MW01235655

First-Order Partial Differential Equations

Volume 2: Theory and Application
of Hyperbolic Systems of
Quasilinear Equations

Hyun-Ku Rhee
Seoul National University
Seoul, Korea

Rutherford Aris
University of Minnesota
Minneapolis, Minnesota

Neal R. Amundson
University of Houston
Houston, Texas

DOVER PUBLICATIONS, INC.
Mineola, New York

Bibliographical Note

This Dover edition, first published in 2001, is an unabridged reprint of the work originally published by Prentice-Hall, Inc., Englewood Cliffs, N.J., in 1989.

Library of Congress Cataloging-in-Publication Data

Rhee, Hyun-Ku, 1939–
    Theory and application of hyperbolic systems of quasilinear equations / Hyun-Ku Rhee, Rutherford Aris, Neal R. Amundson.
        p. cm. — (First-order partial differential equations ; v. 2)
    Originally published: Englewood Cliffs, N.J. : Prentice-Hall, c1989, in series: Prentice-Hall international series in the physical and chemical engineering sciences.
    Includes index.
    ISBN 0-486-41994-0 (pbk.)
    1. Differential equations, Hyperbolic. 2. Quasilinearization. I. Aris, Rutherford. II. Amundson, Neal Russell, 1916– III. Title.

QA374 .R47 2001 vol. 2
[QA377]
515'.353 s—dc21
[515'.353]
                                                                2001041183

Manufactured in the United States of America
Dover Publications, Inc., 31 East 2nd Street, Mineola, N.Y. 11501

To

Junhie, Seungyoon, Sangwoo,
Claire, and Shirley D.

for their forbearance

Contents

Preface

This book is the continuation of *First-Order Partial Differential Equations Volume I: Theory and Application of Single Equations*, published in 1986. As before, there is a variety of physical systems, which are dominated by convective and exchange processes that can usefully be modelled by equations of the first order. The development of models demanding several equations was considered in the first chapter of Volume I and need not be repeated here. Many of the physical situations, such as adsorption, chromatography, chemical kinetics, sedimentation, ultracentrifugation, or oil recovery, call for extension to many components. As soon as there is any interaction between these components, the systems of equations become much more than simply the juxtaposition of single equations.

It is with the new features, introduced by the linking of these sets of partial differential equations, that we are concerned in the second volume. Simple waves and discontinuities will again appear and their interaction will be a major topic. Adsorption and desorption phenomena in flow through a fixed or moving bed again provide a most important example and the Riemann problem, with its piecewise constant boundary values, arises so naturally that it provides a suitable workhorse throughout this book. With the Langmuir isotherm the full range of nonlinear phenomena comes into play, yet there are certain simplifying characteristics that allow the geometry of the situation to be brought out rather clearly. At the same time we have not shunned more pedestrian situations where, to make any progress, one has to resort to numerical calculation at an early stage.

Briefly, Chapter 1 considers pairs of quasi-linear hyperbolic equations of the first order leading off with the example of the chromatography of two solutes. Through a discussion of reducible equations and simple waves, the characteristic initial value problem for two-solute chromatography is discussed and notions of compression waves and the formation of shocks are formulated. In a later section, this is also applied to polymer flooding. Riemann invariants and their applications and the development of singularities and weak solutions are considered. Chapter 2 continues the discussion of two-solute chromatography with Langmuir isotherms and introduces displacement chromatography and shock layers. The problems of two equations with two independent variables allow a very geometrical presentation, which is good preparation for the higher dimensional case. This is considered in Chapter 3, where multicomponent chromatography is discussed. The problem of wave interactions in this context is continued in Chapter 4. In Chapter 5, the additional complications of counter current moving-bed adsorbers are considered. Here, the solid moves in the opposite direction to the fluid phase and continuous separations can be achieved. The adiabatic adsorption column is the subject of Chapter 6, while Chapter 7 introduces the notion of chemical reaction in countercurrent reactors. In general, as has been shown in the previous volume, this means that the characteristics are no longer straight and a new range of interactions comes into play. In particular, it is found that it is possible for a discontinuity to cross the thermodynamic equilibrium line so that complete conversion can be achieved even for a reversible reaction.

The level, at which we have aimed, is again that of graduate study. The text should be accessible to anyone with a thorough grounding in undergraduate mathematics. The earlier volume on single equations is, of course, a good introductory preparation. We hope that it will be as useful to the practicing engineer as it is to the student and we have endeavored, at all points, to relate the mathematical developments to the physical background.

A word is essential about what parts will be useful as a teaching text. The materials in Chapter 1, Sec. 3.1 through Sec. 3.4, and Sec. 6.1 through Sec. 6.5 are general and may be given in any applied mathematics course. The remainder may serve as a source of applications and illustrations, which may prove helpful for the understanding of the subject. For engineering students, Secs. 1.10, 1.11, and 6.5 may be trying and may be omitted.

Those who are mainly interested in the theory of multicomponent chromatography may begin with Chapter 3. If, however, one aims for application to other systems as well, it is important to start from Chapter 1. For a one-semester course for graduate students in engineering, we suggest Sec. 1.1 through Sec. 1.9, Sec. 2.1 through Sec. 2.8, Sec. 3.1 through Sec. 3.9, Secs. 3.11 and 3.12, Sec. 4.1 through Sec. 4.5, Sec. 6.1 through Sec. 6.4, and Sec. 6.6 through Sec. 6.8.

Hyun-Ku Rhee
Rutherford Aris
Neal R. Amundson

Pairs of Quasilinear Hyperbolic Equations of First Order

1

In this chapter the mathematical theory for pairs of quasilinear equations is developed in a rather general fashion. Although we shall be dealing with some physical examples for illustration, extensive application is to be presented in Chapter 2. A number of important and interconnected ideas are introduced here, and since these are sufficiently complex subjects, we shall not avoid repetition of certain ideas but rather, attempt to elucidate the interconnected feature of these ideas in detail. The reader who finds the discussion here insufficient may go to the original sources, to which references are given at the end of the chapter.

Equilibrium chromatography of two solutes is employed as the representative example, for it gives rise to equations of general form. Other examples of conservation equations that are considered are those of polymer flooding, compressible fluid flow, and shallow-water wave theory.

Section 1.1 deals with formulation of the equations for two-solute chromatography and illustrates how a pair of first-order equations is realized in practice. It also discusses how to put the equations in a simplified form. In Sec. 1.2 we consider the general form of two quasilinear equations and introduce the method of characteristics. This then transforms the pair of differential equations into a system of four ordinary differential equations, which are the characteristic differential equations. The solution scheme is also discussed. Section 1.3 is concerned with reducible systems, special in a sense but typical of physical examples, and the hodograph transformation which is the key concept of the solution scheme is introduced. The notion of simple waves is discussed in detail since it assumes an important role in the development of the theory.

In the following three sections, equations of two-solute chromatography are treated in the light of the theoretical development of Secs. 1.2 and 1.3. In Sec. 1.4 the characteristic directions in both the hodograph plane and the (x, τ)-plane are formulated and the directional derivatives along Γ characteristics are introduced. We then proceed to treat the characteristic initial value problem as well as the Riemann problem in Sec. 1.5, showing the structure of the solution in detail and emphasizing the possibility of the necessary existence of discontinuities. In Section 1.6 we are concerned with compression waves and introduce the notion of genuine nonlinearity. The point of shock inception is determined by tracing the envelope of straight characteristics.

In Sec. 1.7 we consider discontinuities in solutions, for which the conservation law is formulated to give the compatibility relation and the jump condition. Important features are discussed, including the entropy condition which is required to ensure the uniqueness of the solution. Based on their different properties, discontinuities are classified into shocks, semishocks, and contact discontinuities. The nature of each discontinuity is fully explored. The equations of polymer flooding are treated in Sec. 1.8 to illustrate how discontinuities come into the solution in a physical system. A numerical example for the Riemann problem is analyzed in detail.

In Sec. 1.9 there is a discussion of the Riemann invariants, whose existence is special to systems of two equations. Their application is illustrated with physical examples. In Sec. 1.10 Riemann invariants are used to show under general circumstances that singularities can develop in solutions after finite time. Then weak solutions are defined, the jump condition is formulated, and the entropy condition is discussed. In particular, Lax's treatment of conservation equations with a convex extension is presented in some detail.

Finally, Sec. 1.11 is concerned with the existence and uniqueness proof for weak solutions as well as their structure and asymptotic behavior. A large number of papers have been published on these subjects and it seems just reasonable to make a survey of these developments rather than going into the detail of specific papers.

1.1 Equations for the Chromatography of Two Solutes

We start with a practical realization of the type of equation we wish to discuss in this chapter, the equilibrium chromatography of two interacting solutes. When two solutes, A_1 and A_2, are present in concentrations c_1 and c_2 in the fluid phase, their equilibrium concentrations n_1 and n_2 in the solid phase will be functions of both concentrations, i.e.,

$$n_1 = n_1(c_1, c_2), \qquad n_2 = n_2(c_1, c_2) \qquad (1.1.1)$$

the nature of which is essentially nonlinear because the mutual influence between the solutes A_1 and A_2 is taken into account. Just what form these functions may take is seen by considering the case of adsorption according to the Langmuir isotherm. If the number of sites at which adsorption can take place is limited, the total adsorbed concentration $(n_1 + n_2)$ cannot exceed an upper bound, say N. In fact, $(N - n_1 - n_2)$ will be proportional to the number of vacant sites and the rate at which A_1 is adsorbed might be expected to be proportional to the product of its concentration above the surface, c_1, and the availability of adsorption sites on the surface, $N - n_1 - n_2$. The rate of adsorption of A_1 might be written as

$$r_{a1} = k_{a1}c_1(N - n_1 - n_2)$$

But at equilibrium this is exactly balanced by the rate of desorption, which we may take to be $k_{d1} n_1$, proportional to the adsorbed concentration, n_1. Thus

$$k_{a1}c_1(N - n_1 - n_2) = k_{d1}n_1$$

or

$$(1 + K_1c_1)n_1 + K_1c_1n_2 = NK_1c_1 \qquad (1.1.2)$$

where

$$K_1 = \frac{k_{a1}}{k_{d1}} \qquad (1.1.3)$$

Compare the rate expressions to those in Eq. (1.3.9) of Vol. I. Similarly, the equilibrium of A_2 gives

$$K_2c_2n_1 + (1 + K_2c_2)n_2 = NK_2c_2 \qquad (1.1.4)$$

where

$$K_2 = \frac{k_{a2}}{k_{d2}} \qquad (1.1.5)$$

Now Eqs. (1.1.2) and (1.1.4) can be solved to give

$$n_1 = \frac{NK_1c_1}{1 + K_1c_1 + K_2c_2}$$

$$\qquad (1.1.6)$$

$$n_2 = \frac{NK_2c_2}{1 + K_1c_1 + K_2c_2}$$

If $K_2 > K_1$, it is reasonable to speak of the solute A_2 as being more strongly adsorbed than A_1. Figure 1.1 shows two typical surfaces of n_1 and n_2 as functions of c_1 and c_2; the surface corresponding to the more strongly adsorbed solute A_2 lies above that of A_1 over most of the plane. Clearly, we can always take

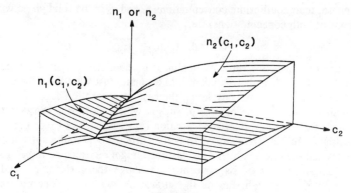

Figure 1.1

$K_2 > K_1$ by correctly numbering the two solutes; the case $K_1 = K_2$ is too special to merit attention. Another way of representing the two surfaces is by contours of n_1/N and n_2/N in the plane of $K_1 c_1$ and $K_2 c_2$. These are the families of straight lines shown in Fig. 1.2. The family of contours for n_1/N radiates from the point $K_1 c_1 = 0$, $K_2 c_2 = -1$; and any line intersects the $K_1 c_1$ axis at $K_1 c_1 = n_1/(N - n_1)$. Similarly, the contours for n_2/N all emanate from the point $K_1 c_1 = -1$, $K_2 c_2 = 0$ and have intercepts $n_2/(N - n_2)$ on the $K_2 c_2$ axis.

We observe that

$$\frac{\partial n_1}{\partial c_1} = \frac{NK_1(1 + K_2 c_2)}{(1 + K_1 c_1 + K_2 c_2)^2}, \qquad \frac{\partial n_1}{\partial c_2} = -\frac{NK_2 K_1 c_1}{(1 + K_1 c_1 + K_2 c_2)^2}$$

$$\frac{\partial n_2}{\partial c_1} = -\frac{NK_1 K_2 c_2}{(1 + K_1 c_1 + K_2 c_2)^2}, \qquad \frac{\partial n_2}{\partial c_2} = \frac{NK_2(1 + K_1 c_1)}{(1 + K_1 c_1 + K_2 c_2)^2} \qquad (1.1.7)$$

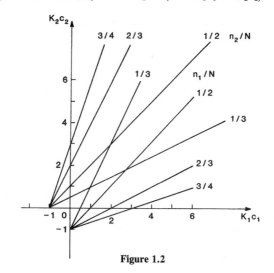

Figure 1.2

Pairs of Quasilinear Hyperbolic Equations of First Order   Chap. 1

The general features of these expressions are shared by all isotherms representing the adsorption of two solutes that compete for the same sites. Thus we expect the derivatives to tend to zero as either c_1 or c_2 tends to infinity and their signs to be

$$\frac{\partial n_1}{\partial c_1} > 0, \qquad \frac{\partial n_1}{\partial c_2} < 0, \qquad \frac{\partial n_2}{\partial c_1} < 0, \qquad \frac{\partial n_2}{\partial c_2} > 0 \qquad (1.1.8)$$

We also expect to have the conditions

$$\left| \frac{\partial n_1}{\partial c_1} \right| > \left| \frac{\partial n_2}{\partial c_1} \right|, \qquad \left| \frac{\partial n_2}{\partial c_2} \right| > \left| \frac{\partial n_1}{\partial c_2} \right| \qquad (1.1.9)$$

which imply that the variation of c_1 (or c_2) has a stronger influence on the adsorption of A_1 (or A_2) than on the adsorption of A_2 (or A_1). The Jacobian of n_1 and n_2 with respect to c_1 and c_2 will generally be positive and in the case of the Langmuir isotherm

$$\frac{\partial(n_1, n_2)}{\partial(c_1, c_2)} = \frac{N^2 K_1 K_2}{(1 + K_1 c_1 + K_2 c_2)^3} \qquad (1.1.10)$$

In the chromatographic column with voidage ϵ and V as the interstitial velocity of the fluid phase we have the usual mass balance equations for A_i, $i = 1, 2$, in terms of $c_i(z, t)$ and $n_i(z, t)$, the two concentrations at position z and time t. These have been derived in Secs. 1.2 and 1.3 of Vol. 1. Thus

$$V\frac{\partial c_1}{\partial z} + \frac{\partial c_1}{\partial t} + \frac{1 - \varepsilon}{\varepsilon} \frac{\partial n_1}{\partial t} = 0$$

and $\qquad\qquad\qquad\qquad\qquad\qquad\qquad\qquad\qquad\qquad\qquad\qquad (1.1.11)$

$$V\frac{\partial c_2}{\partial z} + \frac{\partial c_2}{\partial t} + \frac{1 - \varepsilon}{\varepsilon} \frac{\partial n_2}{\partial t} = 0$$

If we define the dimensionless parameter

$$\nu = \frac{1 - \varepsilon}{\varepsilon} \qquad (1.1.12)$$

and introduce the equilibrium relationships in Eq. (1.1.1), we obtain a pair of first-order equations for $c_1(z, t)$ and $c_2(z, t)$,

$$V\frac{\partial c_1}{\partial z} + \left(1 + \nu\frac{\partial n_1}{\partial c_1}\right)\frac{\partial c_1}{\partial t} + \nu\frac{\partial n_1}{\partial c_2}\frac{\partial c_2}{\partial t} = 0$$
$$\nu\frac{\partial n_2}{\partial c_1}\frac{\partial c_1}{\partial t} + V\frac{\partial c_2}{\partial z} + \left(1 + \nu\frac{\partial n_2}{\partial c_2}\right)\frac{\partial c_2}{\partial t} = 0 \qquad (1.1.13)$$

The natural initial data on these equations would be the specification of c_1 and c_2 at the inlet $z = 0$ and on the column at $t = 0$. These equations are entirely typical of the general pair of equations we shall study in Sec. 1.2, for under suitable conditions the coefficients could be functions of the independent as well as the dependent variables.

To simplify the equations, however, we introduce the equilibrium column isotherms, defined as

$$f_1 = f_1(c_1, c_2) = c_1 + vn_1(c_1, c_2)$$
$$f_2 = f_2(c_1, c_2) = c_2 + vn_2(c_1, c_2)$$
(1.1.14)

[see Eq. (7.1.5) of Vol. 1] and, as usual, make the independent variables dimensionless; i.e.,

$$x = \frac{z}{Z}, \qquad \tau = \frac{Vt}{Z}$$
(1.1.15)

where Z denotes the characteristic length of the system. Thus we can rewrite Eq. (1.1.11) as

$$\frac{\partial c_1}{\partial x} + \frac{\partial}{\partial \tau} f_1(c_1, c_2) = 0$$
$$\frac{\partial c_2}{\partial x} + \frac{\partial}{\partial \tau} f_2(c_1, c_2) = 0$$
(1.1.16)

or put it in the form

$$\frac{\partial c_1}{\partial x} + \frac{\partial f_1}{\partial c_1} \frac{\partial c_1}{\partial \tau} + \frac{\partial f_1}{\partial c_2} \frac{\partial c_2}{\partial \tau} = 0$$
$$\frac{\partial f_2}{\partial c_1} \frac{\partial c_1}{\partial \tau} + \frac{\partial c_2}{\partial x} + \frac{\partial f_2}{\partial c_2} \frac{\partial c_2}{\partial \tau} = 0$$
(1.1.17)

We shall return to this pair of equations in Sec. 1.4.

In the case of Langmuir isotherms we notice that $K_1 c_1$ and $K_2 c_2$ generally keep together so that, setting

$$u = K_1 c_1, \qquad v = K_2 c_2$$
(1.1.18)

and

$$\kappa_1 = vNK_1, \qquad \kappa_2 = vNK_2$$
(1.1.19)

we have the equations

$$\frac{\partial u}{\partial x} + \frac{\partial}{\partial \tau}\left(u + \frac{\kappa_1 u}{1 + u + v}\right) = 0$$
$$\frac{\partial v}{\partial x} + \frac{\partial}{\partial \tau}\left(v + \frac{\kappa_2 v}{1 + u + v}\right) = 0$$

or in the form

$$\frac{\partial u}{\partial x} + \left\{1 + \frac{\kappa_1(1 + v)}{(1 + u + v)^2}\right\}\frac{\partial u}{\partial \tau} - \frac{\kappa_1 u}{(1 + u + v)^2}\frac{\partial v}{\partial \tau} = 0$$
$$-\frac{\kappa_2 v}{(1 + u + v)^2}\frac{\partial u}{\partial \tau} + \frac{\partial v}{\partial x} + \left\{1 + \frac{\kappa_2(1 + u)}{(1 + u + v)^2}\right\}\frac{\partial v}{\partial \tau} = 0$$
(1.1.20)

This will be the form of the equations that we will use to illustrate the general case in the following sections.

Although we shall not pursue it, another way of simplifying equations may be of particular interest. This is accomplished by applying the transformation of independent variables which is essentially the same as the one we treated in Eq. (0.2.8) of Vol. 1. Let

$$x' = v\frac{z}{V}, \qquad y' = t - \frac{z}{V} \tag{1.1.21}$$

then

$$\frac{\partial}{\partial t} = \frac{\partial}{\partial y'}, \qquad \frac{\partial}{\partial z} = \frac{v}{V}\frac{\partial}{\partial x'} - \frac{1}{V}\frac{\partial}{\partial y'}$$

or

$$V\frac{\partial}{\partial z} + \frac{\partial}{\partial t} = v\frac{\partial}{\partial x'} \tag{1.1.22}$$

Thus Eq. (1.1.11) becomes

$$\frac{\partial c_1}{\partial x'} + \frac{\partial}{\partial y'}n_1(c_1, c_2) = 0$$

$$\frac{\partial c_2}{\partial x'} + \frac{\partial}{\partial y'}n_2(c_1, c_2) = 0 \tag{1.1.23}$$

For the case of the Langmuir isotherm we shall again use u and v from Eq. (1.1.18), so that the equations are

$$\frac{1}{K_1}\frac{\partial u}{\partial x'} + N\frac{\partial}{\partial y'}\frac{u}{1 + u + v} = 0$$

$$\frac{1}{K_2}\frac{\partial v}{\partial x'} + N\frac{\partial}{\partial y'}\frac{v}{1 + u + v} = 0$$

This suggests that we put

$$x = NK_1x', \qquad y = y', \qquad \kappa = \frac{K_1}{K_2} \tag{1.1.24}$$

and then

$$\frac{\partial u}{\partial x} + \frac{1 + v}{(1 + u + v)^2}\frac{\partial u}{\partial y} - \frac{u}{(1 + u + v)^2}\frac{\partial v}{\partial y} = 0$$

$$-\frac{v}{(1 + u + v)^2}\frac{\partial u}{\partial y} + \kappa\frac{\partial v}{\partial x} + \frac{1 + u}{(1 + u + v)^2}\frac{\partial v}{\partial y} = 0 \tag{1.1.25}$$

This pair of equations is equivalent to Eqs. (1.1.20). Although the dependent variables are the same, the independent variables are defined in a different manner. Both the independent variables here have the dimensions of time, but because of their origin, we can think of x as space-like and y as time-like. It involves the fewest possible number of parameters, the only one being κ, $0 < \kappa < 1$.

Concerning the initial and boundary conditions, the line $x = 0$ clearly represents the inlet of the feedstream entering the column at $z = 0$ and thus

$$u(0, y) = u^f(y), \qquad v(0, y) = v^f(y) \qquad (1.1.26)$$

the superscript f standing for *feed*. Again the situation on the line $y = 0$ gives the condition at $z = Vt$. Now $t = z/V$ is the instant at which an element of the carrier fluid entering the column at $t = 0$ first reaches the point z. Hence the feed conditions cannot have any influence until $y = 0$ and it has been commonly supposed in the theory of chromatography that u and v can be specified at $y = 0$, say

$$u(x, 0) = u^0(x), \qquad v(x, 0) = v^0(x) \qquad (1.1.27)$$

This argument implies that the system is passive until the adsorption front arrives and is quite valid for an initially constant state, but it would be more accurate to consider *initial* conditions specified on the line $t = 0$, i.e.,

$$y = -\frac{x}{vNK_1} = -\mu x \qquad (1.1.28)$$

In this case we would have

$$u(x, -\mu x) = u^i(x), \qquad v(x, -\mu x) = v^i(x) \qquad (1.1.29)$$

in which the superscript i stands for *initial*. We also observe that if we wish to recover the distribution of solutes in the column from the solution $u(x, y)$, $v(x, y)$ at any instant t, we must take sections by lines

$$y = t - \mu x, \qquad \mu = \frac{1}{vNK_1} \qquad (1.1.30)$$

The geometry of variables is shown in Fig. 1.3.

Exercise

1.1.1 When a single solute adsorbs in an adiabatic adsorption column, we have the usual mass balance equation (1.4.1) of Vol. 1 with $i = 1$ and the energy balance equation (1.4.9) of Vol. 1 with $m = 1$, where c_2 and n_2 are defined by Eqs. (1.4.8a) and (1.4.8b) of Vol. 1, respectively. Consider the Langmuir isotherm for a single solute, i.e., Eq. (1.2.7) of Vol. 1 with $c = c_1$ and $n = n_1$, and note that the parameter $K = K_1$ is expressed as a function of the temperature T by Eq. (1.4.10) of Vol. 1 with $i = 1$. Rewrite the pair of conservation equations in terms of c_1 and T, and arrange them in the form of Eq. (1.1.20) and also in the form of Eq. (1.1.25).

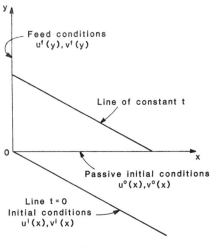

Feed conditions
$u^f(y), v^f(y)$

Line of constant t

0

x

Passive initial conditions
$u^o(x), v^o(x)$

Line t = 0
Initial conditions
$u^i(x), v^i(x)$

Figure 1.3

1.2 Hyperbolic Systems of Two First-Order Equations

We shall consider the general quasilinear system of first-order equations for two dependent variables, u and v, with two independent variables, x and y; i.e.,

$$\mathcal{L}_1 \equiv A_1 \frac{\partial u}{\partial x} + B_1 \frac{\partial u}{\partial y} + C_1 \frac{\partial v}{\partial x} + D_1 \frac{\partial v}{\partial y} + E_1 = 0$$

$$\mathcal{L}_2 \equiv A_2 \frac{\partial u}{\partial x} + B_2 \frac{\partial u}{\partial y} + C_2 \frac{\partial v}{\partial x} + D_2 \frac{\partial v}{\partial y} + E_2 = 0$$

(1.2.1)

where $A_1, A_2, B_1, \ldots, E_1$, and E_2 are given functions of x, y, u, and v and continuous with as many continuous derivatives as may be required. We shall assume that nowhere in the domain do $A_1/A_2 = B_1/B_2 = C_1/C_2 = D_1/D_2$, which asserts that the two equations are independent. These equations are *quasilinear* since the derivatives all enter linearly; they would be called *strictly linear* if the coefficients plus E_1 and E_2 were all functions of x and y only, *linear* if dependence on u and v only appeared in E_1 and E_2 in a linear fashion, and *semilinear* if dependence on u and v only appeared in E_1 and E_2 in some nonlinear manner. Frequently, the equations are called simply *linear* if the coefficients $A_1, A_2, \ldots, D_2$ are all dependent on x and y only.

The equations above are *homogeneous* if $E_1 = E_2 = 0$ and for reasons that will appear later are called *reducible* if they are homogeneous and the coefficients $A_1, \ldots, D_2$ are functions of u and v only. Comparison of Eq.

(1.2.1) and Eq. (1.1.17) or (1.1.20) shows that we shall be dealing with a reducible pair of equations in the equilibrium theory of two-solute chromatography.

There are two ways of working toward ordinary differential equations such as we have seen to be *characteristic* of single equations, and these are extensions of two of the features of characteristics that we discussed previously. The first way is to look for directional derivatives that will simplify the equations; the second is to ask under what conditions the initial data would fail to specify the solution properly. We shall follow each way in turn.

The combination $A_1 \partial u/\partial x + B_1 \partial u/\partial y$ represents the derivative of u in a direction such that $dy/dx = B_1/A_1$ (see Sec. 0.9 of Vol. 1). Similarly, $C_1 \partial v/\partial x + D_1 \partial v/\partial y$ is the derivative of v in the direction $dy/dx = D_1/C_1$. Unless B_1C_1 happens to equal A_1D_1, these directions will not be the same, but we may ask whether it is possible to make a linear combination of the two equations, $\mathcal{L}_1 = 0$ and $\mathcal{L}_2 = 0$, in which these derivatives are taken in the same direction. Let us thus consider the linear combination

$$\mathcal{L} \equiv \lambda_1 \mathcal{L}_1 + \lambda_2 \mathcal{L}_2 = 0 \tag{1.2.2}$$

The derivatives of both u and v in $\mathcal{L} = 0$ will be in the same direction $dy/dx = \sigma$ if

$$\frac{\lambda_1 B_1 + \lambda_2 B_2}{\lambda_1 A_1 + \lambda_2 A_2} = \frac{\lambda_1 D_1 + \lambda_2 D_2}{\lambda_1 C_1 + \lambda_2 C_2} = \sigma \tag{1.2.3}$$

Suppose that at a point (x, y, u, v) on the solution surface, λ_1 and λ_2 are chosen to satisfy these equations; then

$$\begin{aligned} \lambda_1(A_1\sigma - B_1) + \lambda_2(A_2\sigma - B_2) = 0 \\ \lambda_1(C_1\sigma - D_1) + \lambda_2(C_2\sigma - B_2) = 0 \end{aligned} \tag{1.2.4}$$

But if these equations are to give nontrivial values of λ_1 and λ_2, the determinant of the coefficients must vanish, i.e.,

$$\begin{vmatrix} A_1\sigma - B_1 & A_2\sigma - B_2 \\ C_1\sigma - D_1 & C_2\sigma - D_2 \end{vmatrix} = 0 \tag{1.2.5}$$

This may be written as a quadratic,

$$a\sigma^2 - 2b\sigma + c = 0 \tag{1.2.6}$$

where

$$a = A_1 C_2 - A_2 C_1$$
$$2b = A_1 D_2 - A_2 D_1 + B_1 C_2 - B_2 C_1 \tag{1.2.7}$$
$$c = B_1 D_2 - B_2 D_1$$

If the discriminant

$$b^2 - ac > 0 \tag{1.2.8}$$

this quadratic has two real roots and there will be two directions in which it is true that the directional derivatives of u and v in the equation $\lambda_1 \mathcal{L}_1 + \lambda_2 \mathcal{L}_2 = 0$ are aligned. Such systems are called *hyperbolic*, and it is with these that we will be concerned. If $b^2 - ac = 0$, the system is said to be *parabolic*, and if $b^2 - ac < 0$, it is *elliptic*—an obvious classification that we will not pursue further.

Let us denote the two roots of the characteristic equation (1.2.6) by σ_+ and σ_- with $\sigma_+ > \sigma_-$. Since they are real, they will both be positive if a, b, and c are all of the same sign, both negative if b is of the opposite sign to a and c, and of contrary sign if the signs of a and c are different. If we imagine for a moment that the solution surfaces $u(x, y)$, $v(x, y)$ are known, then Eq. (1.2.6) could be solved at every point and two families of curves C_+ and C_- could be drawn in the (x, y)-plane, respectively, satisfying at each point

$$C_+ : \quad \frac{dy}{dx} = \sigma_+$$

$$C_- : \quad \frac{dy}{dx} = \sigma_-$$

(1.2.9)

These two directions will be called the *characteristic directions* and the two families of curves the net of *characteristic curves* or simply *characteristics*, C_+ and C_- in the (x, y)-plane, belonging to the solution $u(x, y)$, $v(x, y)$. In general, they will depend on the solution surface we are looking at, and as with the single equation, they can only be laid down once and for all in the linear or semilinear case. For then both σ_+ and σ_- would be functions only of x and y, and the family of solutions of Eqs. (1.2.9) would form a net of characteristic ground curves.

The two families of characteristics C_+ and C_- may be represented in the form $\beta(x, y) = \text{constant}$ and $\alpha(x, y) = \text{constant}$, respectively, and form a curvilinear coordinate net. We can then introduce new parameters α and β instead of x and y in such a way that β is constant along the C_+ characteristics and α is constant along the C_- characteristics. In other words, α and β are the parameters running along the C_+ and C_- characteristics, respectively, as shown in Fig. 1.4 and we shall call them *characteristic parameters*. To see

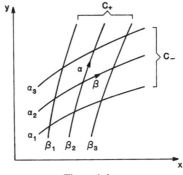

Figure 1.4

how these parameters can be specified, let us consider any curve I, $x = x(\xi)$ and $y = y(\xi)$, that is nowhere tangent to a characteristic, i.e.,

$$a\left(\frac{\partial y}{\partial \xi}\right)^2 - 2b\frac{\partial x}{\partial \xi}\frac{\partial y}{\partial \xi} + c\left(\frac{\partial x}{\partial \xi}\right)^2 \neq 0 \qquad (1.2.10)$$

on the curve I. Through any two points $\xi = \alpha$ and $\xi = \beta$ that are located sufficiently close on I, we draw the C_- and C_+ characteristics, respectively, up to the point of intersection $P(x, y)$ as illustrated in Fig. 1.5. The pair (α, β) is then the set of characteristic parameters for the point $P(x, y)$. Obviously, we can use any monotone functions $\alpha' = A(\alpha)$ and $\beta' = B(\beta)$ as characteristic parameters since such a transformation leaves Eq. (1.2.9) unchanged.

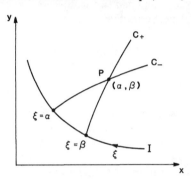

Figure 1.5

Now we can write from Eq. (1.2.9)

$$\frac{\partial y}{\partial \alpha} - \sigma_+ \frac{\partial x}{\partial \alpha} = 0$$

$$\frac{\partial y}{\partial \beta} - \sigma_- \frac{\partial x}{\partial \beta} = 0 \qquad (1.2.11)$$

where $\partial/\partial\alpha$ or $\partial/\partial\beta$ denotes differentiation with respect to α or β along a C_+ or C_- characteristic. In general, we have

$$\frac{\partial}{\partial \alpha} = \frac{\partial x}{\partial \alpha}\left(\frac{\partial}{\partial x} + \sigma_+ \frac{\partial}{\partial y}\right) = \frac{\partial y}{\partial \alpha}\left(\frac{1}{\sigma_+}\frac{\partial}{\partial x} + \frac{\partial}{\partial y}\right) \qquad (1.2.12)$$

or

$$\frac{\partial}{\partial \beta} = \frac{\partial x}{\partial \beta}\left(\frac{\partial}{\partial x} + \sigma_- \frac{\partial}{\partial y}\right) = \frac{\partial y}{\partial \beta}\left(\frac{1}{\sigma_-}\frac{\partial}{\partial x} + \frac{\partial}{\partial y}\right) \qquad (1.2.13)$$

We still have not determined λ_1 and λ_2 to give the equation $\mathscr{L} = 0$. But instead of finding λ_1 and λ_2 directly, we shall proceed in some other way and determine the equation $\mathscr{L} = 0$. Let $\sigma = \sigma_+$ in the first equation of Eq. (1.2.4) and multiply this equation by $\partial x/\partial\alpha$. Then we have

$$\lambda_1(A_1\sigma_+ - B_1)\frac{\partial x}{\partial \alpha} + \lambda_2(A_2\sigma_+ - B_2)\frac{\partial x}{\partial \alpha} = 0 \qquad (1.2.14)$$

Similarly, multiplying Eq. (1.2.2) by $\partial x/\partial \alpha$ gives

$$(\lambda_1 A_1 + \lambda_2 A_2)\frac{\partial u}{\partial x}\frac{\partial x}{\partial \alpha} + (\lambda_1 B_1 + \lambda_2 B_2)\frac{\partial u}{\partial y}\frac{\partial x}{\partial \alpha} + (\lambda_1 C_1 + \lambda_2 C_2)\frac{\partial v}{\partial x}\frac{\partial x}{\partial \alpha}$$

$$+ (\lambda_1 D_1 + \lambda_2 D_2)\frac{\partial v}{\partial y}\frac{\partial x}{\partial \alpha} + (\lambda_1 E_1 + \lambda_2 E_2)\frac{\partial x}{\partial \alpha} = 0$$

and using Eq. (1.2.3) with $\sigma = \sigma_+$ to eliminate the terms in the B's and D's yields

$$(\lambda_1 A_1 + \lambda_2 A_2)\left(\frac{\partial u}{\partial x} + \sigma_+\frac{\partial u}{\partial y}\right)\frac{\partial x}{\partial \alpha} + (\lambda_1 C_1 + \lambda_2 C_2)\left(\frac{\partial v}{\partial x} + \sigma_+\frac{\partial v}{\partial y}\right)\frac{\partial x}{\partial \alpha}$$

$$+ (\lambda_1 E_1 + \lambda_2 E_2)\frac{\partial x}{\partial \alpha} = 0$$

Equation (1.2.12) can now be used to express the derivatives with respect to α; then

$$(\lambda_1 A_1 + \lambda_2 A_2)\frac{\partial u}{\partial \alpha} + (\lambda_1 C_1 + \lambda_2 C_2)\frac{\partial v}{\partial \alpha} + (\lambda_1 E_1 + \lambda_2 E_2)\frac{\partial x}{\partial \alpha} = 0$$

But this can be rearranged to give

$$\lambda_1\left(A_1\frac{\partial u}{\partial \alpha} + C_1\frac{\partial v}{\partial \alpha} + E_1\frac{\partial x}{\partial \alpha}\right) + \lambda_2\left(A_2\frac{\partial u}{\partial \alpha} + C_2\frac{\partial v}{\partial \alpha} + E_2\frac{\partial x}{\partial \alpha}\right) = 0 \qquad (1.2.15)$$

If we had multiplied Eq. (1.2.2) by $\partial y/\partial \alpha$ and eliminated the terms in A's and C's by using Eq. (1.2.3), we would have had the equation

$$\lambda_1\left(B_1\frac{\partial u}{\partial \alpha} + D_1\frac{\partial v}{\partial \alpha} + E_1\frac{\partial y}{\partial \alpha}\right) + \lambda_2\left(B_2\frac{\partial u}{\partial \alpha} + D_2\frac{\partial v}{\partial \alpha} + E_2\frac{\partial y}{\partial \alpha}\right) = 0 \qquad (1.2.16)$$

which can be used as well instead of Eq. (1.2.15).

Consider now Eq. (1.2.15) along with Eq. (1.2.14) as a pair of equations to be solved for λ_1 and λ_2. Since these equations are homogeneous, their determinant must vanish if we are not to have trivial values for λ_1 and λ_2. Thus

$$\begin{vmatrix} (A_1\sigma_+ - B_1)\dfrac{\partial x}{\partial \alpha} & (A_2\sigma_+ - B_2)\dfrac{\partial x}{\partial \alpha} \\[2ex] A_1\dfrac{\partial u}{\partial \alpha} + C_1\dfrac{\partial v}{\partial \alpha} + E_1\dfrac{\partial x}{\partial \alpha} & A_2\dfrac{\partial u}{\partial \alpha} + C_2\dfrac{\partial v}{\partial \alpha} + E_2\dfrac{\partial x}{\partial \alpha} \end{vmatrix} = 0$$

or

$$L\frac{\partial u}{\partial \alpha} + M_+\frac{\partial v}{\partial \alpha} + N_+\frac{\partial x}{\partial \alpha} = 0 \qquad (1.2.17)$$

where

$$L = A_1 B_2 - A_2 B_1$$

$$M_+ = (A_1 C_2 - A_2 C_1)\sigma_+ - (B_1 C_2 - B_2 C_1) \qquad (1.2.18)$$

$$N_+ = (A_1 E_2 - A_2 E_1)\sigma_+ - (B_1 E_2 - B_2 E_1)$$

We could work through the same development with σ_- and the parameter β to obtain a similar equation,

$$L\frac{\partial u}{\partial \beta} + M_-\frac{\partial v}{\partial \beta} + N_-\frac{\partial x}{\partial \beta} = 0 \qquad (1.2.19)$$

where M_- and N_- are given by Eq. (1.2.18) with σ_- in place of σ_+. Then the four equations

$$\frac{\partial y}{\partial \alpha} - \sigma_+\frac{\partial x}{\partial \alpha} = 0$$

$$\frac{\partial y}{\partial \beta} - \sigma_-\frac{\partial x}{\partial \beta} = 0$$

$$L\frac{\partial u}{\partial \alpha} + M_+\frac{\partial v}{\partial \alpha} + N_+\frac{\partial x}{\partial \alpha} = 0 \qquad (1.2.20)$$

$$L\frac{\partial u}{\partial \beta} + M_-\frac{\partial v}{\partial \beta} + N_-\frac{\partial x}{\partial \beta} = 0$$

are called the *characteristic differential equations*. They will be satisfied on the solution surfaces $u(x, y)$, $v(x, y)$ provided that σ_+ and σ_- are roots of the quadratic (1.2.6), and they allow us to learn a good deal about the way the solution can be developed from the initial data. We may think of them as four coupled equations for x, y, u, and v as functions of two parameters α and β. When Eqs. (1.2.20) are solved to give x and y as functions of α and β, $x = x(\alpha, \beta)$, $y = y(\alpha, \beta)$, these functions may be inverted to give α and β in terms of x and y. Finally, when these are substituted in $u(\alpha, \beta)$ and $v(\alpha, \beta)$, one obtains u and v as functions of x and y. In principle this can be done provided that

$$\frac{\partial(x, y)}{\partial(\alpha, \beta)} = (\sigma_+ - \sigma_-)\frac{\partial x}{\partial \alpha}\frac{\partial x}{\partial \beta} \neq 0 \qquad (1.2.21)$$

but this is ensured by the criterion for hyperbolicity.

Before seeing what these equations tell us about the structure of the solution, let us look at the initial data and ask when it seems to be suitably given. Suppose that u and v are given along a curve I in the (x, y)-plane. If ξ is a parameter running along this curve, we may specify the curve and the initial data by giving four functions,

$$I: x = X(\xi), \quad y = Y(\xi), \quad u = U(\xi), \quad v = V(\xi) \qquad (1.2.22)$$

To move off this initial curve I it is essential to be able to calculate the partial derivatives of u and v at any point ξ. These will satisfy the differential equations

$$A_1\frac{\partial u}{\partial x} + B_1\frac{\partial u}{\partial y} + C_1\frac{\partial v}{\partial x} + D_1\frac{\partial v}{\partial y} + E_1 = 0$$
$$A_2\frac{\partial u}{\partial x} + B_2\frac{\partial u}{\partial y} + C_2\frac{\partial v}{\partial x} + D_2\frac{\partial v}{\partial y} + E_2 = 0$$

(1.2.23)

in which $A_1, \ldots, E_2$ have been made functions of ξ by substituting from Eqs. (1.2.22), and also the strip conditions [see Eq. (0.12.2) of Vol. 1].

$$X'\frac{\partial u}{\partial x} + Y'\frac{\partial u}{\partial y} - U' = 0$$
$$X'\frac{\partial v}{\partial x} + Y'\frac{\partial v}{\partial y} - V' = 0$$

(1.2.24)

Thus the four equations (1.2.23) and (1.2.24), as four simultaneous equations for $\partial u/\partial x$, $\partial u/\partial y$, $\partial v/\partial x$, and $\partial v/\partial y$, must be solvable. This will only be the case if

$$\begin{vmatrix} A_1 & B_1 & C_1 & D_1 \\ A_2 & B_2 & C_2 & D_2 \\ X' & Y' & 0 & 0 \\ 0 & 0 & X' & Y' \end{vmatrix} = a(Y')^2 - 2bX'Y' + c(X')^2 \neq 0 \qquad (1.2.25)$$

This implies that the partial derivatives of u and v can be calculated if the initial curve I is nowhere characteristic in the sense defined above. Conversely, this is to say that we shall not be able to determine the partial derivatives uniquely along the curve I if it is characteristic, i.e., if $Y'/X' = \sigma_+$ or σ_-. We have observed this feature with the single equation in Sec. 2.5 of Vol. 1. Another way of putting it is to say that two different solutions can intersect only along a characteristic. Thus the derivatives across a characteristic can be discontinuous, or a sharp edge in a solution surface must be a characteristic. This also is true of a single equation (see Sec. 5.2 of Vol. 1).

On the other hand, if the initial curve I is characteristic, we can only get a nontrivial solution of Eqs. (1.2.23) and (1.2.24) when the matrix of coefficients and its augment,

$$\begin{vmatrix} A_1 & B_1 & C_1 & D_1 & E_1 \\ A_2 & B_2 & C_2 & D_2 & E_2 \\ X' & Y' & 0 & 0 & -U' \\ 0 & 0 & X' & Y' & -V' \end{vmatrix}$$

are of the same rank. But by setting the determinant of the first three columns and the last equal to zero, we recover Eq. (1.2.17) for $Y'/X' = \sigma_+$, $\xi = \alpha$, and Eq. (1.2.19) for $Y'/X' = \sigma_-$, $\xi = \beta$. Thus if data are given on a characteristic curve, they must satisfy the full characteristic differential equations, which is to say that the solution must satisfy the full characteristic differential equations.

Consider next the noncharacteristic data given on the initial curve I shown in Fig. 1.6. We shall show how a step-by-step procedure will allow us to calculate u and v given their values on I. Now α is a parameter running along the characteristics C_+ and β along C_-. Hence two points such as A and E will have the same value of β but different values of α, and points such as B and E the same α but different β's. Let us assume that these points are close together and write the parametric coordinates of the three points as follows (see the diagram on the right of Fig. 1.6):

$$A \text{ is } (\alpha, \beta); \quad B \text{ is } (\alpha + \delta\alpha, \beta - \delta\beta); \quad E \text{ is } (\alpha + \delta\alpha, \beta)$$

Since we know x, y, u, and v at A and B, we can calculate σ_{+A} and σ_{-B}. Then if ξ_A and ξ_B are the values of ξ on I at A and B and if we put $x_A = X(\xi_A)$, $y_A = Y(\xi_A)$, $x_B = X(\xi_B)$, and $y_B = Y(\xi_B)$, the coordinates of E lying at the intersection of the two characteristics must satisfy

$$y_E = y_A + \sigma_{+A}(x_E - x_A)$$
$$y_E = y_B + \sigma_{-B}(x_E - x_B)$$

$$(1.2.26)$$

where the suffixes A and B denote that the quantities must be evaluated at these points. Hence

$$x_E = \frac{(y_B - \sigma_{-B}x_B) - (y_A - \sigma_{+A}x_A)}{\sigma_{+A} - \sigma_{-B}}$$

$$y_E = \frac{\sigma_{+A}(y_B - \sigma_{-B}x_B) - \sigma_{-B}(y_A - \sigma_{+A}x_A)}{\sigma_{+A} - \sigma_{-B}}$$

$$(1.2.27)$$

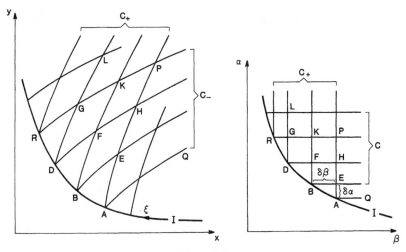

Figure 1.6

　　　Pairs of Quasilinear Hyperbolic Equations of First Order　　Chap. 1

Another way of getting these equations is to multiply the first two equations of Eq. (1.2.20) by $\delta\alpha$ and $\delta\beta$, respectively, and to note that $y_E - y_A = (\partial y/\partial\alpha)\,\delta\alpha$, $y_E - y_B = (\partial y/\partial\beta)\,\delta\beta$, etc. Similarly, multiplying the last two equations of Eq. (1.2.20) by $\delta\alpha$ and $\delta\beta$, respectively, and noting that $u_E - u_A = (\partial u/\partial\alpha)\,\delta\alpha$, $u_E - u_B = (\partial u/\partial\beta)\,\delta\beta$, etc., we obtain

$$L_A(u_E - u_A) + M_{+A}(v_E - v_A) + N_{+A}(x_E - x_A) = 0$$
$$L_B(u_E - u_B) + M_{-B}(v_E - v_B) + N_{-B}(x_E - x_B) = 0$$

(1.2.28)

This provides a pair of equations to be solved for u_E and v_E. Thus, from values of u and v on the initial curve I, the values of u and v at a series of points such as E, F, G can be determined and the location of these points fixed. From this row of points we can go to a row of points such as H, K, L, and so on.

Now, although this is a crude scheme and we would have to indulge in some rather tricky limiting processes to justify it, it does show how the solution can in principle be built up. It shows, for example, that in determining the values of u and v at a point P, all the data on the segment AR of the initial curve I between the C_+ and C_- characteristics through P will be needed (see also Fig. 1.7). In this sense the segment AR is called the *domain of dependence* of the point P. Similarly, the values of u and v at the point A are needed in the calculation of u and v at any point between the two characteristics emanating from A. The angular region PAQ is therefore called the *range of influence* of the point A. In other words, the range of influence of the point A consists of all points whose domains of dependence contain the point A. Compare these definitions here with those for the case of a single equation shown in Fig. 5.5 of Vol. 1.

The existence of such domains of dependence and ranges of influence, which is typical of hyperbolic systems, makes the solution relatively easy to obtain since it allows the construction of the solution by pieces. This further implies that the solution is not necessarily analytic and thus it may consist of analytically different portions in different regions of the (x, y)-plane.

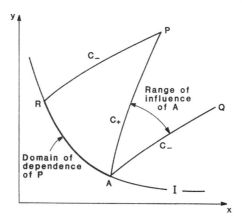

Figure 1.7

Numerical techniques based on this way of discretizing the variables are indeed used and can be refined beyond the crude scheme outlined here. For example, once u_E and v_E are determined by Eqs. (1.2.27) and (1.2.28), we may calculate the average values of σ_+, L, M_+, and N_+ between the two points A and E, and use them as the second estimates for σ_{+A}, L_A, M_{+A}, and N_{+A}. Similarly, the average values of σ_-, L, M_-, and N_- can be determined between the two points B and E, and used as the new estimates for σ_{-B}, L_B, M_{-B}, and N_{-B}. Then Eqs. (1.2.27) and (1.2.28) are solved again to give new values of u_E and v_E, and this procedure can be repeated until a desired convergence is obtained. Various schemes are discussed in the book edited by Ralston and Wilf (1960) and listed at the end of this chapter. It is clear that linearity has the great advantage of making it possible to lay down a network of C_+ and C_- characteristics once and for all.

Before leaving this description we should note that data are sometimes given on a noncharacteristic curve that lies between the two characteristics, as in arc J in Fig. 1.8. Such a curve is called a *time-like boundary*, in contrast to I, which is *space-like*, and on J only one of the dependent variables can be specified. To see this let us take a triad of points A, B, and E just as before, but locate B at the intersection of the curves I and J. The values of x, y, u, and v at E can certainly be determined as before, but now we should ask what can be said of them at D.

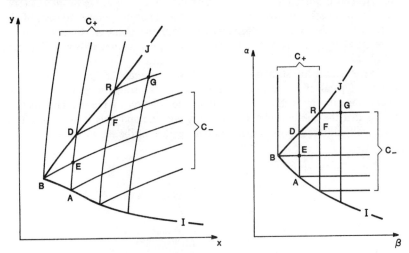

Figure 1.8

Since D lies on the C_+ characteristic through A, we have

$$y_D - y_A = \sigma_{+A}(x_D - x_A) \qquad (1.2.29)$$

But if s_B is the slope of the curve J at B, we also have

$$y_D - y_B = s_B(x_D - x_B) \qquad (1.2.30)$$

This determines the coordinates of D. On the other hand, since D lies on the

Pairs of Quasilinear Hyperbolic Equations of First Order   Chap. 1

C_+ characteristic through A, we can adapt the first equation of Eq. (1.2.28) by replacing E by D; then

$$L_A(u_D - u_A) + M_{+A}(v_D - v_A) + N_{+A}(x_D - x_A) = 0 \qquad (1.2.31)$$

This relation between u_D and v_D means that not both of them can be specified independently on J. Of course, the argument applies only to the infinitesimal triangle ABD, but it can be used again on DFR, etc., and so shows that on a time-like curve only one of the dependent variables can be specified. This becomes even clearer if we consider the picture in the characteristic coordinate system as shown on the right-hand side of Fig. 1.8. The point D on J has the same value of β as the point A on I, which has already been specified, and thus we can specify only α there. This is true for every point on J so that we have one less degree of freedom along the curve J. This implies that we can specify only one of the dependent variables on the time-like boundary J.

Exercises

1.2.1. Consider the spherical isentropic flow of a compressible fluid with spherical symmetry. If x is the distance from the origin, ρ the density of fluid, and u the radial velocity component, show that the governing equations at time t are

$$\frac{\partial \rho}{\partial t} + u\frac{\partial \rho}{\partial x} + \rho\frac{\partial u}{\partial x} + 2\frac{\rho u}{x} = 0$$

$$\rho\left(\frac{\partial u}{\partial t} + u\frac{\partial u}{\partial x}\right) + c^2\frac{\partial \rho}{\partial x} = 0$$

where c^2 is a given function of ρ. By considering a linear combination of these equations, determine the four characteristic differential equations.

1.2.2. For steady irrotational isentropic flow in three dimensions with cylindrical symmetry, let x be the abscissa along the axis and y the distance from the axis, and show that the governing equations are

$$\frac{\partial v}{\partial x} - \frac{\partial u}{\partial y} = 0$$

$$(c^2 - u^2)\frac{\partial u}{\partial x} - uv\left(\frac{\partial u}{\partial y} + \frac{\partial v}{\partial x}\right) + (c^2 - v^2)\frac{\partial v}{\partial y} + \frac{c^2 v}{y} = 0$$

where u and v are, respectively, the x and y components of the velocity and c^2 is a given function of $u^2 + v^2$. Under what conditions is this system hyperbolic? When the system is hyperbolic, find the four characteristic differential equations.

1.3 Reducible Equations and Simple Waves

When the coefficients $A_1, A_2, \ldots, D_2$ in the homogeneous equations

$$A_1\frac{\partial u}{\partial x} + B_1\frac{\partial u}{\partial y} + C_1\frac{\partial v}{\partial x} + D_1\frac{\partial v}{\partial y} = 0$$

$$A_2 \frac{\partial u}{\partial x} + B_2 \frac{\partial u}{\partial y} + C_2 \frac{\partial v}{\partial x} + D_2 \frac{\partial v}{\partial y} = 0 \qquad (1.3.1)$$

are functions only of u and v, the equations are said to be *reducible*. The reason for this is that their quasilinearity can be reduced to strict linearity by interchanging the roles of dependent and independent variables. This transformation, known as the *hodograph transformation*, can be applied whenever the Jacobian

$$j = \frac{\partial(u, v)}{\partial(x, y)} = J^{-1} \qquad (1.3.2)$$

does not vanish or become infinite. Clearly, if we can obtain equations for $x = x(u, v)$ and $y = y(u, v)$ as functions of u and v, they will serve just as well to generate a solution. These functions will answer the question, "Where can a pair of values of u and v be found?", just as the functions $u = u(x, y)$ and $v = v(x, y)$ answer the question, "What are the values of u and v at (x, y)?"

If $x = x(u, v)$, $y = y(u, v)$ are the inverse functions of $u(x, y)$, $v(x, y)$, then the equations

$$x(u(x, y), v(x, y)) = x \qquad (1.3.3)$$

and

$$y(u(x, y), v(x, y)) = y \qquad (1.3.4)$$

must be identities. Thus differentiating Eqs. (1.3.3) and (1.3.4) with respect to x keeping y constant, we have

$$\frac{\partial x}{\partial u} \frac{\partial u}{\partial x} + \frac{\partial x}{\partial v} \frac{\partial v}{\partial x} = 1$$

$$\frac{\partial y}{\partial u} \frac{\partial u}{\partial x} + \frac{\partial y}{\partial v} \frac{\partial v}{\partial x} = 0$$

These may be solved for $\partial u / \partial x$ and $\partial v / \partial x$ to give

$$\frac{\partial u}{\partial x} = j \frac{\partial y}{\partial v}, \qquad \frac{\partial v}{\partial x} = -j \frac{\partial y}{\partial u} \qquad (1.3.5)$$

Similarly, partial differentiation with respect to y gives

$$\frac{\partial x}{\partial u} \frac{\partial u}{\partial y} + \frac{\partial x}{\partial v} \frac{\partial v}{\partial y} = 0$$

$$\frac{\partial y}{\partial u} \frac{\partial u}{\partial y} + \frac{\partial y}{\partial v} \frac{\partial v}{\partial y} = 1$$

and thus

$$\frac{\partial u}{\partial y} = -j \frac{\partial x}{\partial v}, \qquad \frac{\partial v}{\partial y} = j \frac{\partial x}{\partial u} \qquad (1.3.6)$$

Another way of putting this is to say that the Jacobian matrices

$$\begin{bmatrix} \dfrac{\partial u}{\partial x} & \dfrac{\partial u}{\partial y} \\[2ex] \dfrac{\partial v}{\partial x} & \dfrac{\partial v}{\partial y} \end{bmatrix} \quad \text{and} \quad \begin{bmatrix} \dfrac{\partial x}{\partial u} & \dfrac{\partial x}{\partial v} \\[2ex] \dfrac{\partial y}{\partial u} & \dfrac{\partial y}{\partial v} \end{bmatrix}$$

are inverses of each other. Now substituting from Eqs. (1.3.5) and (1.3.6) into Eq. (1.3.1) and dividing through by j, which we must presume does not vanish, we have

$$A_1 \frac{\partial y}{\partial v} - C_1 \frac{\partial y}{\partial u} - B_1 \frac{\partial x}{\partial v} + D_1 \frac{\partial x}{\partial u} = 0$$

$$A_2 \frac{\partial y}{\partial v} - C_2 \frac{\partial y}{\partial u} - B_2 \frac{\partial x}{\partial v} + D_2 \frac{\partial x}{\partial u} = 0$$

(1.3.7)

These equations are linear in the hodograph plane of u and v, in which we can therefore lay down characteristic ground curves.

If we repeat the analysis at the beginning of Sec. 1.2 on the system (1.3.7), we see that

$$\frac{du}{dv} = \zeta$$

(1.3.8)

will be a characteristic direction if

$$a'\zeta^2 - 2b'\zeta + c' = 0$$

(1.3.9)

where

$$a' = -A_1 B_2 + A_2 B_1$$
$$2b' = A_1 D_2 - A_2 D_1 - B_1 C_2 + B_2 C_1$$
$$c' = -C_1 D_2 + C_2 D_1$$

(1.3.10)

It is a matter of algebra to check that the values of ζ given by Eq. (1.3.9) are related to the values of σ given by Eq. (1.2.6), by the equation

$$\zeta_\pm = -\frac{A_1 C_2 - A_2 C_1}{A_1 B_2 - A_2 B_1}\, \sigma_\pm + \frac{B_1 C_2 - B_2 C_1}{A_1 B_2 - A_2 B_1}$$

(1.3.11)

But this is just what we should expect from the last two equations of the four characteristic differential equations (1.2.20). For in the reducible case $N_\pm \equiv 0$, so that

$$\frac{du}{dv} = \frac{\partial u/\partial \alpha}{\partial v/\partial \alpha} = -\frac{M_+}{L} = \zeta_+$$

(1.3.12)

or

$$\frac{du}{dv} = \frac{\partial u/\partial \beta}{\partial v/\partial \beta} = -\frac{M_-}{L} = \zeta_-$$

(1.3.13)

where L, M_+, and M_- are given by Eq. (1.2.18).

Since all the coefficients are functions of only u and v, the equations

$$\frac{du}{dv} = \zeta_+ (u, v) \tag{1.3.14}$$

and

$$\frac{du}{dv} = \zeta_- (u, v) \tag{1.3.15}$$

are two separate ordinary differential equations and can be solved to give two one-parameter families of characteristic curves, Γ_+ and Γ_-, in the hodograph plane of u and v. The advantage of reducibility is that these characteristics can be laid down once and for all and do not depend on the particular solution involved. These two families of Γ_+ and Γ_- may be represented in the form $\beta(u, v) = $ constant and $\alpha(u, v) = $ constant, respectively, and form a curvilinear coordinate net in the hodograph plane. The curvilinear coordinates α and β of the point (u, v) are then the characteristic parameters and there exists a one-to-one correspondence between the set (α, β) and the set (u, v) since ζ_+ and ζ_- do not vanish at the same time.

It is now natural to introduce the characteristic coordinate system (α, β) instead of (u, v) in such a way that β is constant along each of the Γ_+ characteristics and α is constant along each of the Γ_- characteristics. In other words, α and β are two new variables running along the Γ_+ and Γ_- characteristics, respectively. These characteristic parameters can be specified in the same way as we have discussed with the C-characteristics in the (x, y)-plane in the preceding section. Another choice for β and α in this case is the pair of the integration constants from Eqs. (1.3.14) and (1.3.15). More specifically, if we take a curve J given by $u = g(v)$, which nowhere has a characteristic direction, i.e., $g' \neq \zeta_+$ and $g' \neq \zeta_-$, the pair of integration constants can be expressed in terms of the information $u = g(v)$ given along the curve J. This pair certainly meets the requirement for the characteristic parameters. As before, any monotone functions $\alpha' = A(\alpha)$ and $\beta' = B(\beta)$ may be used as characteristic parameters since such a transformation leaves Eqs. (1.3.14) and (1.3.15) invariant and the characteristics Γ_+ and Γ_- unchanged. In Chapter 2 we shall see an example for which this transformation is conveniently used.

If u and v are known as functions of α and β, we can use Eq. (1.3.11) to express $\sigma_\pm$ in terms of α and β, i.e.,

$$\sigma_\pm = -\frac{A_1 B_2 - A_2 B_1}{A_1 C_2 - A_2 C_1} \zeta_\pm + \frac{B_1 C_2 - B_2 C_1}{A_1 C_2 - A_2 C_1} = \sigma_\pm (\alpha, \beta) \tag{1.3.16}$$

and then solve the first two characteristic differential equations of Eq. (1.2.20),

$$\frac{\partial y}{\partial \alpha} - \sigma_+(\alpha, \beta) \frac{\partial x}{\partial \alpha} = 0 \tag{1.3.17}$$

and

$$\frac{\partial y}{\partial \beta} - \sigma_-(\alpha, \beta) \frac{\partial x}{\partial \beta} = 0 \qquad (1.3.18)$$

for x and y as functions of α and β. This is just the sort of process we have outlined in Section 1.2, but it takes place in the hodograph plane of u and v and the initial data provided take the form of the values of x and y for given u and v.

In fact, the solutions of Eqs. (1.3.17) and (1.3.18) will give two families of characteristics, C_+ and C_-, in the (x, y)-plane. Since the parameter β remains constant along the C_+ characteristics and α along the C_- characteristics, we see that the characteristics C_+ and C_- in the (x, y)-plane are, respectively, the images of the characteristics Γ_+ and Γ_- in the hodograph plane, or vice versa, where the correspondence is signified by the subscripts $+$ and $-$. On the other hand, the totality of Γ characteristics is the very image of the solution to be determined. In principle, therefore, the solution can be constructed by mapping the net of Γ characteristics onto the (x, y)-plane by virtue of the correspondence given by Eq. (1.3.16). This process may be regarded as the inverse of the hodograph transformation.

Given the initial curve I as depicted in Fig. 1.9, we can construct the mapping by a step-by-step procedure. Just to show the strategy, let us first take points such as A, B, C, and D on I in the (x, y)-plane that are sufficiently close together and locate the corresponding points A', B', C', and D' on I' in the hodograph plane. We can then determine both Γ_+ and Γ_- characteristics passing through each of these points by integrating Eqs. (1.3.17) and (1.3.18) and this will give the net of Γ characteristics in the hodograph plane. Now the ini-

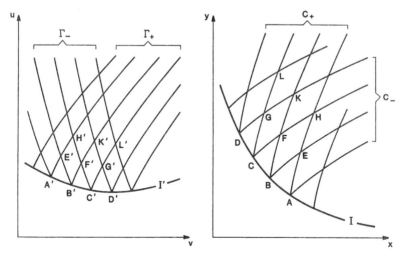

Figure 1.9

tial curve I' separates the hodograph plane into two distinct regions. But there is no confusion on which is the region of interest because with the arrangement as in Fig. 1.9 we expect that the Γ_+ characteristic passing through A' intersects the Γ_- characteristic passing through B' in the region of our interest. To find the map of the point E' we shall calculate σ_+ from Eq. (1.3.16) by using the values of u and v at A' and again by using the values at E', and then take the average value of the two as the slope of the C_+ characteristic emanating from A, which will be approximated by a straight-line segment. Similarly, taking the average value of σ_- between points B' and E', we can draw a straight line for the C_- characteristic from B until it interects the C_+ from A. The intersection E is obviously the map of the point E', so we know u and v at E. This procedure can be repeated to locate the points F and G which are the maps of the points F' and G', respectively. Based on this row of points we can move on to the next row of points, such as H, K, and L, and by continuing this process step by step we can construct the solution over the whole region of interest in the (x, y)-plane. As the points such as A, B, C, and D on the initial curve are taken closer and closer, the solution so determined will certainly converge to the exact solution.

Due to the nonlinear nature of the system (1.3.1) the slope of the C characteristics ($\sigma_\pm$) may vary significantly as u and v change. Under certain circumstances, therefore, we expect to see the C characteristics of the same family cross each other in the (x, y)-plane in the process of mapping. Beyond this point, then, u and v cannot be defined as continuous functions since both u and v have different values along the two characteristics, so that a discontinuity must be introduced. This is the case when the Jacobian J vanishes. Once introduced, the discontinuity is not governed by Eq. (1.3.1) but by another independent conservation principle which we shall discuss in Sec. 1.7. If the initial data include a discontinuity we must examine whether a unique determination of u and v is possible by calculating σ_+ and σ_- for the values of u and v on both sides of the discontinuity. If not, the discontinuity remains sharp in the course of time and here again we have the case of vanishing J. This will also be treated in Sec. 1.7.

Paradoxically, however, we often find the chief value of the reducibility of the equations is most evident when the hodograph transformation itself cannot be used. This will obtain when the Jacobian

$$ j = \frac{\partial(u, v)}{\partial(x, y)} $$

or its inverse vanishes. Let us note that the vanishing of the Jacobian means that we cannot invert the transformation and recover u and v as functions of x and y, but it does not mean that we cannot use the characteristic differential equations (1.2.20), or equivalently Eqs. (1.3.8), (1.3.9), and (1.3.10), for these were derived quite generally before the notion of reducibility was introduced.

Now the Jacobian can vanish in two ways. If the Jacobian matrix is of rank zero, all its terms vanish and $u(x, y)$, $v(x, y)$ are therefore independent of x and y and so constant. Such a *region of constant state* is represented by a single point in the hodograph plane, and the hodograph transformation is singular in the sense that a two-dimensional region is mapped onto a point of zero dimensions. In this case both sets of C characteristics in the (x, y)-plane are straight lines and the same kind are all parallel. Therefore, a region of constant state in the (x, y)-plane is necessarily bounded by straight characteristics.

On the other hand, the Jacobian can vanish if the Jacobian matrix has rank unity; that is, by virtue of u being a function of v throughout a certain region of the (x, y)-plane. If u and v are functionally related, this relation can be represented by a curve in the hodograph plane. But such a curve must be composed of segments of characteristics taken from the families of Γ_+ and Γ_-, as we can easily show. Suppose that $u = u(v)$; then

$$\frac{\partial u}{\partial x} = u'(v) \frac{\partial v}{\partial x} = \frac{du}{dv} \frac{\partial v}{\partial x} \quad \text{and} \quad \frac{\partial u}{\partial y} = \frac{du}{dv} \frac{\partial v}{\partial y}$$

Thus we can rewrite Eq. (1.3.1) in the form

$$\left(\frac{du}{dv} A_1 + C_1 \right) \frac{\partial v}{\partial x} + \left(\frac{du}{dv} B_1 + D_1 \right) \frac{\partial v}{\partial y} = 0$$

$$\left(\frac{du}{dv} A_2 + C_2 \right) \frac{\partial v}{\partial x} + \left(\frac{du}{dv} B_2 + D_2 \right) \frac{\partial v}{\partial y} = 0$$

and because these must be the same, we have

$$\frac{\dfrac{du}{dv} A_1 + C_1}{\dfrac{du}{dv} A_2 + C_2} = \frac{\dfrac{du}{dv} B_1 + D_1}{\dfrac{du}{dv} B_2 + D_2}$$

or

$$(-A_1 B_2 + B_1 A_2) \left(\frac{du}{dv} \right)^2 - (A_1 D_2 - A_2 D_1 - B_1 C_2 + B_2 C_1) \frac{du}{dv}$$
$$- C_1 D_2 + C_2 D_1 = 0 \qquad (1.3.19)$$

which is clearly the same as the combination of Eqs. (1.3.9) and (1.3.10). In this case the region of the (x, y)-plane in which u and v are functions of each other is mapped onto a one-dimensional region in the hodograph plane, and the hodograph transformation is again degenerate through this loss of dimensionality. A region of the (x, y)-plane in which u and v are functionally related, and which therefore maps onto characteristic arcs Γ_+ and Γ_-, is called a *simple wave region* and the solution in such a region is called a *simple wave*. If the image lies on a Γ_+ characteristic, we shall call the simple wave a Γ_+ simple wave, or if on a Γ_- characteristic, a Γ_- *simple wave*. We will show that a simple wave region in the (x, y)-plane is covered by a family of straight C_+ or C_- characteristics.

Consider a Γ_- simple wave region that maps on a particular Γ_- characteristic, say Γ_-^0. Since a C_- characteristic has its image on a Γ_- characteristic in the hodograph plane, it follows that all the C_- characteristics in the simple wave region map onto the Γ_-^0. On the other hand, a C_+ characteristic crosses all the C_- characteristics in the (x, y)-plane and its image falls on a Γ_+ characteristic in the hodograph plane as shown in Fig. 1.10(a). This implies that the image of each C_+ characteristic in the simple wave region must be a single point on the Γ_-^0. Hence both u and v remain constant along a C_+ characteristic, so every C_+ characteristic becomes a straight line. The same argument holds for a Γ_+ simple wave region if the roles of Γ_+ and Γ_- as well as those of C_+ and C_- are interchanged [see Fig. 1.10(b)]. Consequently, a Γ_- (or Γ_+) simple wave region is covered by arcs of the C_+ (or C_-) characteristics, which carry constant values of u and v and hence are straight lines.

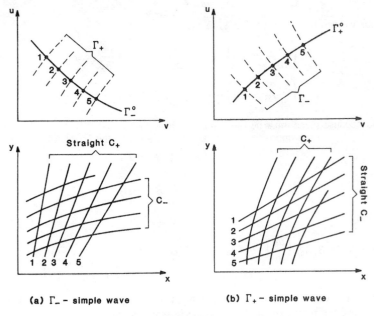

(a) Γ_- - simple wave (b) Γ_+ - simple wave

Figure 1.10

Suppose now that region $\mathcal{R}$ in Fig. 1.11 is one of constant state; then both the characteristics, C_+ and C_-, in this region are straight lines and the region is bounded by straight characteristics. Let us consider a portion of the boundary which is given by a C_+ characteristic, say C_+^0, and examine the solution in the region $\mathcal{Q}$ which is the opposite side of the C_+^0. The image of the region $\mathcal{R}$ is the point $\mathcal{R}'$ in the hodograph plane of u and v, so the boundary characteristic C_+^0 also maps into the point $\mathcal{R}'$. In the region $\mathcal{Q}$ consider the two families of

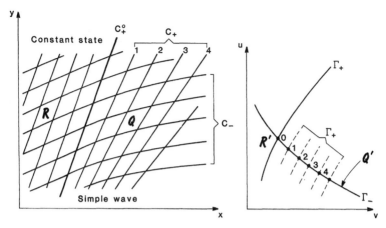

Figure 1.11

characteristics C_+ and C_-. All the C_- characteristics intersect C_+^0 and have their images along the Γ_- characteristics in the hodograph plane. This implies that all the C_- characteristics in the region $\mathfrak{Q}$ map onto the single Γ_- characteristics passing through the point $\mathfrak{R}'$ in the hodograph plane. It then follows that each C_+ in $\mathfrak{Q}$ maps into a point on the Γ_- characteristic passing through $\mathfrak{R}'$, so every C_+ characteristic in $\mathfrak{Q}$ is a straight line. Consequently, the solution in the region $\mathfrak{Q}$ is a Γ_- simple wave, but this is valid only when the C_+ characteristics fan out in the manner as shown on the left of Fig. 1.11. If the C_+ characteristics fan out in the opposite direction and overlap with the C_+ characteristics in the region $\mathfrak{R}$, we need to introduce a discontinuity. This will become clearer when we take a Riemann problem in Sec. 1.6. If the region $\mathfrak{R}$ of constant state is bounded by a C_- characteristic, the solution in the region $\mathfrak{Q}$ adjacent to $\mathfrak{R}$ has its image on the Γ_+ characteristic passing through the point $\mathfrak{R}'$ in the hodograph plane, which is the image of the region $\mathfrak{R}$. Hence the region $\mathfrak{Q}$ is covered by straight C_- characteristics and the solution in $\mathfrak{Q}$ is a Γ_+ simple wave.

Here we see that the solutions in the two regions $\mathfrak{R}$ and $\mathfrak{Q}$ are analytically different and the transition from one region to the other involves discontinuities of derivatives. The two regions are separated by a C characteristic. This again confirms our discussion in the preceding section that two different solutions can intersect only along a characteristic or a sharp edge in a solution surface must be a characteristic.

In summary, we state the following theorem that holds for pairs of reducible equations.

THEOREM 1.1   (on simple waves):

1. The solution in a region adjacent to a region of constant state is a simple wave. If the boundary is a C_+ (or C_-) characteristic, the solution is a Γ_- (or Γ_+) simple wave.

2. A Γ_- (or Γ_+) simple wave region has its image on a Γ_- (or Γ_+) characteristic in the hodograph plane and thus covered by arcs of C_+ (or C_-) characteristics, which carry constant values of u and v and hence are straight lines.

Exercise

1.3.1. If $\sigma_\pm$ are the roots of $a\sigma^2 - 2b\sigma + c = 0$, show that $\zeta_\pm = \mu\sigma_\pm + \nu$ are the roots of $a'\zeta^2 - 2b'\zeta + c' = 0$, provided that

$$\frac{b'}{a'} = \mu\frac{b}{a} + \nu$$

and

$$\frac{c'}{a'} = \mu^2\frac{c}{a} + 2\mu\nu\frac{b}{a} + \nu^2$$

Hence establish Eq. (1.3.11).

1.4 Characteristic Directions for Two-Solute Chromatography

In this section the developments of the general theory in the preceding sections are applied to the equilibrium chromatography of two solutes, which is the system to be treated subsequently in the forthcoming sections. Thus we have, from Eq. (1.1.16), the pair of equations

$$\frac{\partial c_1}{\partial x} + \frac{\partial}{\partial \tau} f_1 (c_1, c_2) = 0$$

$$\frac{\partial c_2}{\partial x} + \frac{\partial}{\partial \tau} f_2 (c_1, c_2) = 0$$

(1.4.1)

or in the form

$$\frac{\partial c_1}{\partial x} + \frac{\partial f_1}{\partial c_1}\frac{\partial c_1}{\partial \tau} + \frac{\partial f_1}{\partial c_2}\frac{\partial c_2}{\partial \tau} = 0$$

$$\frac{\partial f_2}{\partial c_1}\frac{\partial c_1}{\partial \tau} + \frac{\partial c_2}{\partial x} + \frac{\partial f_2}{\partial c_2}\frac{\partial c_2}{\partial \tau} = 0$$

(1.4.2)

where f_1 and f_2 are defined as functions of c_1 and c_2, i.e.,

$$f_1 = c_1 + \nu n_1 (c_1, c_2)$$

$$f_2 = c_2 + \nu n_2 (c_1, c_2)$$

(1.4.3)

The adsorption equilibrium isotherms n_1 and n_2 are continuous functions of c_1 and c_2 with as many derivatives as may be required and satisfy the conditions given by Eqs. (1.1.8) and (1.1.9). The independent variables x and τ are dimensionless and defined by Eq. (1.1.15). Clearly, the system of equations (1.4.2) is reducible so that the characteristic differential equations can be readily obtained by substitution in the general formulas derived in the previous sections. But here we formulate the relevant equations by following closely the procedure of the general theory.

Let us start with the linear combination of the pair of equations in Eq. (1.4.2)

$$\lambda_1 \frac{\partial c_1}{\partial x} + \left(\lambda_1 \frac{\partial f_1}{\partial c_1} + \lambda_2 \frac{\partial f_2}{\partial c_1}\right) \frac{\partial c_1}{\partial \tau} + \lambda_2 \frac{\partial c_2}{\partial x} + \left(\lambda_1 \frac{\partial f_1}{\partial c_2} + \lambda_2 \frac{\partial f_2}{\partial c_2}\right) \frac{\partial c_2}{\partial \tau} = 0$$

If we let ω represent the characteristic parameter (i.e., either α or β), the condition that this linear combination involves the derivatives $\partial c_1/\partial \omega$ and $\partial c_2/\partial \omega$ in the same direction, given by $(\partial x/\partial \omega, \partial \tau/\partial \omega)$, entails the four equations

$$\lambda_1 \frac{\partial \tau}{\partial \omega} - \left(\lambda_1 \frac{\partial f_1}{\partial c_1} + \lambda_2 \frac{\partial f_2}{\partial c_1}\right) \frac{\partial x}{\partial \omega} = 0$$

$$\lambda_2 \frac{\partial \tau}{\partial \omega} - \left(\lambda_1 \frac{\partial f_1}{\partial c_2} + \lambda_2 \frac{\partial f_2}{\partial c_2}\right) \frac{\partial x}{\partial \omega} = 0$$

$$\lambda_1 \frac{\partial c_1}{\partial \omega} + \lambda_2 \frac{\partial c_2}{\partial \omega} = 0$$

$$\left(\lambda_1 \frac{\partial f_1}{\partial c_1} + \lambda_2 \frac{\partial f_2}{\partial c_1}\right) \frac{\partial c_1}{\partial \omega} + \left(\lambda_1 \frac{\partial f_1}{\partial c_2} + \lambda_2 \frac{\partial f_2}{\partial c_2}\right) \frac{\partial c_2}{\partial \omega} = 0$$

(1.4.4)

which correspond to Eqs. (1.2.4), (1.2.15), and (1.2.16). These are homogeneous linear equations for λ_1 and λ_2. If we pick any two equations from Eq. (1.4.4), the determinant of coefficients must vanish because of our insistence on nontrivial values for λ_1 and λ_2. With the hodograph transformation in mind we shall take the last two equations to obtain

$$\frac{\partial f_2}{\partial c_1} \left(\frac{\partial c_1}{\partial \omega}\right)^2 - \left(\frac{\partial f_1}{\partial c_1} - \frac{\partial f_2}{\partial c_2}\right) \frac{\partial c_1}{\partial \omega} \frac{\partial c_2}{\partial \omega} - \frac{\partial f_1}{\partial c_2} \left(\frac{\partial c_2}{\partial \omega}\right)^2 = 0 \qquad (1.4.5)$$

Now we shall put

$$\frac{dc_1}{dc_2} = \frac{\partial c_1/\partial \omega}{\partial c_2/\partial \omega} = \zeta \qquad (1.4.6)$$

Then Eq. (1.4.5) becomes a quadratic equation for ζ,

$$\frac{\partial f_2}{\partial c_1} \zeta^2 - \left(\frac{\partial f_1}{\partial c_1} - \frac{\partial f_2}{\partial c_2}\right) \zeta - \frac{\partial f_1}{\partial c_2} = 0 \qquad (1.4.7)$$

or, by using Eq. (1.4.3),

$$\frac{\partial n_2}{\partial c_1} \zeta^2 - \left(\frac{\partial n_1}{\partial c_1} - \frac{\partial n_2}{\partial c_2}\right) \zeta - \frac{\partial n_1}{\partial c_2} = 0 \qquad (1.4.8)$$

in which n_1 and n_2 are the adsorption equilibrium isotherms and are expressed as functions of c_1 and c_2. Since both $\partial n_2/\partial c_1$ and $\partial n_1/\partial c_2$ are of negative sign [see Eq. (1.1.8)], this equation has two distinct real roots, one being positive and the other negative. Let these be ζ_+ and ζ_-. Then we have

$$\zeta_{\pm} = \frac{1}{2}\left(\frac{\partial n_2}{\partial c_1}\right)^{-1}\left\{\left(\frac{\partial n_1}{\partial c_1} - \frac{\partial n_2}{\partial c_2}\right) \pm \sqrt{\left(\frac{\partial n_1}{\partial c_1} - \frac{\partial n_2}{\partial c_2}\right)^2 + 4\frac{\partial n_1}{\partial c_2}\frac{\partial n_2}{\partial c_1}}\right\} \quad (1.4.9)$$

Accordingly, there exist two different characteristic directions from every point (c_1, c_2) in the hodograph plane, which are given by

$$\frac{dc_1}{dc_2} = \zeta_+ \, (c_1, c_2) > 0 \quad (1.4.10)$$

and

$$\frac{dc_1}{dc_2} = \zeta_- \, (c_1, c_2) < 0 \quad (1.4.11)$$

respectively. This implies that the pair of chromatographic equations (1.4.1) are hyperbolic, and Eqs. (1.4.10) and (1.4.11) represent the mutual interaction between the two solutes A_1 and A_2. The equations above can be integrated to give two families of characteristics, Γ_+ and Γ_-, in the hodograph plane, and as discussed in the preceding section, we can introduce the characteristic parameters α and β running, respectively, along the Γ_+ and Γ_- characteristics.

To find the characteristic directions in the (x, τ)-plane, we can choose any other pair of equations from Eq. (1.4.4), but the proper coordination is derived from the first and third equations, which give

$$\frac{\partial c_2}{\partial \omega}\left(\frac{\partial \tau}{\partial \omega} - \frac{\partial f_1}{\partial c_1}\frac{\partial x}{\partial \omega}\right) + \frac{\partial c_1}{\partial \omega}\frac{\partial f_2}{\partial c_1}\frac{\partial x}{\partial \omega} = 0$$

or

$$\frac{\partial \tau}{\partial \omega} - \left(\frac{\partial f_1}{\partial c_1} - \zeta\frac{\partial f_2}{\partial c_1}\right)\frac{\partial x}{\partial \omega} = 0 \quad (1.4.12)$$

Other combinations will give essentially the same expression. By using Eq. (1.4.3) we can express this equation in terms of the isotherms n_1 and n_2 as

$$\frac{\partial \tau}{\partial \omega} - \left\{1 + \nu\left(\frac{\partial n_1}{\partial c_1} - \zeta\frac{\partial n_2}{\partial c_1}\right)\right\}\frac{\partial x}{\partial \omega} = 0 \quad (1.4.13)$$

Now this equation holds if we identify ζ with ζ_+ and ω with α and, similarly, ζ with ζ_- and ω with β.

Finally, we have arrived at the four characteristic differential equations as follows. In the hodograph plane of c_1 and c_2,

$$\Gamma_+: \quad \left(\frac{\partial c_1}{\partial \alpha}\right)_\beta - \zeta_+\left(\frac{\partial c_2}{\partial \alpha}\right)_\beta = 0 \quad (1.4.14)$$

$$\Gamma_- : \quad \left(\frac{\partial c_1}{\partial \beta}\right)_\alpha - \zeta_- \left(\frac{\partial c_2}{\partial \beta}\right)_\alpha = 0$$

In the (x, τ)-plane,

$$C_+ : \quad \left(\frac{\partial \tau}{\partial \alpha}\right)_\beta - \left\{1 + \nu\left(\frac{\partial n_1}{\partial c_1} - \zeta_+ \frac{\partial n_2}{\partial c_1}\right)\right\}\left(\frac{\partial x}{\partial \alpha}\right)_\beta = 0$$

$$C_- : \quad \left(\frac{\partial \tau}{\partial \beta}\right)_\alpha - \left\{1 + \nu\left(\frac{\partial n_1}{\partial c_1} - \zeta_- \frac{\partial n_2}{\partial c_1}\right)\right\}\left(\frac{\partial x}{\partial \beta}\right)_\alpha = 0 \qquad (1.4.15)$$

Here the suffixes $+$ and $-$ represent the correspondence between the Γ characteristics and the C characteristics; i.e., Γ_+ and C_+ are the images of each other and, similarly, Γ_- and C_- are the images of each other.

Equations (1.4.15) may be put in the form

$$\left(\frac{d\tau}{dx}\right)_\beta = \frac{\partial f_1}{\partial c_1} - \zeta_+ \frac{\partial f_2}{\partial c_1}$$

$$= 1 + \nu\left(\frac{\partial n_1}{\partial c_1} - \zeta_+ \frac{\partial n_2}{\partial c_1}\right) \equiv \sigma_+ \qquad (1.4.16)$$

and

$$\left(\frac{d\tau}{dx}\right)_\alpha = \frac{\partial f_1}{\partial c_1} - \zeta_- \frac{\partial f_2}{\partial c_1}$$

$$= 1 + \nu\left(\frac{\partial n_1}{\partial c_1} - \zeta_- \frac{\partial n_2}{\partial c_1}\right) \equiv \sigma_- \qquad (1.4.17)$$

in which the first expressions are evident from Eq. (1.4.12). Here σ_+ and σ_- represent the slopes of the C_+ and C_- characteristics, respectively. By substituting Eq. (1.4.9), we can show that

$$\sigma_+ = 1 + \frac{\nu}{2}\left\{\frac{\partial n_1}{\partial c_1} + \frac{\partial n_2}{\partial c_2} - \sqrt{\left(\frac{\partial n_1}{\partial c_1} - \frac{\partial n_2}{\partial c_2}\right)^2 + 4\frac{\partial n_1}{\partial c_2}\frac{\partial n_2}{\partial c_1}}\right\} \qquad (1.4.18)$$

$$\sigma_- = 1 + \frac{\nu}{2}\left\{\frac{\partial n_1}{\partial c_1} + \frac{\partial n_2}{\partial c_2} + \sqrt{\left(\frac{\partial n_1}{\partial c_1} + \frac{\partial n_2}{\partial c_2}\right)^2 - 4\left(\frac{\partial n_1}{\partial c_1}\frac{\partial n_2}{\partial c_2} - \frac{\partial n_1}{\partial c_2}\frac{\partial n_2}{\partial c_1}\right)}\right\}$$
$$(1.4.19)$$

whence we deduce by applying Eqs. (1.1.8) and (1.1.9) that

$$1 < \sigma_- < \sigma_+ \qquad (1.4.20)$$

A direct consequence is that the slope of any C characteristic is always greater than 1, which is the direction of the fluid flow [see Eq. (1.1.15)], and that the C_+ characteristic is steeper than the C_- characteristic everywhere in the (x, τ)-plane. Also, we notice that any disturbance in c_1 and c_2 cannot propagate faster than the fluid flow.

Now we shall return to the characteristic quadratic, Eq. (1.4.7), in the hodograph plane and make an important observation. Let us rearrange the equation in the form

$$\frac{\partial f_1}{\partial c_1} + \frac{1}{\zeta}\frac{\partial f_2}{\partial c_1} = \zeta\frac{\partial f_2}{\partial c_1} + \frac{\partial f_2}{\partial c_2} \qquad (1.4.21)$$

The characteristic direction in the hodograph plane is given by $dc_1/dc_2 = \zeta$, so if we use the symbols $\mathcal{D}/\mathcal{D}c_1$ and $\mathcal{D}/\mathcal{D}c_2$ to denote derivatives with respect to c_1 and c_2 along a Γ characteristics, we have

$$\begin{aligned}
\frac{\mathcal{D}}{\mathcal{D}c_1} &= \frac{\partial}{\partial c_1} + \frac{dc_2}{dc_1}\frac{\partial}{\partial c_2} = \frac{\partial}{\partial c_1} + \frac{1}{\zeta}\frac{\partial}{\partial c_2} \\
\frac{\mathcal{D}}{\mathcal{D}c_2} &= \frac{dc_1}{dc_2}\frac{\partial}{\partial c_1} + \frac{\partial}{\partial c_2} = \zeta\frac{\partial}{\partial c_1} + \frac{\partial}{\partial c_2}
\end{aligned} \qquad (1.4.22)$$

Thus Eq. (1.4.21) can be rewritten as

$$\frac{\mathcal{D}f_1}{\mathcal{D}c_1} = \frac{\mathcal{D}f_2}{\mathcal{D}c_2} \qquad (1.4.23)$$

which we shall call the *fundamental differential equation* of Eq. (1.4.1). This equation can also be written in terms of n_1 and n_2 instead of f_1 and f_2 by applying Eq. (1.4.3), i.e.,

$$\frac{\mathcal{D}n_1}{\mathcal{D}c_1} = \frac{\mathcal{D}n_2}{\mathcal{D}c_2} \qquad (1.4.24)$$

This implies that the characteristic direction in the hodograph plane is the one in which the derivatives of n_1 and n_2 with respect to c_1 and c_2, respectively, are identical. What is more, since

$$\sigma_{\pm} = \frac{\partial f_1}{\partial c_1} - \zeta_{\pm}\frac{\partial f_2}{\partial c_1}$$

from Eqs. (1.4.16) and (1.4.17), and

$$\sigma_+ + \sigma_- = \frac{\partial f_1}{\partial c_1} + \frac{\partial f_2}{\partial c_2}$$

by applying Eq. (1.4.7), we can write

$$\sigma_+ = \frac{\partial f_1}{\partial c_1} + \frac{\partial f_2}{\partial c_2} - \sigma_- = \zeta_-\frac{\partial f_2}{\partial c_1} + \frac{\partial f_2}{\partial c_2} = \frac{\mathcal{D}_-f_2}{\mathcal{D}c_2} = \frac{\mathcal{D}_-f_1}{\mathcal{D}c_1} \qquad (1.4.25)$$

where $\mathcal{D}_-$ denotes differentiation along the Γ_- characteristics. Similarly,

$$\sigma_- = \zeta_+\frac{\partial f_2}{\partial c_1} + \frac{\partial f_2}{\partial c_2} = \frac{\mathcal{D}_+f_2}{\mathcal{D}c_2} = \frac{\mathcal{D}_+f_1}{\mathcal{D}c_1} \qquad (1.4.26)$$

These equations can be expressed in terms of n_1 and n_2 as

$$\sigma_+ = 1 + \nu \frac{\mathcal{D}_- n_1}{\mathcal{D}c_1} = 1 + \nu \frac{\mathcal{D}_- n_2}{\mathcal{D}c_2}$$

$$\sigma_- = 1 + \nu \frac{\mathcal{D}_+ n_1}{\mathcal{D}c_1} = 1 + \nu \frac{\mathcal{D}_+ n_2}{\mathcal{D}c_2}$$

(1.4.27)

Although motivated by the assumption of constant initial and feed conditions, Eqs. (1.4.24) and (1.4.27) have been well recognized in the equilibrium theory of two-solute chromatography (see e.g., DeVault, 1943; Glueckauf, 1949; and Offord and Weiss, 1949).

Just as in the case of a single solute the curve along which it is natural to give the initial data is the pair of positive x and τ axes. Thus in general we have

at $\tau = 0$: $c_1 = c_1^0(x)$ and $c_2 = c_2^0(x)$ for $x \geq 0$

at $x = 0$: $c_1 = c_1^f(\tau)$ and $c_2 = c_2^f(\tau)$ for $\tau \geq 0$ (1.4.28)

where the superscripts 0 and f denote the initial and feed conditions, respectively. As usual, we shall parametrize the pair of axes, if necessary, in such a way that $x = \xi$ along the x axis and $\tau = \eta$ along the τ axis. Then we have $c_i^0(\xi)$ and $c_i^f(\eta)$, $i = 1, 2$, for the initial and feed conditions, respectively. Since the initial curve has an edge at the origin of the (x, τ)-plane, the initial data are not continuously differentiable there. Clearly, this edge propagates along the C_+ or C_- characteristics emanating from the origin, so this pair of characteristics are of particular importance in the analysis of chromatographic problems. Another observation is that since $\sigma_+ > \sigma_- > 1$, both the x and τ axes as the initial curve are never tangent to a characteristic direction and they are both space-like.

If the initial and feed conditions are given by piecewise continuously differentiable functions and specified in such a way that their images have nowhere-characteristic directions in the hodograph plane, we can construct the solution by the procedure of mapping briefly discussed in Sec. 1.3. But there are cases when the images of the initial and feed data completely lie on the Γ-characteristics or fall on two different points in the hodograph plane so that the Jacobian j vanishes along the initial curve. These cases will be treated in Sec. 1.5. As remarked in Sec. 1.3, the solution can contain one or more discontinuities in c_1 and c_2. Since c_1 and c_2 are coupled through the equilibrium relations, we note that if one of c_1 and c_2 is discontinuous, the other is also necessarily discontinuous. After discussing the formation of such a discontinuity in Sec. 1.6, we shall treat discontinuous solutions in Sec. 1.7.

Before closing this section, let us discuss an important feature associated with the constant initial data. In the forthcoming sections we shall frequently be dealing with problems for which the initial data (or feed data) are given by constant values of c_1 and c_2 over some portions of the x (or τ)-axis. Consider the portion AB of the x axis on which c_1 and c_2 are specified by constant values c_1^0 and c_2^0, respectively, as shown in Fig. 1.12. Since the characteristic direc-

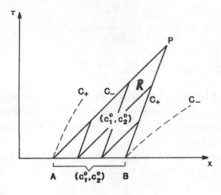

Figure 1.12

tions neither vanish nor diverge at any point in the (x, τ)-plane [see Eq. (1.4.20)], there develops a region $\mathfrak{R}$ adjacent to the portion AB of the x axis such that, for every point in $\mathfrak{R}$, we find its domain of dependence entirely falling on AB. This implies that the solution in $\mathfrak{R}$ is given by constant values c_1^0 and c_2^0. The region $\mathfrak{R}$ is bounded by the C_- characteristic emanating from the point A on the left-hand side by the C_+ from B on the right. If there are discontinuities at A and/or B or in the neighborhood of AB, the region $\mathfrak{R}$ still exists, although its boundaries may become different as we shall see later. Consequently, we have the following:

> **RULE 1.1:** If we specify constant initial or feed data along a portion of the x or τ axis, adjacent to this portion there develops a region of constant state, which is the same as the specified data.

We note that this conclusion is particularly meaningful if related to Theorem 1.1 on simple waves.

Exercises

1.4.1. By applying Eq. (1.3.11) to the system (1.4.2), obtain Eqs. (1.4.16) and (1.4.17).

1.4.2. If we let (for this example alone)

$$\omega_{ij} = \frac{\partial n_i}{\partial c_j}, \quad i, j = 1, 2$$

Show from the inequalities, $\zeta_- < 0 < \zeta_+$, $1 < \sigma_- < \sigma_+$ and other relevant equations of this section that

$$\frac{\omega_{11}}{\omega_{21}} < \zeta_- < \frac{\omega_{12}}{\omega_{22}}$$

and

$$\max\left\{-\frac{\omega_{12}}{\omega_{11}}, \frac{\omega_{22}^2 - \omega_{11}\omega_{22} + \omega_{12}\omega_{21}}{-\omega_{21}\omega_{22}}\right\} < \zeta_+ < \frac{\omega_{22}}{\omega_{21}}$$

1.4.3. Consider the equations for the one-dimensional, isentropic flow of compressible fluids discussed in Sec. 1.12 of Vol. 1 and determine the characteristic differential equations. Can you argue that the propagation speed of a disturbance differs from the particle velocity by the velocity of sound $\pm c$? Compare the results here with those of Exercise 1.2.1.

1.4.4. For steady two-dimensional irrotational isentropic flow, show that the governing equations are

$$\frac{\partial v}{\partial x} - \frac{\partial u}{\partial y} = 0$$

$$(c^2 - u^2)\frac{\partial u}{\partial x} - uv\left(\frac{\partial u}{\partial y} + \frac{\partial v}{\partial x}\right) + (c^2 - v^2)\frac{\partial v}{\partial y} = 0$$

where u and v are, respectively, the x and y components of the velocity and c, the velocity of sound, is a given function of $u^2 + v^2$. Can you argue that the system of equations is hyperbolic provided that the flow is supersonic? By forming a linear combination of these equations, find expressions for the characteristic directions in the hodograph plane as well as for those in the (x, y)-plane. How are the two pairs related? Compare the results here with those of Exercise 1.2.2.

1.4.5. Suppose that two solutes A_1 and A_2 are competing in a countercurrent adsorber under the equilibrium condition, so the conservation equations are given by Eq. (1.5.1) of Vol. 1 for $i = 1, 2$. By imposing the condition (1.1.8) show that this pair of equations is hyperbolic. Determine the four characteristic differential equations. What can you say about the signs of $\sigma_\pm$, the characteristic directions in the (z, t)-plane?

1.4.6. Consider adiabatic adsorption of a single solute in fixed beds for which the conservation equations have been derived in Exercise 1.1.1. Show that the system of equations is hyperbolic and determine the characteristic directions in the hodograph plane as well as in the (z, t)-plane.

1.5 Characteristic Initial Value Problem and Riemann Problem

Here and in Secs. 1.6 and 1.7 we shall take the equilibrium chromatography of two solutes as an example and discuss further developments of the general theory for cases when the Jacobian j or its inverse J vanishes. Thus we shall see how the solutions, being composed of constant states, simple waves, and discontinuities, can be constructed from the initial and feed data prescribed along the x and τ axes, respectively.

Just as the image of a simple wave region lies on a Γ characteristic, we can think of a case when the initial data are specified in such a way that the image of the initial curve falls on a Γ characteristic in the hodograph plane. In other words, the initial values of c_1 and c_2 satisfy the relationship representing the Γ_+ or Γ_- characteristics. We shall call such a class of problems the *characteristic initial value problem*. We notice that in such problems we are at liberty to prescribe only one concentration, e.g. c_2, and, at a single point, the value of the other concentration c_1, while the dependence of one upon the other is given by either Eq. (1.4.10) or (1.4.11). The initial data may consist of several portions which are analytically different. Then the image of the initial curve will be composed of several pieces of different Γ characteristics. But since the construction of the solution by pieces is possible, we shall confine our discussion here to a portion of the x or τ axis whose image falls on a single Γ characteristic. Thus the initial and feed conditions may be expressed as follows[†]:

$$
\begin{aligned}
\text{at } \tau = 0: \quad & c_2 = c_2^0(\xi) & \text{for } 0 \le \xi \le 1 \\
& c_2 = c_2^h = c_2^0(1) & \text{for } \xi > 1 \\
\text{at } x = 0: \quad & c_2 = c_2^t = c_2^0(0) & \text{for } \eta > 0 \\
& c_1 = c_1^t & \text{for } \eta > 0
\end{aligned}
\tag{1.5.1}
$$

and

$$
c_1 = c_1(c_2; c_1^t, c_2^t) \qquad \text{for } \xi, \eta \ge 0 \tag{1.5.2}
$$

where ξ and η are the parameters along the x and τ axes, respectively. Here the superscripts h and t denote the head and tail of a concentration wave and we shall use them in such a way that a quantity, if it carries the superscript h or t, remains constant.

Suppose now that the image of the interval $[0, 1]$ of the x axis lies on a portion of a Γ characteristic, say Γ_-^0, given by Eq. (1.5.2). The end points of this portion of Γ_-^0 are the images of the τ axis and the remainder of the x axis. We know from Rule 1.1 that there is a region of constant state adjacent to the τ axis and also adjacent to the remainder of the x axis as marked by (T) and (H), respectively, in Fig. 1.13, in which the parentheses are used to imply that it is a region of constant state. We shall show here that the solution in the region adjacent to the interval $[0, 1]$ of the x axis is a Γ_- simple wave. Since neither σ_+ nor σ_- vanishes, the x axis intersects every C characteristic appearing in the neighborhood. Consider now the C_- characteristics in the region adjacent to the interval $[0, 1]$ of the x axis. Each also has its image on a Γ_- characteristic. Since the Γ_- characteristics are disjoint in the hodograph plane, it can be inferred that all the images of these C_- characteristics fall on the same Γ_- characteristic, Γ_-^0. Therefore, adjacent to the interval $[0, 1]$ of the x axis there appears a region of Γ_- simple wave.

[†] Note that since $\zeta_\pm = dc_1/dc_2$, it is more convenient to select c_2 as the specified variable. But one could choose c_1 as well.

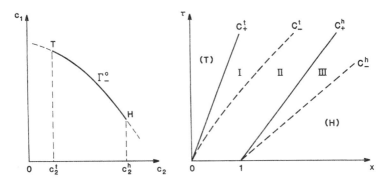

Figure 1.13

It is not difficult to picture the (x, τ)-diagram as shown in Fig. 1.13. The solution in the region II is entirely dependent on the data, $c_1^0(\xi)$ and $c_2^0(\xi)$, so that it is given by the Γ_- simple wave. Since C_+^t maps on T and C_-^t on Γ_-^0, respectively, in the hodograph plane, the region I is also of the Γ_- simple wave. On the other hand, both C_+^h and C_-^h map on H and thus the region III is a part of (H). Hence the Γ_- simple wave region is partitioned by the two characteristics, C_+^t and C_+^h, between the two regions of constant state. Similarly, if the interval $[0, 1]$ of the x axis has its image on a Γ_+ characteristic, we have a Γ_+ simple wave the region of which is bounded by two C_- characteristics. Similar arguments hold for the case of characteristic feed data, $c_1^f(\eta)$ and $c_2^f(\eta)$, since both σ_+ and σ_- never diverge. In conclusion, we propose the following:

RULE 1.2: If the initial or feed data are prescribed in such a way that c_1 and c_2 satisfy the relationship existing along a Γ characteristic, then the solution in the region adjacent to the corresponding portion of the x or τ axis is a simple wave. Further, if it is a Γ_+ simple wave, the region is bounded by the two marginal C_- characteristics and if a Γ_- simple wave, the region is bounded by the two marginal C_+ characteristics.

Let us again consider the initial and feed conditions given by Eqs. (1.5.1) and (1.5.2) so that there develops a region of Γ_- simple wave adjacent to the interval $[0, 1]$ of the x axis as before. Since the parameter α remains constant throughout the simple wave region, β is the only variable, so we can express $c_1^0(\xi)$ and $c_2^0(\xi)$ as functions of β. The inversion of $c_2^0(\xi)$ then defines ξ in terms of β. The C_+ characteristics with intercepts on the interval $[0, 1]$ are all straight, as shown in Fig. 1.14, carrying constant values c_1^0 and c_2^0. Their slopes σ_+ are determined by Eq. (1.4.16) or (1.4.18) with substitution of c_1^0 and c_2^0 for c_1 and c_2, respectively, and here it is assumed that the value of σ_+ decreases monotonically along the Γ_-^0 from T to H. Since σ_+ depends on c_1^0 and c_2^0, it can also be expressed as a function of β alone. Consequently, we obtain a one-parameter representation of the simple wave, i.e.,

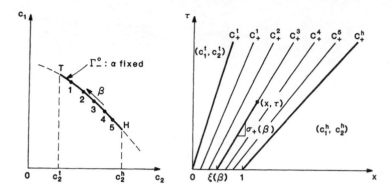

Figure 1.14

$$c_1 = \hat{c}_1(\beta), \qquad c_1 = \hat{c}_2(\beta) \qquad (1.5.3)$$

$$\tau = \sigma_+(\beta)\{x - \xi(\beta)\} \qquad (1.5.4)$$

Since $0 < \sigma_+ < \infty$, Eq. (1.5.4) has an inverse which gives β as a function of x and τ. Substituting this into Eq. (1.5.3), we can express c_1 and c_2 in terms of the original variables x and τ. The functions $c_1(x, \tau)$ and $c_2(x, \tau)$ so determined are the simple wave solution of the system (1.4.1) subject to the initial and feed conditions given by Eqs. (1.5.1) and (1.5.2). The same argument is valid for a Γ_+ simple wave if we use the parameter α in place of β and σ_- instead of σ_+.

Although the process of inversion above may not be amenable in many cases of practical interest, the actual construction of the solution can be carried out in a straightforward manner. First we select as many points as needed in the interval [0, 1] of the x axis and calculate the value of σ_+ for each point by using the values of c_1 and c_2 specified there. Then we draw a straight line of slope σ_+ from each point and this diagram establishes the Γ_- simple wave since we know the values of c_1 and c_2 along each of the C_+ characteristics so located. The profiles of c_1 and c_2 at any fixed time can be drawn by directly reading their values on each of the C_+ characteristics. This procedure is exactly the same as the one we have applied for the case of a single reducible equation.

Next we shall consider a Γ_+ simple wave based on characteristic feed data; i.e.,

$$
\begin{array}{llll}
\text{at } x = 0: & c_2 = c_2^f(\eta) & \text{for } 0 \leq \eta \leq 1 & \\
& c_2 = c_2^i = c_2^f(1) & \text{for } \eta > 1 & \\
\text{at } \tau = 0: & c_2 = c_2^h = c_2^f(0) & \text{for } \xi > 0 & \text{(1.5.5)} \\
& c_1 = c_1^h & \text{for } \xi > 0 &
\end{array}
$$

and

$$c_1 = c_1(c_2; c_1^h, c_2^h) \qquad \text{for } \xi, \eta \geq 0 \qquad (1.5.6)$$

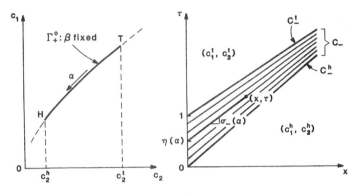

Figure 1.15

where Eq. (1.5.6) represents a Γ_+ characteristic. Now the parameter β remains fixed throughout the simple wave region and the region is covered by a family of straight C_- characteristics with intercepts on the interval $[0, 1]$ of the τ axis as shown in Fig. 1.15. Thus we can again establish a one-parameter representation of the simple wave:

$$c_1 = \hat{c}_1(\alpha), \qquad c_2 = \hat{c}_2(\alpha) \tag{1.5.7}$$

$$\tau - \eta(\alpha) = \sigma_-(\alpha)x \tag{1.5.8}$$

Here $\eta(\alpha)$ is the inverse of the feed data $c_2^f(\eta)$ after it is expressed in terms of α. Obviously, it is assumed in Fig. 1.15 that the value of σ_- increases monotonically along the Γ_+^0 from T to H. But as long as the C_- characteristics do not cross each other, Eq. (1.5.8) has an inverse since $0 < \sigma_- < \infty$, and thus c_1 and c_2 can be expressed as functions of x and τ.

In the special case when the interval $[0, 1]$ collapses to the origin, all the straight C characteristics issue from the same point; i.e., the simple wave is *centered* at the origin. This implies that in this case we can think of all the concentration pairs along the Γ_-^0 or Γ_+^0 between T and H in the hodograph plane being prescribed at the origin. As we shall see in the following, this observation will serve as the basis for the solution of a Riemann problem. Also we notice that the layouts of straight C-characteristics are in contrast between Figs. 1.14 and 1.15. Although it really depends on the nature of the nonlinearity and the initial data specified, the characteristics fan clockwise in the former case and in the opposite direction in the latter case as we proceed in the x direction. If the straight C-characteristics fan clockwise, we see the concentration profiles become more and more diffuse as time increases, so the simple wave is called an *expansion wave* or a *rarefaction wave*.[†] On the other hand, if the characteristics fan counterclockwise, the concentration profiles become steeper with time and eventually the characteristics will cross each other. In this case the simple wave is called a *compression wave* and we see a disconti-

---

[†] This term, originated from the compressible fluid flow, appears to be used more frequently.

nuity must be considered after a finite time. If a compression wave is to be centered, we shall be in a difficult situation from the beginning.

Now we come to a more practical situation in which the initial and feed data are given by two different constant states with a discontinuity at the origin, i.e.,

$$\text{at } \tau = 0: \quad c_i = c_i^0 \ (= \text{constant}) \quad \text{for } \xi > 0$$
$$\text{at } x = 0: \quad c_i = c_i^f \ (= \text{constant}) \quad \text{for } \eta > 0 \tag{1.5.9}$$

for $i = 1, 2$ and

$$c_i^0 \neq c_i^f \quad \text{for } i = 1 \text{ and/or } 2 \tag{1.5.10}$$

Thus the initial state and the feed state map onto two different points in the hodograph plane. We shall call this class of problems the *Riemann problem* just as in the case of a single quasilinear equation. In the practice of chromatography we find this kind of initial and feed conditions are most common: in particular, we have $c_i^0 = 0$ for the saturation of a fresh column and $c_i^f = 0$ for the elution of a presaturated column.

Let the points F and I denote the images of the constant states (c_1^f, c_2^f) and (c_1^0, c_2^0), respectively, in the hodograph plane as shown on the left of Fig. 1.16. According to Rule 1.1, we have a region of constant concentrations (c_1^f, c_2^f) adjacent to the τ axis and another region of constant state (c_1^0, c_2^0) adjacent to the x axis as marked by (F) and (I) on the right of Fig. 1.16. Furthermore, we know from Theorem 1.1 on simple waves that on the other side of the region (F) or (I) there appears a simple wave region, whose image lies on one of the two Γ characteristics passing through the point F or I. On the other side of this simple wave region we expect to have again a region of constant state and this arrangement may be repetitive as we proceed in the x direction. It then follows that the image of the relevant solution is represented by a path along the Γ characteristics connecting the feed point F to the initial point I. At the moment it would seem that there are infinitely many paths of this kind because, for example, the paths *FDRHI* and *FGREI* are certainly can-

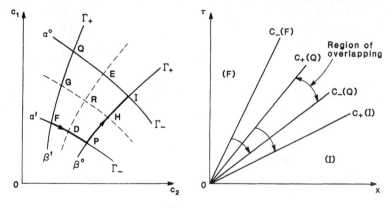

Figure 1.16

Pairs of Quasilinear Hyperbolic Equations of First Order   Chap. 1

didates. But by considering the map of each path onto the (x, τ)-plane we shall show that there is only one possibility.

Let us first take the path FQI and determine the solution in the (x, τ)-plane by mapping. If we assume that σ_- is monotone decreasing along FQ from F to Q, mapping of the portion FQ will give rise to a Γ_+ simple wave region in the (x, τ)-plane, which is covered by a family of straight C_- characteristics. This region is bounded below by the $C_-(Q)$, the C_- characteristic corresponding to the state Q, as shown on the right of Fig. 1.16. We shall assume that σ_+ is monotone decreasing along QI from Q to I so that we have a Γ_- simple wave region corresponding to QI. This region is covered by a family of straight C_+ characteristics and bounded above by the $C_+(Q)$. But we know from Eq. (1.4.20) that $\sigma_{-Q} < \sigma_{+Q}$ and thus the two simple wave regions must overlap as illustrated on the right of Fig. 1.16. Since the state of concentrations cannot be determined in this region of overlapping, the path FQI is not feasible. On the other hand, if we take the path FPI and proceed from F to P and then from P to I, the transition at P is from the $C_+(P)$ to the $C_-(P)$ and the slope of the former is greater. Therefore, there is no overlapping between the two simple wave regions as shown in the upper diagram of Fig. 1.17. This argument fur-

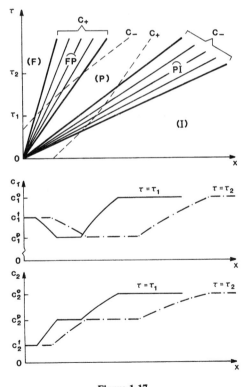

Figure 1.17

ther implies that a feasible path cannot contain more than two segments of Γ_+ and Γ_-, since then the (x, τ)-diagram is necessarily involved with one or more regions of overlapping. Thus we have established the uniqueness of the solution and propose the following:

> **RULE 1.3** For a Riemann problem the image of the solution in the hodograph plane consists of two segments of Γ characteristics of different kinds and connects the point image F of the feed state to the point image I of the initial state. As we proceed from F to I, we must move along the Γ_- first and then along the Γ_+.

The upper diagram of Fig. 1.17 shows that the solution is represented in the (x, τ)-plane by two regions of constant state, (F) and (I), and the range of influence of the discontinuity specified at the origin. This range of influence consists of two simple wave regions, $\widehat{FP}$ and $\widehat{PI}$, and the third region of constant state (P). The image of the region (P) is the point of intersection P between the two Γ characteristics. If we assign the characteristic parameters (α^f, β^f) to F and (α^0, β^0) to I, then the characteristic parameters for the state P is (α^f, β^0). Both simple waves are centered at the origin and this implies that all the concentration pairs along FP and PI are thought to be present at the origin. The dashed lines are drawn to emphasize that the C_- characteristic is curved in the Γ_- simple wave region $\widehat{FP}$ and so is the C_+ characteristic in the region $\widehat{PI}$. In the special case when the two image points F and I fall on the same Γ characteristic, the image of the solution is the portion of the Γ characteristic between the two points, so the solution contains only one centered simple wave. The profiles of c_1 and c_2 are illustrated in the middle and lower parts of Fig. 1.17 at two different times. Here we see that c_1 and c_2 vary in opposite directions in the Γ_- simple wave, whereas the two change in the same direction in the Γ_+ simple wave. As time increases, both profiles tend to be more diffuse and the intermediate plateau of state P also expands.

Let us now concentrate our attention on a specific region of simple wave, e.g., the Γ_- simple wave region $\widehat{FP}$ in Fig. 1.17. Throughout the region the equation of the Γ_- characteristic FP is valid, i.e.,

$$c_1 = c_1(c_2; c_1^f, c_2^f) \tag{1.5.11}$$

In addition, since the region is covered by a family of straight C_+ characteristics all emanating from the origin, we have

$$\frac{\tau}{x} = \sigma_+(c_1, c_2) \tag{1.5.12}$$

We note that σ_+ neither vanishes nor diverges and thus it is always possible to express c_1 and c_2 as functions of τ/x only. The same argument holds for the Γ_+ simple wave. A direct consequence is that the solution of a Riemann problem is determined by a function of τ/x only. This fact was noticed by some of the earlier contributors, such as Walter, Offord and Weiss, and Sillen in the field of chromatography, and played an important role in the development of the conventional theory of chromatography.

Pairs of Quasilinear Hyperbolic Equations of First Order   Chap. 1

If σ_+ is monotone increasing along FP from F to P, the C_+ characteristics in the Γ_- simple wave region $\widehat{FP}$ would fan backward, so the profiles of c_1 and c_2 would become sigmoid. Hence we must introduce a discontinuity in the solution. The same is true if σ_- is monotone increasing along $\widehat{PI}$ from P to I. We shall treat these situations in Sec. 1.7. If the initial and feed data include more than one point of discontinuity, then exactly the same strategy can be applied to each point of discontinuity. Such a separate treatment will lead to the solution as a whole. Therefore, with piecewise constant initial and feed data, the solution is determined by a finite number of centered simple waves, all separated by regions of constant state, unless discontinuities are to be introduced. But as time goes on, the simple waves centered at different points will meet and interact with each other. This is a subject yet to be resolved in the general theory, but a complete analysis can be accomplished in case of the equilibrium chromatography with Langmuir isotherms, as we shall see in Chapter 2.

1.6 Compression Waves and the Formation of Shocks

In Sec. 1.5 we have seen examples of the expansion wave and the compression wave. Although we were concerned with simple waves there, the argument can be extended to the general case by recalling Fig. 1.9, in which the concentration profiles would diffuse with time only if both C_+ and C_- characteristics fan clockwise as we proceed in the x-direction. On the other hand, the profiles tend to steepen if either C_+ or C_- characteristics fan backward. Thus we propose the criterion concerning nonlinear wave propagation in the following form; that is, a concentration wave is expansive if

$$\frac{\partial \sigma_+}{\partial x} < 0 \quad \text{and} \quad \frac{\partial \sigma_-}{\partial x} < 0 \qquad (1.6.1)$$

and compressive if

$$\frac{\partial \sigma_+}{\partial x} > 0 \quad \text{and} \quad \frac{\partial \sigma_-}{\partial x} > 0 \qquad (1.6.2)$$

We understand that the signs of these derivatives are determined by the nonlinear character of the partial differential equations (1.4.1) and the way the initial data are specified, but it is not expected, at the moment, to interpret the conditions (1.6.1) and (1.6.2) in more direct terms.[†] If we confine our attention to simple waves, we always have a one-parameter representation of the solution. Furthermore, the C characteristics of one kind or the other are all straight and each of those carries constant values of c_1 and c_2. Thus in case of a Γ_- simple wave based on the characteristic initial data $c_2^0(\xi)$ [see Eqs. (1.5.1) and (1.5.2)], we can rewrite the condition (1.6.1) as[‡]

[†] See the treatment in Sec. 1.10.

[‡] It is a matter of convenience to choose c_2 here. One may choose c_1 and develop similar arguments.

$$\frac{\partial \sigma_+}{\partial \xi} = \left(\zeta_- \frac{\partial \sigma_+}{\partial c_1} + \frac{\partial \sigma_+}{\partial c_2} \right) \frac{dc_2}{d\xi} < 0 \qquad (1.6.3)$$

in which $dc_2/d\xi$ is to be determined along the initial curve, i.e., the x axis in this case. But from Eq. (1.4.22) we can put

$$\zeta_- \frac{\partial \sigma_+}{\partial c_1} + \frac{\partial \sigma_+}{\partial c_2} = \frac{\mathscr{D}_- \sigma_+}{\mathscr{D} c_2} \qquad (1.6.4)$$

where $\mathscr{D}_-$ denotes differentiation along the Γ_- characteristics. Similarly, $\mathscr{D}_+$ will be used for differentiation along the Γ_+ characteristics [see Eqs. (1.4.25) and (1.4.26)]. Here we note that the derivative $\mathscr{D}_- \sigma_+/\mathscr{D} c_2$ really represents the nonlinearity of the system, while $dc_2/d\xi$ describes the manner in which the initial data are specified. This argument can easily be extended to various cases and gives the following rule.

RULE 1.4:   For a characteristic initial value problem with the image on a Γ_- characteristic, the solution is given by an expansion wave if the equilibrium relations and the initial data satisfy the inequality

$$\frac{d\sigma_+}{d\xi} = \frac{\mathscr{D}_- \sigma_+}{\mathscr{D} c_2} \frac{dc_2^0}{d\xi} < 0 \qquad (1.6.5)$$

and by a compression wave if

$$\frac{d\sigma_+}{d\xi} = \frac{\mathscr{D}_- \sigma_+}{\mathscr{D} c_2} \frac{dc_2^0}{d\xi} > 0 \qquad (1.6.6)$$

If the image falls on a Γ_+ characteristic, the same is true with $\mathscr{D}_+ \sigma_-/\mathscr{D} c_2$ in place of $\mathscr{D}_- \sigma_+/\mathscr{D} c_2$. If we have characteristic feed data, ξ and c_2^0 are replaced by η and c_2^f, respectively, and both the inequalities are reversed.

The derivatives $\mathscr{D}_- \sigma_+/\mathscr{D} c_2$ and $\mathscr{D}_+ \sigma_-/\mathscr{D} c_2$ are worth examining further in the light of the nonlinear character of the system. First, these derivatives may be expressed in terms of the characteristic parameters α and β as

$$\frac{\mathscr{D}_- \sigma_+}{\mathscr{D} c_2} = \frac{d\sigma_+/d\beta}{dc_2/d\beta} \quad \text{and} \quad \frac{\mathscr{D}_+ \sigma_-}{\mathscr{D} c_2} = \frac{d\sigma_-/d\alpha}{dc_2/d\alpha} \qquad (1.6.7)$$

so that if the characteristic parameters are given explicitly, we shall find these expressions more convenient. In Sec. 1.5 we have treated cases when σ_+ is monotone increasing or monotone decreasing along a Γ_- characteristic. Thus if this is true along every Γ_- characteristic, we can write

$$\frac{\mathscr{D}_- \sigma_+}{\mathscr{D} c_2} \neq 0 \qquad (1.6.8)$$

along every Γ_- characteristic. When this condition does hold, we say that the (+) *characteristic field* of the system (1.4.1) is *genuinely nonlinear,* after Lax (1957). Similarly, if we have

$$\frac{\mathcal{D}_+ \sigma_-}{\mathcal{D}c_2} \neq 0 \qquad (1.6.9)$$

along every Γ_+ characteristic, the $(-)$ characteristic field is called genuinely nonlinear. This definition is in accordance with the case of single equations (see Sec. 5.2 of Vol. 1) and bears a particular meaning in relation to the discontinuous solutions which we shall discuss in Sec. 1.7. If both Eqs. (1.6.8) and (1.6.9) hold, we may call the system of equations genuinely nonlinear.

Suppose now that the $(+)$ characteristic field is genuinely nonlinear. In addition, if the initial data are specified in such a way that c_2 is a monotonic function of ξ, then the Γ_- simple wave solution becomes entirely expansive or entirely compressive. A similar argument can be made with respect to the $(-)$ characteristic field. But there are also cases when one or both of the characteristic fields are not genuinely nonlinear (i.e., $\mathcal{D}_-\sigma_+/\mathcal{D}c_2$ and/or $\mathcal{D}_+\sigma_-/\mathcal{D}c_2$ change their signs along Γ characteristics) or when the initial data are such that c_2 is not given by a monotonic function of ξ. Under these situations we expect to see simple waves that are partially expansive and partially compressive depending on the nature of the characteristic field and the way the initial data are prescribed. This feature of the nonlinear wave propagation is essentially the same as in the case of single equations (see Fig. 5.8 of Vol. 1).

As we have seen in Fig. 1.15, the family of straight C characteristics in a compression wave region tend to fan backward as we proceed in the x direction, so they will cross each other after a finite period of time. Since these characteristics are of the same kind and carry different values of c_1 and c_2, it is obvious that beyond this time the concentrations c_1 and c_2 will have more than one pair of values at one position and time. This certainly poses a physically impossible situation and thus here again we must introduce a discontinuity at the earliest moment when the straight C characteristics begin to cross. As we have observed in Sec. 6.1 of Vol. 1, the family of straight C characteristics possess an *envelope* that encloses an angular region as shown in Fig. 1.18.

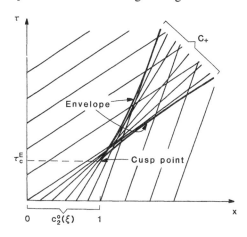

Figure 1.18

This region is covered three times by the straight C characteristics as we proceed in the x direction and all the crossing of the characteristics takes place inside the region. An immediate consequence is that a discontinuity is formed at the point where the envelope takes the lowest ordinate, i.e., the least time in this case. Therefore, we shall first determine the envelope of straight characteristics and thereby locate the point where a discontinuity is to be formed.

Let us now consider the initial and feed conditions given by Eqs. (1.5.1) and (1.5.2) that will lead to a Γ_- simple wave solution and suppose that this is a compression wave. Since the region of this compression wave is covered by a family of straight C_+ characteristics, we can write Eq. (1.5.4), i.e.,

$$\tau = \sigma_+(\beta)\{x - \xi(\beta)\} \tag{1.6.10}$$

where $\xi(\beta)$, denoting the intercept of each characteristic, is given by the inverse of the initial data $c_2^0(\xi)$ with c_2^0 expressed in terms of the characteristic parameter β. Differentiating this equation of C_+ characteristics with respect to β, we have

$$\frac{d\tau}{d\beta} = \sigma_+\left(\frac{dx}{d\beta} - \frac{d\xi}{d\beta}\right) + \frac{d\sigma_+}{d\beta}\{x - \xi(\beta)\}$$

but the term $\sigma_+(dx/d\beta)$ is just equal to $d\tau/d\beta$, so

$$\frac{d\sigma_+}{d\beta}\{x - \xi(\beta)\} = \sigma_+\frac{d\xi}{d\beta} \tag{1.6.11}$$

According to the Theorem of Sec. 0.10 of Vol. 1, the envelope of the characteristics is determined by Eqs. (1.6.10) and (1.6.11), so that we have

$$x^E = \xi(\beta) + \sigma_+(\beta)\frac{d\xi/d\beta}{d\sigma_+/d\beta}$$
$$\tau^E = [\sigma_+(\beta)]^2\frac{d\xi/d\beta}{d\sigma_+/d\beta} \tag{1.6.12}$$

which form a one-parameter representation of the envelope. Clearly, a discontinuity will first appear at $\tau = \min \tau^E$. These equations can be rewritten by using Eq. (1.6.7) in the form

$$x^E = \xi + \sigma_+\left\{\frac{\mathcal{D}_-\sigma_+}{\mathcal{D}c_2}\frac{dc_2^0}{d\xi}\right\}^{-1}$$
$$\tau^E = \sigma_+^2\left\{\frac{\mathcal{D}_-\sigma_+}{\mathcal{D}c_2}\frac{dc_2^0}{d\xi}\right\}^{-1} \tag{1.6.13}$$

which are to be compared with Eq. (6.1.3) of Vol. 1 given for the case of single equations.

If we have a Γ_- simple wave which is compressive and based on the characteristic feed data [see Eqs. (1.5.5) and (1.5.6)], a similar procedure starting from the equation

$$\tau - \eta(\beta) = \sigma_+(\beta)x \tag{1.6.14}$$

leads to the following parametric representation of the envelope:

$$x^E = -\frac{d\eta/d\beta}{d\sigma_+/d\beta}$$

$$\tau^E = \eta(\beta) - \sigma_+(\beta)\frac{d\eta/d\beta}{d\sigma_+/d\beta}$$

(1.6.15)

or to the pair of equations

$$x^E = -\left\{\frac{\mathscr{D}_- \sigma_+}{\mathscr{D}c_2}\frac{dc_2^f}{d\eta}\right\}^{-1}$$

$$\tau^E = -\sigma_+\left\{\frac{\mathscr{D}_- \sigma_+}{\mathscr{D}c_2}\frac{dc_2^f}{d\eta}\right\}^{-1}$$

(1.6.16)

For a Γ_+ simple wave which is compressive, we need only to change σ_+ and β to σ_- and α, respectively, in Eqs. (1.6.12) and (1.6.15) and, in addition, $\mathscr{D}_-$ to $\mathscr{D}_+$ in Eqs. (1.6.13) and (1.6.16).

The envelope of straight characteristics consists of two branches which come together forming a sharp corner, as illustrated in Fig. 1.18. This corner is called a *cusp* point and it is at this cusp point (x_c^E, τ_c^E) that a discontinuity is first formed. Thus we have

$$\tau = \tau_c^E = \min \tau^E$$

$$x = x_c^E$$

(1.6.17)

for the point of inception of a discontinuity. Since we are usually concerned with a finite interval [0, 1] of the x or τ axis, it is quite possible to have local minima at both end points. Among various cases, therefore, there is a possibility that two discontinuities are separately formed on the two boundary characteristics of the compression wave region. The situation here is really analogous to that for the case of a single equation treated in Sec. 6.1 of Vol. 1.

Once it is formed, a discontinuity will grow in size and propagate along a path which would not allow the straight characteristics, say C_+, to pass through. If this path is located outside the angular region of the envelope, we see that the characteristics on one side or the other of the path still intersect each other. Hence the path of a discontinuity must be located inside the angular region enclosed by the envelope. It then follows that to every point along the path there corresponds a pair of C_+ characteristics impinging on the path, one from the left-hand side and the other from the right. A discontinuity of this kind is called a *shock*, as we shall see in Sec. 1.7. This has been observed for a Γ_- simple wave, and obviously a similar argument can be applied to a Γ_+ simple wave. This implies that there are two different kinds of shocks, one originating from a Γ_- simple wave and the other from a Γ_+ simple wave. We shall come back to this subject in Sec. 1.7.

If the characteristic initial or feed data are not prescribed by smooth functions but by numerical values, we can apply the "base curve" method of Sec. 6.1 of Vol. 1 with a slight modification. Suppose, for example, that a Γ_- simple wave is compressive and based on the characteristic initial data specified over the interval [0, 1] of the x-axis [see Eqs. (1.5.1) and (1.5.2)]. Here the C_+ characteristics are straight, so the corresponding base curve is given by the equation

$$(\sigma_+)_B = \frac{\sigma'_+}{1 - (1 - \sigma'_+/\sigma^h_+)\xi}, \qquad 0 \le \xi \le 1 \qquad (1.6.18)$$

which is in accordance with Eq. (6.1.15) of Vol. 1. The superscripts h and t represent that the values are evaluated for $\xi = 1$ and $\xi = 0$, respectively. The further procedure is essentially the same as discussed for single equations in Sec. 6.1 of Vol. 1 and will not be repeated here.

Exercises

1.6.1. Consider two vectors, $\mathbf{r}_-$ and $\mathbf{r}_+$, that are tangent to the Γ_+ and Γ_- characteristics, respectively. Let the second elements be equal to one and show that

$$\frac{\mathscr{D}_-\sigma_+}{\mathscr{D}c_2} = \mathbf{r}^T_+ \, \nabla\sigma_+$$

and

$$\frac{\mathscr{D}_+\sigma_-}{\mathscr{D}c_2} = \mathbf{r}^T_- \, \nabla\sigma_-$$

where ∇ denotes the gradient in the (c_1, c_2)-plane.

1.6.2. For a Γ_- simple wave we can take one of the dependent variables, say c_2, as the parameter and express other variables in terms of c_2 alone. Apply this argument to Eq. (1.6.10) and derive Eq. (1.6.13).

1.7 Discontinuities in Solutions and the Entropy Condition

We should now be convinced that under certain circumstances we must have discontinuities in the dependent variables, and here we examine the nature of such discontinuities. Since the concept of discontinuous solutions is introduced on physical grounds, we shall again consider equilibrium chromatography of two solutes. Thus we have from Eq. (1.4.1)

$$\frac{\partial c_1}{\partial x} + \frac{\partial}{\partial \tau}f_1(c_1, c_2) = 0$$

$$\frac{\partial c_2}{\partial x} + \frac{\partial}{\partial \tau}f_2(c_1, c_2) = 0 \qquad (1.7.1)$$

where f_1 and f_2 are the equilibrium column isotherms defined as

$$f_1(c_1, c_2) = c_1 + v n_1(c_1, c_2)$$
$$f_2(c_1, c_2) = c_2 + v n_2(c_1, c_2)$$

$$(1.7.2)$$

Suppose that there are discontinuities in c_1 and c_2 as illustrated in Fig. 1.19. As usual, we shall adopt the superscripts l and r to denote, respectively, the left- and right-hand sides of a discontinuity and brackets [$\cdot$], to denote the jump in the quantity enclosed across a discontinuity so that, for example, $[c_1] = c_1^l - c_1^r$. The mass balance over the discontinuity in c_1 may be established in exactly the same way as in Sec. 5.4 or 7.2 of Vol. 1 to give the equation

$$\frac{d\tau}{dx} = \frac{f_1^l - f_1^r}{c_1^l - c_1^r} = \frac{[f_1]}{[c_1]} = \tilde{\sigma}_1 \qquad (1.7.3)$$

for the propagation direction of the discontinuity. Similarly, for the discontinuity in c_2 we have

$$\frac{d\tau}{dx} = \frac{f_2^l - f_2^r}{c_2^l - c_2^r} = \frac{[f_2]}{[c_2]} = \tilde{\sigma}_2 \qquad (1.7.4)$$

But the two concentrations are related to each other by the equilibrium relations, $n_1(c_1, c_2)$ and $n_2(c_1, c_2)$, and hence if, for instance, c_1 is discontinuous, both n_1 and n_2 become discontinuous, and so is c_2. This implies that a discontinuity in one of the concentrations will generally be accompanied by a discontinuity in the other. The two are not separate but rather form an entity and must move as such. Thus we have $\tilde{\sigma}_1 = \tilde{\sigma}_2$ or

$$\frac{[f_1]}{[c_1]} = \frac{[f_2]}{[c_2]} \qquad (1.7.5)$$

which is an algebraic equation. If we know both concentrations on one side and one concentration on the other, the unknown concentration can be determined by solving Eq. (1.7.5). As the discontinuity moves, the values of these

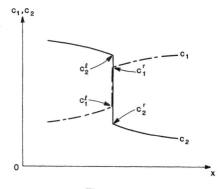

Figure 1.19

concentrations may change, but they must always satisfy Eq. (1.7.5). For this reason Eq. (1.7.5) is called the *compatibility relation*. By using Eq. (1.7.2) we can rewrite the compatibility relation (1.7.5) in an equivalent form,

$$\frac{[n_1]}{[c_1]} = \frac{[n_2]}{[c_2]} \tag{1.7.6}$$

Under the condition that the compatibility relation is satisfied, we have the equation

$$\frac{d\tau}{dx} = \bar{\sigma}(c_1^l, c_2^l, c_1^r, c_2^r) = \frac{[f_i]}{[c_i]} = 1 + \nu\frac{[n_i]}{[c_i]} \tag{1.7.7}$$

for $i = 1$ or $i = 2$ for the propagation direction of a discontinuity or its reciprocal speed of propagation. Equation (1.7.7) is called, in general, the *jump condition* and it is equivalent to the Rankine–Hugoniot relation in compressible fluid flow. Here we notice that by virtue of the compatibility relation the propagation speed of a discontinuity is determined by the concentration state on one side and one condition on the other. Similarly, given the state on one side and the propagation speed of the discontinuity, the unknown state of concentrations on the other side can be determined in terms of known quantities.

There is a striking analogy between the compatibility relation (1.7.5) and the fundamental differential equation (1.4.23) and also between the jump condition (1.7.7) and the equation for the slope of a characteristic [Eq. (1.4.25) or (1.4.26)]. For the latter pair we had similar relations in case of single quasilinear equations. Suppose that we have a discontinuity with the state on its left-hand side (c_1^l, c_2^l) fixed and consider the locus of the possible states on the right-hand side (c_1^r, c_2^r) in the hodograph plane of c_1 and c_2. Clearly, such a locus will be determined by the compatibility relation to give a curve issuing from the point image L of the state (c_1^l, c_2^l). Let this curve be denoted by Σ. By comparing the compatibility relation with the fundamental differential equation (1.4.23) it is apparent that the curve Σ must become tangent to a Γ characteristic passing through L in the limit as the discontinuity becomes infinitely weak in strength, i.e., as $c_1^r \to c_1^l$ and $c_2^r \to c_2^l$. This implies that, in general, there are two different loci of the state $R(c_1^r, c_2^r)$, one being tangent to the Γ_+ at L and the other tangent to the Γ_- at L. We shall denote the former by Σ_+ and the latter by Σ_- as illustrated in Fig. 1.20. This is in accordance with our observation with regard to the formation of shocks in Sec. 11.6. In fact, it can be shown that the curve Σ makes a third-order contact at L with the corresponding Γ characteristic and thus $\Sigma_{\pm}$ and $\Gamma_{\pm}$ have the same curvature, in addition to the common tangent, at L.[†]

In the special case when the Γ characteristics are straight, the argument above implies that the curve Σ would completely fall on a Γ characteristic. In fact, this can be proven as follows. If we substitute the equation for Γ characteristics,

[†] For the formal proof of this, see Keyfitz (1978).

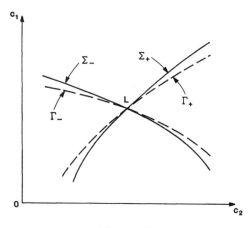

Figure 1.20

$$\zeta = \frac{dc_1}{dc_2} = \text{constant} \tag{1.7.8}$$

into the fundamental differential equation, we obtain

$$\frac{\mathscr{D}f_1}{\mathscr{D}c_1} = \zeta \frac{\mathscr{D}f_2}{\mathscr{D}c_1} \tag{1.7.9}$$

This can be integrated with respect to c_1 along a Γ characteristic from a point (c_1^l, c_2^l) to (c_1^r, c_2^r) to give

$$[f_1] = \zeta[f_2]$$

But from Eq. (1.7.8) we have

$$[c_1] = \zeta[c_2]$$

and hence the compatibility relation (1.7.5) is satisfied. An immediate consequence is that the curve Σ necessarily coincides with a Γ characteristic.

If the states on both sides of a discontinuity fall on a Σ_+, we shall call it a Σ_+ discontinuity and if on a Σ_-, a Σ_- discontinuity. Their propagation directions will also be distinguished in accordance with the direction of the straight characteristics in a simple wave region. Thus for a Σ_- discontinuity we have

$$\bar{\sigma}_+(c_1^l, c_2^l, c_1^r, c_2^r) = \left(\frac{[f_i]}{[c_i]}\right)_{\Sigma_-} \tag{1.7.10}$$

with $i = 1$ or $i = 2$ and for a Σ_+ discontinuity

$$\bar{\sigma}_-(c_1^l, c_2^l, c_1^r, c_2^r) = \left(\frac{[f_i]}{[c_i]}\right)_{\Sigma_+} \tag{1.7.11}$$

Comparing these equations with Eqs. (1.4.25) and (1.4.26), we find that

$$\bar{\sigma}_+(c_1^l, c_2^l, c_1^l, c_2^l) = \sigma_+(c_1^l, c_2^l) = \sigma_+^l$$

$$\bar{\sigma}_-(c_1^l, c_2^l, c_1^l, c_2^l) = \sigma_-(c_1^l, c_2^l) = \sigma_-^l$$

(1.7.12)

If we take the state on the right-hand side (c_1^r, c_2^r) as fixed, Eqs. (1.7.10) and (1.7.11) still hold since they are symmetric with respect to the superscripts l and r. This implies that although the locus of the state (c_1^l, c_2^l) may be different from the previous one, it passes through the point image L of the state (c_1^l, c_2^l). Now taking the limit as $c_1^l \to c_1^r$ and $c_2^l \to c_2^r$ along this new locus, we have

$$\bar{\sigma}_+(c_1^r, c_2^r, c_1^r, c_2^r) = \sigma_+(c_1^r, c_2^r) = \sigma_+^r$$

$$\bar{\sigma}_-(c_1^r, c_2^r, c_1^r, c_2^r) = \sigma_-(c_1^r, c_2^r) = \sigma_-^r$$

(1.7.13)

Although we have the compatibility relation and the jump condition, these are not sufficient to determine a unique solution since an ambiguity still exists concerning the directions of jumps in c_1 and c_2. This is the same situation as we encountered before in case of single equations. Here again we need to introduce an additional condition, the so-called *entropy condition*, that specifies the relevant direction of jumps across a discontinuity. At the moment it is not promising to discuss the entropy condition in general terms, so we shall confine our attention to the characteristic initial value problems and Riemann problems. One way to advance, however, is to treat problems with piecewise constant data and then take the general initial data as the limit of piecewise constant data (see, e.g., Smoller and Johnson, 1969).

Within the scope of our interest here the image of the solution in the hodograph plane consists of segments of Γ characteristics and thus in the (x, τ)-plane the family of characteristics C_+ or C_- are all straight. According to our experience in the past two sections, there is no need to introduce a discontinuity as long as the straight C characteristics fan clockwise as we proceed in the x direction. Only when the C characteristics fan backward in the (x, τ)-plane, do we need to consider a discontinuity. Therefore, it appears plausible to assert that the entropy condition is given by

$$\sigma_+(c_1^l, c_2^l) < \sigma_+(c_1^r, c_2^r)$$

(1.7.14)

for a Σ_- discontinuity and

$$\sigma_-(c_1^l, c_2^l) < \sigma_-(c_1^r, c_2^r)$$

(1.7.15)

for a Σ_+ discontinuity. But the argument above is valid only if the system is genuinely nonlinear and indeed, we can show that this pair of inequalities form the necessary condition for discontinuities in a genuinely nonlinear system. If the system of equations is not genuinely nonlinear, we can easily think of cases when the straight C characteristics fan forward and then backward, or vice versa, or even cases when this pattern is repetitive in a single wave region.

Liu (1974) has extended Oleinik's celebrated Condition E [Eq. (5.4.4) of Vol. 1] to the case of pairs of conservation laws and established the following *extended entropy condition*.

RULE 1.5 (Extended Entropy Condition): Consider a discontinuity connecting the states $L(c_1^l, c_2^l)$ and $R(c_1^r, c_2^r)$, and let the two loci of R passing through L be Σ_+^0 and Σ_-^0, respectively. If R lies on Σ_-^0, we must have[†]

$$\bar{\sigma}_+(c_1^l, c_2^l, c_1^r, c_2^r) \geq \bar{\sigma}_+(c_1^l, c_2^l, c_1, c_2) \qquad (1.7.16)$$

for every (c_1, c_2) on Σ_-^0 between L and R. Similarly, if R falls on Σ_+^0, we must have

$$\bar{\sigma}_-(c_1^l, c_2^l, c_1^r, c_2^r) \geq \bar{\sigma}_-(c_1^l, c_2^l, c_1, c_2) \qquad (1.7.17)$$

for every (c_1, c_2) on Σ_-^0 between L and R. If we take the two loci of L passing through R to be Σ_+^0 and Σ_-^0, we must have the inequality signs reversed with $\bar{\sigma}_\pm(c_1, c_2, c_1^r, c_2^r)$ instead of $\bar{\sigma}_\pm(c_1^l, c_2^l, c_1, c_2)$ in Eqs. (1.7.16) and (1.7.17).

With this entropy condition Liu has proved that there exists a unique solution to any Riemann problem in the class of simple waves, discontinuities, and constant states. We shall not attempt to present the proof here but interpret the implication of the entropy condition in relationship to various types of nonlinearity. In fact, we shall see that discontinuities can be classified into shocks, semishocks, and contact discontinuities, in accordance with the case of single equations.

Shocks

Suppose that the $(+)$ characteristic field is genuinely nonlinear, i.e.,

$$\frac{\mathcal{D}_-\sigma_+}{\mathcal{D}c_2} \neq 0 \qquad (1.7.18)$$

along every Γ_- characteristic, so that the characteristic direction σ_+ varies in a strictly monotone fashion along every Γ_- characteristic. Let us now consider a region of constant state (c_1^l, c_2^l) adjacent to the τ axis in the (x, τ)-plane and its image L in the hodograph plane. If this region is to be connected to another region of constant state adjacent to the x axis for which only c_2 is specified by

---

[†] In general, the extended entropy condition is expressed in terms of the propagation speed, so that Eqs. (1.7.16) and (1.7.17) should be written as

$$1/\bar{\sigma}_+(c_1^l, c_2^l, c_1^r, c_2^r) \leq 1/\bar{\sigma}_+(c_1^l, c_2^l, c_1, c_2) \qquad (1.7.16a)$$

and

$$1/\bar{\sigma}_-(c_1^l, c_2^l, c_1^r, c_2^r) \leq 1/\bar{\sigma}_-(c_1^l, c_2^l, c_1, c_2) \qquad (1.7.17a)$$

respectively. As long as $\bar{\sigma}_+$, σ_+, $\bar{\sigma}_-$, and σ_- all have values of a common sign, the two sets of inequalities are equivalent. If, however, $\bar{\sigma}_+$, σ_+, $\bar{\sigma}_-$, and σ_- can have positive as well as negative values, we must use Eqs. (1.7.16a) and (1.7.17a). For examples of the latter case, see Chapters 5 and 6.

c_2^r, it is only natural to think of the connection by one of the two simple waves. We shall consider the case of a Γ_- simple wave so that the portrait in the hodograph plane will be as shown in the upper part of Fig. 1.21. If σ_+ decreases from L to R' along the Γ_-^0 passing through L, the connection is actually made by the Γ_- simple wave and c_1 is given by $c_1^{r'}$.

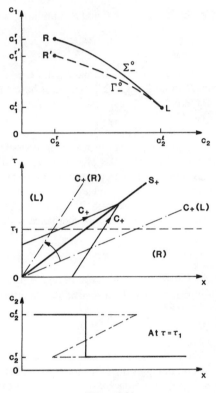

Figure 1.21

Our interest here is in the case when σ_+ increases from L to R' so that the straight characteristics C_+ fan backward as illustrated in the middle part of Fig. 1.21. Now we must introduce a discontinuity in such a way that the Rule 1.5 is satisfied. Thus we shall locate the curve Σ_-^0 passing through L and this gives c_1^r as well as $\bar{\sigma}_+(c_1^l, c_2^l, c_1^r, c_2^r)$. Recalling the fact that Σ_-^0 and Γ_-^0 have the same curvature at L and σ_+ is monotone increasing from L and R' along Γ_-^0, we claim that

$$\sigma_+(c_1^l, c_2^l) < \bar{\sigma}_+(c_1^l, c_2^l, c_1^r, c_2^r) < \sigma_+(c_1^r, c_2^r) \qquad (1.7.19)$$

on the basis of Eqs. (1.7.12) and (1.7.13). Thus we can draw the propagation path S_+ of the discontinuity having slope $\bar{\sigma}_+$ as shown in the middle part of Fig. 1.21 and the profile of c_2 at fixed τ is illustrated in the lower part of Fig. 1.21.

Pairs of Quasilinear Hyperbolic Equations of First Order   Chap. 1

It can easily be seen that Eq. (1.7.19) satisfies the entropy condition (Rule 1.5). Liu (1974) has added another condition,

$$\bar{\sigma}_+(c_1^l, c_2^l, c_1^r, c_2^r) > \sigma_-(c_1^r, c_2^r) \qquad (1.7.20)$$

and proved that if the $(+)$ characteristic field is genuinely nonlinear, the entropy condition reduces to Eqs. (1.7.19) and (1.7.20). According to Lax (1957), a discontinuity satisfying these conditions is called Σ_- *shock,* and for this reason Eqs. (1.7.19) and (1.7.20) are frequently referred to as the *shock condition* and the curve Σ is called a *shock curve.*

A similar argument can be applied when the $(-)$ characteristic field is genuinely nonlinear, i.e.,

$$\frac{\mathcal{D}_+ \sigma_-}{\mathcal{D}c_2} \neq 0 \qquad (1.7.21)$$

along every Γ_+ characteristic to give

$$\bar{\sigma}_-(c_1^l, c_2^l) < \bar{\sigma}_-(c_1^l, c_2^l, c_1^r, c_2^r) < \sigma_-(c_1^r, c_2^r) \qquad (1.7.22)$$

But the additional condition is

$$\bar{\sigma}_-(c_1^l, c_2^l, c_1^r, c_2^r) < \sigma_+(c_1^l, c_2^l) \qquad (1.7.23)$$

A discontinuity satisfying these conditions will be called a Σ_+ shock, and Eqs. (1.7.22) and (1.7.23) form the shock condition for a Σ_+ shock. Therefore, we conclude that the discontinuities encountered in genuinely nonlinear systems of equations are all shocks.

Some remarks are worth mentioning here in regard to the shock condition. First, Eq. (1.7.19) [or Eq. (1.7.22)] shows that the corresponding characteristics, C_+ [or C_-], must be impinging on the shock path, S_+ [or S_-], from both sides in the τ direction (see Fig. 1.21). Hence the shock tends to overtake the state on the right-hand side and to be overtaken by the state on the left so that it bears a *self-sharpening tendency.* Second, Eq. (1.7.20) [or Eq. (1.7.23)] allows the characteristics of the other kind, C_- [or C_+], to pass through the shock path, S_+ [or S_-]. This then guarantees that the shock path does not overlap with the simple wave region of different kind. It also implies that if two shocks of different kinds are to propagate side by side with a constant state in between, we have

$$\bar{\sigma}_+ > \bar{\sigma}_- \qquad (1.7.24)$$

To get a better understanding we shall consider a Riemann problem in which the initial and feed data are specified in the manner reversed from the case of Figs. 1.16 and 1.17. Thus the portrait in the hodograph plane will be as shown in Fig. 1.22. Here again we assume that σ_+ is monotone increasing from F to Q' along the Γ_- and σ_- is also monotone increasing from Q' to I along the Γ_+. This implies that the system is genuinely nonlinear and we must consider two shocks of different kinds. To locate the appropriate shock curve Σ we first take the state F as the state (c_1^l, c_2^l) and solve the compatibility relation (1.7.5) for the state (c_1^r, c_2^r). This gives the curve Σ_- passing through F.

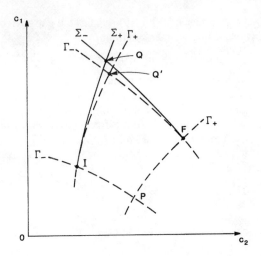

Figure 1.22

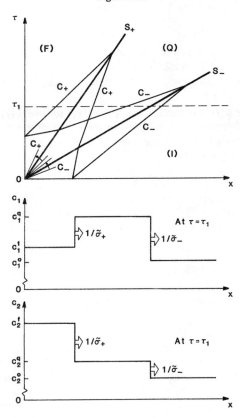

Figure 1.23

Next we take the state I as the state (c_1^r, c_2^r) and solve Eq. (1.7.5) for the state (c_1^l, c_2^l) to obtain the curve Σ_+ passing through I. The two curves meet at Q, which represents the constant state to appear between the two shocks. With the state Q so determined, we can evaluate $\bar{\sigma}_+$ between the states F and Q and $\bar{\sigma}_-$ between Q and I. Thus the two shock paths, S_+ and S_-, are located as shown in the upper part of Fig. 1.23. Here we clearly discern the impinging characteristics and the transmitting characteristics with respect to each shock path so as to show that the shock condition is satisfied. The profiles of c_1 and c_2 at a fixed time are illustrated in the lower parts of Fig. 1.23. As the two shocks propagate, the region of constant state Q expands, so the intermediate plateau of the profile tends to grow.

Semishocks

If the system of equations is not genuinely nonlinear, we can think of a variety of situations. Here we shall concentrate on the $(+)$ characteristic field and assume that the derivative $\mathscr{D}_-\sigma_+/\mathscr{D}c_2$ changes its sign only once along every Γ_- characteristic. Consider a situation similar to the one shown in Fig. 1.21 but the constant state in the region adjacent to the x axis is denoted by Q for which only c_2 is specified by c_2^0 as depicted in Fig. 1.24. We shall assume that along the Γ_-^0 passing through L, σ_+ increases from L to P and then decreases from P to R''. Thus the C_+ characteristics would first fan backward and then forward as we pass in the x direction, the transition taking place at $C_+(P)$. Since a discontinuity must be introduced, we first locate the shock curve Σ_-^0 passing through L and determine $\sigma_+(c_1, c_2)$ and $\bar{\sigma}_+(c_1^l, c_2^l, c_1, c_2)$ for every (c_1, c_2) on Σ_-^0 between L and R'. Although it depends on the nature of the $(+)$ characteristic field, we can think of two possibilities. In one case we have the inequality

$$\sigma_+(c_1, c_2) > \bar{\sigma}_+(c_1^l, c_2^l, c_1, c_2) \qquad (1.7.25)$$

satisfied at R', so the shock condition (1.7.19) is satisfied with the state R' on the right-hand side. This implies that the state L is connected to the state R' by a Σ_- shock in the same manner as before.

Another possibility is that there exists a point $R(c_1^r, c_2^r)$ on Σ_-^0 between L and R' such that at R,

$$\sigma_+(c_1^r, c_2^r) = \bar{\sigma}_+(c_1^l, c_2^l, c_1^r, c_2^r) \qquad (1.7.26)$$

but for every (c_1, c_2) beyond the point R up to R',

$$\sigma_+(c_1, c_2) < \bar{\sigma}_+(c_1^l, c_2^l, c_1, c_2) \qquad (1.7.27)$$

Hence the two states L and R can be connected by a discontinuity but beyond the state R the solution must be continued by a simple wave. At the point R we locate the Γ_- characteristic and follow this to find the point Q on the line $c_2 = c_2^0$. The portion RQ of this Γ_- characteristic is the image of the simple wave region. Consequently, the solution is constructed in the (x, τ)-plane as shown in the middle part of Fig. 1.24 while the profile of c_2 at a fixed time τ_1 is illustrated in the lower part of Fig. 1.24.

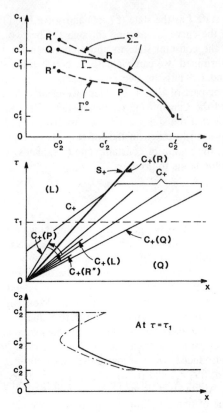

Figure 1.24

Along the path of the discontinuity S_+ we have

$$\sigma_+(c_1^l, c_2^l) < \bar{\sigma}_+(c_1^l, c_2^l, c_1^r, c_2^r) = \sigma_+(c_1^r, c_2^r) \qquad (1.7.28)$$

and

$$\bar{\sigma}_+(c_1^l, c_2^l, c_1^r, c_2^r) > \sigma_-(c_1^r, c_2^r) \qquad (1.7.29)$$

so that the entropy condition (1.7.16) is satisfied. The path S_+ coincides with the $C_+(R)$ and thus the discontinuity propagates with the same speed as the state R on its right-hand side. But the state on its left-hand side L still tends to overtake the discontinuity. Since it is *shock-like* only on one side, a discontinuity of this type will be called a *semishock*.

If σ_+ varies in the reverse manner along the Γ_-^0, i.e., first decreases and then increases starting from L, we shall consider the state $R(c_1^r, c_2^r)$ as fixed and apply an entirely similar procedure to recognize the possibility of a semishock for which we have the condition

$$\sigma_+(c_1^l, c_2^l) = \bar{\sigma}_+(c_1^l, c_2^l, c_1^r, c_2^r) < \sigma_+(c_1^r, c_2^r) \qquad (1.7.30)$$

Pairs of Quasilinear Hyperbolic Equations of First Order   Chap. 1

plus Eq. (1.7.29). In this case we have a simple wave joined to the left-hand side of the semishock. An analogous argument can be applied to the $(-)$ characteristic field if it is not genuinely nonlinear. Thus we expect to have a Σ_+ semishock for which we have Eq. (1.7.28) or (1.7.30) with the subscript $+$ replaced by $-$ everywhere. The second condition is given by Eq. (1.7.23). We note that in any case a semishock is joined by a simple wave of the same family on one side. The totality of the two waves joined together, which are obviously of the same family, is referred to as a *combined wave*.

Now looking at Fig. 1.23, we can easily imagine a Riemann problem whose solution may include a semishock. Suppose that the shock path S_+ in Fig. 1.23 is replaced by the combined wave of Fig. 1.24. The fast propagating wave may be a simple wave, a shock, or even a combined wave of different family. Between the two waves we still have the region of constant state Q.

Contact discontinuities

We shall finally consider the case when the system of equations is not genuinely nonlinear and the derivative $\mathcal{D}_-\sigma_+/\mathcal{D}c_2$, for example, changes its sign more than once along every Γ_- characteristic in a certain region of the hodograph plane. Typical of the situation here will be the case when the sign of $\mathcal{D}_-\sigma_+/\mathcal{D}c_2$ changes twice along a Γ_- characteristic.

Let us consider a region of constant state $A(c_1^a, c_2^a)$ adjacent to the τ axis and another region of constant state B adjacent to the x axis but for the state B only c_2 is specified by c_2^b. The other concentration c_1^b is to be determined in such a way that the states A and B are connected by a single wave. Taking the point image of the state A and the line $c_2 = c_2^b$ in the hodograph plane as depicted in Fig. 1.25, we immediately look for the Γ characteristic passing through A. Between the two choices we shall discuss the case with the Γ_- characteristic passing through A, say Γ_-^0, and assume that, along the Γ_-^0, σ_+ decreases from A to P, increases from P to Q, and again decreases from Q to B''. If the image of the state B falls on the portion AP' of Γ_-^0, where P' is some point in the neighborhood to the right of P, the two states will be connected by a Γ_- simple wave.

But if the state B is to fall on a point to the right of P', we know that the simple wave must be joined by a discontinuity from the right. Thus for every point $L(c_1^l, c_2^l)$ along the Γ_-^0 from the point A on, we find the value of σ_+ to be denoted by σ_+^l. Note that the lowercase superscripts on c_i's or σ's are used here to indicate that the quantities represent the state itself or are to be evaluated for the state denoted by the corresponding capital letters. Next we use c_1^l, c_2^l, and c_2^b in the compatibility relation (1.7.5) and the jump condition (1.7.10) to determine c_1^b and $\tilde{\sigma}_+$. Now we can evaluate σ_+^b to make the following test. If there exists a point L for which we have the particular condition

$$\sigma_+^l = \tilde{\sigma}_+ < \sigma_+^b \qquad (1.7.31)$$

satisfied, the states L and B are connected by a Σ_- semishock so that the state A is connected to the state B by a combined wave, the semishock being joined by a simple wave on the left-hand side.

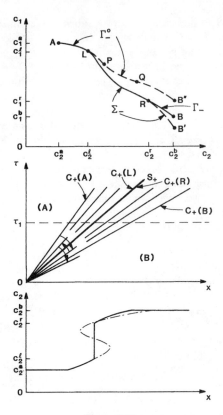

Figure 1.25

If the condition (1.7.31) is not satisfied (i.e., $\sigma_+^l \neq \tilde{\sigma}_+$ or $\tilde{\sigma}_+ > \sigma_+^b$) at every point L on Γ_-^0 from A to B'', we expect the discontinuity to be joined by another simple wave on the right-hand side. For every point $L(c_1^l, c_2^l)$ on the portion AP of Γ_-^0, we construct the shock curve Σ_- and see if there is a point R on this curve Σ_- such that

$$\sigma_+^l = \tilde{\sigma}_+ = \sigma_+^r \tag{1.7.32}$$

According to Liu (1974), there exists a unique pair of such points L and R satisfying the entropy condition (1.7.16). After locating the points L and R, we follow the Γ_- characteristic passing through R to the line $c_2 = c_2^b$ as illustrated in the upper part of Fig. 1.25. Of course, we must check whether σ_+ decreases along the Γ_- characteristic from R to B, but this is guaranteed by the nature of the $(+)$ characteristic field. Hence the solution consists of a Γ_- simple wave with image on AL, a discontinuity $L \to R$, and another Γ_- simple wave with image on RB which are ordered in the x direction. The middle part of Fig. 1.25 shows the portrait of the solution in the (x, τ)-plane, whereas the profile of c_2 at a fixed time is illustrated in the lower part.

Pairs of Quasilinear Hyperbolic Equations of First Order   Chap. 1

At every point along the path of the discontinuity S_+ we have the condition

$$\sigma_+(c_1^l, c_2^l) = \bar{\sigma}_+(c_1^l, c_2^l, c_1^r, c_2^r) = \sigma_+(c_1^r, c_2^r) \tag{1.7.33}$$

so the states on both sides propagate with the same speed as the discontinuity itself. There is no self-sharpening tendency. A discontinuity of this nature is called a *contact discontinuity*.

If the roles of the states A and B are interchanged in Fig.1.25, we can picture the solution in the form of two semishocks connected by a simple wave. Exactly the same argument can be extended to the $(-)$ characteristic field.

A special case may be worth discussing here. Suppose that the derivative $\mathcal{D}_-\sigma_+/\mathcal{D}c_2$ identically vanishes along every Γ_- characteristic in a certain region of the hodograph plane. Let us take a Γ_- characteristic, say Γ_-^0, belonging to this region and consider two constant states L and R whose images fall on Γ_-^0. We shall examine how these states can be connected in the (x,τ)-plane. Since σ_+ is constant along Γ_-^0, every point on Γ_-^0 between L and R will map on a single, straight C_+ characteristic in the (x, τ)-plane. This implies that the states L and R are connected by a discontinuity irrespective of which state is on which side. Clearly, Eq. (1.7.33) is satisfied and so the discontinuity propagates with the same speed as the states L and R. It is immediately apparent that the entropy condition (1.7.16) is satisfied with the equality sign. On the other hand, from Eq. (1.4.25) we have

$$\frac{\mathcal{D}_-f_1}{\mathcal{D}c_1} = \sigma_+ = \frac{\mathcal{D}_-f_2}{\mathcal{D}c_2} = \text{constant} \tag{1.7.34}$$

and by integrating the first equation with respect to c_1 and the second with respect to c_2 along Γ_-^0, we obtain

$$\frac{[f_1]}{[c_1]} = \sigma_+ = \frac{[f_2]}{[c_2]} \tag{1.7.35}$$

Hence both the compatibility relation and the jump condition are satisfied. It follows that the discontinuity above is a contact discontinuity.

If we have the equality

$$\frac{\mathcal{D}_-\sigma_+}{\mathcal{D}c_2} = 0 \tag{1.7.36}$$

for all values of (c_1, c_2) in question, the $(+)$ characteristic field is called *linearly degenerate* and we can say the same for the $(-)$ characteristic field. Thus discontinuities encountered in a linearly degenerate system are all contact discontinuities. In the limiting case when the equations are linear, we have the same situation. In contrast to quasilinear systems, linear or linearly degenerate systems of equations have discontinuous solutions only when the initial and feed data include discontinuities. The directions of the jumps are strictly given by those in the data.

In conclusion, we make a summary in the following form. The solution of a Riemann problem consists of two wave regions of different families separated by a region of constant state. The region of the fast-moving wave is separated from the x axis by a region of the initial state, whereas the region of the slowly moving wave is separated from the τ axis by a region of the feed state. If the system is genuinely nonlinear, each wave is given either by a simple wave or by a shock. If the system is linear or linearly degenerate, each wave is given by a contact discontinuity. If the system is neither linearly degenerate nor genuinely nonlinear, each wave may still be given by a simple wave or a shock but, in general, it consists of a set of semishocks, simple waves, and contact discontinuities all of the same family.

A more rigorous treatment of semishocks and contact discontinuities has been given by Liu (1974), who used the terms "one-sided contact discontinuity" and "two-sided contact discontinuity." These have also been discussed by Lax (1957), Jeffrey and Taniuti (1964), and Jeffrey (1976). The latter authors used the term "intermediate discontinuity" for the semishock.

We shall treat a physical example involved with a shock, a semishock, and a contact discontinuity in the next section, while an exhaustive discussion on shocks will be given in Chapter 2 when we deal with two solute chromatography with the Langmuir isotherm. More examples involving shocks, semishocks, and contact discontinuities are treated in Chapter 6.

Exercises

1.7.1. Consider the countercurrent adsorber of Exercise 1.4.4. If there is a discontinuity in the solution, formulate the mass balances for A_1 and A_2 over the discontinuity and arrange them in the form of the jump condition and the compatibility relation.

1.7.2. Suppose that the adsorption equilibrium follows the Langmuir isotherm given by Eq. (1.1.6). If the state on the left-hand side of a discontinuity (c_1^l, c_2^l) is fixed, show that there are two distinct loci of (c_1^r, c_2^r), the state on the right-hand side.

1.7.3. For the adiabatic adsorption of a single solute in a fixed bed, we have the usual mass conservation law, Eq. (1.4.1) of Vol. 1 with $i = 1$, and the energy conservation law, Eq. (1.4.9) of Vol. 1 with $m = 1$ (see Exercise 1.1.1). If there is a discontinuity in the solution, show that the mass and energy conservation laws applied to the discontinuity give Eqs. (1.7.6) and (1.7.7) in which c_2 and n_2 are given by Eqs. (1.4.8a) and (1.4.8b) of Vol. 1, respectively. Consider now the Langmuir isotherm given by Eq. (1.2.7) of Vol. 1 with $n = n_1$ and $c = c_1$, and note that the parameter $K = K_1$ is expressed as a function of the temperature by Eq. (1.4.10) of Vol. 1 with $i = 1$. Taking the state on the left-hand side of a discontinuity as fixed, show that the locus of the state on the right-hand side consists of two different branches.

1.7.4. Consider a shock in the one-dimensional flow of a compressible fluid. Using the notation of Sec. 1.12 of Vol. 1 and following the procedure of Sec. 7.2 of Vol. 1 (see Fig. 7.1 of Vol. 1), formulate the conservation of mass and the conservation of momentum over the shock. Since the flow across a shock is essentially an irreversible process, it is necessary to consider the conservation of energy as well as the increase or conservation of entropy. From these four equations, can you derive the jump condition, the compatibility relation, and the entropy condition?

1.8 Analysis of Polymer Flooding[†]

While we shall deal with applications of the general theory in Chapter 2 and again in Chapter 6, it is instructive to see here how discontinuities come into the solution in physical systems. To illustrate this we consider the equations of polymer flooding, which have been set up in Sec. 1.10 of Vol. 1 (see also Sec. 5.6 of Vol. 1). Thus if we let

ϕ = fractional void volume in the porous medium

A = cross-sectional area of the porous medium

q_T = total flow rate

f = fractional flow rate of water

s = volume fraction of water (or water saturation) in the pore space

c = polymer concentration in the mobile phase

c_s = polymer concentration adsorbed in the immobile phase

the mass conservation equations for water and polymer at a distance z from the inlet at time t can be written as

$$q_T \frac{\partial f}{\partial z} + \phi A \frac{\partial s}{\partial t} = 0$$

$$q_T \frac{\partial}{\partial z}(fc) + \phi A \frac{\partial}{\partial t}(sc + c_s) = 0$$

(1.8.1)

respectively, from Eqs. (1.10.1) and (1.10.7) of Vol. 1. The subscript w used in the original equations has been deleted for brevity. Let us put the independent variables in dimensionless form

$$x = \frac{z}{L}, \qquad \tau = \frac{q_T t}{\phi A L}$$

(1.8.2)

where L denotes the length of the reservoir core, and obtain

$$\frac{\partial f}{\partial x} + \frac{\partial s}{\partial \tau} = 0$$

$$\frac{\partial}{\partial x}(fc) + \frac{\partial}{\partial \tau}(sc + c_s) = 0$$

(1.8.3)

Now if we neglect the effects of gravity and capillarity, the fractional flow rate f may be expressed as a function of s and c. Also, we shall assume local equilibrium for the adsorption of polymer so that c_s is related to c by the adsorption isotherm. Hence we have

---

[†] Since this section is concerned with a rather particular example, it may be omitted without loss of continuity. The reader may return to this section after studying the adiabatic adsorption of single solutes in Chapter 6.

$$f = f(s, c) \tag{1.8.4}$$

$$c_s = g(c) \tag{1.8.5}$$

and Eq. (1.8.3) can be rewritten as

$$\frac{\partial f}{\partial s}\frac{\partial s}{\partial x} + \frac{\partial s}{\partial \tau} + \frac{\partial f}{\partial c}\frac{\partial c}{\partial x} = 0$$
$$f\frac{\partial c}{\partial x} + \{s + g'(c)\}\frac{\partial c}{\partial \tau} = 0 \tag{1.8.6}$$

in which the first equation has been used to put the second equation in this form.

Clearly, the pair of equations (1.8.6) form a reducible system and take such a particularly simple form that we can immediately find the characteristic directions $\sigma_\pm$, one from each equation, by examining them. To apply the hodograph transformation, however, we shall start with a linear combination of the equations

$$\lambda_1\frac{\partial f}{\partial s}\frac{\partial s}{\partial x} + \lambda_1\frac{\partial s}{\partial \tau} + \left(\lambda_1\frac{\partial f}{\partial c} + \lambda_2 f\right)\frac{\partial c}{\partial x} + \lambda_2(s + g')\frac{\partial c}{\partial \tau} = 0$$

and by requiring that the derivatives $\partial s/\partial \omega$ and $\partial c/\partial \omega$ be in a common direction given by $(\partial x/\partial \omega, \partial \tau/\partial \omega)$, we obtain the four equations

$$\lambda_1\frac{\partial f}{\partial s}\frac{\partial \tau}{\partial \omega} - \lambda_1\frac{\partial x}{\partial \omega} = 0$$

$$\left(\lambda_1\frac{\partial f}{\partial c} + \lambda_2 f\right)\frac{\partial \tau}{\partial \omega} - \lambda_2(s + g')\frac{\partial x}{\partial \omega} = 0$$

$$\lambda_1\frac{\partial f}{\partial s}\frac{\partial s}{\partial \omega} + \left(\lambda_1\frac{\partial f}{\partial c} + \lambda_2 f\right)\frac{\partial c}{\partial \omega} = 0 \tag{1.8.7}$$

$$\lambda_1\frac{\partial s}{\partial \omega} + \lambda_2(s + g')\frac{\partial c}{\partial \omega} = 0$$

Taking the last two equations and applying the condition for nontrivial values of λ_1 and λ_2, we get

$$\frac{dc}{ds}\left\{\frac{\partial f}{\partial c}(s + g')\frac{dc}{ds} - f + \frac{\partial f}{\partial s}(s + g')\right\}\left(\frac{\partial s}{\partial \omega}\right)^2 = 0 \tag{1.8.8}$$

where

$$\frac{dc}{ds} = \frac{\partial c/\partial \omega}{\partial s/\partial \omega} \tag{1.8.9}$$

Since $\partial s/\partial \omega$ should not vanish, we have the equations for the Γ characteristics in the form

$$\frac{dc}{ds} = 0 = \zeta_1 \tag{1.8.10}$$

$$\frac{dc}{ds} = \frac{f - \dfrac{\partial f}{\partial s}\{s + g'(c)\}}{\dfrac{\partial f}{\partial c}\{s + g'(c)\}} = \zeta_2$$

This implies that the system (1.8.6) is hyperbolic. We shall not assign the subscripts $+$ and $-$ at the moment because it should be done on the basis of the requirement $\sigma_+ > \sigma_-$. In any case the Γ characteristics of one kind are all straight and parallel to the s axis in the hodograph plane and thus a simplification is expected when we construct the solution.

Next, we choose the second and the fourth equations of Eq. (1.8.7) and ask for nontrivial values of λ_1 and λ_2 to obtain

$$\left\{ f - \frac{\partial f}{\partial c}(s + g')\zeta \right\} \frac{\partial \tau}{\partial \omega} - (s + g')\frac{\partial x}{\partial \omega} = 0$$

or

$$\frac{d\tau}{dx} = \frac{\partial \tau/\partial \omega}{\partial x/\partial \omega} = \frac{s + g'}{f - \dfrac{\partial f}{\partial c}(s + g')\zeta} \tag{1.8.11}$$

Substituting Eq. (1.8.10) into this gives

$$\frac{d\tau}{dx} = \frac{s + g'(c)}{f} = \sigma_1$$

$$\frac{d\tau}{dx} = \left(\frac{\partial f}{\partial s}\right)^{-1} = \sigma_2 \tag{11.8.12}$$

It is not obvious which of σ_1 and σ_2 would be larger. Although it seems necessary to have specific forms of the functions $f(s, c)$ and $g(c)$ for the comparison between σ_1 and σ_2, we can make a general argument to some extent.

Let us recall the shape of the fractional flow curve from Fig. 5.28 of Vol. 1 (see also Fig. 1.26 in the present volume). We note that the curve has zero slope both at $s = s_{wc}$ and $s = 1 - s_{or}$, where s_{wc} is the connate water saturation and s_{or} the residual oil saturation, and the slope of the curve changes rather rapidly between the two limits. This implies that σ_2 diverges for $s = s_{wc}$ and for $s = 1 - s_{or}$. But for $s = s_{wc}$, σ_1 also diverges and it does faster than σ_2. Putting these arguments together, we expect to have $\sigma_1 < \sigma_2$ in the neighborhood of the line $s = 1 - s_{or}$ in the hodograph plane but $\sigma_1 > \sigma_2$ outside the region. We shall call the former region I and the latter region II. Then inside region I, where $\sigma_1 < \sigma_2$, we must have

$$\sigma_+ = \sigma_2, \qquad \sigma_- = \sigma_1 \tag{1.8.13}$$

$$\zeta_+ = \zeta_2, \qquad \zeta_- = \zeta_1 = 0 \tag{1.8.14}$$

whereas in region II, where $\sigma_1 > \sigma_2$, we have

$$\sigma_+ = \sigma_1, \qquad \sigma_- = \sigma_2 \qquad (1.8.15)$$

$$\zeta_+ = \zeta_1 = 0, \qquad \zeta_- = \zeta_2 \qquad (1.8.16)$$

This implies that a Γ_- characteristic remains straight and horizontal in region I but becomes curved in region II. On the other hand, a Γ_+ characteristic is straight and horizontal in region II and becomes curved in region I. This feature will become clearer when we deal with a numerical example later in this section.

If there is a discontinuity in the solution, we can apply the conservation law expressing the fact that the discontinuity must propagate with such speed that there is no accumulation of water or polymer at the discontinuity. We also note that both discontinuities in s and c coexist and propagate together. Thus we obtain the jump condition as well as the compatability relation

$$\frac{d\tau}{dx} = \frac{[s]}{[f]} = \frac{[sc + c_s]}{[fc]} \qquad (1.8.17)$$

Given the state $L(s^l, c^l)$, the compatibility relation gives two branches for the locus of the state $R(s^r, c^r)$ in the hodograph plane. These are the shock curves Σ_+ and Σ_- but since the Γ_- (or Γ_+) characteristics are straight in region I (or II), we know that the shock curve Σ_- (or Σ_+) coincides with the Γ_- (or Γ_+) characteristic in region I (or II). There are discontinuities of two different kinds and for each kind we have the jump condition

$$\frac{d\tau}{dx} = \left(\frac{[s]}{[f]}\right)_{\Sigma_-} = \tilde{\sigma}_+$$

$$\frac{d\tau}{dx} = \left(\frac{[s]}{[f]}\right)_{\Sigma_+} = \tilde{\sigma}_- \qquad (1.8.18)$$

To proceed further, let us suppose that the relative permeability data are correlated as

$$K_{ro}(s) = (1 - \psi)^2(1 + 2\psi)$$

$$K_{rw}(s) = 0.25\psi^3 \qquad (1.8.19)$$

where

$$\psi = \frac{s - s_{wc}}{1 - s_{or} - s_{wc}} \qquad (1.8.20)$$

These expressions were used by Claridge and Bondor (1974) and are to be compared with Eqs. (1.10.3) and (5.6.10) of Vol. 1. We shall further assume that the viscosity of the polymer solution is a linear function of c within the range of interest, i.e.,

$$\mu_w = \mu_w^0(1 + \beta c) \qquad (1.8.21)$$

while the viscosity of oil, μ_o, is constant. Thus the fractional flow rate of water, f, is expressed from Eq. (1.10.2) of Vol. 1 in the form

$$f = f(s, c) = \frac{\psi^3}{\psi^3 + \alpha(1 + \beta c)(1 - \psi)^2(1 + 2\psi)} \qquad (1.8.22)$$

in which

$$\alpha = \frac{4\mu_w^0}{\mu_o} \qquad (1.8.23)$$

The derivatives needed to evaluate the characteristic directions are given by

$$\frac{\partial f}{\partial s} = \frac{3\alpha(1 + \beta c)\psi^2(1 - \psi^2)}{(1 - s_{or} - s_{ow})\{\psi^3 + \alpha(1 + \beta c)(1 - \psi)^2(1 + 2\psi)\}^2} \qquad (1.8.24)$$

$$\frac{\partial f}{\partial c} = \frac{-\alpha\beta\psi^3(1 - \psi)^2(1 + 2\psi)}{\{\psi^3 + \alpha(1 + \beta c)(1 - \psi)^2(1 + 2\psi)\}^2} \qquad (1.8.25)$$

The equilibrium relation for the polymer adsorption on rock surfaces is often given by the Langmuir isotherm

$$c_s = g(c) = \frac{NKc}{1 + Kc} \qquad (1.8.26)$$

where N and K are constants, so

$$g'(c) = \frac{NK}{(1 + Kc)^2} \qquad (1.8.27)$$

All in all this is a rather complex system, and thus we shall resort to numerical treatment. As a numerical example, let us consider the saturation values

$$s_{wc} = 0.2, \qquad s_{or} = 0.3$$

and the viscosity data

$$\mu_o = 40 \text{ cp}, \qquad \mu_w^0 = 1 \text{ cp}$$

so that

$$\alpha = 0.1$$

Also, we shall assume that

$$\beta = 4 \text{ liters/g}, \quad N = 0.6 \text{ g/liter}, \quad K = 10.0 \text{ liters/g}$$

which are taken to apply to polyacrylamide solutions up to a concentration of 1 g/liter.

The fractional flow curves for $c = 0$ and $c = 1$ g/liter are shown in Fig. 1.26. We observe the rapid change in the slope of each curve. For the same value of s the fractional flow of water f is substantially lower when $c = 1$ g/liter than when $c = 0$ over a wide range of s. This implies that the mobility of oil is significantly enhanced due to the presence of polymer.

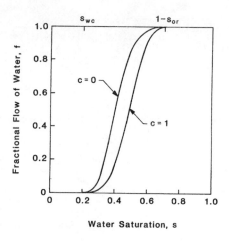

Figure 1.26

To construct the Γ characteristics we integrate Eq. (1.8.10) numerically from various points in the hodograph plane. Along each of the integral curves we compare the two values σ_1 and σ_2 to discern the Γ_+ and Γ_- characteristics. The result is depicted in Fig. 1.27, in which we clearly see the two regions as discussed in the above. Thus the Γ_- characteristic emanating from point A consists of the horizontal segment TA in region I and the curved portion PT in

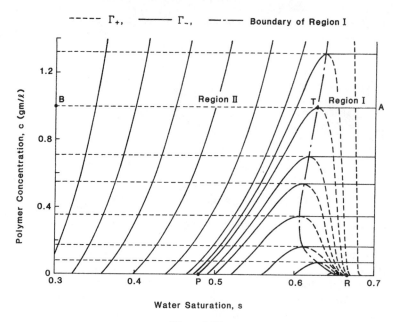

Figure 1.27

Pairs of Quasilinear Hyperbolic Equations of First Order   Chap. 1

region II, whereas the Γ_+ characteristic passing through point T is composed of the horizontal segment BT in region II and the curve TR in region I. The single broken line represents the boundary between the two regions.

Let us now consider the actual situation of polymer flooding. We shall assume that water flooding has been applied to the oil reservoir so that the initial water saturation is relatively high (see Sec. 5.6 of Vol. 1). The polymer solution to be fed has a constant concentration of 1 g/liter. Thus we shall take the initial and feed conditions as

$$\text{at } \tau = 0: \quad s = 0.6 \quad \text{and} \quad c = 0$$
$$\text{at } x = 0: \quad s = 1.0 \quad \text{and} \quad c = 1.0$$

(1.8.28)

so we have a Riemann problem.

The feed and initial conditions have their images on the points F and I, respectively, in the hodograph plane as marked in Fig. 1.28. To determine the image of the solution connecting the points F and I we take the Γ_- characteristic emanating from F, which consists of the horizontal segment $A \rightarrow T$ and the curved portion $T \rightarrow P$, and then the Γ_+ characteristic $P \rightarrow I$. The abscissa of the point P is 0.4794.

If we calculate $\sigma_+ = \sigma_2$ over the portion $F \rightarrow T$ of the Γ_- characteristic, σ_+ remains at infinity from F to A and decreases continuously to 0.7079 at T, where $s = 0.6263$. At the point T both $\sigma_+ = \sigma_2$ and $\sigma_- = \sigma_1$ have the same value 0.7079. From this point on we have $\sigma_+ = \sigma_1$ and this increases along the Γ_+ characteristic to 8.0028 at P. Hence the C_+ characteristics for the state on $F \rightarrow A$ are all vertical and give rise to a contact discontinuity remaining stationary at the inlet. The C_+ characteristics for the states on $A \rightarrow T$ fan clockwise but those for the state on $T \rightarrow P$ fan counterclockwise and thus we expect to have a semishock. Along the Γ_+ characteristic $P \rightarrow I$ we have

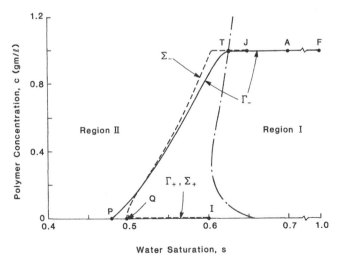

Figure 1.28

$\sigma_- = \sigma_1$ and this value increases from 0.3611 at P to 1.9741 at I. This implies that the C_- characteristics for the states on $P \rightarrow I$ fan counterclockwise, so we must have a Σ_+ shock.

For the Σ_+ shock we note that the shock curve Σ_+ is given by the horizontal line $c = 0$ because the latter is a straight Γ_+ characteristic. Therefore, the constant state that will appear between the two waves has its image on the line $c = 0$. To determine the semishock we calculate σ_+ and $\bar{\sigma}_+$ at every point along the Γ_- characteristic starting from the point A, where for $\bar{\sigma}_+$ we put $c^r = 0$ The two values become identical to give $\sigma_+ = \bar{\sigma}_+ = 1.2184$ at the point J, for which we have $s = 0.6507$ and $c = 1.0$, with the state on the right-hand side given by the point Q, for which we have $s = 0.4987$ and $c = 0$. In fact, the shock curve Σ_- can be constructed as shown by the dashed

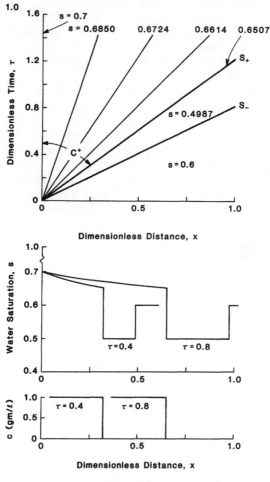

Figure 1.29

line $J \rightarrow T \rightarrow Q$ in Fig. 1.28. Here the crossing between the Γ_- characteristic and the shock curve Σ_- is not unexpected since the system (1.8.1) is not genuinely nonlinear.

Now we can construct the solution as shown in Fig. 1.29. In the physical plane, we have a contact discontinuity overlapping with the τ axis across which s drops from 1.0 to 0.7 and then a portion of Γ_- simple wave through which s varies continuously from 0.7 to 0.6507. In the course of these variations in s the polymer concentration c remains at 1.0 because we are moving along the horizontal line $c = 1$ from F to A and again to J. The simple wave is joined on the right-hand side by the semishock that propagates side by side with the state J, $s = 0.6507$ and $c = 1.0$. This semishock connects the state J to the state Q, $s = 0.4987$ and $c = 0$, and thus both s and c drop accordingly. The constant state Q is then connected to the initial state I by the Σ_+ shock that propagates along the shock path S_- of slope 0.8229.

In the lower part of Fig. 1.29 the profiles of s and c are presented at $\tau = 0.4$ and $\tau = 0.8$. We can clearly observe that the oil is being displaced by the polymer in the form of an *oil bank*. Profiles of this type have been reported by Patton et al. (1971). The breakthrough curves for s and c can be readily constructed from the physical plane portrait of the solution. In an actual situation, however, we would observe the fractional flows of water and oil at the exit rather than the water and oil saturations. Hence, we present the curves f and $(1 - f)$ versus τ observed at $x = 1$ in Fig. 1.30. Until $\tau = 0.8229$, the time when the Σ_+ shock reaches the exit, we have little oil $(1 - f = 0.02)$. However, the Σ_+ shock brings the breakthrough of the oil bank so that the fractional flow of oil jumps up to 0.147 and is maintained till $\tau = 1.2184$, when the semishock arrives at the exit. At this instant the polymer makes its breakthrough and the fractional flow of oil drops to 0.0183. The flow of oil gradually decreases further as time goes on.

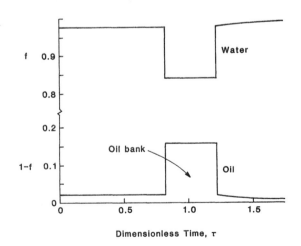

Figure 1.30

We note that the dimensionless time τ defined by Eq. (1.8.2) represents the cumulative pore volume and so the volume of oil recovered V_o, when expressed in terms of the pore volume, is given by the area under the breakthrough curve of oil, $(1 - f)$ versus τ. The pore volume of oil recovered V_o is shown as a function of time in Fig. 1.31, which shows that we can recover 0.073 pore volume of oil until the breakthrough of polymer. This amounts to 14.6% of the total available oil, which is 0.5 pore volume in this case.

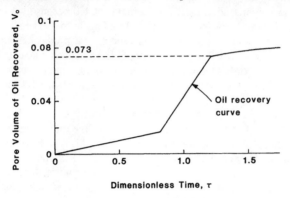

Figure 1.31

Exercises

1.8.1. For the numerical example treated in this section, change the polymer concentration of the feed solution to 0.5 g/liter and repeat the analysis to determine the pore volume of oil recovered.

1.8.2. Consider the numerical example treated in this section and repeat the analysis for each case.
(a) at $\tau = 0$, $s = 0.55$ (b) at $\tau = 0$, $s = 0.50$ (c) at $\tau = 0$, $s = s_{wc} = 0.20$
Compare the oil recovery curve in each case with that of Fig. 1.31.

1.9 Riemann Invariants and Their Application

A pair of quasilinear first-order partial differential equations can always be written in matrix form as

$$\frac{\partial \mathbf{w}}{\partial t} + \mathbf{A} \frac{\partial \mathbf{w}}{\partial x} + \mathbf{b} = \mathbf{0} \qquad (1.9.1)$$

where

$$\mathbf{w} = \begin{bmatrix} u \\ v \end{bmatrix}, \qquad \frac{\partial \mathbf{w}}{\partial t} = \begin{matrix} \dfrac{\partial u}{\partial t} \\ \dfrac{\partial v}{\partial t} \end{matrix}, \qquad \frac{\partial \mathbf{w}}{\partial x} = \begin{matrix} \dfrac{\partial u}{\partial x} \\ \dfrac{\partial v}{\partial x} \end{matrix} \qquad (1.9.2)$$

The symbols **A** and **b** represent 2×2 and 2×1 matrices, respectively, whose elements are functions of x, t, u, and v. For example, the chromatographic equations (1.1.20) take this form if the independent variables x and τ are renamed t and x, respectively, and

$$\mathbf{A} = \frac{1}{D^2}\begin{bmatrix} D^2 + \kappa_1(1 + v) & -\kappa_1 u \\ -\kappa_2 v & D^2 + \kappa_2(1 + u) \end{bmatrix}, \qquad \mathbf{b} = \begin{bmatrix} 0 \\ 0 \end{bmatrix} \qquad (1.9.3)$$

where

$$D = 1 + u + v$$

If we let $\mathbf{w} = \begin{bmatrix} c_1 \\ c_2 \end{bmatrix}$ and again rename x and τ, respectively, as t and x, Eq.

(1.1.17) can be put in the form of Eq. (1.9.1), where

$$\mathbf{A} = \begin{bmatrix} \dfrac{\partial f_1}{\partial c_1} & \dfrac{\partial f_1}{\partial c_2} \\[2mm] \dfrac{\partial f_2}{\partial c_1} & \dfrac{\partial f_2}{\partial c_2} \end{bmatrix}, \qquad \mathbf{b} = \begin{bmatrix} 0 \\ 0 \end{bmatrix} \qquad (1.9.4)$$

Another example is afforded by the equations of polymer flooding, Eq. (1.8.6), which also take the form of Eq. (1.9.1) if we put $t = \tau$ and

$$\mathbf{w} = \begin{bmatrix} s \\ c \end{bmatrix}, \qquad \mathbf{A} = \begin{bmatrix} \dfrac{\partial f}{\partial s} & \dfrac{\partial f}{\partial c} \\[2mm] 0 & \dfrac{f}{s + g'(c)} \end{bmatrix}, \qquad \mathbf{b} = \begin{bmatrix} 0 \\ 0 \end{bmatrix} \qquad (1.9.5)$$

The characteristic differential equations may again be derived by asking for the initial curve on which the partial derivatives would be indeterminate. If the initial data are prescribed as

$$t = T(\xi), \qquad x = X(\xi), \qquad u = U(\xi), \qquad v = V(\xi) \qquad (1.9.6)$$

the strip conditions to be satisfied are

$$T'(\xi)\frac{\partial u}{\partial t} + X'(\xi)\frac{\partial u}{\partial x} = U'(\xi)$$
$$T'(\xi)\frac{\partial v}{\partial t} + X'(\xi)\frac{\partial v}{\partial x} = V'(\xi) \qquad (1.9.7)$$

These may be written in matrix form

$$T'\mathbf{I}\frac{\partial \mathbf{w}}{\partial t} + X'\mathbf{I}\frac{\partial \mathbf{w}}{\partial x} - \mathbf{W}' = \mathbf{0} \qquad (1.9.8)$$

where **I** is the unit matrix and $\mathbf{W}'$ denotes $\begin{bmatrix} U' \\ V' \end{bmatrix}$. We shall put

$$\lambda = \frac{X'(\xi)}{T'(\xi)} \tag{1.9.9}$$

so that λ denotes the reciprocal slope of the initial curve. Now substituting $\lambda T'$ for X' in Eq. (1.9.8) and combining it with Eq. (1.9.1), we see that

$$\begin{bmatrix} \mathbf{I} & \mathbf{A} \\ T'\mathbf{I} & \lambda T'\mathbf{I} \end{bmatrix} \begin{bmatrix} \dfrac{\partial \mathbf{w}}{\partial t} \\ \dfrac{\partial \mathbf{w}}{\partial x} \end{bmatrix} = \begin{bmatrix} -\mathbf{b} \\ \mathbf{W}' \end{bmatrix} \tag{1.9.10}$$

The determinant of the 4×4 matrix on the left is $T'(\xi) | \lambda \mathbf{I} - \mathbf{A} |$, and since $T'(\xi) \neq 0$, the condition for the indeterminacy is

$$|\lambda \mathbf{I} - \mathbf{A}| = 0 \tag{1.9.11}$$

This is a quadratic and if the system of equations (1.9.1) is hyperbolic, there will be two real roots for λ. Let these be denoted by λ_+ and λ_-. Of course, λ_+ and λ_- are both functions of x, t, u, and v. As we shall see in the following, λ is equivalent to the reciprocal of σ of previous sections.[†] In other words, λ_+ and λ_- represent the reciprocal directions of the C_+ and C_- characteristics in the (x, t)-plane. They are frequently referred to as the *characteristic speed*.

Corresponding to the two eigenvalues λ_+ and λ_- of $\mathbf{A}$, there will be two left eigenvectors $\boldsymbol{\ell}_+^T$ and $\boldsymbol{\ell}_-^T$, such that

$$\boldsymbol{\ell}_+^T \mathbf{A} = \lambda_- \boldsymbol{\ell}_+^T \quad \text{and} \quad \boldsymbol{\ell}_-^T \mathbf{A} = \lambda_- \boldsymbol{\ell}_-^T \tag{1.9.12}$$

The eigenvectors, too, depend on x, t, u, and v. Thus multiplying Eq. (1.9.1) on the left by $\boldsymbol{\ell}_+^T$ or $\boldsymbol{\ell}_-^T$, we have

$$\boldsymbol{\ell}_+^T \frac{\partial \mathbf{w}}{\partial t} + \lambda_+ \boldsymbol{\ell}_+^T \frac{\partial \mathbf{w}}{\partial x} + \boldsymbol{\ell}_+^T \mathbf{b} = 0$$
$$\boldsymbol{\ell}_-^T \frac{\partial \mathbf{w}}{\partial t} + \lambda_- \boldsymbol{\ell}_-^T \frac{\partial \mathbf{w}}{\partial x} + \boldsymbol{\ell}_-^T \mathbf{b} = 0 \tag{1.9.13}$$

We note that the curves having directions $1/\lambda_\pm$ in the (x, t)-plane are the $C_\pm$ characteristics. If we introduce parameters α and β along the C_+ and C_- characteristics, respectively, so that

$$\frac{\partial}{\partial \alpha} = \left(\frac{\partial}{\partial t} + \lambda_+ \frac{\partial}{\partial x} \right) \frac{\partial t}{\partial \alpha}$$
$$\frac{\partial}{\partial \beta} = \left(\frac{\partial}{\partial t} + \lambda_- \frac{\partial}{\partial x} \right) \frac{\partial t}{\partial \beta} \tag{1.9.14}$$

the two equations in Eq. (1.9.13) can be written as

$$\boldsymbol{\ell}_+^T \left(\frac{\partial \mathbf{w}}{\partial \alpha} + \mathbf{b} \frac{\partial t}{\partial \alpha} \right) = 0 \tag{1.9.15}$$

[†] Note that the roles of x and t, the space and time variables, are interchanged in Eq. (1.9.1) in comparison to the pair of chromatographic equations (1.1.17).

and

$$\boldsymbol{\ell}_-^T \left(\frac{\partial \mathbf{w}}{\partial \beta} + \mathbf{b} \frac{\partial t}{\partial \beta} \right) = 0 \qquad (1.9.16)$$

respectively. This is equivalent to the second pair of equations in the set of the characteristic differential equations (1.2.20).

If the equations are homogeneous, i.e., $\mathbf{b} = \mathbf{0}$, we have

$$\boldsymbol{\ell}_+^T \frac{\partial \mathbf{w}}{\partial \alpha} = 0 \quad \text{and} \quad \boldsymbol{\ell}_-^T \frac{\partial \mathbf{w}}{\partial \beta} = 0 \qquad (1.9.17)$$

Thus remembering that the components of $\boldsymbol{\ell}_\pm$ are functions of x, t, u, and v and writing them as

$$\boldsymbol{\ell}_+^T = [m_+, \, n_+] \quad \text{and} \quad \boldsymbol{\ell}_-^T = [m_-, \, n_-] \qquad (1.9.18)$$

we see that

$$m_+ \frac{\partial u}{\partial \alpha} + n_+ \frac{\partial v}{\partial \alpha} = 0 \qquad (1.9.19)$$

along a C_+ characteristic or by integrating along a C_+ characteristic,

$$\int m_+ \, du + \int n_+ dv = \text{constant}$$

Since β is constant on C_+, we can write

$$\int m_+ du + \int n_+ dv = J_+(\beta) \qquad (1.9.20)$$

Similarly, on a C_- characteristic we have

$$m_- \frac{\partial u}{\partial \beta} + n_- \frac{\partial v}{\partial \beta} = 0 \qquad (1.9.21)$$

so

$$\int m_- du + \int n_- dv = J_-(\alpha) \qquad (1.9.22)$$

where the integrations are along a C_- characteristic. Thus the functions $J_+(\beta)$ and $J_-(\alpha)$ can be calculated. As functions of x, t, u, and v, J_+ is constant along the C_+ characteristics and J_- remains invariant along the C_- characteristics. It is immediately apparent that J_+ and J_- can be used as the characteristic parameters β and α, respectively. After their discoverer, the functions J_+ and J_- are called the *Riemann invariants*.

In the reducible case the eigenvalues and eigenvectors are all functions of only u and v. Since then m and n are functions of only u and v, we can write from Eqs. (1.9.19) and (1.9.21) the following equations:

$$-\frac{n_+}{m_+} = \frac{du}{dv} = \zeta_+ = -\frac{M_+}{L} \qquad (1.9.23)$$

and

$$-\frac{n_-}{m_-} = \frac{du}{dv} = \zeta_- = -\frac{M_-}{L} \tag{1.9.24}$$

where L, M_+, and M_- are defined by Eqs. (1.2.18). This pair of equations gives the Γ_+ and Γ_- characteristics in the hodograph plane, so J_+ remains invariant along a Γ_+ characteristic and J_- along a Γ_- characteristic. An immediate consequence is the following:

RULE 1.6: In a Γ_- simple wave region the Riemann invariant J_- is constant and, similarly, in a Γ_+ simple wave region J_+ remains constant.

Clearly, J_+ has distinct values along distinct Γ_+ characteristics and J_- also has distinct values along distinct Γ_- characteristics. Moreover, since in a simply connected domain of the hodograph plane a Γ_+ characteristic intersects a Γ_- characteristic in at most one point, it follows that the mapping $(u, v) \rightarrow (J_+, J_-)$ is one-to-one over any simply connected domain. This implies that the Riemann invariants J_+ and J_- can be used as the characteristic parameters β and α, respectively.

We also note the fact that the left and right eigenvectors of $\mathbf{A}$ are biorthogonal. Thus if we let $\mathbf{r}_+$ and $\mathbf{r}_-$ be the right eigenvectors corresponding to λ_+ and λ_-, respectively, and write them as

$$\mathbf{r}_+ = \begin{bmatrix} \mu_+ \\ \nu_+ \end{bmatrix} \quad \text{and} \quad \mathbf{r}_- = \begin{bmatrix} \mu_- \\ \nu_- \end{bmatrix} \tag{1.9.25}$$

we have the equations

$$\boldsymbol{\ell}_+^T \mathbf{r}_- = 0 \quad \text{and} \quad \boldsymbol{\ell}_-^T \mathbf{r}_+ = 0$$

or, in terms of their elements,

$$\begin{aligned} m_+ \mu_- + n_+ \nu_- &= 0 \\ m_- \mu_+ + n_- \nu_+ &= 0 \end{aligned} \tag{1.9.26}$$

Comparing these with Eqs. (1.9.19) and (1.9.21), we find that

$$\begin{aligned} -\frac{n_+}{m_+} &= \frac{\mu_-}{\nu_-} = \frac{du}{dv} = \zeta_+ \\ -\frac{n_-}{m_-} &= \frac{\mu_+}{\nu_+} = \frac{du}{dv} = \zeta_- \end{aligned} \tag{1.9.27}$$

These equations imply that Γ_+ characteristics are everywhere tangent to $\mathbf{r}_-$ and Γ_- to $\mathbf{r}_+$. Thus the fact that J_+ remains invariant along a Γ_+ characteristic may be expressed in the form

$$\mathbf{r}_-^T \nabla J_+ = 0 \tag{1.9.28}$$

Pairs of Quasilinear Hyperbolic Equations of First Order   Chap. 1

and, similarly, for J_-,

$$\mathbf{r}_+^T \nabla J_- = 0 \qquad (1.9.29)$$

where ∇ represents the gradient in the hodograph plane of u and v.

Genuine nonlinearity has played an important role with respect to discontinuous solutions in Sec. 1.7. Now if σ_+ changes in a strictly monotone fashion along a Γ_- characteristic, clearly we can write

$$\mathbf{r}_+^T \nabla \sigma_+ \neq 0 \quad \text{or equivalently} \quad \mathbf{r}_+^T \nabla \lambda_+ \neq 0 \qquad (1.9.30)$$

so Eq. (1.9.30) implies that the $(+)$ characteristic field is genuinely nonlinear. Similarly, if the $(-)$ characteristic field is genuinely nonlinear, we have $\mathbf{r}_-^T \nabla \sigma_- \neq 0$ or equivalently $\mathbf{r}_-^T \nabla \lambda_- \neq 0$. If $\mathbf{r}_+^T \nabla \sigma_+ = 0$ for every (u, v) along a Γ_- characteristic, the $(+)$ characteristic field is linearly degenerate. To show that this is consistent with the criterion based on the derivative $\mathcal{D}_-/\mathcal{D}c_2$ or $\mathcal{D}_+/\mathcal{D}c_2$, we rewrite $\mathbf{r}_+^T \nabla \sigma_+$ in terms of the elements:

$$\mathbf{r}_+^T \nabla \sigma_+ = \mu_+ \frac{\partial \sigma_+}{\partial c_1} + \nu_+ \frac{\partial \sigma_+}{\partial c_2}$$

in which c_1 and c_2 are used for u and v, respectively. But the right-hand side can be rearranged as

$$\nu_+ \left(\frac{\mu_+}{\nu_+} \frac{\partial \sigma_+}{\partial c_1} + \frac{\partial \sigma_+}{\partial c_2} \right) = \nu_+ \left(\zeta_- \frac{\partial \sigma_+}{\partial c_1} + \frac{\partial \sigma_+}{\partial c_2} \right) = \nu_+ \frac{\mathcal{D}_- \sigma_+}{\mathcal{D}c_2}$$

by applying the second equation of Eq. (1.9.27). Since the choice of μ_+ and ν_+ is arbitrary up to a multiplicative constant, we can always make ν_+ equal to 1, and thus we see that $\mathbf{r}_+^T \nabla \sigma_+$ is indeed equivalent to $\mathcal{D}_- \sigma_+/\mathcal{D}c_2$ and, similarly, $\mathbf{r}_-^T \nabla \sigma_-$ is equivalent to $\mathcal{D}_+ \sigma_-/\mathcal{D}c_2$.

Let us illustrate the use of Riemann invariants by the equations of isentropic one-dimensional flow. These have been obtained in Sec. 1.12 of Vol. 1 and need only be quoted here:

$$\frac{\partial \rho}{\partial t} + u \frac{\partial \rho}{\partial x} + \rho \frac{\partial u}{\partial x} = 0$$

$$\frac{\partial u}{\partial t} + u \frac{\partial u}{\partial x} + \frac{c^2}{\rho} \frac{\partial \rho}{\partial x} = 0 \qquad (1.9.31)$$

The term

$$\frac{c^2}{\rho} = \frac{1}{\rho} \frac{\partial p}{\partial \rho} \qquad (1.9.32)$$

is a function of ρ and the equation can be written in matrix form as

$$\frac{\partial \mathbf{w}}{\partial t} + \mathbf{A} \frac{\partial \mathbf{w}}{\partial x} = \mathbf{0} \qquad (1.9.33)$$

where

$$\mathbf{w} = \begin{bmatrix} \rho \\ u \end{bmatrix} \quad \text{and} \quad \mathbf{A} = \begin{bmatrix} u & \rho \\ \dfrac{c^2}{\rho} & u \end{bmatrix} \tag{1.9.34}$$

The quadratic (1.9.11) is

$$|\lambda \mathbf{I} - \mathbf{A}| = (\lambda - u)^2 - c^2 = 0$$

so that

$$\lambda_+ = u + c, \qquad \lambda_- = u - c \tag{1.9.35}$$

Furthermore, the left eigenvectors are $[m_\pm , n_\pm]$ such that

$$[m_\pm, n_\pm] \begin{bmatrix} \pm c & -\rho \\ -\dfrac{c^2}{\rho} & \pm c \end{bmatrix} = [0, 0]$$

The choice of $m_\pm$ and $n_\pm$ is arbitrary up to a multiplicative constant and we shall take

$$m_\pm = \frac{c}{2\rho} \quad \text{and} \quad n_\pm = \pm \frac{1}{2} \tag{1.8.36}$$

Then by Eqs. (1.9.20) and (1.9.22), the Riemann invariants are

$$J_+(\beta) = \frac{1}{2} L(\rho) + \frac{1}{2} u \tag{1.8.37}$$

and

$$J_-(\alpha) = \frac{1}{2} L(\rho) - \frac{1}{2} u \tag{1.9.38}$$

where

$$L(\rho) = \int_0^\rho \frac{c}{\rho} \, d\rho = \int_0^\rho \frac{1}{\rho c} \frac{\partial p}{\partial \rho} \, d\rho = \int_0^\rho \frac{dp}{\rho c} \tag{1.9.39}$$

For a polytropic gas

$$p = k\rho^\gamma \tag{1.9.40}$$

so

$$c^2 = \gamma k \rho^{\gamma - 1}$$

and

$$L(\rho) = \int_0^\rho (\gamma k)^{1/2} \rho^{(\gamma - 3)/2} d\rho = \frac{2(\gamma k)^{1/2}}{\gamma - 1} \rho^{(\gamma - 1)/2} = \frac{2c}{\gamma - 1} \tag{1.9.41}$$

Hence the Riemann invariants are

$$J_+ = \frac{c}{\gamma - 1} + \frac{u}{2}, \qquad J_- = \frac{c}{\gamma - 1} - \frac{u}{2} \tag{1.9.42}$$

This suggests that u and c might be taken as dependent variables, since $\rho = (c^2/\gamma k)^{1/(\gamma - 1)}$, and then, in the hodograph plane of u and c, the Γ characteristics become straight lines. It is also interesting that if $\gamma = 3$, the characteristic speeds are $\lambda_+ = u + c = 2J_+$ and $\lambda_- = u - c = 2J_-$, which are constant along the C_+ and C_- characteristics, respectively. This implies that the C characteristics in the (x, t)-plane are straight lines in the special case when $\gamma = 3$.

In any case

$$u = J_+ - J_-, \qquad c = \frac{\gamma - 1}{2} (J_+ + J_-) \qquad (1.9.43)$$

and the equation for the C_+ characteristic is

$$\frac{\partial x}{\partial \alpha} = (u + c) \frac{\partial t}{\partial \alpha} \qquad (1.9.44)$$

We could change the parameter α along C_+ to J_-, since J_- is a function of α, and write

$$C_+ : \quad \frac{\partial x}{\partial J_-} = (u + c) \frac{\partial t}{\partial J_-} = \frac{1}{2} \{ (\gamma + 1)J_+ + (\gamma - 3)J_- \} \frac{\partial t}{\partial J_-} \qquad (1.9.45)$$

Similarly, for the C_- characteristic,

$$C_- : \quad \frac{\partial x}{\partial J_+} = (u - c) \frac{\partial t}{\partial J_+} = -\frac{1}{2} \{ (\gamma - 3)J_+ + (\gamma + 1)J_- \} \frac{\partial t}{\partial J_+} \qquad (1.9.46)$$

Differentiating the first with respect to J_+ and the second with respect to J_- and equating $\partial^2 x / \partial J_+ \, \partial J_-$ and $\partial^2 x / \partial J_- \, \partial J_+$, we get

$$2c \frac{\partial^2 t}{\partial J_+ \partial J_-} + \frac{\partial (u + c)}{\partial J_+} \frac{\partial t}{\partial J_-} - \frac{\partial (u - c)}{\partial J_-} \frac{\partial t}{\partial J_+} = 0$$

or

$$2(\gamma - 1) (J_+ + J_-) \frac{\partial^2 t}{\partial J_+ \partial J_-} + (\gamma + 1) \left(\frac{\partial t}{\partial J_+} + \frac{\partial t}{\partial J_-} \right) = 0 \qquad (1.9.47)$$

If this equation can be solved for t, then x can be obtained from Eqs. (1.9.45) and (1.9.46) (see Exercises 1.9.5 and 1.9.6).

As the second example we shall consider the equations for a shallow-water wave (see, e.g., Jeffrey and Taniuti, 1964). Let $y = y^*(x, t)$ and $y = y_*(x)$ denote the elevation of the free surface of water and the depth of the seabed, respectively, as shown in Fig. 1.32, where x is the horizontal distance taken downstream. If u and v are the horizontal and vertical components of water velocity $\mathbf{v}$, respectively, the appropriate boundary conditions are

$$\text{at } y = -y_* : \quad u \frac{\partial y_*}{\partial x} + v = 0 \qquad (1.9.48)$$

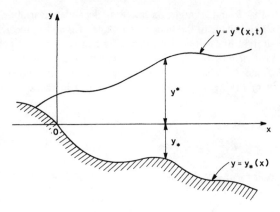

Figure 1.32

and

$$\text{at } y = y^*: \quad \frac{\partial}{\partial t}(y - y^*) + \mathbf{v} \cdot \nabla(y - y^*) = 0$$

or

$$\text{at } y = y^*: \quad \frac{\partial y^*}{\partial t} + u\frac{\partial y^*}{\partial x} - v = 0 \tag{1.9.49}$$

The latter condition expresses the fact that a fluid particle which is once at the free surface will stay there for all time. By applying these conditions to the continuity equation, $\nabla.\mathbf{v} = 0$, we obtain

$$\frac{\partial y^*}{\partial t} + \frac{\partial}{\partial x}\{(y^* + y_*)u\} = 0 \tag{1.9.50}$$

which is equivalent to the floodwater equation discussed in Sec. 1.11 of Vol. 1. We shall assume that the viscosity has negligible effect and the pressure of water is determined by the hydrostatic pressure alone. Thus the x component of the Navier–Stokes equation can be written as

$$\frac{\partial u}{\partial t} + u\frac{\partial u}{\partial x} = -g\frac{\partial y^*}{\partial x} \tag{1.9.51}$$

in which g is the acceleration due to gravity. By letting

$$c = \sqrt{g(y^* + y_*)}$$
$$H = gy_* \tag{1.9.52}$$

we can rewrite Eqs. (1.9.50) and (1.9.51) in the form

$$\frac{\partial u}{\partial t} + u\frac{\partial u}{\partial x} + 2c\frac{\partial c}{\partial x} - \frac{\partial H}{\partial x} = 0 \tag{1.9.53}$$

Pairs of Quasilinear Hyperbolic Equations of First Order   Chap. 1

$$2\frac{\partial c}{\partial t} + c\frac{\partial u}{\partial x} + 2u\frac{\partial c}{\partial x} = 0$$

This pair of equations represents the so-called shallow-water wave approximation or the long-wave approximation. As we shall see in the following, the variable c represents the propagation speed of a surface disturbance relative to the fluid particle.

In the matrix form the equations become

$$\frac{\partial \mathbf{w}}{\partial t} + \mathbf{A}\frac{\partial \mathbf{w}}{\partial x} - \mathbf{b} = \mathbf{0} \qquad (1.9.54)$$

where

$$\mathbf{w} = \begin{bmatrix} u \\ c \end{bmatrix}, \qquad \mathbf{A} = \begin{bmatrix} u & 2c \\ c/2 & u \end{bmatrix}, \qquad \mathbf{b} = \begin{bmatrix} -\dfrac{\partial H}{\partial x} \\ 0 \end{bmatrix} \qquad (1.9.55)$$

Now the characteristic quadratic (1.9.11) gives the eigenvalues of $\mathbf{A}$,

$$\lambda_+ = u + c, \qquad \lambda_- = u - c \qquad (1.9.56)$$

which imply that a surface disturbance would propagate with a speed c relative to the fluid particle. Since we have

$$\text{adj } (\lambda_\pm \mathbf{I} - \mathbf{A}) = \begin{bmatrix} \pm c & 2c \\ c/2 & \pm c \end{bmatrix}$$

the left and right eigenvectors corresponding to $\lambda_\pm$ are given by

$$\boldsymbol{\ell}_\pm = \begin{bmatrix} 1 \\ \pm 2 \end{bmatrix} \quad \text{and} \quad \mathbf{r}_\pm = \begin{bmatrix} \pm 2 \\ 1 \end{bmatrix} \qquad (1.9.57)$$

respectively.

For the case when the seabed is flat, i.e., $\partial H/\partial x = 0$, we find from Eqs. (1.9.20) and (1.9.22) that the Riemann invariants are simply[†]

$$J_+(\beta) = u + 2c, \qquad J_-(\alpha) = u - 2c \qquad (1.9.58)$$

and thus the Γ characteristics in the hodograph plane of u and c are all straight, the same kind being parallel to each other. We can now express the variables u and c in terms of J_+ and J_-, i.e.,

$$u = \frac{1}{2}(J_+ + J_-), \qquad c = \frac{1}{4}(J_+ - J_-) \qquad (1.9.59)$$

and by making the identifications $\alpha \to J_-$ and $\beta \to J_+$ in the first pair of equations of the set of four characteristic differential equations (1.2.20), we obtain

---

[†] Note that Eq. (1.9.58) is essentially identical to Eq.(1.9.42) for $\gamma = 2$. In fact, Eq. (1.9.53) with $\partial H/\partial x = 0$ is the same as Eq. (1.9.31) for a polytropic gas with $\gamma = 2$.

$$C_+: \quad \frac{\partial x}{\partial J_-} = (u + c)\frac{\partial t}{\partial J_-} = \frac{1}{4}(3J_+ + J_-)\frac{\partial t}{\partial J_-}$$

$$C_-: \quad \frac{\partial x}{\partial J_+} = (u - c)\frac{\partial t}{\partial J_+} = \frac{1}{4}(J_+ + 3J_-)\frac{\partial t}{\partial J_+} \qquad (1.9.60)$$

As in the previous example, equating the mixed derivatives $\partial^2 x/\partial J_+ \, \partial J_-$ and $\partial^2 x/\partial J_- \, \partial J_+$ gives the linear second-order equation

$$2(J_+ - J_-)\frac{\partial^2 t}{\partial J_+ \, \partial J_-} + 3\frac{\partial t}{\partial J_+} - 3\frac{\partial t}{\partial J_-} = 0 \qquad (1.9.61)$$

Solving this equation for t and then using Eq. (1.9.60) to determine x, we can find the solution to the shallow-water wave approximation for water above a flat seabed.

Finally, let us take the pair of chromatographic equations (1.4.2) and put them into matrix form as

$$\frac{\partial \mathbf{w}}{\partial x} + \mathbf{A}\frac{\partial \mathbf{w}}{\partial \tau} = 0 \qquad (1.9.62)$$

where

$$\mathbf{w} = \begin{bmatrix} c_1 \\ c_2 \end{bmatrix}, \qquad \mathbf{A} = \begin{bmatrix} \dfrac{\partial f_1}{\partial c_1} & \dfrac{\partial f_1}{\partial c_2} \\ \dfrac{\partial f_2}{\partial c_1} & \dfrac{\partial f_2}{\partial c_2} \end{bmatrix} \qquad (1.9.63)$$

The eigenvalues of $\mathbf{A}$ are given by the characteristic quadratic

$$|\sigma \mathbf{I} - \mathbf{A}| = \sigma^2 - \left(\frac{\partial f_1}{\partial c_1} + \frac{\partial f_2}{\partial c_2}\right)\sigma + \frac{\partial f_1}{\partial c_1}\frac{\partial f_2}{\partial c_2} - \frac{\partial f_1}{\partial c_2}\frac{\partial f_2}{\partial c_1} = 0$$

or

$$2\sigma_\pm = \frac{\partial f_1}{\partial c_1} + \frac{\partial f_2}{\partial c_2} \pm \sqrt{\left(\frac{\partial f_1}{\partial c_1} + \frac{\partial f_2}{\partial c_2}\right)^2 - 4\left(\frac{\partial f_1}{\partial c_1}\frac{\partial f_2}{\partial c_2} - \frac{\partial f_1}{\partial c_2}\frac{\partial f_2}{\partial c_1}\right)} \qquad (1.9.64)$$

and these are clearly identical to those given by Eqs. (1.4.18) and (1.4.19), respectively. Thus the characteristic directions in the (x, τ)-plane are given by the eigenvalues of $\mathbf{A}$ in Eq. (1.9.62). This is so because here the matrix $\mathbf{A}$ is associated with the time derivative $\partial \mathbf{w}/\partial \tau$ in contrast to the case of Eq. (1.9.1).

From Sec. 1.4 we know that the characteristic directions in the hodograph plane are

$$\Gamma_+: \quad \frac{dc_1}{dc_2} = \zeta_+(c_1, c_2)$$

$$\Gamma_-: \quad \frac{dc_1}{dc_2} = \zeta_-(c_1, c_2) \qquad (1.9.65)$$

where ζ_+ and ζ_- are the positive and negative roots, respectively, of the quadratic equation

$$\frac{\partial f_2}{\partial c_1} \zeta^2 - \left(\frac{\partial f_1}{\partial c_1} - \frac{\partial f_2}{\partial c_2}\right)\zeta - \frac{\partial f_1}{\partial c_2} = 0 \qquad (1.9.66)$$

Since the Γ_+ characteristics are tangent to the right eigenvector $\mathbf{r}_-$ corresponding to σ_- and the Γ_- characteristics to $\mathbf{r}_+$, the right eigenvector corresponding to σ_+, we can write

$$\mathbf{r}_+ = \begin{bmatrix} \zeta_- \\ 1 \end{bmatrix} \quad \text{and} \quad \mathbf{r}_- = \begin{bmatrix} \zeta_+ \\ 1 \end{bmatrix} \qquad (1.9.67)$$

By applying the biorthogonality property of the left and right eigenvectors, we find that

$$\boldsymbol{\ell}_+ = \begin{bmatrix} -1 \\ \zeta_+ \end{bmatrix} \quad \text{and} \quad \boldsymbol{\ell}_- = \begin{bmatrix} -1 \\ \zeta_- \end{bmatrix} \qquad (1.9.68)$$

Hence the Riemann invariants are expressed as

$$\begin{aligned}
J_+(\beta) &= -c_1 + \int \zeta_+(c_1, c_2)\, dc_2 \\
J_-(\alpha) &= -c_1 + \int \zeta_-(c_1, c_2)\, dc_2
\end{aligned} \qquad (1.9.69)$$

If the adsorption equilibrium is described by Langmuir isotherms, these expressions take explicit forms in terms of the characteristic parameters α and β as we shall see in Chapter 2.

Exercises

1.9.1. Put the pair of equations (1.2.1) into matrix form. Show that Eq. (1.9.11) gives the characteristic quadratic (1.2.6) and Eqs. (1.9.15) and (1.9.16) represent the last two equations in the set of the characteristic differential equations (1.2.20).

1.9.2. Consider the equations of polymer flooding (1.8.6) and put them in the form

$$\mathbf{w} = \begin{bmatrix} s \\ c \end{bmatrix}, \qquad \frac{\partial \mathbf{w}}{\partial x} + \mathbf{A}\frac{\partial \mathbf{w}}{\partial \tau} = 0$$

How is $\mathbf{A}$ defined? Determine the eigenvalues of $\mathbf{A}$ and corresponding left and right eigenvectors. By using this information, show that the characteristic directions given by Eqs. (1.8.10) and (1.8.12) are correct. Also, find expressions for the Riemann invariants.

1.9.3. For the system of equations (1.9.31) with the equation of state (1.9.40), find the right eigenvectors corresponding to λ_+ and λ_-, and show that the system is genuinely nonlinear.

1.9.4. Consider the steady, irrotational, isentropic flow in three dimensions with cylindrical symmetry for which the conservation equations are given in Exercise 1.2.2. If we put the equations in matrix form

$$\mathbf{w} = \begin{bmatrix} u \\ v \end{bmatrix}, \qquad \frac{\partial \mathbf{w}}{\partial x} + \mathbf{A}\frac{\partial \mathbf{w}}{\partial y} + \mathbf{b} = 0$$

how are $\mathbf{A}$ and $\mathbf{b}$ defined? From this equation determine the characteristic directions $dy/dx = \lambda_\pm$ and the corresponding left eigenvectors, and thereby find the characteristic directions in the hodograph plane.

1.9.5. Solve Eq. (1.9.47) for $\gamma = 3$ and show that the general solution is given by

$$ct = f(u + c) + g(u - c)$$
$$cx = (u - c)f(u + c) + (u + c)g(u - c) + h$$

where f and g are arbitrary functions and h an arbitrary constant.

1.9.6. If $\gamma = (2N + 3)/(2N + 1)$, $N = 1, 2, 3, \ldots$, show that the solution of Eq. (1.9.47) is

$$t = \left(\frac{\partial}{\partial J_+}\right)^N \frac{f(J_+)}{(J_+ + J_-)^{N+1}} + \left(\frac{\partial}{\partial J_-}\right)^N \frac{g(J_-)}{(J_+ + J_-)^{N+1}} + h$$

and find x, where f and g are arbitrary functions and h an arbitrary constant.

1.10 Development of Singularities, Weak Solution, and the Entropy Condition

In Sec. 6.1 of Vol. 1 we showed for single conservation laws a geometric argument that under certain circumstances the solution cannot be determined by a smooth function beyond a certain time. This has also been proven from a different viewpoint in Sec. 7.8 of Vol. 1; that is, if $f''(c)\Gamma'(\tau - xf''(c)) < 0$ in Eqs. (7.8.15) and (7.8.16) of Vol. 1 the derivatives $\partial c/\partial x$ and $\partial c/\partial \tau$ become infinite in finite time. In Sec. 1.6 we have treated the characteristic initial value problems for pairs of conservation laws, and in this case we have been able to give the same geometric argument for the nonexistence of continuous solutions after a finite time. In both cases the goemetric argument was based on the fact that the solution was constant along characteristics C and the characeristics were straight lines.

For pairs of conservation laws, in general, we can make use of the Riemann invariants, J_+ and J_-, each of which remains constant along characteristics of one family. But the characteristics are not straight lines, so that the geometric argument cannot be extended to this case. Thus we shall take a different approach for a single equation that can be generalized to pairs of conservation laws. Let us consider the equation

$$\frac{\partial u}{\partial t} + f'(u)\frac{\partial u}{\partial x} = 0 \tag{1.10.1}$$

and differentiate it with respect to x to obtain

$$\frac{\partial^2 u}{\partial t\, \partial x} + f'(u)\frac{\partial^2 u}{\partial x^2} + f''(u)\left(\frac{\partial u}{\partial x}\right)^2 = 0$$

Let

$$p = \frac{\partial u}{\partial x} \qquad (1.10.2)$$

and d/dt denote differentiation with respect to t in the characteristic direction, i.e.,

$$\frac{d}{dt} = \frac{\partial}{\partial t} + f'(u)\frac{\partial}{\partial x} \qquad (1.10.3)$$

Then we have

$$\frac{dp}{dt} + f''(u)p^2 = 0 \qquad (1.10.4)$$

which is an ordinary differential equation for p along the characteristic. Since $f''(u)$ remains fixed along the characteristic, integration gives

$$p(x, t) = \frac{p_0(x)}{1 + p_0(x)f''(u)t} \qquad (1.10.5)$$

where $p_0(x) = p(x, 0) = \frac{\partial u}{\partial x}(x, 0)$. Thus if $p_0(x)f''(u) > 0$, $p(x, t)$ is bounded for all $t > 0$ while if $p_0(x)f''(u) < 0$, $p(x, t)$ becomes infinite at $t = -\{p_0(x)f''(u)\}^{-1}$. [Compare this with Eq. (7.8.19) of Vol. 1.]

We shall now extend this approach to the case of a pair of reducible equations. To do this we shall follow the development due to Lax (1964, 1973). Thus let us consider the system of equations

$$\mathbf{w} = \begin{bmatrix} u \\ v \end{bmatrix}, \qquad \frac{\partial \mathbf{w}}{\partial t} + \mathbf{A}\frac{\partial \mathbf{w}}{\partial x} = 0 \qquad (1.10.6)$$

where $\mathbf{A}$ is a 2×2 matrix whose elements are functions of u and v. Here we have the Riemann invariants, J_+ and J_-, and if we let d_+/dt denote differentiation with respect to t in the direction of a C_+ characteristic, we find that J_+ satisfies the equation

$$\frac{d_+ J_+}{dt} = \frac{\partial J_+}{\partial t} + \lambda_+ \frac{\partial J_+}{\partial x} = 0 \qquad (1.10.7)$$

Similarly, if d_-/dt denotes differentiation with respect to t in the direction of a C_- characteristic, we have

$$\frac{d_- J_-}{dt} = \frac{\partial J_-}{\partial t} + \lambda_- \frac{\partial J_-}{\partial x} = 0 \qquad (1.10.8)$$

Differentiating Eq. (1.10.7) with respect to x, we obtain

$$\frac{\partial^2 J_+}{\partial t\, \partial x} + \lambda_+ \frac{\partial^2 J_+}{\partial x^2} + \frac{\partial \lambda_+}{\partial J_+}\left(\frac{\partial J_+}{\partial x}\right)^2 + \frac{\partial \lambda_+}{\partial J_-}\frac{\partial J_+}{\partial x}\frac{\partial J_-}{\partial x} = 0$$

or, by letting

$$p = \frac{\partial J_+}{\partial x} \tag{1.10.9}$$

we have the equation

$$\frac{d_+p}{dt} + \frac{\partial \lambda_+}{\partial J_+}p^2 + \frac{\partial \lambda_+}{\partial J_-}\frac{\partial J_-}{\partial x}p = 0 \tag{1.10.10}$$

On the other hand, we can write

$$\frac{d_+ J_-}{dt} = \frac{\partial J_-}{\partial t} + \lambda_+\frac{\partial J_-}{\partial x} = (\lambda_+ - \lambda_-)\frac{\partial J_-}{\partial x}$$

in which Eq. (1.10.8) has been applied, so that

$$\frac{\partial J_-}{\partial x} = \frac{1}{\lambda_+ - \lambda_-}\frac{d_+ J_-}{dt}$$

Substituting this into Eq. (1.10.10) gives

$$\frac{d_+p}{dt} + \frac{\partial \lambda_+}{\partial J_+}p^2 + \frac{1}{\lambda_+ - \lambda_-}\frac{\partial \lambda_+}{\partial J_-}\frac{d_+ J_-}{dt}p = 0 \tag{1.10.11}$$

Let $h = h(J_+, J_-)$ be a function of J_+ and J_- that satisfies

$$\frac{\partial h}{\partial J_-} = \frac{1}{\lambda_+ - \lambda_-}\frac{\partial \lambda_+}{\partial J_-}$$

Since $d_+ J_+/dt = 0$, we find

$$\frac{d_+h}{dt} = \frac{\partial h}{\partial J_+}\frac{d_+ J_+}{dt} + \frac{\partial h}{\partial J_-}\frac{d_+ J_-}{dt} = \frac{1}{\lambda_+ - \lambda_-}\frac{\partial \lambda_+}{\partial J_-}\frac{d_+ J_-}{dt}$$

and substitution of this into Eq. (1.10.11) gives

$$\frac{d_+p}{dt} + \frac{\partial \lambda_+}{\partial J_+}p^2 + \frac{d_+h}{dt}p = 0$$

We then multiply this equation through by e^h and let

$$q = e^h p = e^h \frac{\partial J_+}{\partial x} \tag{1.10.12}$$

to obtain

$$e^h\frac{d_+p}{dt} + e^h\frac{d_+h}{dt}p + \frac{\partial \lambda_+}{\partial J_+}e^h p^2 = \frac{d_+q}{dt} + \frac{\partial \lambda_+}{\partial J_+}e^{-h}q^2 = 0$$

If we let

$$k = e^{-h} \frac{\partial \lambda_+}{\partial J_+} \tag{1.10.13}$$

this equation becomes

$$\frac{d_+ q}{dt} + kq^2 = 0 \tag{1.10.14}$$

which is an ordinary differential equation for q along each C_+ characteristic, similar to Eq. (1.10.4) except that the coefficient of q^2 is not constant. But the solution is explicitly given by

$$q(x, t) = \frac{q_0(x)}{1 + q_0(x)K(t)} \tag{1.10.15}$$

where $q_0(x) = q(x, 0)$ and $K(t) = \int_0^t k \, dt$, the integration being along the C_+ characteristic.

The question is now whether $q_0(x)K(t)$ takes on the value -1. Suppose that the $(+)$ characteristic field is genuinely nonlinear so that $\mathbf{r}_+^T \nabla \lambda_+ \neq 0$, which can be expressed equivalently as $\partial \lambda_+ / \partial J_+ \neq 0$ since J_+ may be regarded as the parameter along the Γ_- characteristic (see the discussion around Rule 1.6 in Sec. 1.9). The sign of J_+ is arbitrary, so we may as well assume that $\partial \lambda_+ / \partial J_+ > 0$. If J_+ and J_- are bounded at $t = 0$, they satisfy the same bounds for all $t > 0$, since the value of J_+ or J_- at any point P equals the value of J_+ or J_- at that point on the initial line to which P can be connected by a C_+ or C_- characteristic. When both J_+ and J_- are bounded for all time, the function k defined by Eq. (1.10.13) is bounded from below for all t and x by a positive constant k_0 so that $K(t) \geq k_0 t$ for all $t \geq 0$. Applying this to Eq. (1.10.15), we find that if $q_0(x) > 0$, $q(x, t)$ is bounded, but if $q_0(x) < 0$, $q(x, t)$ becomes unbounded after a finite time. Since q is defined as in Eq. (1.10.12), the sign of $q_0(x)$ is the same as that of the initial value of $\partial J_+ / \partial x$. Clearly, a similar argument holds with respect to the other Riemann invariant J_- (see Exercise 1.10.1) and thus we draw the following conclusion:

1. Suppose that the $(+)$ [or $(-)$] characteristic field of Eq. (1.10.6) is genuinely nonlinear, i.e., $\partial \lambda_+ / \partial J_+ > 0$ [or $\partial \lambda_- / \partial J_- > 0$], and both initial values $u(x, 0)$ and $v(x, 0)$ are bounded. Then if $\partial J_+(x, 0)/\partial x < 0$ [or $\partial J_-(x, 0)/\partial x < 0$] for some x, the derivatives of the solution become unbounded after a finite time.

The converse of this also holds, so we have;

2. Suppose that Eq. (1.10.6) is genuinely nonlinear, i.e., $\partial \lambda_+ / \partial J_+ > 0$ and $\partial \lambda_- / \partial J_- > 0$, and that both the initial values $u(x, 0)$ and $v(x, 0)$ are bounded. If both $J_+(x, 0)$ and $J_-(x, 0)$ are increasing functions of x, then the first derivatives of the solution remain uniformly bounded, and the solution exists and is differentiable for all $t > 0$.

We have now shown in a general manner under what conditions the solution of Eq. (1.10.6) cannot be determined by a continuous function. In such a case we must consider a discontinuity in the solution. Thus it is obvious that we need to introduce the concept of *weak solutions* as we did for single conservation laws in Sec. 7.4 of Vol. 1 so as to accommodate all solutions that are not continuously differentiable. The definition of weak solutions is based on the conservation laws rather than the quasilinear equations in the form of Eq.(1.10.6), so we shall begin with the pair of conservation laws

$$\frac{\partial u}{\partial t} + \frac{\partial}{\partial x} f_1(u, v) = 0$$

$$\frac{\partial v}{\partial t} + \frac{\partial}{\partial x} f_2(u, v) = 0 \qquad (1.10.16)$$

and the initial values

$$u(x, 0) = u_0(x), \qquad v(x, 0) = v_0(x) \quad \text{for } -\infty < x < \infty \qquad (1.10.17)$$

If we let

$$\mathbf{w} = \begin{bmatrix} u \\ v \end{bmatrix} \quad \text{and} \quad \mathbf{f} = \begin{bmatrix} f_1(u, v) \\ f_2(u, v) \end{bmatrix} \qquad (1.10.18)$$

Eqs. (1.10.16) and (1.10.17) can be written in vector form as

$$\frac{\partial \mathbf{w}}{\partial t} + \frac{\partial}{\partial x} \mathbf{f}(\mathbf{w}) = 0 \qquad (1.10.19)$$

$$\mathbf{w}(x, 0) = \mathbf{w}_0(x), \qquad -\infty < x < \infty \qquad (1.10.20)$$

which is typical of conservation laws.

Some remarks are worth mentioning with regard to the connection between Eqs. (1.10.6) and (1.10.16). Clearly, we can put Eq. (1.10.16) in the form of Eq. (1.10.6), i.e.,

$$\frac{\partial \mathbf{w}}{\partial t} + \mathbf{A}(\mathbf{w}) \frac{\partial \mathbf{w}}{\partial x} = 0 \qquad (1.10.21)$$

with the matrix defined as

$$\mathbf{A}(\mathbf{w}) = \begin{bmatrix} \dfrac{\partial f_1}{\partial u} & \dfrac{\partial f_1}{\partial v} \\ \dfrac{\partial f_2}{\partial u} & \dfrac{\partial f_2}{\partial v} \end{bmatrix} = d\mathbf{f} \qquad (1.10.22)$$

Thus $\mathbf{A}$ is given by the Jacobian matrix of the vector-valued function $\mathbf{f}$. This matrix is frequently referred to as the *Fréchet derivative* of $\mathbf{f}$ and denoted by $d\mathbf{f}$. Although the reverse process may not always be an obvious one, it has been proven that pairs of quasilinear equations (1.10.21) can be put in the form of conservation laws (1.10.16). For example, the equations of isentropic one-dimensional flow (1.9.3) can be written in the form of Eq. (1.10.19) by defining

$$\mathbf{w} = \begin{bmatrix} \rho \\ \rho u \end{bmatrix} \quad \text{and} \quad \mathbf{f} = \begin{bmatrix} \rho u \\ \rho u + p \end{bmatrix} \tag{1.10.23}$$

The equations of the shallow-water wave approximation $(1.9.53)$ may also be expressed in the conservation form if we let

$$\mathbf{w} = \begin{bmatrix} u \\ c^2 \end{bmatrix} \quad \text{and} \quad \mathbf{f} = \begin{bmatrix} \dfrac{u^2}{2} + c^2 - H(x) \\ uc^2 \end{bmatrix} \tag{1.10.24}$$

Here we note that *the equations of two-solute chromatography $(1.4.1)$ and $(1.4.2)$ are exactly of the same type as Eqs. $(1.10.16)$ and $(1.10.21)$ if we identify c_1, c_2, x, and τ with u, v, t, and x, respectively. Hence the discussion of this section is equally valid for the chromatographic equations with the correspondence of σ_+ to $1/\lambda_+$ and σ_- to $1/\lambda_-$, respectively.*

Returning to Eqs. $(1.10.16)$ and $(1.10.17)$, we define a weak solution as follows: If $u(x, t)$ and $v(x, t)$ are bounded, measurable functions and satisfy the integral relations

$$\int_0^\infty \int_{-\infty}^\infty \left\{ u \frac{\partial \phi}{\partial t} + f_1(u, v) \frac{\partial \phi}{\partial x} \right\} dx \, dt + \int_{-\infty}^\infty u_0(x)\phi(x, 0) \, dx = 0$$
$$\int_0^\infty \int_{-\infty}^\infty \left\{ v \frac{\partial \phi}{\partial t} + f_2(u, v) \frac{\partial \phi}{\partial x} \right\} dx \, dt + \int_{-\infty}^\infty v_0(x)\phi(x, 0) \, dx = 0 \tag{1.10.25}$$

for all smooth functions $\phi(x, t)$ which are identically zero outside some bounded set (i.e., having compact support), the pair of functions $u(x, t)$ and $v(x, t)$ is called a weak solution of Eqs. $(1.10.16)$ and $(1.10.17)$.

This definition is entirely analogous to that for a single conservation law (see Sec. 7.4 of Vol. 1), so that by applying integration by parts to each equation of Eq. $(1.10.25)$ we can show that if u and v are smooth in any region of the (x, t)-plane and satisfy Eq. $(1.10.25)$, then u and v satisfy Eq. $(1.10.16)$. On the other hand, if u and v are discontinuous, we consider a contour including a portion of the propagation path of discontinuity in the (x, t)-plane and apply Green's theorem as in Sec. 7.4 of Vol. 1 (see Exercise 1.10.4). This leads us to the jump condition as well as the compatibility relation to be satisfied on the path of discontinuity, i.e.,

$$\frac{dx}{dt} = \frac{[f_1]}{[u]} = \frac{[f_2]}{[v]} = \tilde{\lambda} \tag{1.10.26}$$

where the brackets [ ] represent the jump in the quantity enclosed across the discontinuity, e.g., $[u] = u^l - u^r$ and $[f_1] = f_1^l - f_1^r = f_1(u^l, v^l) - f_1(u^r, v^r)$. This may be rewritten in vector form as

$$\tilde{\lambda}[\mathbf{w}] = [\mathbf{f}] \tag{1.10.27}$$

which is called the generalized Rankine–Hugoniot condition.

With the state on one side, say $L(u^l, v^l)$, fixed Eq. $(1.10.26)$ gives the locus of the state on the other side $R(u^r, v^r)$. This locus is called the *shock curve* and denoted by Σ. For hyperbolic systems it has been shown that the locus consists of two branches passing through L (see Fig. 1.20) and further that if the system (1.10.16) is genuinely nonlinear, the shock curve Σ makes a third-order contact at L with the Γ characteristic passing through L; i.e., Σ and Γ have the same tangent and curvature at L (see, e.g., Keyfitz, 1978). Following the treatment in Sec. 1.7, we let Σ_+ and Σ_- denote the shock curves that are tangent to the Γ_+ and Γ_- characteristics, respectively. If the states L and R fall on a Σ_-, we put $\tilde{\lambda} = \tilde{\lambda}_+$, and if L and R fall on a Σ_+, $\tilde{\lambda} = \tilde{\lambda}_-$.

Just as in the case of a single equation, so also with pairs of equations it is necessary to impose an additional condition to ensure the uniqueness of the weak solution. Being called the *entropy condition*, such an additional condition was first given by Lax (1957) for a genuinely nonlinear system, i.e.,

$$\mathbf{r}_+^T \nabla \lambda_+ \neq 0 \quad \text{and} \quad \mathbf{r}_-^T \nabla \lambda_- \neq 0 \qquad (1.10.28)$$

Based on the stability consideration Lax's condition requires the C characteristics of the same family to impinge on the path of discontinuity from both sides as t increases. Thus we must have the inequalities

$$\lambda_+(u^l, v^l) > \tilde{\lambda}_+ > \lambda_+(u^r, v^r), \lambda_-(u^r, v^r) > \tilde{\lambda}_+ \qquad (1.10.29)$$

or

$$\lambda_-(u^l, v^l) > \tilde{\lambda}_- > \lambda_-(u^r, v^r), \tilde{\lambda}_- > \lambda_+(u^l, v^l) \qquad (1.10.30)$$

satisfied on the path of discontinuity. Compare these with Eqs. (1.7.19) and (1.7.20), and Eqs. (1.7.22) and (1.7.23). The two situations are illustrated in Fig. 1.33, in which the slopes of curves are given by the reciprocal values of the λ's or $\tilde{\lambda}$'s marked on them. The inequalities (1.10.29) and (1.10.30) are indeed the *shock condition* defining *shocks*. Thus a discontinuity satisfying Eq. (1.10.29) is called a Σ_- shock and the one satisfying Eq. (1.10.30) a Σ_+ shock. By imposing the shock condition Lax established the existence and uniqueness of the solution to the Riemann problem.[†] The solution consists of three constant states connected by two centered waves, either shocks or simple

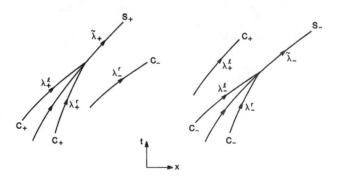

Figure 1.33

[†] In fact, Lax's proof was given for a system of n equations.

Pairs of Quasilinear Hyperbolic Equations of First Order   Chap. 1

waves, satisfying Eq. (1.10.26) across shocks and satisfying Eq. (1.10.16) in the usual sense at all other points. This structure of the solution is consistent with that described at the end of Sec. 1.7. However, Lax's proof was restricted to the case when the feed state as well as the intermediate constant state lies in a neighborhood of the initial state.

Lax (1971, 1973) has expressed the entropy condition in more general terms which both generalizes and unifies all of the previous notions. Let us consider the system (1.10.16) in vector form, i.e., Eq. (1.10.19). We may assume that $\mathbf{f}(0) = 0$. If $\mathbf{w}$ is a solution of Eq. (1.10.19) which is zero for $|x|$ sufficiently large, we can integrate Eq. (1.10.19) with respect to x to obtain

$$\int_{-\infty}^{\infty} \frac{\partial \mathbf{w}}{\partial t} \, dx = 0$$

which implies that the integral of $\mathbf{w}$ is conserved; i.e.,

$$\int_{-\infty}^{\infty} \mathbf{w}(x, t) \, dx = \int_{-\infty}^{\infty} \mathbf{w}(x, 0) \, dx, \qquad t > 0 \qquad (1.10.31)$$

To see this phenomenon more generally, let W be some function of u and v, and ask the following question: When does W satisfy a conservation law; i.e., a law of the form

$$\frac{\partial W}{\partial t} + \frac{\partial F}{\partial x} = 0 \qquad (1.10.32)$$

where F is some function of u and v? To find the answer to this question, we carry out the differentiation in Eq. (1.10.32) to find

$$(\nabla W)^T \frac{\partial \mathbf{w}}{\partial t} + (\nabla F)^T \frac{\partial \mathbf{w}}{\partial x} = 0$$

On the other hand, by multiplying Eq. (1.10.21) on the left by $(\nabla W)^T$, we obtain

$$(\nabla W)^T \frac{\partial \mathbf{w}}{\partial t} + (\nabla W)^T \mathbf{A} \frac{\partial \mathbf{w}}{\partial x} = 0$$

so that Eq. (1.10.32) results if and only if the relation

$$(\nabla W)^T \mathbf{A} = (\nabla F)^T \qquad (1.10.33)$$

holds. This is a pair of partial differential equations for W and F and called the *compatibility equation*.

It is not obvious whether this compatibility equation has a solution in general, but when $\mathbf{A}$ is a symmetric matrix, we can easily find the solution. In this case we have the relation

$$\frac{\partial f_1}{\partial v} = \frac{\partial f_2}{\partial u} \qquad (1.10.34)$$

and thus there exists a function $\phi(u, v)$ such that

$$\mathbf{f} = \nabla \phi$$

If we let

$$W = \frac{1}{2}(u^2 + v^2)$$

$$F = uf_1(u, v) + vf_2(u, v) - \phi(u, v) \qquad (1.10.35)$$

or, in vector form,

$$W = \frac{1}{2}\mathbf{w}^T\mathbf{w}$$

$$F = \mathbf{w}^T\mathbf{f} - \phi = \mathbf{w}^T\nabla\phi - \phi \qquad (1.10.36)$$

then

$$\frac{\partial F}{\partial x} = \frac{\partial \mathbf{w}^T}{\partial x}\nabla\phi + \mathbf{w}^T\frac{\partial(\nabla\phi)}{\partial x} - (\nabla\phi)^T\frac{\partial \mathbf{w}}{\partial x}$$

$$= \mathbf{w}^T\frac{\partial(\nabla\phi)}{\partial x} = \mathbf{w}^T\frac{\partial\mathbf{f}}{\partial x} = -\mathbf{w}^T\frac{\partial\mathbf{w}}{\partial t}$$

where Eq. (1.10.19) has been used to get the last expression. But from Eq. (1.10.36), we have

$$\frac{\partial W}{\partial t} = \mathbf{w}^T\frac{\partial\mathbf{w}}{\partial t}$$

and hence W and F defined by Eq. (1.10.35) satisfy the conservation law (1.10.32). We note that here the function W is given by a convex function.

Now we recall that the role of the entropy condition is to distinguish those discontinuous solutions which are physically realizable from those which are not. Another way in which the physically realizable solutions can be characterized is to identify them as limits of solutions of equations with some dissipation. Specifically, we consider the equation containing the artificial viscous term. Thus from Eq. (1.10.21) we have

$$\frac{\partial\mathbf{w}}{\partial t} + \mathbf{A}\frac{\partial\mathbf{w}}{\partial x} = \mu\frac{\partial^2\mathbf{w}}{\partial x^2}, \qquad \mu > 0 \qquad (1.10.37)$$

Multiplying this on the left by $(\nabla W)^T$ and applying the compatibility equation (1.10.33), we obtain

$$(\nabla W)^T\frac{\partial\mathbf{w}}{\partial t} + (\nabla F)^T\frac{\partial\mathbf{w}}{\partial x} = \mu(\nabla W)^T\frac{\partial^2\mathbf{w}}{\partial x^2}$$

or

$$\frac{\partial W}{\partial t} + \frac{\partial F}{\partial x} = \mu(\nabla W)^T\frac{\partial^2\mathbf{w}}{\partial x^2} \qquad (1.10.38)$$

On the other hand,

$$\frac{\partial^2 W}{\partial x^2} = (\nabla W)^T\frac{\partial^2\mathbf{w}}{\partial x^2} + \frac{\partial\mathbf{w}^T}{\partial x}\mathbf{H}(W)\frac{\partial\mathbf{w}}{\partial x}$$

where $\mathbf{H}(W)$ is the Hessian matrix of W; i.e.,

$$\mathbf{H}(W) = \begin{bmatrix} \dfrac{\partial^2 W}{\partial u^2} & \dfrac{\partial^2 W}{\partial u\,\partial v} \\[2ex] \dfrac{\partial^2 W}{\partial u\,\partial v} & \dfrac{\partial^2 W}{\partial v^2} \end{bmatrix}$$

Suppose that W is strictly convex; then the matrix $\mathbf{H}(W)$ is positive definite, so that

$$\frac{\partial^2 W}{\partial x^2} \geq (\nabla W)^T \frac{\partial^2 \mathbf{w}}{\partial x^2}$$

and substituting this into Eq. (1.10.38) gives

$$\frac{\partial W}{\partial t} + \frac{\partial F}{\partial x} \leq \mu \frac{\partial^2 W}{\partial x^2}$$

since $\mu > 0$. In the limit as $\mu \to 0$, the right-hand side tends to zero in the sense of distributions and hence

$$\frac{\partial W}{\partial t} + \frac{\partial F}{\partial x} \leq 0 \tag{1.10.39}$$

Consequently, we deduce the following theorem.

THEOREM 1.2: Let Eq. (1.10.19) be a pair of conservation laws which implies an additional conservation law (1.10.32); suppose that W is strictly convex. Let $\mathbf{w}(x, t)$ be a weak solution of Eq. (1.10.19) which is the limit of solutions of the viscosity equation (1.10.37). Then $\mathbf{w}$ satisfies the inequality (1.10.39).

Since we may assume that F vanishes for $\mathbf{w} = 0$, if $\mathbf{w}$ is a weak solution of Eq. (1.10.19) which is zero for $|x|$ sufficiently large, we can integrate Eq. (1.10.39) with respect to x and obtain

$$\int_{-\infty}^{\infty} \frac{\partial W}{\partial t} \, dx \leq 0$$

which implies that the integral

$$\int_{-\infty}^{\infty} W \, dx \tag{1.10.40}$$

is a decreasing function of t. Suppose that $\mathbf{w}$ is piecewise continuous. Then by applying the same procedure as the one giving the jump condition (1.10.27), we can show that at a point of discontinuity $\mathbf{w}$ satisfies the condition

$$\tilde{\lambda}[W] - [F] \leq 0 \tag{1.10.41}$$

(see Exercise 1.10.5).

In the case of compressible fluid flow Eq. (1.10.40) corresponds to the increase of total negative entropy and Eq. (1.10.41) indicates that the entropy of particles increases upon crossing a shock. For this reason we shall call W an *entropy* for Eq. (1.10.19) with *entropy flux F*. Also, Eqs. (1.10.39) and (1.10.41) will be referred to as the *entropy conditions*. Indeed, Lax has shown that for genuinely nonlinear systems, Eq. (1.10.41) is equivalent to the shock condition [Eqs. (1.10.29) and (1.10.30)]. For single equations it can be easily shown that Eq. (1.10.41) implies the generalized entropy condition [Eqs. (7.4.18) and (7.4.19) of Vol. 1].

Returning to the compatibility equation (1.10.33), we can rewrite it in two partial differential equations

$$\frac{\partial f_1}{\partial u}\frac{\partial W}{\partial u} + \frac{\partial f_2}{\partial u}\frac{\partial W}{\partial v} = \frac{\partial F}{\partial u}$$

$$\frac{\partial f_1}{\partial v}\frac{\partial W}{\partial u} + \frac{\partial f_2}{\partial v}\frac{\partial W}{\partial v} = \frac{\partial F}{\partial v}$$

Now differentiating the first equation with respect to v and the second with respect to u, and subtracting one from the other, we obtain a homogeneous second-order equation for W:

$$-\frac{\partial f_1}{\partial v}\frac{\partial^2 W}{\partial u^2} + \left(\frac{\partial f_1}{\partial u} - \frac{\partial f_2}{\partial v}\right)\frac{\partial^2 W}{\partial u\,\partial v} + \frac{\partial f_2}{\partial u}\frac{\partial^2 W}{\partial v^2} = 0 \qquad (1.10.42)$$

It is obvious that if the original conservation law (1.10.19) is hyperbolic, so is the derived equation (1.10.42). We then ask the question: Does the second-order equation (1.10.42) have convex solutions? Lax has proven that Eq. (1.10.42) has a convex solution in the neighborhood of every point and further asserted that the compatibility equation (1.10.33) has solutions with W convex in any domain of the (u, v)-plane.

Dafermos (1973) has used the term "total entropy" for the integral of Eq. (1.10.40) and postulated that for the admissible solution the total entropy decreases with the highest possible rate. This entropy rate admissibility criterion has been proven valid in case of the isentropic flow of compressible fluid, where the system of equations is not genuinely nonlinear.

Exercises

1.10.1. Suppose that the $(-)$ characteristic field of Eq. (1.10.6) is genuinely nonlinear and the initial values are bounded. If $\partial J_-(x, 0)/\partial x < 0$ for some x, show that the derivatives of the solution become unbounded after a finite time.

1.10.2. Put the equations of polymer flooding [Eq. (1.8.6)], in the conservation form, i.e., in the form of Eq. (1.10.16).

1.10.3. Consider a weak solution of Eq. (1.10.16) containing one discontinuity that moves along a smooth curve $x = y(t)$. As shown in Fig. 1.34, choose a and b

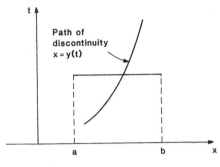

Figure 1.34

so that the curve $x = y(t)$ intersects the interval $a \leq x \leq b$ at time t, and then apply the conservation law for u over the interval $a \leq x \leq b$, regarding $f_1(u, v)$ as the flux of u. By using the resulting equation, derive the jump condition for u. Repeat the same procedure for v and thus obtain Eq. (1.10.26).

1.10.4. Suppose that we take a domain D large enough to occupy the whole of the plane $t \geq 0$ as shown in Fig. 1.35. If S is the path of discontinuity, carry out the integration in each equation of Eq. (1.10.25) by applying Green's theorem, i.e.,

$$\iint_D \left\{ \frac{\partial}{\partial t}(u\phi) + \frac{\partial}{\partial x}(f\phi) \right\} dx \, dt = \oint \phi(-u \, dx + f \, dt)$$

where the single integral is taken around the boundary of D. Noting that the boundary must include the upper and lower sides of S, show that we obtain Eq. (1.10.26) (see Sec. 7.4 of Vol. 1).

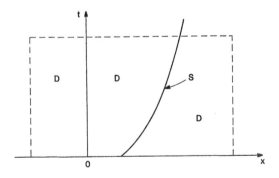

Figure 1.35

1.10.5. Regarding W as an entropy for Eq. (1.10.19) and F as its flux and noting that the integral of W is a nonincreasing function of t [see Eq. (1.10.40)], apply the scheme of Exercise 1.10.3 to formulate Eq. (1.10.41).

1.11 Existence, Uniqueness, Structure, and Asymptotic Behavior of Weak Solutions

Since the pioneering work of Lax (1957), numerous articles have been published concerning the existence and uniqueness of weak solutions, their structure, and their asymptotic behavior. The scope ranges from some special forms of equations to the general form of equations, from genuinely nonlinear systems to nongenuinely nonlinear systems, and from Riemann problems to arbitrary initial value problems with various classes of initial data. Rather than going into the detail of particular articles, we would like to make a survey in the literature so as to introduce some of the features that have been accomplished. In doing this we admit that we are not being exhaustive.

One of the major contributions is attributed to Glimm (1965), who has proven that the initial value problem,

$$\frac{\partial u}{\partial t} + \frac{\partial}{\partial x} f_1(u, v) = 0$$
$$\frac{\partial v}{\partial t} + \frac{\partial}{\partial x} f_2(u, v) = 0$$

(1.11.1)

and

$$u(x, 0) = u_0(x), \qquad v(x, 0) = v_0(x) \qquad \text{for} \quad -\infty < x < \infty \qquad (1.11.2)$$

has a global weak solution provided that the system (1.11.1) is genuinely nonlinear, i.e.,

$$\mathbf{r}_+^T \nabla \lambda_+ \neq 0 \quad \text{and} \quad \mathbf{r}_-^T \nabla \lambda_- \neq 0 \qquad (1.11.3)$$

and the initial data (1.11.2) have a small total variation. His solutions are given by the limits of difference approximations which involve a random choice and thus have a stochastic feature. In Eq. (1.11.3), λ_+, λ_-, $\mathbf{r}_+$, and $\mathbf{r}_-$ all have the usual meanings defined in Sec. 1.10. Thus if we let

$$\mathbf{w} = \begin{bmatrix} u \\ v \end{bmatrix} \quad \text{and} \quad \mathbf{f}(\mathbf{w}) = \begin{bmatrix} f_1 \\ f_2 \end{bmatrix} \qquad (1.11.4)$$

Eq. (1.11.1) can be written in the form

$$\frac{\partial \mathbf{w}}{\partial t} + \frac{\partial \mathbf{f}}{\partial x} = \mathbf{0} \qquad (1.11.5)$$

or

$$\frac{\partial \mathbf{w}}{\partial t} + \mathbf{A}(\mathbf{w}) \frac{\partial \mathbf{w}}{\partial x} = \mathbf{0} \qquad (1.11.6)$$

where $\mathbf{A}(\mathbf{w})$ is the Jacobian matrix of $\mathbf{f}(\mathbf{w})$ [see Eq. (1.10.22)]. Now λ_+ and λ_- are the eigenvalues of $\mathbf{A}$ and $\mathbf{r}_+$ and $\mathbf{r}_-$ are the right eigenvectors corresponding to λ_+ and λ_-, respectively.

Another line of study has been motivated from a consideration of a special form of equations that arise in gas dynamics. Thus the equations take the form

$$\frac{\partial u}{\partial t} - \frac{\partial v}{\partial x} = 0$$

$$\frac{\partial v}{\partial t} + \frac{\partial}{\partial x} g(u) = 0 \tag{1.11.7}$$

where g is an increasing function of u. When this system is genuinely nonlinear [i.e., $g''(u) \neq 0$], Godunov (1956), Oleinik (1957), Douglis (1961), and Hurd (1970) have established the uniqueness of weak solutions under certain smoothness and ordering conditions on the solutions. In particular, Oleinik has considered the class of functions sufficiently broad to include centered simple waves. Zhang and Guo (1965) have discussed the existence of solutions in the large. They first approximate the initial data (continuous almost everywhere) by a sequence of step functions and construct the solution in the same way as for Riemann problems. After treating the interactions between waves they prove that the solution so constructed is everywhere convergent with respect to x for any fixed t and thus establish the existence theorem. This approach has been extended by Johnson and Smoller (1967) to the less restricted systems

$$\frac{\partial u}{\partial t} + \frac{\partial}{\partial x} f(v) = 0$$

$$\frac{\partial v}{\partial t} + \frac{\partial}{\partial x} g(u) = 0 \tag{1.11.8}$$

For the general form of equations [Eq. (1.11.1)], Smoller and Johnson (1969) have discussed the existence of global solutions with bounded, measurable initial data satisfying a monotonicity condition by adapting the same approach as above. In addition to the genuine nonlinearity, they also assume that the system satisfies the shock condition as well as the "shock interaction condition," which states that the interaction of two shocks of the same family produces a shock of that family and a simple wave of the opposite family. They first consider the Riemann problem and finally take the general initial data as the limit of a sequence of piecewise constant data. It is further shown that under the same hypothesis, a solution can be obtained by using Glimm's difference scheme. This implies that Glimm's scheme can be used to solve certain initial value problems where the variation of the initial data is arbitrarily large.

In their paper Smoller and Johnson have expressed the genuine nonlinearity in an alternate form; i.e., if the $(+)$ characteristic field is genuinely nonlinear, the condition

$$\ell_+^T d^2 \mathbf{f}(\mathbf{r}_+, \mathbf{r}_+) \neq 0 \tag{1.11.9}$$

holds, where ℓ_+ is the left eigenvector of $\mathbf{A}$ corresponding to λ_+ and $d^2\mathbf{f}(\mathbf{r}_+, \mathbf{r}_+)$ denote the second Fréchet derivative of the vector-valued function $\mathbf{f}$ in the direction of $\mathbf{r}_+$. Similarly, if the $(-)$ characteristic field is genuinely nonlinear, we have the condition

$$\boldsymbol{\ell}_-^T \, d^2\mathbf{f}(\mathbf{r}_-, \, \mathbf{r}_-) \neq 0 \qquad (1.11.10)$$

where $\mathbf{l}_-$ and $d^2\mathbf{f}(\mathbf{r}_-, \, \mathbf{r}_-)$ have similar meanings. Let $\mathbf{H}(f_i)$ denote the Hessian matrix of f_i,

$$\mathbf{H}(f_i) = \begin{bmatrix} \dfrac{\partial^2 f_i}{\partial u^2} & \dfrac{\partial^2 f_i}{\partial u \, \partial v} \\[2ex] \dfrac{\partial^2 f_i}{\partial u \, \partial v} & \dfrac{\partial^2 f_i}{\partial v^2} \end{bmatrix}$$

then $d^2\mathbf{f}(\mathbf{r}_+, \, \mathbf{r}_+)$ is a column vector defined by

$$d^2\mathbf{f}(\mathbf{r}_+, \, \mathbf{r}_+) = \begin{bmatrix} \mathbf{r}_+^T \, \mathbf{H}(f_1)\mathbf{r}_+ \\[1ex] \mathbf{r}_+^T \, \mathbf{H}(f_2)\mathbf{r}_+ \end{bmatrix} \qquad (1.11.11)$$

Therefore, with

$$\boldsymbol{\ell}_+ = \begin{bmatrix} m_+ \\ n_+ \end{bmatrix} \quad \text{and} \quad \mathbf{r}_+ = \begin{bmatrix} \mu_+ \\ \nu_+ \end{bmatrix} \qquad (1.11.12)$$

we can write

$$\boldsymbol{\ell}_+^T \, d^2\mathbf{f}(\mathbf{r}_+, \, \mathbf{r}_+) = \mu_+^2 \left(m_+ \frac{\partial^2 f_1}{\partial u^2} + n_+ \frac{\partial^2 f_2}{\partial u^2} \right) + 2\mu_+ \nu_+ \left(m_+ \frac{\partial^2 f_1}{\partial u \, \partial v} \right.$$
$$\left. + \, n_+ \frac{\partial^2 f_2}{\partial u \, \partial v} \right) + \nu_+^2 \left(m_+ \frac{\partial^2 f_1}{\partial v^2} + n_+ \frac{\partial^2 f_2}{\partial v^2} \right) \qquad (1.11.13)$$

Expressions in Eqs. (1.11.9) and (1.11.10) can be seen to be equivalent to those in Eq.(1.11.3) as follows. We shall take the case of the (+) characteristic field. By the definition of the right eigenvector we have

$$(\mathbf{A} - \lambda_+ \mathbf{I})\mathbf{r}_+ = 0$$

We note that differentiation in the direction of $\mathbf{r}_+$ is equivalent to taking the gradient and multiplying on the left by $\mathbf{r}_+^T$. Thus differentiating the equation above in the direction of $\mathbf{r}_+$ gives

$$[\mathbf{r}_+^T \, \nabla(\mathbf{A} - \lambda_+ \mathbf{I})]\mathbf{r}_+ + (\mathbf{A} - \lambda_+ \mathbf{I})(\mathbf{r}_+^T \, \nabla \mathbf{r}_+) = 0$$

or

$$\mathbf{r}_+^T \, (\nabla \mathbf{A})\mathbf{r}_+ = (\mathbf{r}_+^T \, \nabla \lambda_+)\mathbf{r}_+ - (\mathbf{A} - \lambda_+ \mathbf{I})(\mathbf{r}_+^T \, \nabla \mathbf{r}_+)$$

But the left-hand side is just identical to $d^2\mathbf{f}(\mathbf{r}_+, \, \mathbf{r}_+)$, and if we multiply this equation on the left by $\boldsymbol{\ell}_+^T$, the last term vanishes by the definition of the left eigenvector so that we obtain

$$\boldsymbol{\ell}_+^T \, d^2\mathbf{f}(\mathbf{r}_+, \, \mathbf{r}_+) = (\mathbf{r}_+^T \, \nabla \lambda_+)(\boldsymbol{\ell}_+^T \, \mathbf{r}_+)$$

Since the choice of $\boldsymbol{\ell}_+$ and $\mathbf{r}_+$ is arbitrary up to a multiplicative constant, we can always make $\boldsymbol{\ell}_+^T \mathbf{r}_+ > 0$, so that $\boldsymbol{\ell}_+^T d^2\mathbf{f}(\mathbf{r}_+, \, \mathbf{r}_+)$ and $\mathbf{r}_+^T \nabla \lambda_+$ have a common sign. A similar argument holds for the (−) characteristic field, and we see that

Eqs. (1.11.9) and (1.11.10) are indeed equivalent to Eq. (1.11.3). Also, it can be shown that for sufficiently weak shocks the shock interaction condition is satisfied if

$$\boldsymbol{\ell}_-^T \, d^2\mathbf{f}(\mathbf{r}_+,\mathbf{r}_+) > 0 \quad \text{and} \quad \boldsymbol{\ell}_+^T \, d^2\mathbf{f}(\mathbf{r}_-, \mathbf{r}_-) > 0 \qquad (1.11.14)$$

The Riemann problem for the system (1.11.1) has been treated by Smoller (1969a, b, 1970a). Under the conditions that the system is genuinely nonlinear and satisfies the shock interaction condition, he shows that the Riemann problem is solvable with two arbitrary constant states of u and v, not necessarily close. It is also proven that the shock condition [Eqs. (1.10.29) and (1.10.30)] is necessary and sufficient for uniqueness of the solution. Indeed, it has been shown by Conway and Smoller (1973) that weak solutions which violate the shock condition are unstable relative to smoothing of the initial data. The existence and uniqueness proof has been extended by Smoller (1970b) to systems which are not genuinely nonlinear and thus admit contact discontinuities. Moler and Smoller (1970) have considered the Riemann problem with two jumps of arbitrary size such that each initial discontinuity produces only a single wave and the two waves necessarily meet and interact. When the system is genuinely nonlinear and satisfies the shock interaction condition, there exists a solution which is bounded and has uniformly bounded total variation on each line $t = $ constant. More recently, Keyfitz (1978) has established the existence and uniqueness proof under less restrictive hypotheses; i.e., she has replaced the condition $(\partial f_1/\partial v)(\partial f_2/\partial u) > 0$ of Smoller by an apparently weaker *half-plane condition*. This half-plane condition requires that the eigenvectors of $\mathbf{A}$ point into opposite fixed half-planes, and thus it is satisfied if $\mathbf{r}_+(\mathbf{w}_1) \neq \mathbf{r}_-(\mathbf{w}_2)$ for every pair of points $\mathbf{w}_1$ and $\mathbf{w}_2$ in the (u, v)-plane.

Existence of global solutions for arbitrary initial values was first undertaken by Nishida (1968) for a specific form for the equations, i.e.,

$$\frac{\partial u}{\partial t} - \frac{\partial v}{\partial x} = 0$$

$$\frac{\partial v}{\partial t} + \frac{\partial}{\partial x}\left(\frac{a^2}{u}\right) = 0 \qquad (1.11.15)$$

which is a special case of Eq. (1.11.7) with $g(u) = a^2/u$, where a is a constant. By using a modified Glimm difference scheme he proved that Eq. (1.11.15) can be solved for arbitrary initial values $u_0(x)$, $v_0(x)$ such that $u_0(x) \geq 0$. Nishida's work has been extended to a more general form of equations; i.e., Eq.(1.11.7), by several authors (see Bakhvalov, 1970; DiPerna, 1973a; Greenberg, 1973a; and Nishida and Smoller, 1973). One method for the proof is to obtain estimates for the interaction of shocks and simple waves (or rarefraction waves) and apply them to the Glimm difference approximations. Nishida (1977) has also considered Eq.(1.11.7) with $g(u) = K^2 u^{-\gamma}$ under the initial and boundary conditions

$$u(x, 0) = u_0(x) > 0, \qquad v(x, 0) = v_0(x) \qquad \text{for } 0 < x < 1$$

$$u(0, t) = u_1(t) \qquad\qquad\qquad\qquad\quad \text{for } t \geq 0 \qquad\qquad (1.11.16)$$

$$u(1, t) = u_2(t) \qquad\qquad\qquad\qquad\quad \text{for } t \geq 0$$

He first discusses the problem in the semi-infinite domain $x \geq 0$ with the condition $u(0, t) = u_1(t)$, which is equivalent to the piston problem in gas dynamics, and then the problem with Eq. (1.11.16), the double-piston problem. The technique is to use a modified functional in the Glimm difference scheme and the generalized Riemann invariants introduced by DiPerna (1973a). Thus it is established that there exists a $\gamma_0 > 0$, depending only on the variation of the data, such that for any γ, $1 < \gamma \leq \gamma_0$, the mixed problem of Eqs. (1.11.7) and (1.11.16) has a weak solution defined for all time. For the piston problem the same holds with $\gamma_0 = 2$.

For the general system (1.11.1), strictly hyperbolic and genuinely nonlinear, DiPerna (1973b) has proved existence of global solutions with prescribed data, $u_0(x)$ and $v_0(x)$, having finite total variation. The key to the existence proof is the global geometry of the shock curves, which leads to the construction of approximate solutions using the Glimm difference scheme. It then follows by a compactness argument that a weak solution exists as the pointwise limit of a sequence of difference approximations. Uniqueness of the weak solution to this problem has been established by Liu (1976) and Diperna (1979). While Liu imposes the shock condition [Eqs. (1.10.29) and (1.10.30)], DiPerna requires the entropy inequality (1.10.41) and further conjectures that the result generalizes to genuinely nonlinear systems of n equations. Both authors also consider the system (1.11.7) and establish the uniqueness theorem for a not necessarily genuinely nonlinear system (Liu) as well as for solutions with arbitrarily large oscillation (DiPerna).

Discussions on systems without the genuine nonlinearity condition have been restricted to the Riemann problem. For the system (1.11.7) Wendroff (1972) and Leibovich (1974) have established the existence of weak solutions with two arbitrary constant states under the generalized Condition E of Oleinik (see Sec. 5.4 of Vol. 1) and discussed the use of vanishing viscosity to eliminate certain weak solutions which violate the Condition E. A more important contribution is due to Liu (1974), who has considered the general system (1.11.1) and extended Oleinik's Condition E to formulate the extended entropy condition (see the Rule 1.5). By imposing this condition he classifies the discontinuities into shocks, semishocks (or one-sided contact discontinuities), and contact discontinuities (or two-sided contact discontinuities), and proves that there exists a unique solution to the Riemann problem with two arbitrary constant states in the class of simple waves (or rarefaction waves), shocks, semishocks, and contact discontinuities (see Sec. 1.7). In addition to the condition, $\partial f_1/\partial v < 0$ and $\partial f_2/\partial u < 0$, he first requires a somewhat unusual condition, i.e., $\partial f_1/\partial u \geq 0$ and $\partial f_2/\partial v \leq 0$, but later replaces this by the second inequalities of the shock condition, which is to say that the shock speed is not equal to the characteristic speed of the opposite family (see Liu, 1975).

Keyfitz and Kranzer (1980) have recently considered the Riemann problem for a pair of nonstrictly hyperbolic conservation laws of the form

$$\frac{\partial u}{\partial t} + \frac{\partial}{\partial x}\{u\phi(u, v)\} = 0$$

$$\frac{\partial v}{\partial t} + \frac{\partial}{\partial x}\{v\phi(u, v)\} = 0$$
(1.11.17)

which describe the propagation of forward longitudinal and transverse waves in a stretched elastic string which moves in a plane. Here the characteristic speeds λ_+ and λ_- of Eq. (1.11.17) may coalesce on some subset of the (u, v)-plane so that the system is not strictly hyperbolic. She has extended the theory to this nonstrictly hyperbolic system and proved the existence of a weak solution to the Riemann problem in the class of functions containing appropriately generalized shocks and rarefaction waves.

Just as in the case of single equations (see Sec. 7.6 of Vol. 1), the viscosity method has been proven instrumental for the study of the structure of solutions of the Riemann problem. In particular, the solutions containing discontinuities are of interest here. Following Conley and Smoller (1970), let us add the standard viscosity operator to the right-hand side of the system (1.11.1) or, in vector form, Eq. (1.11.5), so that

$$\frac{\partial \mathbf{w}}{\partial t} + \frac{\partial \mathbf{f}}{\partial x} = \mu \frac{\partial}{\partial x}\left(\mathbf{P}\frac{\partial \mathbf{w}}{\partial x}\right), \qquad \mu > 0$$
(1.11.18)

where $\mathbf{P}$ is some 2×2 matrix of constant elements, whose physical significance, if any, is generally that of viscosity or diffusivity. Looking for a solution of the form $\mathbf{w} = \mathbf{w}(\xi)$, where

$$\xi = \frac{x - \tilde{\lambda}t}{\mu}$$

we see that $\mathbf{w}(\xi)$ must satisfy the pair of ordinary differential equations

$$-\tilde{\lambda}\frac{d\mathbf{w}}{d\xi} + \frac{d\mathbf{f}}{d\xi} = \mathbf{P}\frac{d^2\mathbf{w}}{d\xi}$$

which, upon integration with respect to ξ, can be rewritten as

$$-\tilde{\lambda}\mathbf{w} + \mathbf{f} + \mathbf{C} = \mathbf{P}\frac{d\mathbf{w}}{d\xi}$$
(1.11.19)

where $\mathbf{C}$ is the constant of integration. Now, if $\mathbf{w}(\xi)$ is to converge to a given shock arising from Eq. (1.11.5), we must have

$$\lim_{\xi \to -\infty} \mathbf{w}(\xi) = \mathbf{w}^l, \qquad \lim_{\xi \to +\infty} \mathbf{w}(\xi) = \mathbf{w}^r$$
(1.11.20)

It follows that the left-hand side of Eq.(1.11.19) must vanish for $\mathbf{w} = \mathbf{w}^l$ and also for $\mathbf{w} = \mathbf{w}^r$. Using $\mathbf{w} = \mathbf{w}^l$, we obtain $\mathbf{C} = \tilde{\lambda}\mathbf{w}^l - \mathbf{f}(\mathbf{w}^l)$, and thus Eq. (1.11.19) becomes

$$\mathbf{p}\, \frac{d\mathbf{w}}{d\xi} = -\tilde{\lambda}(\mathbf{w} - \mathbf{w}^l) + \mathbf{f}(\mathbf{w}) - \mathbf{f}(\mathbf{w}^l)$$

which has two singular points, one at $\mathbf{w} = \mathbf{w}^l$ and the other at $\mathbf{w} = \mathbf{w}^r$. Applying the second condition of Eq. (1.11.20) gives

$$\tilde{\lambda}[\mathbf{w}] = [\mathbf{f}] \qquad (1.11.21)$$

so that we recover the jump condition. If we let

$$\mathbf{V}(\mathbf{w}) = -\tilde{\lambda}(\mathbf{w} - \mathbf{w}^l) + \mathbf{f}(\mathbf{w}) - \mathbf{f}(\mathbf{w}^l) \qquad (1.11.22)$$

the problem is now to solve the pair of ordinary differential equations

$$\frac{d\mathbf{w}}{d\xi} = \mathbf{P}^{-1}\mathbf{V}(\mathbf{w}) \qquad (1.11.23)$$

subject to the boundary conditions of Eq. (1.11.20).

Without loss of generality let us restrict attention to a Σ_+ shock so that the shock condition (1.10.30) is satisfied. Since $d\mathbf{V} = d\mathbf{f} - \tilde{\lambda}\mathbf{I}$, the eigenvalues of $d\mathbf{V}$ are $\lambda_+ - \tilde{\lambda}_-$ and $\lambda_- - \tilde{\lambda}_-$, so that the singularity at $\mathbf{w} = \mathbf{w}^l$ is a saddle while the one at $\mathbf{w} = \mathbf{w}^r$ is a stable node. Conley and Smoller show that these are the only singularities of the vector field $\mathbf{V}(\mathbf{w})$ and there is a unique orbit of $\mathbf{V}(\mathbf{w})$ connecting these singular points. This solution is called the *standing-wave solution* of Eq. (1.11.23) and denoted by $\mathbf{w}(\xi; \mu)$. If the sequence of standing-wave solutions converge to the given shock $(\mathbf{w}^l, \mathbf{w}^r)$ as $\mu \to 0$, the matrix $\mathbf{P}$ is called *suitable* for the Riemann problem of Eq. (1.11.5) with $\mathbf{w} = \mathbf{w}^l$ for $x < 0$ and $\mathbf{w} = \mathbf{w}^r$ for $x > 0$. According to Conley and Smoller, the identity matrix is suitable whatever the form of the particular equations may be or no matter what the initial conditions may be. They also give sufficient conditions for $\mathbf{P}$ to be suitable. Incidentally, it can be shown that a necessary and sufficient condition for all matrices $\mathbf{P}$ with $\mathbf{P}^{-1} + (\mathbf{P}^{-1})^T > 0$ to be suitable for all shocks is that the vector field $\mathbf{f}$ is a gradient field; i.e., there exists a function $\phi(u, v)$ such that $\mathbf{f} = \nabla\phi$.

Smoller and Conley (1972) have extended the argument above to the case when the viscosity terms depend on several parameters and the matrix $\mathbf{P}$ also depends on $\mathbf{w}$ as well as to the cases with singular viscosity matrices. The same authors (1971) have also considered modifications of Eq. (1.11.5) obtained by the addition of *dissipative* ($\partial^2\mathbf{w}/\partial x^2$) and *dispersive* ($\partial^3\mathbf{w}/\partial x^3$) terms:

$$\frac{\partial\mathbf{w}}{\partial t} + \frac{\partial\mathbf{f}}{\partial x} = \alpha\mu\, \frac{\partial^2\mathbf{w}}{\partial x^2} + \beta\mu^2\, \frac{\partial^3\mathbf{w}}{\partial x^3} \qquad (1.11.24)$$

Here α and β are fixed positive constants while μ is a small positive parameter. The problem concerns the behavior of *progressive wave solutions* of

Eq.(1.11.24) as $\mu \rightarrow 0$, where a progressive wave solution of Eq. (1.11.24) means a solution $w(x, t; \mu)$ of the form

$$\mathbf{w}(x, t; \mu) = \mathbf{w}\left(\frac{x - \bar{\lambda}t}{\mu}\right)$$

with a constant $\bar{\lambda}$. If $\beta = 0$, Eq. (1.11.24) corresponds to Eq. (1.11.18) with the identity matrix for $\mathbf{P}$ and the standing-wave solution there is indeed the progressive wave solution in the (x, t)-coordinate system. Thus it is obvious that shocks can be realized as limits of progressive wave solutions of the associated dissipative system. In contrast to this they have shown that if $\alpha = 0$ and $\mathbf{f} = \nabla\phi$, shocks of Eq. (1.11.5) cannot be obtained as limits of progressive wave solutions of the associated dispersive system.

By applying Eq.(1.11.24) with $\alpha = 1$, $\beta = 0$ to systems that are not necessarily genuinely nonlinear, Liu (1976) has justified his extended entropy condition [Eqs. (1.7.16) and (1.7.17)]. Another approach due to Dafermos (1973) has been motivated by the invariance property of the system (1.11.5) under the transformation $(x, t) \rightarrow (\alpha x, \alpha t)$, $\alpha > 0$. In other words, by adding the unconventional form of the viscosity operator to the right-hand side of Eq. (1.11.5), we have

$$\frac{\partial \mathbf{w}}{\partial t} + \frac{\partial \mathbf{f}}{\partial x} = \mu t \frac{\partial^2 \mathbf{w}}{\partial x^2} \qquad (1.11.25)$$

which is invariant under the transformation above and therefore admits solutions that are functions of the single variable

$$\eta = \frac{x}{t}$$

For a solution of the form $\mathbf{w}(x, t) = \mathbf{w}(x/t) = \mathbf{w}(\eta)$, Eq. (1.11.25) reduces to the ordinary differential equation

$$\mu \frac{d^2\mathbf{w}}{d\eta^2} = -\frac{d\mathbf{w}}{d\eta} + \frac{d\mathbf{f}}{d\eta} \qquad (1.11.26)$$

and we shall require this equation to be subject to the boundary conditions

$$\lim_{\eta \to -\infty} \mathbf{w}(\eta) = \mathbf{w}^l, \qquad \lim_{\eta \to +\infty} \mathbf{w}(\eta) = \mathbf{w}^r \qquad (1.11.27)$$

for we are interested in the convergence of $\mathbf{w}(\eta)$ to a shock arising from Eq. (1.11.5).

Dafermos shows that for every fixed $\mu > 0$, Eq. (1.11.26) with Eq. (1.11.27) has a solution $\mathbf{w}(\eta; \mu)$ whose total variation is bounded independently of μ and further that there is a sequence $\mathbf{w}(\eta; \mu_n)$, $\mu_n < 0+$ as $n \rightarrow \infty$, converging boundedly to a function $\mathbf{w}(\eta)$ of bounded variation so that $\mathbf{w}(x/t)$ is a weak solution of the Riemann problem for Eq. (1.11.5) with $\mathbf{w} = \mathbf{w}^l$ for $x < 0$ and $\mathbf{w} = \mathbf{w}^r$ for $x > 0$. He also verifies that this weak solution is admissible by proving that $\mathbf{w}(\eta)$ satisfies the inequality condition

$$-\xi \frac{dW}{d\eta} + \frac{dF}{d\eta} \leq 0 \qquad (1.11.28)$$

in the sense of distributions where $W(\mathbf{w})$ is an entropy for Eq. (1.11.5) with entropy flux $F(\mathbf{w})$ [see Eq. (1.10.39)]. Thus the result applies to systems that are not necessarily genuinely nonlinear. The proof was established first (1973) under rather stringent assumptions on $f_1(u, v)$ and $f_2(u, v)$ and then reestablished by Dafermos and Diperna (1976) under considerably weaker assumptions, which read

$$|f_1(u, v)| \to \infty \quad \text{as } |v| \to \infty, \quad \text{uniformly in } u \text{ for } u \text{ in compact sets}$$

$$|f_2(u, v)| \to \infty \quad \text{as } |u| \to \infty, \quad \text{uniformly in } v \text{ for } v \text{ in compact sets}$$

Dafermos (1974) has applied this approach to obtain a complete picture of the structure of solutions of the Riemann problem for systems that are not necessarily genuinely nonlinear. The result is essentially the same as that of Liu (1974, 1975).

The theory of decay of solutions for genuinely nonlinear systems of conservation laws has been developed by Glimm and Lax (1970) with the aid of notions of approximate characteristic curves and approximate conservation laws. The Glimm–Lax theory is developed in the context of pairs of conservation laws (1.11.1) satisfying the shock interaction condition and treats the Cauchy problem with the initial data (1.11.2) having small oscillation. The results established may be summarized as follows. (1) If the oscillation of the initial data is small enough, the solution of the initial value problem [Eqs. (1.11.1) and (1.11.2)] exists for all positive time. This is an improvement on the result of Glimm (1965). (2) If the initial data are periodic, the solution decays like $1/t$. (3) If the initial data are constant outside a compact interval, the solution decays like $1/\sqrt{t}$. This feature of the decay of solutions is essentially the same as that for the case of single equations (see Sec. 7.9 of Vol. 1). Furthermore, the Glimm–Lax theory demonstrates that the primary mechanisms of decay in genuinely nonlinear systems are the spreading of rarefaction waves and the interaction between shocks and rarefaction waves of the same family.

For the simpler system (1.11.7) Greenberg (1970, 1971) has treated the elementary interactions between two waves of the same family as well as between two waves of different families. To obtain each of these situations he considers special arrangements of the initial data, not necessarily close, and extends the Glimm and Glimm–Lax approach to establish the existence of weak solutions. Greenberg (1973b, 1975) has also considered a modified system of Eq. (1.11.7); i.e.,

$$\frac{1}{g'(u)} \frac{\partial u}{\partial t} - \frac{\partial v}{\partial x} = 0$$

$$\frac{\partial v}{\partial t} - g'(u) \frac{\partial u}{\partial x} = 0 \qquad (1.11.29)$$

for which the Riemann invariants are readily found as

$$J_+ = u + v, \qquad J_- = -u + v$$

When the initial and boundary data are specified as

$$u(x, 0) = 0, \qquad v(x, 0) = 0 \qquad \text{for } x > 0$$

$$u(0, t) = \begin{cases} u_1 \neq 0, & 0 < t < T \\ 0, & t > T \end{cases}$$

it is shown that J_+ and J_- decay as $1/t$ and $1/\sqrt{t}$, respectively.

Ballou (1974) has discussed the structure of compression waves arising from the system(1.11.7). He first shows that if the initial data outside some finite interval consist of two different constant states, then after finite time the solution becomes essentially the same as the solution to the Riemann problem determined by these constant states. This implies that the asymptotic behavior is unaffected by the data prescribed over the finite interval. This result is then applied to prove that in the case of a forward (or Γ_+) compression wave, the solution after the completion of all interactions consists of a weak backward (or Γ_-) rarefaction wave and a strong forward (or Σ_+) shock. The structure of an arbitrary solution to the general system (1.11.5) has been treated by DiPerna (1975a). The main concern here is the local decomposition of the solution into elementary waves and the classification and propagation of singularities in the solution as well as the limiting behavior at singularities. While the analysis is carried out for pairs of equations, he claims that a natural generalization to systems of n equations can be expected because his treatment does not make essential use of the existence of Riemann invariants, a feature that is special to systems of two equations.

By using equations for Riemann invariants, DiPerna (1975b) has shown that if the initial data have a common constant value $\mathbf{w}_0$ outside a finite interval, the solution of the system (1.11.5) decays uniformly to a waveform that consists of the superposition of two N-waves,[†] N_+ and N_-, propagating at the characteristic speeds, $\lambda_+(\mathbf{w}_0)$ and $\lambda_-(\mathbf{w}_0)$, respectively. This feature of the asymptotic behavior of solutions was conjectured by Lax (1963) for systems of n equations. DiPerna further establishes that the solution decays to the superposition of N-waves uniformly at the rate $t^{-1/6}$ in contrast to the rate $t^{-1/2}$ in the case of single equations and attributes the difference in the rates to the interaction of the characteristic fields in systems, a phenomenon that is not present with a single conservation law. Liu (1978) has treated the asymptotic behavior of solutions of the system (1.11.5) when the initial data outside some finite interval take two different constant states, $\mathbf{w}^l$ and $\mathbf{w}^r$. Based on Glimm's scheme he first proves that the error term for wave interactions is of third order on the strength of the shock waves before interaction, where the strength of a shock or a centered rarefaction wave is measured by the differences either in Riemann invariants or in characteristic speeds. Finally, it is shown that the so-

---

[†] For N-waves, see Secs. 7.6 and 7.9 of Vol. 1, in particular, Fig. 7.23.

lution converges to the solution of the Riemann problem with the two constant states, $\mathbf{w}^l$ and $\mathbf{w}^r$, at an algebraic rate t^n provided that the solution contains only weak shocks. The strength of rarefaction waves is allowed to be large because the effect of interactions is of the third order of the shock strength only. The exponent n takes one of the values $-1/2$, $-3/2$, and $-2 + \epsilon$, depending on the pattern of the solution of the Riemann problem, where ϵ is a small constant that goes to zero as the total variation of the initial data goes to zero.

Exercises

1.11.1. Determine the values of $\ell_+^T d^2 \mathbf{f}(\mathbf{r}_+, \mathbf{r}_+)$ and $\ell_-^T d^2 \mathbf{f}(\mathbf{r}_-, \mathbf{r}_-)$ at the point $u = 0$, $v = 0$ for each case.

$$\text{(a)} \quad f_1(u, v) = u - v + \tfrac{1}{6} u^2 + \tfrac{1}{12} v^2$$
$$f_2(u, v) = -2u + \tfrac{11}{20} u^2 + \tfrac{1}{20} v^2$$
$$\text{(a)} \quad f_1(u, v) = u - v + \tfrac{1}{3} u^2 + \tfrac{1}{6} v^2$$
$$f_2(u, v) = -2u + \tfrac{1}{12} u^2 + \tfrac{17}{48} v^2$$

1.11.2. Consider Eq. (1.11.23) when the states $\mathbf{w}^l$ and $\mathbf{w}^r$ satisfy the shock condition (1.10.29) for a Σ_- shock, and show that the singularity at $\mathbf{w} = \mathbf{w}^l$ is an unstable node while the singularity at $\mathbf{w} = \mathbf{w}^r$ is a saddle. Based on this observation, can you argue that there exists a unique trajectory connecting the two singular points in the (u, v)-plane?

REFERENCES

1.1. The chromatographic equations were first viewed in this connection by

D. DeVault, "The theory of chromatography," *J. Amer. Chem. Soc.* **65,** 532–540 (1943).

The transformation of independent variables given by Eq. (1.1.21) has been the usual practice in the field of chromatography starting from the earlier works to be found in

H. C. Thomas, "Heterogeneous ion exchange in a flowing system," *J. Amer. Chem. Soc.* **66,** 1664–1666 (1944).

N. R. Amundson, "A note on the mathematics of adsorption in beds," *J. Phys. Colloid Chem.* 52, 1153–1157 (1948).

L. G. Sillén, "On filtration through a sorbent layer: IV. The ψ condition, a simple approach to the theory of sorption columns," *Ark. Kemi* **2,** 477–498 (1950).

1.2 and **1.3.** Hyperbolic pairs of equations are treated fairly fully in Chapter 2 of

R. Courant and K. O. Friedrichs, *Supersonic Flow and Shock Waves*, Interscience, New York, 1948.

Numerical techniques using the characteristics are to be found in Part IV of

A. Ralston and H. S. Wilf (eds.), *Mathematical Methods for Digital Computers*, Wiley, New York, 1960.

It is also considered in

> G. D. Smith, *Numerical Solution of Partial Differential Equations,* Oxford University Press, London, 1965.

> W. F. Ames, *Nonlinear Partial Differential Equations in Engineering,* Academic Press, New York, 1965.

> R. L. Street, *The Analysis and Solution of Partial Differential Equations,* Brooks/Cole, Monterey, Calif., 1973.

1.4–1.6. The treatment of two-solute chromatography is found in

> H.-K. Rhee, Studies on the theory of chromatography, Ph.D. thesis, University of Minnesota, 1968.

See also the book by Courant and Friedrichs cited above.

The notion of genuine nonlinearity was introduced by Lax in his 1957 paper cited below for Sec. 1.7.

The fundamental differential equations as well as the compatibility relation was first recognized in DeVault's 1943 paper cited above and discussed further in

> G. G. Bayle and A. Klinkenberg, "On the theory of adsorption chromatography for liquid mixtures," *Rec. Trav. Chim. Pays Bas* **73,** 1037–1057 (1954).

> E. Glueckauf, "Theory of chromatography: VII. The general theory of two solutes following non-linear isotherms," *Discuss Faraday Soc.* **7,** 12–25 (1949).

> A. C. Offord and J. Weiss, "Chromatography with several solutes," *Discuss Faraday Soc.* **7,** 26–34 (1949).

In the theory of chromatography, solution of the Riemann problem was first expressed as a function of τ/x only by

> J. E. Walter, "Multiple adsorption from solutions," *J. Chem. Phys.* **13,** 229–234 (1945).

This was further extended in the papers by Offord and Weiss (1949) and Sillén (1950) cited above.

For applications to specific isotherms, see Chapter 2 and the references cited there.

1.7. For the proof of the fact that the shock curve Σ makes third-order contact with the corresponding Γ-characteristic at $L,$ see

> B. L. Keyfitz, "Existence and uniqueness of entropy solutions for hyperbolic systems of two nonlinear conservation laws," *J. Differential Equations* **27,** 444–476 (1978).

The solution with piecewise constant data is proven to converge in the limit to the solution with the general initial data in

> J. A. Smoller and J. L. Johnson, "Global solutions for an extended class of hyperbolic systems of conservation laws," *Arch. Rational Mech. Anal.* **32,** 169–189 (1969).

The extended entropy condition and discontinuous solutions are discussed in

> T. P. Liu, "The Riemann problem for general 2×2 conservation laws," *Trans. Amer. Math. Soc.* **199,** 89–112 (1974).

> T. P. Liu, "Existence and uniqueness theorems for Riemann problems," *Trans. Amer. Math. Soc.* **212,** 375–382 (1975).

The definition of shocks was first given in this form by

P. D. LAX, "Hyperbolic systems of conservation laws: II," *Comm. Pure Appl. Math.* **10,** 537–566 (1957).

Semishocks and contact discontinuities in physical systems are to be found in

H.-K. RHEE and N. R. AMUNDSON, "An analysis of an adiabatic adsorption column: Part I. Theoretical development," *Chem. Engrg. J.* **1,** 241–254 (1970).

H.-K. RHEE, E. D. HEERDT, and N. R. AMUNDSON, "An analysis of an adiabatic adsorption column: Part II. Adsorption of a single solute," *Chem. Engrg. J.* **1,** 279–290 (1970).

H.-K. RHEE and N. R. AMUNDSON, "An analysis of an adiabatic adsorption column: Part VI. Adsorption in the high temperature range," *Chem. Engrg. J.* **3,** 121–135 (1972).

For Jeffrey and Taniuti (1964) and Jeffrey (1976), see the references cited below for Sec. 1.9.

1.8. The treatment here follows somewhat on the lines laid down in

J. T. PATTON, K. H. COATES, and G. T. COLEGROVE, "Prediction of polymer flood performance," *Soc. Pet. Engrs. J.* **11,** 72–84 (Mar. 1971).

K. Y. CHOI and H.-K. RHEE, "A study of the tertiary oil recovery by polymer flooding," *Engrg. Res. Rep. Seoul Nat. Univ.* **9,** 1–12 (1977).

R. G. LARSON and G. J. HIRASAKI, "Analysis of the physical mechanisms in surfactant flooding," *Soc. Pet. Engrs. J.* **18,** 42–58 (Feb. 1978).

See further

B. L. BONDOR, G. J. HIRASAKI, and M. J. THAM, "Mathematical simulation of polymer flooding in complex reservoirs," *Soc. Pet. Engrs. J.* **12,** 369–382 (Oct. 1972).

E. L. CLARIDGE and P. L. BONDOR, "A graphical method for calculating linear displacement with mass transfer and continuously changing mobilities," *Soc. Pet. Engrs. J.* **14,** 609–618 (Dec. 1974).

The former deals with the finite difference simulator, while the latter is concerned with a graphical scheme.

For more recent development in multicomponent, multiphase displacement, see

G. A. POPE, L. W. LAKE, and F. G. HELFFERICH, "Cation exchange in chemical flooding: Part 1. Basic theory without dispersion," *Soc. Pet. Engrs. J.* **18,** 418–456 (Dec. 1978).

F. G. HELFFERICH, "Theory of multicomponent, multiphase displacement in porous media," *Soc. Pet. Engrs. J.* **21,** 51–62 (1981).

G. J. HIRASAKI, "Application of the theory of multicomponent, multiphase displacement to three-component, two-phase surfactant flooding," *Soc. Pet. Engrs. J.* **21,** 191–204 (Apr. 1981).

1.9. The matrix formulation introduced here is used by many authors. See, for example,

A. JEFFREY and T. TANIUTI, *Nonlinear Wave Propagation,* Academic Press, New York, 1964.

A. JEFFREY, *Quasilinear Hyperbolic Systems and Waves,* Pitman, London, 1976.

J. SMOLLER, *Shock Waves and Reaction-Diffusion Equations,* Springer-Verlag, New York, 1983.

Discussions and applications of the Riemann invariants are to be found in the book by Courant and Friedrichs (1984) cited above for Secs. 1.2 and 1.3, and also in the three books cited just above for this section.

For the notion of genuine nonlinearity, see Lax's 1957 paper cited above for Sec. 1.7.

The shallow-water wave approximation is formulated in the book by Jeffrey and Taniuti (1964) and treated in the book by Jeffrey (1976), both cited above. For a two-dimensional description, see the book by Courant and Friedrichs (1948) cited above.

Various problems arising from river flow are treated in

> G. B. WHITHAM, *Linear and Nonlinear Waves,* Wiley-Interscience, New York, 1974.

1.10. This treatment of the development of singularities is due to

> P. D. LAX, "Development of singularities of solutions of nonlinear hyperbolic partial differential equations," *J. Math. Phys.* **5,** 611–613 (1964).

> P. D. LAX, "Hyperbolic systems of conservation laws and the mathematical theory of shock waves," *Regional Conference Series in Applied Mathematics,* No. 11, SIAM, Philadelphia, 1973.

See also

> B. L. ROZDESTVENSKII, "On the discontinuities of solutions of quasilinear equations," *Mat. Sb.* **47**(89), 485–494 (1959); English translation in *Amer. Math. Soc. Transl.,* Ser. 2, **101,** 239–250 (1973).

> A. JEFFREY, "The development of jump discontinuities in nonlinear hyperbolic systems of equations in two independent variables," *Arch. Rational Mech. Anal.* **14,** 27–39 (1963).

> A. JEFFREY, "The development of singularities of solutions of nonlinear hyperbolic equations of order greater than one," *J. Math. Mech.* **15,** 585–598 (1966).

> A. JEFFREY, "The evolution of discontinuities in solutions of homogeneous nonlinear hyperbolic equations having smooth initial data," *J. Math. Mech.* **17,** 331–352 (1967).

For the transformation of systems of quasilinear equations to conservative form (1.10.16), see

> B. L. ROZDESTVENSKII, "Systems of quasilinear equations," *Dokl. Akad. Nauk SSSR* **115,** 454–457 (1957); English translation in *Amer. Math. Soc. Transl.,* Ser. 2, **42,** 13–17 (1964).

> S. K. GODUNOV, "An interesting class of quasilinear systems," *Dokl. Akad. Nauk SSSR* **139,** 521–523 (1961); English translation in *Soviet Math. Dokl.* **2,** 947–949 (1961).

For the definition of weak solutions and derivation of the generalized Rankine–Hugoniot condition, see the three books by Jeffrey and Taniuti (1965), Jeffrey (1976), and Smoller (1983), all cited above for Sec. 1.9.

The structure of the shock curve Σ is discussed in the 1978 paper by Keyfitz cited above for Sec. 1.7.

The shock condition as well as the uniqueness proof for the solution to the Riemann problem is to be found in the 1957 paper by Lax cited above for Sec. 1.7.

This concept of entropy is treated in detail in

K. O. Friedrichs and P. D. Lax, "Systems of conservation equations with a convex extension," *Proc. Nat. Acad. Sci. USA* **68,** 1686–1688 (1971).

P. D. Lax, "Shock waves and entropy," in *Contributions to Nonlinear Functional Analysis,* edited by E. H. Zarantonello, Academic Press, New York, 1971, pp. 603–634.

See also Lax's 1973 paper cited above.

For the entropy rate admissibility criterion, see

C. M. Dafermos, "The entropy rate admissibility criterion for solutions of hyperbolic conservation laws," *J. Differential Equations* **14,** 202–212 (1973).

1.11. The first existence proof of global solutions to quasilinear systems is to be found in

J. Glimm, "Solutions in the large for nonlinear hyperbolic systems of equations," *Comm. Pure Appl. Math.* **18,** 697–715 (1965).

Existence and uniqueness for the system (1.11.7) is discussed in

S. K. Godunov, "On the uniqueness of the solution of the equations of hydrodynamics," *Mat. Sb.* **82** (N.S. 40), 467–478 (1956), in Russian.

O. A. Oleinik, "On the uniqueness of the generalized solution of Cauchy problem for a nonlinear system of equations occurring in mechanics," *Uspehi Mat. Nauk* (*N.S.*) **12**(6), 169–176 (1957), in Russian.

A. Douglis, "The continuous dependence of generalized solutions of nonlinear partial differential equations upon initial data," *Comm. Pure Appl. Math.* **14,** 267–284 (1961).

A. E. Hurd, "A uniqueness theorem for second-order quasilinear hyperbolic equations," *Pacific J. Math.* **32,** 415–427 (1970).

T. Zhang and Y.-H. Guo, "A class of initial value problems for systems of aerodynamic equations," *Acta Math. Sinica* **15,** 386–396 (1965); English translation in *Chinese Math.* **7,** 90–101 (1965).

J. L. Johnson and J. A. Smoller, "Global solutions for certain systems of quasilinear hyperbolic equations," *J. Math. Mech.* **17,** 561–576 (1967).

The last paper is concerned with the system (1.11.8).

An extension to the system (1.11.1) is treated by Smoller and Johnson in their 1969 paper cited above for Sec. 1.7.

The Riemann problem for the system (1.11.1) under the condition of genuine nonlinearity is treated in

J. A. Smoller, "On the solution of the Riemann problem with general step data for an extended class of hyperbolic systems," *Michigan Math. J.* **16,** 201–210 (1969a).

J. A. Smoller, "A uniqueness theorem for Riemann problems," *Arch. Rational Mech. Anal.* **33,** 110–115 (1969b).

E. D. Conway and J. A. Smoller, "Shocks violating Lax's condition are unstable," *Proc. Amer. Math. Soc.* **39,** 353–356 (1973).

J. A. Smoller, "Contact discontinuities in quasilinear hyperbolic systems," *Comm. Pure Appl. Math.* **23,** 791–801 (1970a).

C. Moler and J. A. Smoller, "Elementary interactions in quasilinear hyperbolic systems," *Arch Rational Mech. Anal.* **37,** 307–322 (1970).

Keyfitz's 1978 paper cited above for Sec. 1.7 also discusses the problem.
See also

J. S. SMOLLER, "A survey of hyperbolic systems of conservation laws in two dependent variables," in *Hyperbolic Equations and Waves*, edited by M. Foissart, Springer-Verlag, Berlin, 1970b.

For Nishida's work on the system (1.11.15) and its extension to the system (1.11.7), see

T. NISHIDA, "Global solution for an initial boundary value problem of a quasilinear hyperbolic system," *Proc. Japan Acad.* **44**, 642–646 (1968).

N. BAKHVALOV, "On the existence of regular solutions in the large for quasilinear hyperbolic systems," *Z. Vychisl. Mat. i Mat. Fiz.* **10**, 969–980 (1970), in Russian.

R. J. DIPERNA, "Existence in the large for quasilinear hyperbolic conservation laws," *Arch. Rational Mech. Anal.* **52**, 244–257 (1973a).

J. M. GREENBERG, "Estimates for fully developed shock solutions to the equation $\frac{\partial u}{\partial t} - \frac{\partial v}{\partial x} = 0$ and $\frac{\partial v}{\partial t} - \frac{\partial \sigma(u)}{\partial x} = 0$," *Indiana Univ. Math. J.* **22**, 989–1003 (1973a).

T. NISHIDA and J. A. SMOLLER, "Solutions in the large for some nonlinear hyperbolic conservations laws," *Comm. Pure Appl. Math.* **26**, 183–200 (1973).

T. NISHIDA, "Mixed problems for conservation laws," *J. Differential Equations* **23**, 244–269 (1977).

Existence and uniqueness for global solutions to the system (1.11.1) is established in

R. J. DIPERNA, "Global solutions to a class of nonlinear hyperbolic systems of equations," *Comm. Pure Appl. Math.* **26**, 1–28 (1973b).

T.-P. LIU, "Uniqueness of weak solutions of the Cauchy problem for general 2×2 conservations laws," *J. Differential Equations.* **20**, 369–388 (1976).

R. J. DIPERNA, "Uniqueness of solutions to hyperbolic conservation laws," *Indiana Univ. Math. J.* **28**, 137–188 (1979).

The Riemann problem for the system (1.11.7), not necessarily genuinely nonlinear, is treated in

B. WENDROFF, "The Riemann problem for materials with non-convex equations of state I. Isentropic flow." *J. Math. Anal. Appl.* **38**, 454–466 (1972).

L. LEIBOVICH, "Solutions of the Riemann problem for hyperbolic systems of quasilinear equations without convexity conditions," *J. Math. Anal. Appl.* **45**, 81–90 (1974).

The same problem for the system (1.11.1) is discussed in detail by Liu in his 1974 and 1975 papers cited above for Sec. 1.7.

Concerning the Riemann problem for a nonstrictly hyperbolic system, see

B. L. KEYFITZ and H. C. KRANZER, "A system of non-strictly hyperbolic conservation laws arising in elasticity theory," *Arch. Rational Mech. Anal.* **72**, 219–241 (1980).

Application of the viscosity method is fully discussed in

C. C. CONLEY and J. A. SMOLLER, "Viscosity matrices for two-dimensional nonlinear hyperbolic systems," *Comm. Pure Appl. Math.* **23**, 867–884 (1970).

C. C. CONLEY and J. A. SMOLLER, "Shock waves as limits of progressive wave solutions of higher order equations," *Comm. Pure Appl. Math.* **24**, 459–472 (1971).

J. A. SMOLLER and C. C. CONLEY, "Viscosity matrices for two-dimensional nonlinear hyperbolic systems: II," *Amer. J. Math.* **94**, 631–650 (1972).

T.-P. LIU, "The entropy condition and the admissibility of shocks," *J. Math. Anal. Appl.* **53**, 78–88 (1976).

For the unconventional form of the viscosity operator, see the treatment in

C. M. DAFERMOS, "Solution of the Riemann problem for a class of hyperbolic systems of conservation laws by the viscosity method," *Arch. Rational Mech. Anal.* **52**, 1–9 (1973).

C. M. DAFERMOS, "Structure of solutions of the Riemann problem for hyperbolic systems of conservation laws," *Arch. Rational Mech. Anal.* **53**, 203–217 (1974).

C. M. DAFERMOS and R. J. DIPERNA, "The Riemann problem for a certain class of hyperbolic systems of conservation laws," *J. Differential Equations* **20**, 90–114 (1976).

The theory of decay of solutions to genuinely nonlinear systems was first developed by

J. GLIMM and P. D. LAX, "Decay of solutions of systems of nonlinear hyperbolic conservation laws," *Mem. Amer. Math. Soc.* No. 101, 1–111 (1970).

Elementary interactions for simpler systems (1.11.7) and (1.11.29) are the subject of

J. M. GREENBERG, "On the interaction of shocks and simple waves of the same family," *Arch. Rational Mech. Anal.* **37**, 136–160 (1970).

J. M. GREENBERG, "On the elementary interactions for the quasilinear wave equations," *Arch. Rational Mech. Anal.* **43**, 325–340 (1971).

J. M. GREENBERG, "On the interaction of shocks and simple waves of the same family II," *Arch. Rational Mech. Anal.* **51**, 209–217 (1973b).

J. M. GREENBERG, "Decay theorems for stopping-shock problems," *J. Math. Anal. Appl.* **50**, 314–324 (1975).

See also Greenberg's 1973a paper and Moler and Smoller's 1970 paper cited above, where the latter deals with the system (1.11.1).

About the structure of solutions to the system (1.11.7), see

D. P. BALLOU, "The structure and asymptotic behavior of compression waves," *Arch. Rational Mech. Anal.* **56**, 170–182 (1974).

For the general system (1.11.1), the structure, decay, and asymptotic behavior of solutions are discussed in

R. J. DIPERNA, "Singularities of solutions of nonlinear hyperbolic systems of conservation laws," *Arch. Rational Mech. Anal.* **60**, 75–100 (1975a).

R. J. DIPERNA, "Decay and asymptotic behavior of solutions to nonlinear hyperbolic systems of conservation laws," *Indiana Univ. Math. J.* **24**, 1047–1071 (1975b).

T.-P. LIU, "Asymptotic behavior of solutions of general systems of nonlinear hyperbolic conservation laws," *Indiana Univ. Math. J.* **27**, 211–253 (1978).

Lax's conjecture on the asymptotic behavior of solutions is to be found in

P. D. LAX, "Nonlinear hyperbolic systems of conservation laws," in *Nonlinear Problems,* edited by R. L. Langer, University of Wisconsin Press, Madison, Wis., 1963, pp. 3–12.

Two-Solute Chromatography with the Langmuir Isotherm

2

The mathematical theory developed in Chapter 1 is applied here to treat the equilibrium chromatography of two solutes subject to the Langmuir isotherm. This isotherm is highly nonlinear, but owing to its particular form, the transformation between the concentration pair and the characteristic parameters can be expressed explicitly. Furthermore, the Γ characteristics are straight lines. These features enable us to develop a complete discussion on simple waves and shocks and a rigorous analysis of their interaction. Every stage of the development is elucidated in great detail and in some sections numerical examples are treated for the purpose of illustration. We believe that this is rather important for those who might be interested in applications. In Sec. 2.10 we also consider the situation under nonequilibrium conditions.

In Sec. 2.1 we determine the solution to the fundamental differential equation and this naturally gives the equations of the Γ characteristics. The transformation between the concentrations and the characteristic parameters is explicitly formulated. In Sec. 2.2 we are concerned with the connection between the hodograph plane of c_1 and c_2 and the (x, τ)-plane. Thus we determine the directions of the C characteristics and the shock paths in terms of the characteristic parameters. The nature of the characteristic fields is examined and the entropy condition is interpreted in terms of various physical quantities.

Riemann problems are treated in Sec. 2.3. Typical of these are the elution problem and the saturation problem. After treating the two problems in general terms, the method of solution is illustrated in some detail by using numerical examples. In Sec. 2.4 the characteristic initial (or feed) value problems are considered with a particular emphasis on compression waves, which lead to the formation of shocks. The point of shock inception is determined and an illustrative example is treated along with the application of the base curve method.

When the initial and feed data are such that waves develop from two or more points on or portions of the x and τ axes, these waves are bound to meet each other and interact as the time increases. While the fundamental features of wave interaction are discussed and summarized in Sec. 2.5, the following two sections deal with the analysis of wave interaction, in which the characteristic parameters play the key role. The interactions between waves of the same family are treated in Sec. 2.6 and those between waves of different families in Sec. 2.7. The development in these sections find application in the subsequent two sections.

In Sec. 2.8 we consider the separation of two solutes by successive application of the saturation process and the elution process, which is known as the *chromatographic cycle*. Here the saturation waves (shocks) and the elution waves (simple waves) are inevitably involved in interactions, which can be completely analyzed by applying the development in the previous sections. It is clearly understood that the interaction between waves of different families is responsible for the separation of the two solutes. Another case involved with wave interaction is *displacement development,* in which the column, after being saturated over a finite period of time, is developed by a stream containing a solute more strongly adsorbed than any solute in the column. This gives rise to a sequence of pure solute bands with sharp boundaries. Although limited to the development of single-solute chromatograms, characteristic features are investigated in Sec. 2.9 and this will serve as the precursor for the analysis of the general problem in Chapter 4.

In Sec. 2.10 we deal with the system in which axial dispersion or mass transfer resistance has a significant effect. The shock is then smeared out to some extent to give the shock layer. It is proven that such a shock layer exists if and only if the shock condition is satisfied by the end states and that it is unique if it exists. The case of equal Peclet numbers turns out to be equivalent to the case of a single solute and the shock layer has an analytic expression. A numerical example is treated not only to examine the effects of axial dispersion and mass transfer resistance but also to demonstrate the validity and usefulness of the shock layer analysis.

2.1 Langmuir Isotherm and Characteristic Parameters

In this section we introduce the Langmuir isotherm and develop the characteristic parameters by applying the treatment of Secs. 1.3 and 1.4. Since the development here is very much a continuation of Chapter 1, we will keep the notation consistent.

The Langmuir isotherm has been derived in Sec. 1.1 under the assumption that the maximum value N would be the same for both solutes A_1 and A_2, and this is consistent with Gibbs' adsorption isotherm (see Sec. 1.3 of Vol. 1). In practice, however, the experimental value of N varies from one solute to the other so that we shall develop the formulation as such. If we use the subscripts 1 and 2 to represent the concentration variables and related parameters for solutes A_1 and A_2, respectively, the fraction of vacant sites will be $1 - (n_1/N_1) - (n_2/N_2)$ and the rate of adsorption of A_1 may be expressed as

$$r_{a1} = k'_{a1} c_1 \left(1 - \frac{n_1}{N_1} - \frac{n_2}{N_2} \right)$$

The rate of desorption of A_1 may be taken to be $k'_{d1} n_1/N_1$, and since the two rates must be the same at equilibrium, we have

$$(1 + K_1 c_1) \frac{n_1}{N_1} + K_1 c_1 \frac{n_2}{N_2} = K_1 c_1 \tag{2.1.1}$$

where

$$K_1 = \frac{k'_{a1}}{k'_{d1}} \tag{2.1.2}$$

Similarly, the adsorption equilibrium of A_2 requires that

$$K_2 c_2 \frac{n_1}{N_1} + (1 + K_2 c_2) \frac{n_2}{N_2} = K_2 c_2 \tag{2.1.3}$$

where

$$k_2 = \frac{k'_{a2}}{k'_{d2}} \tag{2.1.4}$$

Solving Eqs. (2.1.1) and (2.1.3) simultaneously, we obtain the Langmuir isotherm in the form

$$n_1 = \frac{N_1 K_1 c_1}{1 + K_1 c_1 + K_2 c_2}$$

$$n_2 = \frac{N_2 K_2 c_2}{1 + K_1 c_1 + K_2 c_2} \tag{2.1.5}$$

The geometrical representation of these equations has been given in Sec. 1.1.

Whenever convenient, we shall denote the denominator of these equations by δ; i.e.,

$$\delta = 1 + K_1 c_1 + K_2 c_2 \qquad (2.1.6)$$

The total coverage Θ of adsorption sites can be expressed in terms of δ; i.e., from Eq. (2.1.5), we have

$$\Theta = \frac{n_1}{N_1} + \frac{n_2}{N_2} = 1 - \frac{1}{\delta} \qquad (2.1.7)$$

The effectiveness of an adsorbent for separating a mixture of solutes A_1 and A_2 into its components may be measured by the *separation factor*, or *relative adsorptivity*, γ, which is defined as

$$\gamma = \frac{n_2/c_2}{n_1/c_1} \qquad (2.1.8)$$

corresponding to the relative volatility of distillation or the selectivity of extraction. For the Langmuir isotherm (2.1.5) we immediately find that $\gamma = N_2 K_2/(N_1 K_1)$, and without loss of generality, we shall assume that the solute A_2 is more strongly adsorbed than A_1; i.e., $N_2 K_2 > N_1 K_1$, so that we have

$$\gamma = \frac{N_2 K_2}{N_1 K_1} > 1 \qquad (2.1.9)$$

One of the special features of the Langmuir isotherm is that for isothermal systems, the relative adsorptivity γ is a constant. In case of nonisothermal systems γ is a strong function of the temperature and the adsorptivities may be reversed between two solutes at higher temperatures, as we shall see in Chapter 6.

Let us now consider the equations of two-solute chromatography. From Eq. (1.1.16) we have

$$\frac{\partial c_1}{\partial x} + \frac{\partial}{\partial \tau} f_1(c_1, c_2) = 0$$
$$\frac{\partial c_2}{\partial x} + \frac{\partial}{\partial \tau} f_2(c_1, c_2) = 0 \qquad (2.1.10)$$

where x and τ denote the dimensionless independent variables defined by Eq. (1.1.15), while f_1 and f_2 are the equilibrium column isotherms defined as

$$f_1(c_1, c_2) = c_1 + \nu n_1(c_1, c_2)$$
$$f_2(c_1, c_2) = c_2 + \nu n_2(c_1, c_2) \qquad (2.1.11)$$

Here ν represents the volume ratio of the solid phase to the fluid phase [see Eq. (1.1.12)].

First, we shall find the solution to the fundamental differential equation, which may be written in the form [see Eq. (1.4.8)]

$$\frac{\partial n_2}{\partial c_1}\left(\frac{dc_1}{dc_2}\right)^2 - \left(\frac{\partial n_1}{\partial c_1} - \frac{\partial n_2}{\partial c_2}\right)\frac{dc_1}{dc_2} - \frac{\partial n_1}{\partial c_2} = 0 \qquad (2.1.12)$$

and thus determine the equations of the Γ characteristics and also define the characteristic parameters. By substituting the partial derivatives from Eq. (2.1.5) into Eq. (2.1.12), we obtain

$$-\frac{N_2 K_2 K_1 c_2}{\delta^2}\left(\frac{dc_1}{dc_2}\right)^2 - \frac{1}{\delta^2}\{N_1 K_1(1 + K_2 c_2)$$

$$- N_2 K_2(1 + K_1 c_1)\}\frac{dc_1}{dc_2} + \frac{N_1 K_1 K_2 c_1}{\delta^2} = 0$$

or, after dividing through by $N_2 K_2 K_1/\delta^2$ (which never vanishes), the equation

$$c_2\left(\frac{dc_1}{dc_2}\right)^2 - \left\{\frac{1}{K_1}\left(1 - \frac{N_1 K_1}{N_2 K_2}\right) + c_1 - \frac{N_1}{N_2}c_2\right\}\frac{dc_1}{dc_2} - \frac{N_1}{N_2}c_1 = 0 \qquad (2.1.13)$$

For brevity we shall put

$$h = \frac{N_1}{N_2} \qquad (2.1.14)$$

and

$$k = \frac{1}{K_1}\left(1 - \frac{N_1 K_1}{N_2 K_2}\right) = \frac{\gamma - 1}{K_1 \gamma} \qquad (2.1.15)$$

so that we have a particularly simple equation

$$c_2\left(\frac{dc_1}{dc_2}\right)^2 - (k + c_1 - hc_2)\frac{dc_1}{dc_2} - hc_1 = 0 \qquad (2.1.16)$$

which, if solved, gives the equation of the Γ characteristics in the hodograph plane of c_1 and c_2.

If we let $\zeta = dc_1/dc_2$, Eq. (2.1.16) can be rewritten as

$$(h + \zeta)c_1 = \zeta(h + \zeta)c_2 - k\zeta$$

or

$$c_1 = \zeta c_2 - \frac{k\zeta}{\zeta + h} \qquad (2.1.17)$$

We may regard this as a nonlinear differential equation of Clairaut's form $c_1 = \zeta c_2 + \phi(\zeta)$ (see Sec. 2.4 of Vol. 1), for which there exists a general solution and a singular solution. The former is, in fact, the family of straight lines given by regarding ζ as a constant in the differential equation itself and the latter gives the envelope of the former.

We shall confirm the observation above by determining the solution of Eq. (2.1.16). Although this solution scheme is not uncommon in the literature (see, e.g., E. L. Ince, *Ordinary Differential Equations,* Dover, New York, 1956, p. 39), Glueckauf (1949) was the first to apply it to Eq. (2.1.16). Let us differentiate Eq. (2.1.16) with respect to c_2 and thus obtain

$$\frac{d^2c_1}{dc_2^2}\{2c_2\frac{dc_1}{dc_2} - (k + c_1 - hc_2)\} = 0 \qquad (2.1.18)$$

which can be satisfied in two ways. In the first case we have

$$\frac{d^2c_1}{dc_2^2} = 0 \qquad (2.1.19)$$

or

$$c_1 = \zeta c_2 + \mu \qquad (2.1.20)$$

where ζ and μ are the two constants of integration. Substituting this back into the original differential equation (2.1.16), we have

$$\mu = -\frac{k\zeta}{\zeta + h}$$

and thus

$$c_1 = \zeta c_2 - \frac{k\zeta}{\zeta + h} \qquad (2.1.21)$$

which is identical to Eq. (2.1.17). Since ζ is constant, Eq. (2.1.21) represents a family of straight lines and this is the general solution mentioned above.

On the other hand, Eq. (2.1.18) can also be satisfied by requiring that

$$2c_2\frac{dc_1}{dc_2} = (k + c_1 - hc_2) \qquad (2.1.22)$$

Substituting this into Eq. (2.1.16) gives

$$(k + c_1 - hc_2)^2 + 4hc_1c_2 = 0 \qquad (2.1.23)$$

which is physically irrelevant since it implies either $c_1 < 0$ or $c_2 < 0$. Equation (2.1.23) can be rewritten as

$$(k + c_1 + hc_2)^2 - 4hkc_2 = 0 \qquad (2.1.24)$$

so we have $c_1 < 0$ and $c_2 > 0$. Obviously, Eq. (2.1.24) is the singular solution and gives the envelope of the family of straight lines of Eq. (2.1.21). This envelope is a parabola that lies in the fourth quadrant and is tangent to the c_1 and c_2 axes at $c_1 = -k = (1 - \gamma)/(K_1\gamma)$ and $c_2 = k/h = (\gamma - 1)/K_2$, respectively, as shown in Fig. 2.1.

The straight lines of Eq. (2.1.21) are all tangent to the parabola and solving Eqs. (2.1.21) and (2.1.24) simultaneously gives the point of tangency as

$$c_1 = -k\left(\frac{\zeta}{\zeta + h}\right)^2, \qquad c_2 = \frac{hk}{(\zeta + h)^2} \qquad (2.1.25)$$

Since the straight lines are tangent to a convex curve, no two different lines possess the same value of ζ. There are two different families of these lines,

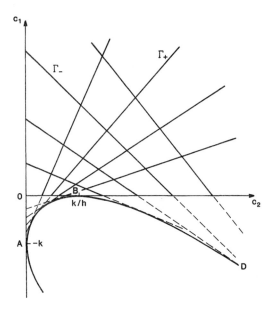

Figure 2.1

one having positive slopes and the other negative slopes (see Fig. 2.1). Those with positive slopes are tangent to the portion AB of the parabola, for which we have $0 \leq c_2 \leq k/h$, and thus from Eq. (2.1.25) we find that

$$0 \leq \zeta_+ \leq \infty \qquad (2.1.26)$$

where ζ_+ denotes the positive slope dc_1/dc_2. On the other hand, the straight lines with negative slopes are tangent to the portion BD of the parabola, over which c_2 varies from k/h to ∞, so from Eq. (2.1.25) we have

$$-h < \zeta_- \leq 0 \qquad (2.1.27)$$

where ζ_- represents the negative slope dc_1/dc_2.

These two families of straight lines represent the solution to Eq. (2.1.16) and are indeed the Γ_+ and Γ_- characteristics in the hodograph plane of c_1 and c_2. Their equations are given by Eq. (2.1.21) with $\zeta = \zeta_+$ for the Γ_+ characteristics and with $\zeta = \zeta_-$ for the Γ_- characteristics, respectively. We note that ζ_+ or ζ_- remains fixed along a Γ_+ or Γ_- characteristic, respectively, and thus we may directly take ζ_+ and ζ_- as the characteristic parameters β and α, respectively. Consequently, we can write

$$\Gamma_+: \quad c_1 = \beta c_2 - \frac{k\beta}{\beta + h} \qquad (2.1.28)$$

$$\Gamma_-: \quad c_1 = \alpha c_2 - \frac{k\alpha}{\alpha + h} \qquad (2.1.29)$$

where

$$-h < \alpha \leq 0, \qquad 0 \leq \beta \leq \infty \qquad (2.1.30)$$

It should be pointed out that the c_1 axis itself is the Γ_+ characteristic for $\beta = \infty$ and the portion $k/h < c_2 < \infty$, of the c_2 axis corresponds to the Γ_+ characteristic for $\beta = 0$ while the portion, $0 \leq c_2 < k/h$, of the c_2 axis is the Γ_- characteristic for $\alpha = 0$. Given the values of h and k, the characteristic net in the hodograph plane can be constructed by applying Eqs. (2.1.28) and (2.1.29) for various values of β and α, respectively, and an example is shown for $h = 1$ in Fig. 2.2.

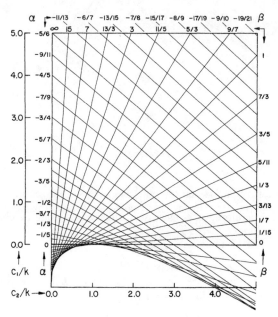

Characteristic net in the hodograph plane when h = 1.

Figure 2.2

For a given state of concentrations (c_1, c_2), the set of corresponding parameters (α, β) is determined by the two roots of the quadratic equation [see Eq. (2.1.16)]

$$c_2\zeta^2 - (k + c_1 - hc_2)\zeta - hc_1 = 0 \qquad (2.1.31)$$

so

$$\begin{Bmatrix} \beta \\ \alpha \end{Bmatrix} = \frac{1}{2c_2}\{k + c_1 - hc_2 \pm \sqrt{(k + c_1 - hc_2)^2 + 4hc_1c_2}\} \qquad (2.1.32)$$

Conversely, with the set (α, β) given, the corresponding state of concentrations (c_1, c_2) can be determined by solving Eqs. (2.1.28) and (2.1.29) simulta-

Two-Solute Chromatography with the Langmuir Isotherm   Chap. 2

neously and thus we have

$$c_1 = \frac{-k\alpha\beta}{(\alpha + h)(\beta + h)} \qquad (2.1.33)$$

$$c_2 = \frac{hk}{(\alpha + h)(\beta + h)} \qquad (2.1.34)$$

Therefore, a one-to-one correspondence is established between the pair of concentrations (c_1, c_2) and the set of characteristic parameters (α, β). We also note that

$$\frac{\partial c_1}{\partial \alpha} = -\frac{hk\beta}{(\alpha + h)^2(\beta + h)} < 0,$$

$$\frac{\partial c_1}{\partial \beta} = -\frac{hk\alpha}{(\alpha + h)(\beta + h)^2} > 0 \qquad (2.1.35)$$

and

$$\frac{\partial c_2}{\partial \alpha} = -\frac{hk}{(\alpha + h)^2(\beta + h)} < 0,$$

$$\frac{\partial c_2}{\partial \beta} = -\frac{hk}{(\alpha + h)(\beta + h)^2} < 0 \qquad (2.1.36)$$

Clearly, Eq. (2.1.36) implies that the characteristic parameters α and β both vary in the opposite direction to c_2, the concentration of more strongly adsorbed solute. Therefore, as c_2 increases, α decreases along Γ_+ characteristics and β decreases along Γ_- characteristics.

The equilibrium concentrations n_1 and n_2 in the solid phase can also be expressed in terms of α and β. For this let us first consider δ and substitute Eqs. (2.1.33) and (2.1.34) into Eq. (2.1.6) to obtain

$$\delta = 1 + K_1c_1 + K_2c_2 = 1 + \frac{-\alpha\beta(\gamma - 1)/\gamma + (\gamma - 1)h^2}{(\alpha + h)(\beta + h)}$$

$$= \frac{1}{\gamma}\left(\frac{\alpha + h\gamma}{\alpha + h}\right)\left(\frac{\beta + h\gamma}{\beta + h}\right) \qquad (2.1.37)$$

Here we note that

$$h\gamma = \frac{K_2}{K_1} \qquad (2.1.38)$$

Now using Eqs. (2.1.33), (2.1.34), and (2.1.37) in Eq. (2.1.5), we find that

$$\frac{n_1}{N_1} = \frac{-(\gamma - 1)\alpha\beta}{(\alpha + h\gamma)(\beta + h\gamma)} \qquad (2.1.39)$$

and

$$\frac{n_2}{N_2} = \frac{h^2\gamma(\gamma - 1)}{(\alpha + h\gamma)(\beta + h\gamma)} \qquad (2.1.40)$$

From Eq. (2.1.7) we can express the surface coverage in terms of α and β, i.e.,

$$\Theta = \frac{(\gamma - 1)(h^2\gamma - \alpha\beta)}{(\alpha + h\gamma)(\beta + h\gamma)} \tag{2.1.41}$$

Let us now recall the discussion of Sec. 1.9 and consider the Jacobian matrix $[\partial f_i/\partial c_j]$ of the system. Since the Γ_+ and Γ_- characteristics are everywhere tangent to the right eigenvectors $\mathbf{r}_-$ and $\mathbf{r}_+$, respectively, we can put

$$\mathbf{r}_- = \begin{bmatrix} \beta \\ 1 \end{bmatrix}, \qquad \mathbf{r}_+ = \begin{bmatrix} \alpha \\ 1 \end{bmatrix} \tag{2.1.42}$$

and by the biorthogonality property, we know that the left eigenvectors are given by

$$\boldsymbol{\ell}_+ = \begin{bmatrix} -1 \\ \beta \end{bmatrix}, \qquad \boldsymbol{\ell}_- = \begin{bmatrix} -1 \\ \alpha \end{bmatrix} \tag{2.1.43}$$

From Eqs. (1.9.20) and (1.9.22), therefore, we obtain the Riemann invariants as

$$J_+(\beta) = -\int dc_1 + \int \beta dc_2 = c_1 + \beta c_2 = \frac{k\beta}{\beta + h} \tag{2.1.44}$$

$$J_-(\alpha) = -\int dc_1 + \int \alpha dc_2 = -c_1 + \alpha c_2 = \frac{k\alpha}{\alpha + h} \tag{2.1.45}$$

and these are to be compared with Eq. (1.9.69). Since, however, the parameters β and α satisfy the requirements of the Riemann invariants J_+ and J_-, we may take β and α for J_+ and J_-, respectively, as well if necessary.

Exercises

2.1.1. Rearrange Eq. (2.1.22) so that it can be integrated directly and show that, when the constant of integration is determined properly, the solution is given by

$$k + c_1 + hc_2 = \pm 2\sqrt{hkc_2}$$

which is the same as Eq. (2.1.24).

2.1.2. Equation (2.1.21) may be rearranged as

$$(\zeta c_2 - c_1)(\zeta + h) = k\zeta$$

In addition to this equation, apply a geometrical argument to show that for large values of c_1 and c_2, we have

$$\frac{c_1}{c_2} < \zeta_+ < \frac{c_1 + k}{c_2}$$

$$-h < \zeta_- < 0$$

Two-Solute Chromatography with the Langmuir Isotherm   Chap. 2

and for small values of c_1 and c_2,

$$\frac{hc_1}{hc_1 - k} < \zeta_-$$

2.2 Directions of C Characteristics and Shock Paths

Once we have determined the equations of Γ characteristics, we need to establish the connection between the hodograph plane and the (x, τ)-plane, and this can be accomplished by expressing the characteristic directions $\sigma_\pm$ in the (x, τ)-plane in terms of c_1 and c_2 or more conveniently in terms of the characteristic parameters α and β.

From Eqs. (1.4.16) and (1.4.17) we have

$$\sigma_\pm = 1 + \nu\left(\frac{\partial n_1}{\partial c_1} - \frac{\partial n_2}{\partial c_1}\zeta_\pm\right) \tag{2.2.1}$$

and thus substitution of the partial derivatives from Eq. (2.1.5) gives

$$\sigma_\pm = 1 + \frac{\nu\{N_1 K_1(1 + K_2 c_2) + N_2 K_2 K_1 c_2 \zeta_\pm\}}{\delta^2}$$

$$= 1 + \frac{\nu N_1 K_1\{1 + (K_2 c_2/h)(\zeta_\pm + h)\}}{\delta^2} \tag{2.2.2}$$

But from Eq. (2.1.34), we see that

$$\frac{K_2 c_2}{h} = \frac{h(\gamma - 1)}{(\alpha + h)(\beta + h)}$$

and $\zeta_+ = \beta$ and $\zeta_- = \alpha$, so that

$$1 + \frac{K_2 c_2}{h}(\zeta_\pm + h) = \begin{cases} \dfrac{\alpha + h\gamma}{\alpha + h} & \text{for } \sigma_+ \\[2mm] \dfrac{\beta + h\gamma}{\beta + h} & \text{for } \sigma_- \end{cases}$$

Hence, by substituting this together with Eq. (2.1.37) for δ into Eq. (2.2.2), we obtain

$$C_+: \quad \sigma_+ = 1 + \rho\left(\frac{\alpha + h}{\alpha + h\gamma}\right)\left(\frac{\beta + h}{\beta + h\gamma}\right)^2 \tag{2.2.3}$$

$$C_-: \quad \sigma_- = 1 + \rho\left(\frac{\alpha + h}{\alpha + h\gamma}\right)^2\left(\frac{\beta + h}{\beta + h\gamma}\right) \tag{2.2.4}$$

where ρ is a positive constant defined as

$$\rho = \nu N_2 K_2 \gamma = \frac{(1 - \epsilon) N_2^2 K_2^2}{\epsilon N_1 K_1} \qquad (2.2.5)$$

For the sake of convenience we shall adopt the notation

$$a = a(\alpha) = \frac{\alpha + h}{\alpha + h\gamma} \qquad (2.2.6)$$

and

$$b = b(\beta) = \frac{\beta + h}{\beta + h\gamma} \qquad (2.2.7)$$

so that we can write

$$C_+: \quad \sigma_+ = 1 + \rho ab^2 \qquad (2.2.8)$$

$$C_-: \quad \sigma_- = 1 + \rho a^2 b \qquad (2.2.9)$$

Since we have $-h < \alpha \le 0$ and $0 \le \beta \le \infty$ from Eq. (2.1.30) and $\gamma > 1$, it is obvious that a and b are both positive and indeed bounded as

$$0 < a(\alpha) \le \gamma^{-1} \le b(\beta) \le 1 \qquad (2.2.10)$$

We also notice that

$$\frac{da}{d\alpha} = \frac{h(\gamma - 1)}{(\alpha + h\gamma)^2} > 0, \qquad \frac{db}{d\beta} = \frac{h(\gamma - 1)}{(\beta + h\gamma)^2} > 0 \qquad (2.2.11)$$

and hence $a(\alpha)$ and $b(\beta)$ are monotone increasing functions of α and β, respectively. An immediate consequence is that we may regard a and b as another set of characteristic parameters (see Sec. 1.3), and this will occur frequently in the following.

Using the bounds on a and b given by Eq. (2.2.10) and the definition (2.2.5) of ρ, we can find bounds on σ_+ and σ_-, i.e.,

$$1 < \sigma_+ < 1 + \nu N_2 K_2$$
$$1 < \sigma_- < 1 + \nu N_1 K_1 \qquad (2.2.12)$$

the upper bounds corresponding to the state $c_1 = c_2 = 0$. Furthermore, we have $a < b$ for any pair (c_1, c_2) and thus

$$1 < \sigma_- < \sigma_+ \qquad (2.2.13)$$

which confirms Eq. (1.4.20). The only exception is at the point B of Fig. 2.1, where $c_1 = 0$ and $c_2 = k/h = (\gamma - 1)/K_2$. At this point we have $\alpha = \beta = 0$ and $a = b = \gamma^{-1}$, so that $\sigma_+ = \sigma_- = 1 + \nu N_2 K_2 / \gamma^2$. This is a peculiar point of the hodograph plane. We see the Γ_+ characteristics all have their c_2 intercepts located to the left of the point B, whereas the Γ_- characteristics have their c_2 intercepts positioned to the right of B. The point B is referred to by some authors as the *watershed point* (see, e.g., Helfferich and Klein, 1970).

Let us now examine the nature of the characteristic fields. From Eqs. (2.2.8) and (2.2.9) we have

$$\frac{\partial \sigma_+}{\partial \beta} = 2\rho ab \frac{db}{d\beta} = \frac{2\rho h (\gamma - 1)ab}{(\beta + h\gamma)^2}$$

$$\frac{\partial \sigma_-}{\partial \alpha} = 2\rho ab \frac{da}{d\alpha} = \frac{2\rho h (\gamma - 1)ab}{(\alpha + h\gamma)^2}$$

(2.2.14)

after substitution of Eq. (2.2.11). By using these together with Eq. (2.1.36) in Eq. (1.6.7) we find that

$$\frac{\mathcal{D}_- \sigma_+}{\mathcal{D}c_2} = \frac{\partial \sigma_+ / \partial \beta}{\partial c_2 / \partial \beta} = -2\rho K_1 \gamma (\alpha + h)ab^3$$

$$\frac{\mathcal{D}_+ \sigma_-}{\mathcal{D}c_2} = \frac{\partial \sigma_- / \partial \alpha}{\partial c_2 / \partial \alpha} = -2\rho K_1 \gamma (\beta + h)a^3 b$$

(2.2.15)

both of which are negative, and this implies that both the $(+)$ and $(-)$ characteristic fields are genuinely nonlinear. Hence the system (2.1.10) is genuinely nonlinear. A similar calculation gives

$$\frac{\mathcal{D}_+ \sigma_+}{\mathcal{D}c_2} = \frac{\partial \sigma_+ / \partial \alpha}{\partial c_2 / \partial \alpha} = -\rho K_1 \gamma (\beta + h)a^2 b^2$$

$$\frac{\mathcal{D}_- \sigma_-}{\mathcal{D}c_2} = \frac{\partial \sigma_- / \partial \beta}{\partial c_2 / \partial \beta} = -\rho K_1 \gamma (\alpha + h)a^2 b^2$$

(2.2.16)

and both of these are negative. It is also a matter of algebraic manipulation to show that the second derivatives of σ_+ and σ_- taken with respect to c_2 along either Γ_+ or Γ_- characteristics are all positive.

From Eqs. (2.2.15) and (2.2.16) we see that along either the Γ_+ or Γ_- characteristics both σ_+ and σ_- decrease in the direction of increasing c_2. This conclusion, which is important in building up a proper picture of the solution, may be summarized by associating a direction of decrease in σ_+ and σ_- with the Γ characteristics as in Fig. 2.3. Also, we draw the following rule.

RULE 2.1: Along either the Γ_+ or Γ_- characteristics, both σ_+ and σ_- decrease in the direction of increasing c_2, the concentration of the more strongly adsorbed solute.

It is now clear that if c_2 decreases in the x direction in a simple wave region, the simple wave is a compression wave and will give birth to a shock after a finite time. If we have $c_2^\ell > c_2^0$ for a Riemann problem [see Eq. (1.5.9)], the solution will contain one or two shocks. We are assured here that the discontinuities we may encounter in this problem are all shocks because the system (2.1.10) is genuinely nonlinear.

Suppose now that the solution contains a shock. Then we must have the compatibility relation

$$\frac{[n_1]}{[c_1]} = \frac{[n_2]}{[c_2]}$$

(2.2.17)

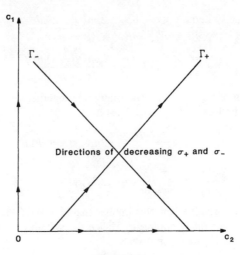

Figure 2.3

satisfied by the states on both sides of the shock. If the state on one side, say (c_1^l, c_2^l), is given, the compatibility relation gives the locus of the state on the other side, i.e., (c_1^r, c_2^r). From the discussion of Sec. 1.7 we know that the locus consists of two branches which are called the shock curves Σ_+ and Σ_-. Furthermore, it has been proven that if the Γ characteristics are straight, the shock curves necessarily coincide with the Γ characteristics; i.e., the curves Σ_+ and Σ_- emanating from the point (c_1^l, c_2^l) coincide with the Γ_+ and Γ_- characteristics, respectively, passing through the point (c_1^l, c_2^l). It then follows that the solution of the compatibility relation (2.2.17) is given by the Γ characteristics, so the states on both sides of a shock must fall on the same Γ characteristic in the hodograph plane (see also Exercise 2.2.3). If the images of the states on both sides of a shock fall on a Γ_+ characteristic, we shall call it a Γ_+ shock, and similarly, if the images fall on a Γ_- characteristic, it is referred to as a Γ_- shock. Note that these are equivalent to the Σ_+ and Σ_- shocks of Sec. 1.7.

A Γ_+ shock propagates in the direction of $\tilde{\sigma}_-$ while a Γ_- shock propagates in the direction of $\tilde{\sigma}_+$. Now we need to express $\tilde{\sigma}_+$ and $\tilde{\sigma}_-$ in terms of the characteristic parameters α and β. First, we shall introduce the Langmuir isotherm (2.1.5) in Eq. (2.2.17) to have

$$
\begin{aligned}
\frac{[n_1]}{[c_1]} = \frac{n_1^l - n_1^r}{c_1^l - c_1^r} &= \frac{1}{c_1^l - c_1^r}\left(\frac{N_1 K_1 c_1^l}{\delta^l} - \frac{N_1 K_1 c_1^r}{\delta^r}\right) \\
&= \frac{N_1 K_1}{\delta^l \delta^r}\left\{1 + K_2 c_2^l - K_2 c_1^l \frac{c_2^l - c_2^r}{c_1^l - c_1^r}\right\} \\
&= \frac{N_1 K_1}{\delta^l \delta^r}\left\{1 + K_2\left(c_2^l - \frac{c_1^l}{\zeta_\pm}\right)\right\}
\end{aligned}
\tag{2.2.18}
$$

where the relation $(c_1^l - c_1^r)/(c_2^l - c_2^r) = \zeta_\pm$ along Γ characteristics has been substituted to obtain the last expression. Using Eqs. (2.1.33) and (2.1.34) for

Two-Solute Chromatography with the Langmuir Isotherm   Chap. 2

c_1^l and c_2^l, and taking $\zeta_+ = \beta = \beta^r = \beta^l$ for a Γ_+ shock and $\zeta_- = \alpha = \alpha^r = \alpha^l$ for a Γ_- shock, we can show that

$$1 + K_2\left(c_2^l - \frac{c_1^l}{\zeta_\pm}\right) = \begin{cases} \dfrac{\beta^l + h\gamma}{\beta^l + h} & \text{for a } \Gamma_+ \text{ shock} \\[2ex] \dfrac{\alpha^l + h\gamma}{\alpha^l + h} & \text{for a } \Gamma_- \text{ shock} \end{cases}$$

Substituting this and δ^l and δ^r from Eq. (2.1.37) into Eq. (2.2.18) and then using the resulting expressions in Eqs. (1.7.10) and (1.7.11), we finally obtain

$$S_+: \quad \tilde{\sigma}_+(\beta^l, \beta^r; \alpha) = 1 + \nu\left(\frac{[n_i]}{[c_i]}\right)_{\Gamma_-}$$
$$= 1 + \rho\left(\frac{\alpha + h}{\alpha + h\gamma}\right)\left(\frac{\beta^l + h}{\beta^l + h\gamma}\right)\left(\frac{\beta^r + h}{\beta^r + h\gamma}\right) \qquad (2.2.19)$$

for a Γ_- shock and

$$S_-: \quad \tilde{\sigma}_-(\alpha^l, \alpha^r; \beta) = 1 + \nu\left(\frac{[n_i]}{[c_i]}\right)_{\Gamma_+}$$
$$= 1 + \rho\left(\frac{\beta + h}{\beta + h\gamma}\right)\left(\frac{\alpha^l + h}{\alpha^l + h\gamma}\right)\left(\frac{\alpha^r + h}{\alpha^r + h\gamma}\right) \qquad (2.2.20)$$

for a Γ_+ shock. Note that α is invariant across a Γ_- shock while β is unchanged across a Γ_+ shock. Of course, we can use a and b defined by Eqs. (2.2.6) and (2.2.7), and then Eqs. (2.2.19) and (2.2.20) are written as

$$S_+: \quad \tilde{\sigma}_+(b^l, b^r; a) = 1 + \rho a b^l b^r \qquad (2.2.21)$$

$$S_-: \quad \tilde{\sigma}_-(a^l, a^r; b) = 1 + \rho b a^l a^r \qquad (2.2.22)$$

If a Γ_- shock and a Γ_+ shock have the state (c_1^0, c_2^0) in common, we can write

$$\tilde{\sigma}_+(b^0, b; a^0) = \rho a^0 b^0 b$$
$$\tilde{\sigma}_-(a^0, a; b^0) = 1 + \rho b^0 a^0 a$$

and since $a < b$, it follows that

$$\tilde{\sigma}_- < \tilde{\sigma}_+ \qquad (2.2.23)$$

This implies that the Γ_+ shock propagates faster than the Γ_- shock.

Based on Rule 2.1, it appears that the entropy condition is simply given by the inequality $c_2^l > c_2^r$. We confirm this observation by applying the extended entropy condition (see Rule 1.5). In terms of the characteristic parameters, Eq. (1.7.16) reads

$$\tilde{\sigma}_+(b^l, b^r; a) \geq \tilde{\sigma}_+(b^l, b; a) \qquad (2.2.24)$$

which is to be satisfied by every b between b^l and b^r. Thus we know from Eq. (2.2.22) that it is required to have $b^r \geq b$. But since b varies monotonously along a Γ_- characteristic, we conclude that the extended entropy condition is satisfied if $b^l < b^r$. This condition may be expressed in terms of β or c_2 because $db/d\beta > 0$ and $\partial c_2/\partial \beta < 0$. Thus we have $\beta^l < \beta^r$ or $c_2^l > c_2^r$ to be satisfied across a Γ_- shock. Also, we see from Eq. (2.1.37) that $\delta^l > \delta^r$ and from Eq. (2.1.7) that $\Theta^l > \Theta^r$. On the other hand, Eq. (1.7.17) reads

$$\tilde{\sigma}_-(a^l, a^r; b) \geq \tilde{\sigma}_-(a^l, a; b) \tag{2.2.25}$$

which is to be satisfied by every a between a^l and a^r. By applying a similar argument we can interpret the entropy condition as $a^l < a^r$, $\alpha^l < \alpha^r$, $c_2^l > c_2^r$, $\delta^l > \delta^r$ or $\Theta^l > \Theta^r$. Consequently, we draw the following conclusion.

RULE 2.2 (Entropy Condition):

1. If two states L and R are to be connected by a Γ_- shock, we must have

$$\beta^l < \beta^r \quad \text{so} \quad b^l < b^r \tag{2.2.26}$$

and if they are to be connected by a Γ_+ shock, we must have

$$\alpha^l < \alpha^r \quad \text{so} \quad a^l < a^r \tag{2.2.27}$$

2. If two states L and R are to be connected by a shock, we must have

$$c_2^l > c_2^r \tag{2.2.28}$$

and equivalently

$$\delta^l < \delta^r \quad \text{or} \quad \Theta^l < \Theta^r \tag{2.2.29}$$

While statement 2 is more general, we shall frequently find that statement 1 is more convenient in application. It is to be noticed that the concentration c_2 of the more strongly adsorbed solute decreases across a shock and the same is true for the surface coverage Θ.

Let us now consider a Γ_- shock. The states on both sides of this shock fall on the same Γ_- characteristic. Since the parameter α remains invariant along a Γ_- characteristic, we can write from Eqs. (2.2.8), (2.2.9), and (2.2.21) the following:

$$\tilde{\sigma}_+^l = 1 + \rho a(b^l)^2$$
$$\tilde{\sigma}_+^r = 1 + \rho a(b^r)^2$$
$$\tilde{\sigma}_-^r = 1 + \rho a^2 b^r$$
$$\tilde{\sigma}_+ = 1 + \rho a b^l b^r$$

From the entropy condition we have $b^l < b^r$, so

$$\sigma_+^l < \tilde{\sigma}_+ < \sigma_+^r \tag{2.2.30}$$

We also have $a < b$ from Eq. (2.2.10), so that

$$\tilde{\sigma}_+ > \sigma_-^r \tag{2.2.31}$$

Similarly, for a Γ_+ shock we have $a^l < a^r$ and again $a < b$ and hence from Eqs. (2.2.8), (2.2.9), and (2.2.22), we can show that

$$\sigma_-^l < \bar{\sigma}_- < \sigma_-^r \qquad (2.2.32)$$

$$\bar{\sigma}_- < \sigma_+^l \qquad (2.2.33)$$

The inequalities (2.2.30)–(2.2.33) are identical to the shock condition, so we again confirm that discontinuities which are admissible in the solution are all shocks.

Exercises

2.2.1. Since $\zeta_+ \zeta_- = -hc_1/c_2$ from Eq. (2.1.16), we may rewrite Eq. (2.2.2) in the form

$$\sigma_\pm = 1 + \frac{\nu N_1 K_1 \{1 + K_2 c_2 - K_2 c_1/\zeta_\pm\}}{\delta^2}$$

By differentiating σ_+ with respect to c_2 along a Γ_- characteristic, show that

$$\frac{\mathcal{D}_- \sigma_+}{\mathcal{D} c_2} = -\frac{2}{\delta}(\sigma_+ - 1)K_1\left(\zeta_- + \frac{K_2}{K_1}\right) < 0$$

and similarly, show that

$$\frac{\mathcal{D}_+ \sigma_-}{\mathcal{D} c_2} = -\frac{2}{\delta}(\sigma_- - 1)K_1\left(\zeta_+ + \frac{K_2}{K_1}\right) < 0$$

so that the system (2.1.10) is genuinely nonlinear.

2.2.2. Show that the second derivatives of σ_+ and σ_- taken with respect to c_2 along Γ_- and Γ_+ characteristics, respectively, are both positive.

2.2.3. Using Eq. (2.2.18), put the compatibility relation (2.2.17) in the form

$$c_2\left(\frac{[c_1]}{[c_2]}\right)^2 - (k + c_1^l - hc_2^l)\frac{[c_1]}{[c_2]} - hc_1^l = 0$$

where h and k are defined as in Eqs. (2.1.14) and (2.1.15). If the state (c_1^l, c_2^l) is given, can you argue that there are two distinct branches of the locus for the state (c_1^r, c_2^r) in the hodograph plane? Show that the equation above is satisfied along Γ characteristics and thus the shock curve Σ necessarily coincides with a Γ characteristic.

2.2.4. From Eq. (2.1.21) we have

$$c_2(\zeta + h) = k + c_1\left(1 + \frac{h}{\zeta}\right)$$

and thus Eq. (2.2.2) may be rewritten as

$$\frac{\delta^2}{\nu N_1 K_1}(\sigma_\pm - 1) = 1 + \frac{K_2 c_2}{h}(\zeta_\pm + h) = \gamma + \frac{K_2 c_1}{h}\left(1 + \frac{h}{\zeta_\pm}\right)$$

By using the fact that $\sigma_- > 1$, show that

$$-\frac{1 + hK_2c_2}{K_2c_2} < \zeta_- < -\frac{hK_2c_2}{h\gamma + K_2c_2}$$

Similarly, note that $\zeta_- > -h$ and thus show that

$$\frac{1}{\delta^2} < \frac{\sigma_- - 1}{\nu N_1 K_1} < \frac{\gamma}{\delta^2}$$

Finally, using the bounds on ζ_+ given in Exercise 2.1.2, show that

$$\max\left[\frac{1 + K_2c_1/h + K_2c_2}{\delta^2}, \frac{(h\gamma + K_2c_1)(c_1 + k) + K_2c_1c_2}{h(c_1 + k)}\right]$$

$$< \frac{\sigma_+ - 1}{\nu N_1 K_1} < \frac{\gamma + K_2c_1/h + K_2c_2}{\delta^2}$$

2.2.5. Consider Eq. (2.1.16) in the form

$$c_2\zeta^2 - (k + c_1 - hc_2)\zeta - hc_1 = 0$$

When $\zeta = \zeta_+$, differentiate with respect to c_2 along a Γ_- characteristic and show that

$$\frac{\mathcal{D}_-\zeta_+}{\mathcal{D}c_2} = -\frac{\zeta_+ + h}{c_2}$$

Similarly, show that

$$\frac{\mathcal{D}_+\zeta_-}{\mathcal{D}c_2} = -\frac{\zeta_- + h}{c_2}$$

Compare these results with Eq. (2.1.36).

2.2.6. Using the relations $c_2(\zeta + h) = k + c_1(1 + h/\zeta)$ and $\zeta_+\zeta_- = -hc_1/c_2$, we can rewrite Eq. (2.2.2) in the form

$$\sigma_\pm = 1 + \nu N_1 K_1\left\{\gamma + \frac{K_2c_1}{h} + \frac{K_2c_1}{\zeta_\pm}\right\}\frac{1}{\delta^2}$$

$$= 1 + \nu N_1 K_1\left\{\gamma + \frac{K_2c_1}{h} - \frac{K_2c_2}{h}\zeta_\mp\right\}\frac{1}{\delta^2}$$

By differentiating with respect to c_2 along Γ characteristics and making some manipulations with the result of Exercise 2.2.5, show that

$$\frac{\mathcal{D}_+\sigma_+}{\mathcal{D}c_2} = -2\nu N_2 K_2 K_1(\zeta_+ + h\gamma)\frac{1}{\delta^2}$$

and

$$\frac{\mathcal{D}_-\sigma_-}{\mathcal{D}c_2} = -2\nu N_2 K_2 K_1(\zeta_- + h\gamma)\frac{1}{\delta^2}$$

Also, show that the second derivatives $\mathcal{D}_+^2\sigma_+/\mathcal{D}c_2^2$ and $\mathcal{D}_-^2\sigma_-/\mathcal{D}c_2^2$ are both positive.

2.3 Riemann Problems

We are now in a position to construct the solutions of various problems. As a first step we shall treat here Riemann problems, i.e., those problems with constant initial and feed data. Typical of these cases are the elution of a column initially saturated with constant concentrations of the two solutes and the saturation of an initially clean column with a feedstream having constant concentrations of the two solutes.

Let us first consider the case of elution and see how the solution can be obtained in terms of simple waves. If c_1^0 and c_2^0 are the constant initial concentrations of the two solutes, our problem is to solve the conservation equations (2.1.10) subject to the initial and feed conditions

$$\text{at } \tau = 0: \quad c_1 = c_1^0 \quad \text{and} \quad c_2 = c_2^0$$
$$\text{at } x = 0: \quad c_1 = 0 \quad \text{and} \quad c_2 = 0 \tag{2.3.1}$$

while the state of the pure solvent feed has its image at the origin O of the hodograph plane of c_1 and c_2, the initial state of the column maps into a single point A as shown in Fig. 2.4.

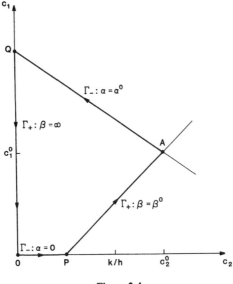

Figure 2.4

According to Rule 1.3, the image of the solution consists of two segments of Γ characteristics connecting points O and A. Furthermore, if we start from the point O of the feed state, we must take the Γ_- characteristic first. Thus the image of the solution is given by the path $O \to P \to A$, where the point P represents the new constant state to appear in the solution.

Now Rule 2.1 ensures that along the path $O \to P$, which is a part of the Γ_-

characteristic, σ_+ decreases and along the path $P \to A$, which is a portion of the Γ_+ characteristic, σ_- decreases. It then follows that the path $O \to A$ gives a Γ_- simple wave and the path $P \to A$ a Γ_+ simple wave. Just as in the case of a single solute, there is no problem with the overlapping of characteristics. Hence the portrait of the solution in the (x, τ)-plane can be pictured as shown in the upper part of Fig. 2.5.

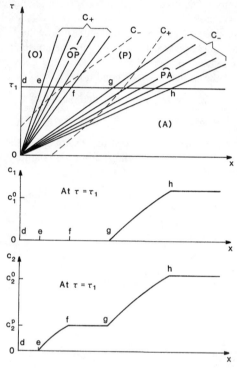

Figure 2.5

It is clear that the characteristic parameters (α, β) corresponding to the feed state O are $\alpha = 0$ and $\beta = \infty$. By using c_1^0 and c_2^0 in Eq. (2.1.32), we obtain the pair of characteristic parameters (α^0, β^0) for the initial state A. From Fig. 2.4 we know that the state P takes $\alpha = 0$ and $\beta = \beta^0$. Along the path $O \to P$ we have $\alpha = 0$ while β decreases from ∞ to β^0 so that from Eqs. (2.1.33) and (2.1.34) we can write

$$c_1 = 0, \qquad c_2 = \frac{k}{\beta + h} \qquad (2.3.2)$$

Thus c_2 increases from 0 to $k/(\beta^0 + h)$ and the constant state P is determined by

$$c_1^p = 0, \qquad c_2^p = \frac{k}{\beta^0 + h} \qquad (2.3.3)$$

Two-Solute Chromatography with the Langmuir Isotherm   Chap. 2

Note that here and in the following the lowercase letters will be used as superscripts to indicate that the values correspond to the state designated by the corresponding capital letters. Along the path $P \rightarrow A$ we have $\beta = \beta^0$ and α decreases from 0 to α^0. Both c_1 and c_2 increase according to the equations

$$c_1 = \frac{-k\alpha\beta^0}{(\alpha + h)(\beta^0 + h)}$$

$$c_2 = \frac{hk}{(\alpha + h)(\beta^0 + h)} \tag{2.3.4}$$

respectively, as α decreases from 0 to α^0. The two concentrations are related by the equation of the Γ_+ characteristic (PA), i.e.,

$$c_1 = c_1^0 + \beta^0(c_2 - c_2^0) \tag{2.3.5}$$

and thus the concentration c_2^p is also determined from this equation as

$$c_2^p = c_2^0 - \frac{c_1^0}{\beta^0} \tag{2.3.6}$$

The argument above is summarized in Table 2.1.

TABLE 2-1

Constant State	Concentrations		Characteristic Parameters		Connecting Wave
	c_1	c_2	α	β	
O	0	0	0	∞	$\}\,\Gamma_-$ simple wave
P	0	$c_2^0 - c_1^0/\beta^0$	0	β^0	$\}\,\Gamma_+$ simple wave
A	c_1^0	c_2^0	α^0	β^0	

Let us then construct the simple waves. From Theorem 1.1 on simple waves we know that to every point on the path $O \rightarrow P$ there corresponds a straight-line C_+ characteristic in the (x, τ)-plane, whose direction is given by[†]

$$\frac{\tau}{x} = \sigma_+ = 1 + \rho a(0)\{b(\beta)\}^2 \tag{2.3.7}$$

$$= 1 + \rho\gamma^{-1}\{b(\beta)\}^2, \qquad \beta^0 \le \beta \le \infty$$

since $a(0) = \gamma^{-1}$. In particular, the uppermost characteristic $C_+(0)$, corresponding to the feed state O (see Fig. 2.5), carries $\beta = \infty$ [i.e., $b(\infty) = 1$] and so has a slope

$$\sigma_+^0 = 1 + \rho\gamma^{-1} = 1 + \nu N_2 K_2 \tag{2.3.8}$$

The Γ_- simple wave region is bounded below by the characteristic $C_+(P)$, which has a slope

[†] Note that A_2 is the only solute present here, so we may apply the development of Chapter 5 of Vol. 1.

$$\sigma^p_+ = 1 + \nu N_2 K_2 \{b(\beta^0)\}^2 \qquad (2.3.9)$$

Between the two bounds ∞ and β^0 we choose any value $\bar{\beta}$ to calculate $\bar{c}_2$ by Eq. (2.3.2) and $\bar{\sigma}_+$ by Eq. (2.3.7). If we draw a straight line of slope $\bar{\sigma}_+$ passing through the origin, this is the corresponding C_+ characteristic that carries the concentration values $c_1 = 0$ and $c_2 = \bar{c}_2$. Repeating this for various values of β will complete the Γ_- simple wave.

Similarly, corresponding to every point on the path $P \rightarrow A$ there is a straight line C_- characteristic in the (x, τ)-plane and its slope is given by

$$\frac{\tau}{x} = \sigma_- = 1 + \rho\{a(\alpha)\}^2 b(\beta^0), \qquad \alpha^0 \le \alpha \le 0 \qquad (2.3.10)$$

Thus the Γ_+ simple wave region is bounded above by the $C_-(P)$ of slope

$$\sigma^p_- = 1 + \rho\gamma^{-2}b(\beta^0) = 1 + \nu N_1 K_1 b(\beta^0) \qquad (2.3.11)$$

and below by the $C_-(A)$ of slope

$$\sigma^a_- = 1 + \rho\{a(\alpha^0)\}^2 b(\beta^0) \qquad (2.3.12)$$

Clearly, we see from Eqs. (2.3.9) and (2.3.11) that $\sigma^p_- < \sigma^p_+$, so the region of the constant state P expands with time. Here again choosing any value $\bar{\alpha}$ between the two bounds 0 and α^0, we obtain $\bar{c}_1$ and $\bar{c}_2$ from Eq. (2.3.4) and $\bar{\sigma}_-$ from Eq. (2.3.10). Then the straight line of slope $\bar{\sigma}_-$ passing through the origin is the corresponding C_- characteristic and bears the concentration values $c_1 = \bar{c}_1$ and $c_2 = \bar{c}_2$. Iterating this procedure for various values of α, we determine the Γ_+ simple wave.

Consequently, the solution is completed in the (x, τ)-plane as shown in the upper part of Fig. 2.5. The sector adjacent to the τ axis is a region of constant state O; a Γ_- simple wave region $\widehat{OP}$ leads to another region of constant state P. Again a simple wave region $\widehat{PA}$ connects the plateau to the constant state A in a region adjacent to the x axis. The dashed lines are drawn to illustrate how the nonstraight C characteristics are curved in simple wave regions. Compare this feature with Rule 2.1.

Since both the simple waves are centered at the origin, we should be able to express the solution in terms of τ/x only. For the Γ_- simple wave we have from Eq. (2.3.7),

$$\beta = h\frac{\gamma\sqrt{\rho^{-1}\gamma(\tau/x - 1)} - 1}{1 - \sqrt{\rho^{-1}\gamma(\tau/x - 1)}}$$

and substituting this in Eq. (2.3.2) and making rearrangement, we obtain

$$c_1 = 0, \qquad K_2 c_2 = -1 + \sqrt{\frac{\nu N_2 K_2}{\tau/x - 1}} \qquad (2.3.13)$$

which is a function of τ/x alone. [Compare this with Eq. (5.3.19) of Vol. 1] For the Γ_+ simple wave Eq. (2.3.10) gives

$$\alpha = -h\frac{1 - \gamma\sqrt{(\tau/x - 1)/\{\rho b(\beta^0)\}}}{1 - \sqrt{(\tau/x - 1)/\{\rho b(\beta^0)\}}}$$

By substituting this in Eq. (2.3.4) and making some manipulations we can show that

$$K_1 c_1 = \frac{\beta^0}{\beta^0 + h}\left(\frac{1}{\gamma}\sqrt{\frac{\rho b(\beta^0)}{\tau/x - 1}} - 1\right)$$

$$K_2 c_2 = \frac{h}{\beta^0 + h}\left(\sqrt{\frac{\rho b(\beta^0)}{\tau/x - 1}} - 1\right)$$

(2.3.14)

and thus the solution is determined in terms of τ/x only.

From the (x, τ)-diagram we can take any cross section by a horizontal line of constant time and get the typical distribution of solute on the column. This is shown in the lower parts of Fig. 2.5. The line of constant time at $\tau = \tau_1$ intersects the important boundaries in points d, e, f, g, and h. Thus at $\tau = \tau_1$ the first part of the bed d–e is swept completely free of solute. The concentration c_1 remains zero in the simple wave region $\overline{OP}$ but c_2 increases to c_1^q over the interval e–f. On the plateau f–g, $c_1 = 0$ and $c_2 = c_2^q$, and after this there is g–h, over which the two concentrations increase together according to relation (2.3.5). Beyond the point h the initial state is unaffected. It is interesting to see that it is the concentration c_1 of the solute we referred to as the less strongly adsorbed of the two, which is most readily removed from the bed. This is true no matter how much larger the initial concentration of this solute might be, so it is sensible to refer to it as the less strongly adsorbed and not merely as the less abundant.

Next we shall consider the saturation of an initially clean column. Here we notice that the initial state and the feed state are interchanged from the previous case so that the problem is to solve the conservation equations (2.1.10) subject to the initial and feed conditions

$$\text{at } \tau = 0: \quad c_1 = 0 \quad \text{and} \quad c_2 = 0$$

$$\text{at } x = 0: \quad c_1 = c_1^0 \quad \text{and} \quad c_2 = c_2^0$$

(2.3.15)

In the hodograph plane of Fig. 2.4, the initial state maps into the origin O, whereas the feed state has its image at the point A. Thus the image of the solution is given by the path $A \to Q \to O$. However, β increases along the path $A \to Q$ and α increases along the path $Q \to O$. This implies that the entropy condition (Rule 2.2) is satisfied at both transitions from A to Q and from Q to O, so that two shocks must be considered in the solution. Since the Γ characteristics are straight, the shocks that we put in must have their states on both sides on the same Γ characteristic. Hence one shock has its end states at A and Q, while the other has its end states at Q and O. In other words, the path $A \to Q$ gives a Γ_- shock and the path $Q \to O$ a Γ_+ shock, and the point Q represents the new constant state to appear between the two shocks. Thus we can picture the portrait of the solution in the (x, τ)-plane as shown in the upper part of Fig. 2.6.

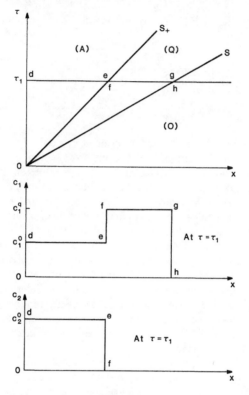

Figure 2.6

The characteristic parameters corresponding to the feed state A are $\alpha = \alpha^0$ and $\beta = \beta^0$. Across the Γ_- shock the parameter β jumps up from β^0 to ∞ while α remains invariant. Thus the constant state Q takes $\alpha = \alpha^0$ and $\beta = \infty$ so that we have, from Eqs. (2.1.33) and (2.1.34),

$$c_1^q = \frac{-k\alpha^0}{\alpha^0 + h}, \qquad c_2^q = 0 \tag{2.3.16}$$

Also, we can apply the equation of the Γ_- characteristic (AQ) to relate the states on both sides of the Γ_- shock, so that

$$c_1^0 - c_1^q = \alpha^0(c_2^0 - c_2^q)$$

or

$$c_1^q = c_1^0 - \alpha^0 c_2^0 \tag{2.3.17}$$

Across the Γ_+ shock the parameter β is fixed at infinity and α jumps up from

Two-Solute Chromatography with the Langmuir Isotherm   Chap. 2

α^0 to 0 so that the state on the right-hand side is the initial state O. A summary of the argument here is given in Table 2.2.

TABLE 2-2

Constant State	Concentrations		Characteristic Parameters		Connecting Wave
	c_1	c_2	α	β	
A	c_1^0	c_2^0	α^0	β^0	$\Big\}$ Γ_- shock
Q	$c_1^0 - \alpha^0 c_2^0$	0	α^0	∞	$\Big\}$ Γ_+ shock
O	0	0	0	∞	

The path of the Γ_+ shock S_+ has a direction given by Eq. (2.2.19) with $\alpha = \alpha^0$, $\beta^l = \beta^0$, and $\beta^r = \infty$, so that we have

$$\bar{\sigma}_+ = 1 + \rho a(\alpha^0) b(\beta^0) \qquad (2.3.18)$$

Similarly, for the path of the Γ_- shock S_- we have $\beta = \infty$, $\alpha^l = a^0$, and $\alpha^r = 0$, so that Eq. (2.2.20) gives

$$\bar{\sigma}_- = 1 + \rho \gamma^{-1} a(\alpha^0) = 1 + \nu N_2 K_2 a(\alpha^0) \qquad (2.3.19)$$

Since $b(\beta^0) > \gamma^{-1}$, we see that $\bar{\sigma}_- < \bar{\sigma}_+$. Hence we draw two straight lines passing through the origin of the (x, τ)-plane, one of slope $\bar{\sigma}_+$ and the other of slope $\bar{\sigma}_-$. These are the two shock paths, S_+ and S_-, having the region of constant state R in between as shown in the upper part of Fig. 2.6.

Taking a section by a line of constant time from the (x, τ)-diagram, we obtain the distribution of the two-solute concentrations. A typical one taken at $\tau = \tau_1$ is shown in the lower parts of Fig. 2.6. Again it makes very good sense to refer to the second solute as the more strongly adsorbed. It is, in fact, so strongly adsorbed that the first solute cannot be fully adsorbed and pushed out in the front. The two shocks at time τ_1 are at $x = \tau_1/\bar{\sigma}_+$ and $x = \tau_1/\bar{\sigma}_-$, respectively. If the length of the column is X, the two shocks reach the exit at $\tau = \bar{\sigma}_- X$ and $\tau = \bar{\sigma}_+ X$, respectively. During the period between $\bar{\sigma}_- X$ and $\bar{\sigma}_+ X$ the outgoing stream contains only the first solute A_1, so partial separation is achieved. Beyond $\tau = \bar{\sigma}_+ X$ the column is completely saturated, so that the state of the outgoing stream is the same as the feed state A.

To get a better understanding we shall consider a numerical example, for which we have

$$\epsilon = 0.4$$

$$N_1 = N_2 = 1.0 \text{ mol/liter}$$

$$K_1 = 5.0 \text{ liters/mol}, \quad K_2 = 7.5 \text{ liters/mol}$$

$$c_1^0 = \frac{2}{30} \text{ mol/liter}, \quad c_2^0 = \frac{1}{10} \text{ mol/liter}$$

It follows that

$$\nu = 1.5, \qquad h = 1, \qquad \gamma = 1.5, \qquad k = \frac{1}{15}, \qquad \rho = 16.875$$

and Eq. (2.1.31) gives

$$\alpha^0 = -\frac{2}{3}, \qquad \beta^0 = 1$$

so that

$$a(\alpha^0) = 0.4, \qquad b(\beta^0) = 0.8$$

For the elution problem we have two simple waves separated by the new constant state P, where

$$c_1^p = 0, \qquad c_2^p = \frac{1}{30} \text{ mol/liter}$$

from Eq. (2.3.3) or (2.3.6). The Γ_- simple wave is given by the family of straight C_+ characteristics of slope

$$\begin{aligned} \sigma_+ &= 1 + \nu N_2 K_2 \{b(\beta)\}^2 \\ &= 1 + 11.25 \left(\frac{\beta + 1}{\beta + 1.5}\right)^2, \qquad 1 \le \beta \le \infty \end{aligned} \tag{2.3.20}$$

and thus the region is bounded above by the $C_+(O)$ of slope $\sigma_+^0 = 12.25$ and below by the $C_+(P)$ of slope $\sigma_+^p = 8.2$ (see Fig. 2.5). For the Γ_+ simple wave we have a family of straight C_- characteristics of slope

$$\begin{aligned} \sigma_- &= 1 + \rho b(\beta^0)\{a(\alpha)\}^2 \\ &= 1 + 13.5 \left(\frac{\alpha + 1}{\alpha + 1.5}\right)^2, \qquad -\frac{2}{3} \le \alpha \le 0 \end{aligned} \tag{2.3.21}$$

and the region is bounded by the $C_-(P)$ of slope $\sigma_-^p = 7.0$ and the $C_-(A)$ of slope $\sigma_-^a = 3.16$. These results are summarized in Table 2.3. To complete the simple wave solutions we take various values of β in the range $1 < \beta < \infty$

TABLE 2-3

Constant State	Concentrations (mol/liter)		Characteristic Parameters[a]		Characteristic Directions[a]	
	c_1	c_2	α	β	σ_+	σ_-
O	0	0	0	∞	12.25	(8.5)
				$\downarrow$	$\downarrow$	
P	0	$\frac{1}{30}$	0	1	8.2	7.0
			$\downarrow$			$\downarrow$
A	$\frac{2}{30}$	$\frac{1}{10}$	$-\frac{2}{3}$	1	(5.32)	3.16

[a] Arrows indicate the directions of change as we proceed in the x direction.

and locate the straight C_+ characteristics by Eq. (2.3.20); then on each of these lines we assign the concentration values given by Eq. (2.3.2), i.e.,

$$c_1 = 0, \qquad c_2 = \frac{1}{15(\beta + 1)} \qquad (2.3.22)$$

This establishes the Γ_- simple wave. Similarly, for the Γ_+ simple wave we choose various values of α in the range $-\frac{2}{3} < \alpha < 0$ and locate the straight C_- characteristics by Eq. (2.3.21). Then the concentration values borne on each of these characteristics are determined by Eq.(2.3.4), which now reads

$$c_1 = \frac{-\alpha}{30(\alpha + 1)}, \qquad c_2 = \frac{1}{30(\alpha + 1)} \qquad (2.3.23)$$

By using Eqs. (2.3.13) and (2.3.14), we can express the solution as follows:

$$
\begin{aligned}
&c_1 = 0, \quad c_2 = 0 && \text{for } 12.25 < \frac{\tau}{x} \\[2mm]
&c_1 = 0, \quad K_2 c_2 = -1 + \frac{3.354}{\sqrt{\tau/x - 1}} && \text{for } 8.2 < \frac{\tau}{x} < 12.25 \\[2mm]
&c_1 = 0, \quad c_2 = \frac{1}{30} && \text{for } 7.0 < \frac{\tau}{x} < 8.2 \\[2mm]
&K_1 c_1 = -\frac{1}{2} + \frac{1.225}{\sqrt{\tau/x - 1}}, \quad K_2 c_2 = -\frac{1}{2} + \frac{1.837}{\sqrt{\tau/x - 1}} && \\[1mm]
& && \text{for } 3.16 < \frac{\tau}{x} < 7.0 \\[2mm]
&c_1 = \frac{2}{30}, \quad c_2 = \frac{1}{10} && \text{for } 0 < \frac{\tau}{x} < 3.16
\end{aligned}
$$

$$(2.3.24)$$

where the unit of concentration is moles per liter.

The solution to the saturation problem consists of two shocks with the new constant state Q in between (see Fig. 2.6), where

$$c_1^q = \frac{2}{15} \text{ mol/liter}, \qquad c_2^q = 0$$

from Eq. (2.3.16) or (2.3.17). The Γ_- shock propagates in the direction

$$\tilde{\sigma}_+ = 1 + \rho a(\alpha^0) b(\beta^0) = 6.4$$

while the Γ_+ shock propagates in the direction

$$\tilde{\sigma}_- = 1 + \rho \gamma^{-1} a(\alpha^0) = 5.5$$

Hence we can write the solution as

$$c_1 = \frac{2}{30}, \qquad c_2 = \frac{1}{10} \qquad \text{for } 6.4 < \frac{\tau}{x}$$

$$c_1 = \frac{2}{15}, \qquad c_2 = 0 \qquad \text{for } 5.5 < \frac{\tau}{x} < 6.4 \qquad (2.3.25)$$

$$c_1 = 0, \qquad c_2 = 0 \qquad \text{for } 0 < \frac{\tau}{x} < 5.5$$

These are summarized in Table 2.4, where the arrow indicates the direction of jump when we proceed in the x direction.

TABLE 2-4

Constant State	Concentrations (mol/liter)		Characteristic Parameters[a]		Slope of the Shock Path
	c_1	c_2	α	β	
A	$\frac{2}{30}$	$\frac{1}{10}$	$-\frac{2}{3}$	1	
				$\downarrow$	$\bar{\sigma}_+ = 6.4$
Q	$\frac{2}{15}$	0	$-\frac{2}{3}$	∞	
			$\downarrow$		$\bar{\sigma}_- = 5.5$
O	0	0	0	∞	

[a] Arrows indicate the direction of jump as we proceed in the x direction.

We shall next examine a more general situation. Suppose that the voidage and the Langmuir isotherm parameters are the same as in the preceding example, but the initial and feed conditions are prescribed as

$$\text{at } \tau = 0: \quad c_1 = 0.10 \text{ mol/liter} \quad \text{and} \quad c_2 = 0.08 \text{ mol/liter}$$
$$\text{at } x = 0: \quad c_1 = 0.02 \text{ mol/liter} \quad \text{and} \quad c_2 = 0.06 \text{ mol/liter} \qquad (2.3.26)$$

Although this example may not have much physical meaning, it will illustrate the structure of the solution in a more general context. For the feed state F (0.02, 0.06), Eq. (2.1.31) with $h = 1$ and $k = \frac{1}{15}$ gives the characteristic parameters

$$\alpha^f = -0.396, \qquad \beta^f = 0.841$$

and similarly, for the initial state I (0.10, 0.08) we obtain

$$\alpha^0 = -0.701, \qquad \beta^0 = 1.784$$

The image of the solution in the hodograph plane is given by the path $F \rightarrow P \rightarrow I$ as shown in Fig. 2.7. Along the path $F \rightarrow P$, which is a portion of the Γ_- characteristic with $\alpha = -0.396$, the parameter β increases (or c_2 decreases), so by the entropy condition (see Rule 2.2) we have a Γ_- shock corresponding to this path. The path $P \rightarrow I$ is a portion of the Γ_+ characteristic with $\beta = 1.784$, and along this path α decreases (or c_2 increases) so that it gives a Γ_+ simple wave. Between the two waves we have a new constant state P for which

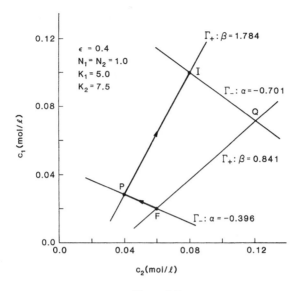

Figure 2.7

$$\alpha = \alpha^f = -0.396, \qquad \beta = \beta^0 = 1.784$$

and thus from Eqs. (2.1.33) and (2.1.34) we obtain

$$c_1^p = 0.028 \text{ mol/liter}, \qquad c_2^p = 0.040 \text{ mol/liter}$$

These results are tabulated in Table 2.5.

TABLE 2-5

Constant State	Concentrations (mol/liter)		Characteristic Parameters[a]		Connecting Wave
	c_1	c_2	α	β	
F	0.02	0.06	−0.396	0.841	$\Big\}\,\Gamma_-$ shock
P	0.028	0.040	−0.396	↓ 1.784	
I	0.10	0.08	−0.701	1.784	$\Big\}\,\Gamma_-$ simple wave

[a] Arrows indicate the direction of change as we proceed in the x direction.

The Γ_- shock propagates in the direction

$$\tilde{\sigma}_+ = 1 + \rho a(\alpha^f)b(\beta^f)b(\beta^0)$$

$$= 1 + (16.875)\left(\frac{0.604}{1.104}\right)\left(\frac{1.841}{2.341}\right)\left(\frac{2.784}{3.284}\right)$$

$$= 7.155$$

The region of the Γ_+ simple wave is covered by a family of straight C_- characteristics, each of which has a slope given by

$$\sigma_- = 1 + \rho b(\beta^0)\{a(\alpha)\}^2 = 1 + (16.875)\left(\frac{2.784}{3.284}\right)\left(\frac{\alpha + 1}{\alpha + 1.5}\right)^2$$

$$= 1 + 14.306\left(\frac{\alpha + 1}{\alpha + 1.5}\right)^2 \qquad \text{for } -0.396 \geq \alpha \geq -0.701 \qquad (2.3.27)$$

Thus the region is bounded above by the $C_-(P)$ of slope 5.282 and below by the $C_-(I)$ of slope 3.003. The relation above can be inverted to give

$$\alpha = -\frac{1.5 - 3.782\sqrt{\tau/x - 1}}{1 - 3.782\sqrt{\tau/x - 1}} \qquad (2.3.28)$$

since $\sigma_- = \tau/x$. The concentrations along each characteristic are determined by Eqs. (2.1.33) and (2.1.34), i.e.,

$$c_1 = \frac{-k\alpha\beta^0}{(\alpha + h)(\beta^0 + h)} = \frac{-(1/15)(1.784)\alpha}{(2.784)(\alpha + 1)} = \frac{-\alpha}{(23.408)(\alpha + 1)}$$

$$c_2 = \frac{hk}{(\alpha + h)(\beta^0 + h)} = \lambda\frac{1/15}{(2.784)(\alpha + 1)} = \frac{1}{(41.760)(\alpha + 1)} \qquad (2.3.29)$$

Substituting α from Eq. (2.3.28), we can express the Γ_+ simple wave solution as a function of τ/x only:

$$c_1 = \frac{0.323}{\sqrt{\tau/x - 1}} - 0.128, \qquad c_2 = \frac{0.181}{\sqrt{\tau/x - 1}} - 0.048 \qquad (2.3.30)$$

Hence the solution can be written as

$$c_1 = 0.02, \qquad c_2 = 0.06 \qquad\qquad \text{for } 7.155 < \frac{\tau}{x}$$

$$c_1 = 0.028, \qquad c_2 = 0.04 \qquad\qquad \text{for } 5.282 < \frac{\tau}{x} < 7.155$$

$$c_1 = \frac{0.323}{\sqrt{\tau/x - 1}} - 0.128, \qquad c_2 = \frac{0.181}{\sqrt{\tau/x - 1}} - 0.048 \qquad (2.3.31)$$

$$\text{for } 3.003 < \frac{\tau}{x} < 5.282$$

$$c_1 = 0.10, \qquad c_2 = 0.08 \qquad\qquad \text{for } 0 < \frac{\tau}{x} < 3.003$$

where the concentration values are in mol/liter.

For the purpose of illustration we shall construct the solution in the (x, τ)-plane and also the concentration profiles. To establish the Γ_+ simple wave we take several values of α in the interval $-0.396 \leq \alpha \leq -0.701$ and calculate σ_- by Eq. (2.3.27) and the concentrations by Eq. (2.3.29). These values are presented in Table 2.6. Note that the concentration values can also be determined by Eq. (2.3.30) with $\sigma_- = \tau/x$. The (x, τ)-diagram consists of a straight-line shock path S_+ of slope 7.155 for the Γ_- shock and a family of straight C_- characteristics for the Γ_+ simple wave as shown in the upper part of Fig. 2.8, where the numbers marked on the C_- characteristics correspond to those of Table 2.6. Each of the C_- characteristics bears constant concentrations whose values are listed in the corresponding row of Table 2.6. Hence it is a straightforward matter to construct the concentration profiles as illustrated at two different times in the lower part of Fig. 2.8.

TABLE 2-6

Number	α	σ_-	$c_1(mol/liter)$	$c_2(mol/liter)$
1	−0.396	5.282	0.028	0.040
2	−0.45	4.925	0.035	0.044
3	−0.50	4.576	0.043	0.048
4	−0.55	4.210	0.052	0.053
5	−0.60	3.826	0.064	0.060
6	−0.65	3.426	0.079	0.068
7	−0.701	3.003	0.100	0.080

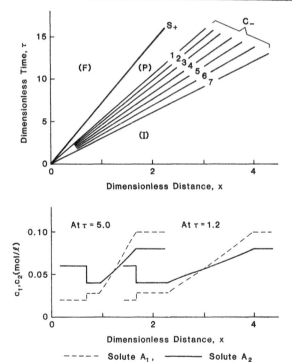

Figure 2.8

Exercises

2.3.1. Show that the solution of the saturation problem [i.e., the problem with the data given by Eq. (2.3.15)] accounts for the total amounts of the two solutes fed to the column.

2.3.2. By using Eqs. (2.3.20)–(2.3.23), construct the solution of the elution problem in the (x, τ)-plane and thereby draw the concentration profiles at two successive times. Confirm that your solution satisfies Eq. (2.3.24).

2.3.3. Consider the problem for which the initial and feed conditions given by Eq. (2.3.26) are interchanged and determine the solution as much in detail as you can.

2.4 The Formation of Shocks

Since the solutions are represented by nonlinear waves, the concentration profiles become distorted in the course of time. In general, it is necessary to examine the variations of both σ_+ and σ_- in the x direction in order to tell whether the concentration profiles would tend to steepen or to become diffuse [see Eqs. (1.6.1) and (1.6.2)]. Here we shall be concerned with characteristic initial value problems and thus the solutions are determined by simple waves. In this case one of the characteristic parameters is invariant in the simple wave region so that the slope of the straight C characteristic is expressed as a function of one parameter. More specifically, from Eq. (2.2.14) we have

$$\frac{\partial \sigma_+}{\partial \beta} < 0 \tag{2.4.1}$$

for a Γ_- simple wave and

$$\frac{\partial \sigma_-}{\partial \alpha} > 0 \tag{2.4.2}$$

for a Γ_+ simple wave. Hence we can determine the nature of the wave form by examining the variation of β or α in the x direction.

Furthermore, we have from Eqs. (2.1.35) and (2.1.36)

$$\frac{\partial c_1}{\partial \alpha} < 0, \qquad \frac{\partial c_1}{\partial \beta} > 0 \tag{2.4.3}$$

$$\frac{\partial c_2}{\partial \alpha} < 0, \qquad \frac{\partial c_2}{\partial \beta} < 0 \tag{2.4.4}$$

and notice that the variation in c_2 is consistent in sign with the variation in α as well as that in β. This implies that we need only to check the variation of c_2 to see whether the wave form is expansive or compressive. The argument above is embodied in the following rule.

RULE 2.3:

1. A Γ_- (or Γ_+) simple wave is expansive if β (or α) decreases in the x direction, whereas it is compressive if β (or α) increases in the x direction.

2. A simple wave is expansive if c_2 increases in the x-direction, whereas it is compressive if c_2 decreases in the x direction.

Application of this rule is explained in detail in Table 2.7.

TABLE 2-7

	Characteristic Parameters[a]		Concentrations[a]		Characteristics		Concentration Profiles
	α	β	c_1	c_2	C_+	C_-	
Constant state	$\rightarrow$	$\rightarrow$	$\rightarrow$	$\rightarrow$	Straight and parallel	Straight and parallel	—
Γ_- simple wave	$\rightarrow$	$\searrow$	$\searrow$	$\nearrow$	Straight and diffuse	—	Expansive
	$\rightarrow$	$\nearrow$	$\nearrow$	$\searrow$	Straight and steepening	—	Compressive
Γ_+ simple wave	$\searrow$	$\rightarrow$	$\nearrow$	$\nearrow$	—	Straight and diffuse	Expansive
	$\nearrow$	$\rightarrow$	$\searrow$	$\searrow$	—	Straight and steepening	Compressive

[a] Arrows indicate the directions of change as we proceed in the x direction.

Let us now confine attention to compression waves and investigate when and where shocks are to be formed. For this we shall consider the characteristic initial data the image of which falls on a Γ_- characteristic in the hodograph plane. From Eq. (1.5.1) we have

$$\text{at } \tau = 0: \quad c_2 = c_2^0(\xi) \qquad \text{for } 0 \le \xi \le 1$$
$$c_2 = c_2^h = c_2^0(1) \qquad \text{for } \xi > 1$$
$$\text{at } x = 0: \quad c_2 = c_2^i = c_2^0(0) \qquad \text{for } \eta > 0 \qquad (2.4.5)$$
$$c_1 = c_1^i \qquad \text{for } \eta > 0$$

and

$$c_1 = c_1(c_2; c_1^i, c_2^i) \qquad \text{for } \xi, \eta \ge 0 \qquad (2.4.6)$$

where ξ and η are the parameters running along the x and τ axes, respectively, and Eq. (2.4.6) represents the Γ_- characteristic. The initial condition is also shown in Fig. 2.9. Clearly, there develops a region of Γ_- simple wave adjacent to the portion [0, 1] of the x axis.

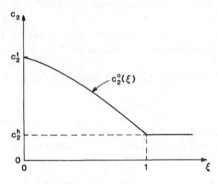

Figure 2.9

According to the Rule 2.3, this Γ_- simple wave is compressive if c_2 decreases in the x direction, i.e.,

$$\frac{dc_2^0}{d\xi} < 0 \qquad (2.4.7)$$

Throughout the region of this compression wave the parameter α remains constant. We shall denote the fixed value by α^0. From Eq. (2.1.34) we have

$$c_2^0(\xi) = \frac{hk}{(\alpha^0 + h)(\beta + h)} \qquad (2.4.8)$$

which defines a relationship between ξ and β. Differentiating this with respect to β gives

$$\frac{dc_2^0}{d\beta} = \frac{dc_2^0}{d\xi}\frac{d\xi}{d\beta} = -\frac{hk}{(\alpha^0 + h)(\beta + h)^2}$$

or

$$\frac{d\xi}{d\beta} = \frac{hk}{(\alpha^0 + h)(\beta + h)^2}\left(-\frac{dc_2^0}{d\xi}\right)^{-1} \qquad (2.4.9)$$

On the other hand, the Γ_- simple wave region is covered by a family of straight C_+ characteristics having slope

$$\sigma_+ = 1 + \rho a(\alpha^0)b^2 \qquad (2.4.10)$$

These characteristics possess an envelope as was shown in Fig. 1.18 and the equation of the envelope is given in parametric form by Eq. (1.6.12). Hence, substituting Eqs. (2.4.9), (2.4.10), and

$$\frac{d\sigma_+}{d\beta} = 2\rho a(\alpha^0)b\frac{db}{d\beta} = 2\rho h(\gamma - 1)a(\alpha^0)\frac{b}{(\beta + h\gamma)^2} \qquad (2.4.11)$$

into Eq. (1.6.12), we obtain

$$x^E(\beta) = \xi(\beta) + \frac{1 + \rho a(\alpha^0)b^2}{2\rho K_1 \gamma(\alpha^0 + h)a(\alpha^0)b^3}\left(-\frac{dc_2^0}{d\xi}\right)^{-1}$$
$$\tau^E(\beta) = \frac{\{1 + \rho a(\alpha^0)b^2\}^2}{2\rho K_1 \gamma(\alpha^0 + h)a(\alpha^0)b^3}\left(-\frac{dc_2^0}{d\xi}\right)^{-1} \qquad (2.4.12)$$

in which α^0 is fixed and $\xi(\beta)$ denotes the x intercept of the C_+ characteristic.

A shock is to be formed at the cusp point of the envelope for which we have $x = \min x^E$ and $\tau = \min \tau^E$. Since the compression wave is based on a finite interval of the x axis, the cusp point can be positioned on one of the characteristics emanating from end points of the interval (see Fig. 6.3 of Vol. 1).

As an illustrative example, we shall consider a linear distribution of the initial concentrations such that

$$c_2^0(\xi) = c_2^l - (c_2^l - c_2^h)\xi, \qquad c_2^l > c_2^h \qquad (2.4.13)$$

and

$$c_1^0 = \alpha^0 c_2^0 - \frac{k\alpha^0}{\alpha^0 + 1} \qquad (2.4.14)$$

Thus $-dc_2^0/d\xi = c_2^l - c_2^h = $ constant. Since τ^E is expressed explicitly in terms of b and $db/d\beta > 0$, we shall differentiate τ^E with respect to b to locate the cusp point. This gives

$$\frac{d\tau^E}{db} = \frac{3\{1 + \rho a(\alpha^0)b^2\}}{2\rho k_1 \gamma(c_2^l - c_2^h)(\alpha^0 + h)a(\alpha^0)b^2}\left\{\frac{\rho a(\alpha^0)}{3} - \frac{1}{b^2}\right\} \qquad (2.4.15)$$

The sign of this derivative is determined by that of the factor

$$\frac{\rho a(\alpha^0)}{3} - \frac{1}{b^2} \qquad \text{for } b^l \leq b \leq b^h$$

where $b^l = b(\beta^l)$ and $b^h = b(\beta^h)$ correspond to the concentration values c_2^l and c_2^h, respectively. By using Eq. (2.4.8) we can show that

$$\frac{1}{b} = \frac{\beta + h\gamma}{\beta + h} = 1 + \frac{\gamma - 1}{k}(\alpha^0 + h)c_2^0(\xi)$$

or

$$c_2^0(\xi) = \frac{-1 + 1/b}{K_1 \gamma(\alpha^0 + h)} \qquad (2.4.16)$$

Since $dc_2^0/db < 0$ and $c_2^l > c_2^h$, we have three different cases concerning the sign of $d\tau^E/db$.

1. If

$$b' > \left\{ \frac{\rho a(\alpha^0)}{3} \right\}^{-1/2}$$

or

$$c_2^t < \frac{-1 + \sqrt{\rho a(\alpha^0)/3}}{K_1 \gamma (\alpha^0 + h)} \qquad (2.4.17)$$

then $d\tau^E/db > 0$, so τ^E obtains a minimum for $b = b'$ or $c_2 = c_2^t$. This implies that a shock is formed on the C_+ characteristic emanating from the origin ($\xi = 0$) and bearing the concentration c_2^t.

2. If

$$b^h < \left\{ \frac{\rho a(\alpha^0)}{3} \right\}^{-1/2}$$

or

$$c_2^h > \frac{-1 + \sqrt{\rho a(\alpha^0)/3}}{K_1 \gamma (\alpha^0 + h)} \qquad (2.4.18)$$

we have $d\tau^E/db < 0$ so that a shock is to appear first on the C_+ characteristic emanating from the point $\xi = 1$ and carrying the concentration c_2^h.

3. Finally, if

$$b' < \left\{ \frac{\rho a(\alpha^0)}{3} \right\}^{-1/2} < b^h$$

or

$$c_2^t > \frac{-1 + \sqrt{\rho a(\alpha^0)/3}}{K_1 \gamma (\alpha^0 + h)} > c_2^h \qquad (2.4.19)$$

then τ^E becomes minimum on the C_+ characteristic bearing the concentration c_2^* given by

$$c_2^* = \frac{-1 + \sqrt{\rho a(\alpha^0)/3}}{K_1 \gamma (\alpha^0 + h)} \qquad (2.4.20)$$

and the x intercept of this characteristic is

$$\xi = \xi^* = \frac{1}{c_2^t - c_2^h} \left\{ c_2^t - \frac{-1 + \sqrt{\rho a(\alpha^0)/3}}{K_1 \gamma (\alpha^0 + h)} \right\} \qquad (2.4.21)$$

Hence the point of shock inception lies on this characteristic.

To demonstrate the use of the base curve method discussed in Sec. 1.6 (see also Sec. 6.1 of Vol. 1), let us consider a numerical example for which we have

$$N_1 = N_2 = 1.0 \text{ mol/liter}$$

$$K_1 = 5.0 \text{ mol/liter}, \qquad K_2 = 7.5 \text{ mol/liter}$$

so that $h = 1$, $\gamma = 1.5$, and $k = 1/15$. The initial concentrations are specified by Eqs. (2.4.13) and (2.4.14) with

$$c_2^i = 0.15 \text{ mol/liter}, \qquad c_2^h = 0.05 \text{ mol/liter}, \qquad c_1^i = 0.01 \text{ mol/liter}$$

For the state (c_1^i, c_2^i) we find from Eq. (2.1.31) that

$$\alpha^i = \alpha^0 = -\frac{3}{5}, \qquad \beta^i = \frac{1}{9}$$

and thus Eq. (2.4.14) gives

$$c_1 = -0.6c_2 + 0.1$$

This relation holds throughout the Γ_- simple wave region, so we have $c_1^h = 0.07$ mol/liter. The initial data are shown in Fig. 2.10.

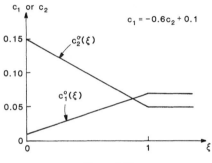

c_1 or c_2

$c_1 = -0.6c_2 + 0.1$

0.15

$c_2^0(\xi)$

0.10

$c_1^0(\xi)$

0.05

0

0   1 ξ

Figure 2.10

For the state (c_1^h, c_2^h) we obtain again from Eq. (2.1.31)

$$\alpha^h = \alpha^0 = -\frac{3}{5}, \qquad \beta^h = \frac{7}{3}$$

Also, we determine

$$a(\alpha^0) = \frac{4}{9}, \qquad b^i = b(\beta^i) = \frac{20}{29}, \qquad b^h = b(\beta^h) = \frac{20}{23}$$

and

$$\rho = \nu N_2 K_2 \gamma = \frac{45}{4}\nu$$

Suppose now that we have a column with $\epsilon = \frac{1}{3}$, so that $\rho a(\alpha^0) = 10$. Then

$$b^i = \frac{20}{29} > \sqrt{\frac{3}{10}} = \left\{ \frac{\rho a(\alpha^0)}{3} \right\}^{-1/2}$$

and we have case 1. If $\epsilon = \frac{2}{3}$, we see that $\rho a(\alpha^0) = 5.2$ and

$$b^h = \frac{20}{23} < \sqrt{\frac{6}{5}} = \left\{ \frac{\rho a(\alpha^0)}{3} \right\}^{-1/2}$$

so that we have case 2. If we take $\epsilon = \frac{1}{2}$, $\rho a(\alpha^0) = 5$, so $\{\rho a(\alpha^0)/3\}^{-1/2} = 0.795$ falls between b^t and b^h. Hence we have case 3 and the C_+ characteristic on which a shock is to form carries the concentrations, $c_2^* = 0.097$ mol/liter and $c_1^* = 0.043$ mol/liter, and has its x intercept at $\xi^* = 0.53$.

Now we shall show that the result above can also be obtained by applying the base curve method. From Eq. (1.6.18) we have

$$(\sigma_+)_B = \frac{\sigma_+^t}{1 - (1 - \sigma_+^t/\sigma_+^h)\xi}, \qquad 0 \le \xi \le 1 \qquad (2.4.22)$$

where

$$\sigma_+^t = 1 + \rho a(\alpha^0)(b^t)^2$$
$$\sigma_+^h = 1 + \rho a(\alpha^0)(b^h)^2 \qquad (2.4.23)$$

For three different values of ϵ, i.e., $\frac{1}{3}$, $\frac{1}{2}$, and $\frac{2}{3}$, $(\sigma_+)_B$ is calculated as a function of ξ and plotted as marked by the dashed lines in Fig. 2.11. These are the

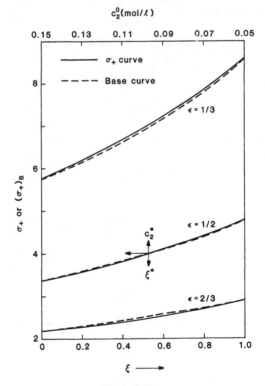

Figure 2.11

Two-Solute Chromatography with the Langmuir Isotherm   Chap. 2

base curves for the three cases. The actual value of the characteristic direction is given by

$$\sigma_+ = 1 + \rho a(a^0)b^2 \qquad \text{for } b' \le b \le b^h \qquad (2.4.24)$$

where b is related to $c_2^0(\xi)$ by Eq. (2.4.16) and again to ξ by Eq. (2.4.13). For the present example we have

$$b = \frac{1}{1.45 - 0.3\xi}$$

Hence we can compute σ_+ as a function of ξ for the given values of ϵ. These are plotted to give the solid lines of Fig. 2.11. When $\epsilon = \frac{1}{3}$, the σ_+ curve lies entirely above the base curve, so a shock will first form on the characteristic emanating from the point $\xi = 0$. In the case of $\epsilon = \frac{1}{2}$, the two curves intersect each other at $\xi = 0.53$ or $c_2 = 0.97$ and for low values of ξ the base curve lies above while it comes below the σ_+ curve for large values of ξ. This implies that a shock is to appear first on the characteristic emanating from the point $\xi = 0.53$ and carrying the concentration $c_2 = 0.97$. For $\epsilon = \frac{2}{3}$ the σ_+ curve lies above the base curve for all values of ξ so that the point of shock inception lies on the characteristic emanating from the point $\xi = 1$. These are exactly what we have observed from the previous analysis, and hence we have demonstrated the validity of the base curve method.

Exercises

2.4.1. Repeat the analysis of this section for a Γ_+ simple wave.

2.4.2. Consider the characteristic feed data

$$\text{at } x = 0: \quad c_2 = c_2^f(\eta) \qquad \text{for } 0 \le \eta \le 1$$
$$c_2 = c_2^f = c_2^f(1) \qquad \text{for } \eta > 1$$
$$\text{at } \tau = 0: \quad c_2 = c_2^h = c_2^f(0) \qquad \text{for } \xi > 0$$
$$c_1 = c_1^h \qquad \text{for } \xi > 0$$

and

$$c_1 = c_1(c_2; c_1^h, c_2^h)$$

where the last equations represent a Γ_- characteristic. Assuming that $dc_2^f/d\eta > 0$, determine the equation of the envelope of the straight C_+ characteristics. If $c_2^f(\eta)$ is linear, show that a shock is always formed on the characteristic emanating from the origin of the (x, τ)-plane.

2.5 Fundamentals of Wave Interaction

We note that the fundamental differential equation (2.1.12) is expressed in terms of the dependent variables, c_1 and c_2, only and does not explicitly depend on the independent variables, x and τ. Thus the characteristic parameters are determined in terms of c_1 and c_2 only. This implies that the image of the solution in the hodograph plane is solely determined by the concentration values of the initial data and remains invariant under the translation of the position of the initial data. Hence the theory of simple waves and shocks can be applied equally well no matter where the characteristic initial data and/or the initial discontinuities may be prescribed.

The development in previous sections then finds its natural extension to a class of problems associated with piecewise constant initial and/or feed data. The solution as a whole can be established by constructing the centered wave solutions separately from each point of discontinuity. This is the characteristic feature of hyperbolic systems of equations. After a finite period of time, however, any two wave solutions centered at two different, but adjacent, points of discontinuity will meet each other so that an overlapped region appears in which the solution is influenced by two different sets of data at the same time. The two confronting waves will be involved in an *interaction* between themselves, so the solution is to be determined by the analysis of such interactions. If we have characteristic initial data in the neighborhood of initial discontinuities, the overall features will be the same because the solution still consists of simple waves, shocks, and constant states.

In this section we identify the pairs of interacting waves and discuss the basic principles of interaction. The analysis of interaction will be the subject of the Secs. 2.6 and 2.7, whereas application will be treated in Secs. 2.8 and 2.9.

If two simple waves of the same family are side by side, the boundary characteristics facing each other are parallel, so the two waves are noninteractive. On the other hand, the shock condition [Eq.(2.2.30) or (2.2.32)] clearly shows that not only two shocks but also a shock and a simple wave will interact with each other if the two are of the same family. The shock condition also indicates that a Γ_+ shock (or a Γ_+ simple wave) will interact with a Γ_- simple wave (or a Γ_- shock) traveling ahead. Finally, from Eqs. (2.2.13) and (2.2.23) we notice that a Γ_+ simple wave or a Γ_+ shock will interact with a Γ_- simple wave or a Γ_- shock, respectively, traveling ahead. The observation above can be summarized as in the following rule.

RULE 2.4 (Pairs of Interacting Waves):

1. Two shocks of the same family necessarily interact with each other.
2. A simple wave and a shock of the same family necessarily interact with each other.
3. A Γ_+ wave (either a simple wave or a shock) necessarily interacts with a Γ_- wave traveling ahead.

Excluded from this rule is the case of two simple waves of the same family and the case of a Γ_- wave traveling behind a Γ_+ wave. In these cases the two waves do not meet each other.

Let us now consider the images of the interacting wave pairs in the hodograph plane. When the waves are of the same family, their images fall on a single Γ characteristic, either nonoverlapping (in case of two shocks) or overlapping (in case of a simple wave and a shock). These are shown in Fig. 2.12(a) and (b), respectively. In the first case the shock $L \rightarrow P$ meets the shock $P \rightarrow R$ from behind to give a new Riemann problem with the constant states L on the left and the state R on the right. It is immediately apparent that the solution to this new problem will be given by a single shock $L \rightarrow P$ since both states fall on the same Γ characteristic. Hence two shocks of the same family are simply superposed upon interaction. This phenomenon is called the *confluence* (or *union*) of two shocks.

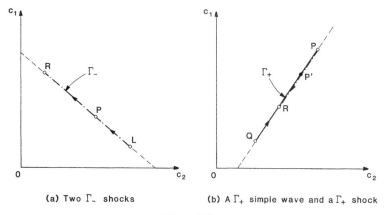

(a) Two Γ_- shocks   (b) A Γ_+ simple wave and a Γ_+ shock

Figure 2.12

In the second case [see Fig. 2.12(b)], suppose that the simple wave $Q \rightarrow P$ meets the shock $P \rightarrow R$ from behind. As soon as the interaction starts, the constant state P disappears and the state on the left-hand side of the shock moves toward the point R along the Γ characteristic. Thus at a particular instant during the interaction the images of the simple wave and the shock are given by $Q \rightarrow P'$ and $P' \rightarrow R$, respectively. The portion $P' \rightarrow P$ of the original simple wave and the portion $P \rightarrow P'$ of the original shock are canceled due to the interaction. This phenomenon is known as the *cancellation* of a simple wave and a shock or *decay* of a shock. If the state R falls between Q and P as depicted in Fig. 2.12(b), the state P' approaches the point P only asymptotically, for in the limit as $P' \rightarrow R$ the shock speed approaches the characteristic speed of the state R.[†] Hence the shock is not completely canceled in a finite time and the interaction goes on indefinitely. If the state Q fell between P and R, the interaction would be over when the state P' reaches the point Q. At this stage the simple wave would be completely canceled to give a weaker shock $Q \rightarrow R$.

[†] This feature will become clearer as we analyze the interaction in Sec. 2.6.

Next we shall consider the interaction between waves of different families. The images of the two waves fall on a Γ_+ and a Γ_- characteristic, respectively, and the two meet together at the point of common state as shown in Fig. 2.13. In case of two shocks the Γ_+ shock $A \rightarrow B$ will meet the Γ_- shock $B \rightarrow D$ from behind, so we have a new Riemann problem with the state A on the left and the state D on the right. The solution to this problem is given by the Γ_- shock $A \rightarrow E$ and the Γ_+ shock $E \rightarrow D$. This implies that the Γ_+ shock $A \rightarrow D$ is transmitted through the Γ_- shock to give the new Γ_+ shock $E \rightarrow D$, while the Γ_- shock $B \rightarrow D$ is transmitted to give the new Γ_- shock $A \rightarrow E$. This phenomenon will be referred to as the *transmission* between waves. In this case the interaction is instantaneous. If a Γ_+ simple wave follows a Γ_- shock, the image of the solution will be given by $B \rightarrow A$ and $A \rightarrow E$ as we proceed in the x direction. When the two waves meet, the constant state A disappears and the Γ_- shock $A \rightarrow E$ recedes gradually toward $B \rightarrow D$ with the states on both sides tracing AB and ED, respectively. Hence when the interaction is over, we see a new shock $B \rightarrow D$ trailing the transmitted simple wave $D \rightarrow E$. Between the two waves we have the new constant state D. Finally, suppose that the Γ_+ simple wave $D \rightarrow E$ follows the Γ_- simple wave $E \rightarrow A$. The two are bound to meet and interact. When the interaction is over, we have the constant state D to be connected to the constant state A in the x direction. Clearly, the proper connection is achieved by the Γ_- simple wave $D \rightarrow B$, the constant state B, and the Γ_+ simple wave $B \rightarrow A$. This implies that both simple waves are transmitted through each other while interacting and give two transmitted simple waves $D \rightarrow B$ and $B \rightarrow A$ separated by the new constant state B. The interaction takes place in a finite region of the (x, τ)-plane in which the solution is not determined by simple waves because its image would occupy the region $ABDE$ of the hodograph plane.

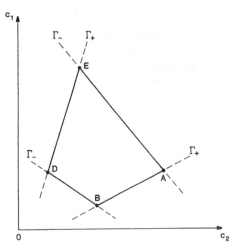

Figure 2.13

Two-Solute Chromatography with the Langmuir Isotherm   Chap. 2

The argument above concerning the nature of wave interactions leads us to the following rule.

RULE 2.5 (Basic Principles of Interaction):

1. *Confluence of shocks.* Two shocks of the same family are instantly superposed when interacting.
2. *Cancellation of simple waves and shocks.* A simple wave and a shock of the same family are continuously canceled while interacting.
3. *Transmission between waves.* Two waves of different families are transmitted through each other while interacting.

Confluence and cancellation have their counterparts in the case of single equations but transmission is a new phenomenon that we encounter here.

During the process of interaction not only the shock path but also the characteristics involved will be curved gradually or refracted instantly because the state of concentrations no longer remains constant. This may be regarded as the *accelerated motion* of a shock or a disturbance. When a Γ_+ simple wave interacts with a Γ_- wave from behind, the parameter $a(\alpha)$ remains invariant along each C_- characteristic and thus we can write

$$\frac{d\sigma_-}{db} = \rho a^2 \tag{2.5.1}$$

Similarly, if a Γ_- simple wave interacts with a Γ_+ wave, the parameter $b(\beta)$ is constant along each C_+ characteristic so that

$$\frac{d\sigma_+}{da} = \rho b^2 \tag{2.5.2}$$

But from Eqs. (2.1.36) and (2.2.11) we notice that

$$\frac{\partial a}{\partial c_2} < 0, \qquad \frac{\partial b}{\partial c_2} < 0 \tag{2.5.3}$$

An immediate consequence is that the slope of a C_+ or C_- characteristic decreases as c_2 increases in the x-direction, so a disturbance is accelerated if c_2 increases as it propagates. When a shock interacts with another wave of the same family or of the opposite family, we can write from Eqs. (2.2.21) and (2.2.22) that

$$\frac{\partial \bar{\sigma}_+}{\partial a} > 0, \qquad \frac{\partial \bar{\sigma}_+}{\partial b^l} > 0, \qquad \frac{\partial \bar{\sigma}_+}{\partial b^r} > 0 \tag{2.5.4}$$

and

$$\frac{\partial \bar{\sigma}_-}{\partial b} > 0, \qquad \frac{\partial \bar{\sigma}_-}{\partial a^l} > 0, \qquad \frac{\partial \bar{\sigma}_-}{\partial a^r} > 0 \tag{2.5.5}$$

These, if combined with Eq. (2.5.3), show that a shock is accelerated if c_2^l and/or c_2^r increases as it propagates. Consequently, we draw the following rule.

RULE 2.6:

1. A disturbance, propagating along a C_+ or C_- characteristic, is accelerated (or decelerated) if c_2 increases (or decreases) as it propagates.

2. A shock is accelerated (or decelerated) if c_2^1 and/or c_2^2 increases (or decreases) as it propagates.

2.6 Interactions between Waves of the Same Family

For a piecewise constant data problem the wave propagation inevitably leads to interactions between waves issuing from different points of discontinuity. If we have characteristic initial data specified over a finite portion of the initial curve, we expect to have a simple wave in the region adjacent to that portion and this wave may interact with other waves traveling ahead or behind. To obtain the complete solution, it is necessary to analyze each interaction that may be involved. In the present case of two-solute chromatography with the Langmuir isotherm the wave interaction is of particularly simple nature. First, since the Γ characteristics are straight, both simple waves and shocks have their images on Γ characteristics in the hodograph plane, so their interaction does not produce a third wave as we have discussed in the preceding section. Moreover, one of the two characteristic parameters remains invariant along each of the Γ characteristics, and this enables us to perform a complete analysis of every interaction. Here we will be concerned with the analysis of interactions between waves of the same family, while interactions between waves of different families will be treated in the next section.

Confluence of two shocks

Let us consider the situation when two Γ_- shocks propagate facing each other. This is shown in Fig. 2.14, one shock emanating from the origin of the (x,τ) plane and the other from the point $A(0, \eta)$. The image of the solution in the hodograph plane will be just as shown in Fig. 2.12(a). The parameter α is constant here and so is $a(\alpha)$. Hence we can write

$$\tilde{\sigma}_{+,1} = 1 + \rho a b^p b^r \tag{2.6.1}$$

$$\tilde{\sigma}_{+,2} = 1 + \rho a b^l b^p \tag{2.6.2}$$

where the lowercase superscripts are used to indicate that the values are for the states designated by the corresponding capital letters. This scheme of notation will be adopted throughout this section whenever convenient.

The two shocks will meet at $B(x_0, \tau_0)$, where

$$x_0 = \frac{\eta}{\tilde{\sigma}_{+,1} - \tilde{\sigma}_{+,2}} = \frac{\eta}{\rho a b^p (b^r - b^l)} \tag{2.6.3}$$

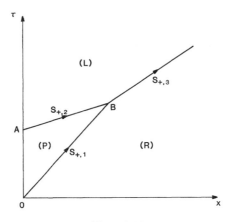

Figure 2.14

$$\tau_0 = \tilde{\sigma}_{+,1}x_0 = \frac{1 + \rho ab^p b^r}{\rho ab^p (b^r - b^l)} \eta$$

and instantly superposed to give a new shock with the states L and R on both sides. The state P disappears and the new shock propagates in the direction

$$\tilde{\sigma}_{+,3} = 1 + \rho ab^l b^r \qquad (2.6.4)$$

Since we have $b^l < b^p < b^r$ from the entropy condition (2.2.26), it follows that

$$\tilde{\sigma}_{+,1} < \tilde{\sigma}_{+,3} < \tilde{\sigma}_{+,2} \qquad (2.6.5)$$

so the new shock takes an intermediate speed of propagation.

In case of two Γ_+ shocks a similar argument can be developed. We also note that the situation here is entirely analogous to the interaction between two shocks in case of single conservation laws (see Secs. 6.5 and 6.7 of Vol. 1).

Cancellation of a simple wave and a shock

Here we shall consider a Γ_- simple wave interacting with a Γ_- shock from behind. For a more general discussion we assume that the simple wave is not centered but based on the characteristic feed data, while the shock originates from the origin. The situation is depicted in Fig. 2.15(a) together with the image in the hodograph plane shown in Fig. 2.15(b). Thus the initial and feed data may be expressed as

$$\text{at } \tau = 0: \quad c_2 = c_2^r \qquad \text{for } \xi > 0$$

$$\text{at } x = 0: \quad c_2 = \begin{cases} c_2^p = c_2^f(\eta_1) & \text{for } 0 < \eta \le \eta_1 \\ c_2^f(\eta) & \text{for } \eta_1 < \eta < \eta_2 \\ c_2^q = c_2^f(\eta_2) & \text{for } \eta \ge \eta_2 \end{cases} \qquad (2.6.6)$$

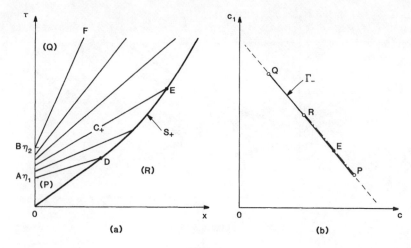

Figure 2.15

and

$$c_1 = \alpha c_2 - \frac{k\alpha}{\alpha + h} \tag{2.6.7}$$

where α is fixed here. For the given situation we have $\beta^p < \beta^r < \beta^q$, so

$$b^p < b^r < b^q \tag{2.6.8}$$

whereas a is constant.

For the C_+ characteristics in the simple wave region we can write the equation

$$\tau - \eta(b) = \sigma_+(b)x = (1 + \rho ab^2)x \tag{2.6.9}$$

where $\eta(b)$ denotes the inverse of the feed data

$$c_2 = c_2^f(\eta) = \frac{-1 + 1/b}{K_1\gamma(\alpha + h)} \tag{2.6.10}$$

Thus the lowermost characteristic AD has the equation

$$\tau - \eta_1 = \{1 + \rho a(b^p)^2\}x$$

and solving this simultaneously with the equation of the shock path OD,

$$\tau = (1 + \rho ab^pb^r)x$$

we determine the point of intersection $D(x_0, \tau_0)$, i.e.,

$$
\begin{aligned}
x_0 &= \frac{\eta_1}{\rho ab^p(b^r - b^p)} \\
\tau_0 &= \frac{1 + \rho ab^pb^r}{\rho ab^p(b^r - b^p)}\eta_1
\end{aligned}
\tag{2.6.11}
$$

Beyond the point D the two waves gradually cancel each other by interaction. The shock decelerates continuously because c_1' decreases while c_2' remains constant (see Rule 2.6). To an arbitrary point E on the shock path there corresponds a straight C_+ characteristic bearing the parameter value $b(\beta)$ up to the left-hand side of the shock, so we can write

$$\tau - \eta(b) = \sigma_+(b)x = (1 + \rho ab^2)x$$

$$\frac{d\tau}{dx} = \bar{\sigma}_+(b) = 1 + \rho abb' \tag{2.6.12}$$

This pair of equations gives a parametric representation of the shock path beyond the point D. When the shock arrives at the point E, the image of the simple wave becomes $Q \to E$, whereas the image of the shock is now $E \to R$ and the point E tends to approach R as the interaction goes on [see Fig. 2.15(b)].

Since a and b' remain constant, b is the only variable parameter in Eq. (2.6.12). Differentiating the first equation with respect to b gives

$$\frac{d\tau}{db} - \frac{d\eta}{db} = \sigma_+(b)\frac{dx}{db} + \frac{d\sigma_+}{db} x$$

and the second equation can be rewritten as

$$\frac{d\tau}{db} = \bar{\sigma}_+(b)\frac{dx}{db}$$

Now combining these equations we obtain

$$\{\bar{\sigma}_+(b) - \sigma_+(b)\}\frac{dx}{db} - \frac{d\sigma_+}{db} x = \frac{d\eta}{db} \tag{2.6.13}$$

or, by substituting the expressions for $\bar{\sigma}_+(b)$ and $\sigma_+(b)$,

$$(b' - b)\frac{dx}{db} - 2x = \frac{1}{\rho ab}\frac{d\eta}{db} \tag{2.6.14}$$

This is an ordinary differential equation which, if solved subject to the initial condition

$$x = x_0 \quad \text{at} \quad b = b^p \tag{2.6.15}$$

will give x as a function of b.

Equation (2.6.14) can be made exact by multiplying throughout by $(b' - b)$, so that

$$\frac{d}{db}\{(b' - b)^2 x\} = \frac{1}{\rho a}\left(\frac{b'}{b} - 1\right)\frac{d\eta}{db}$$

or

$$x(b) = x_0 \left(\frac{b' - b^p}{b' - b}\right)^2 + \frac{1}{\rho a(b' - b)^2}\int_{b^p}^{b}\left(\frac{b'}{b} - 1\right)\frac{d\eta}{db}\,db \tag{2.6.16}$$

Substituting this in the first equation of Eq. (2.6.12), we can express τ in terms of b and thus determine the shock path S_+. Here we note that $x(b) \to \infty$ as $b \to b^r$, so the interaction will go on indefinitely unless $b^q < b^r$. If $b^q < b^r$, the interaction will be over when $b = b^q$. This implies that the simple wave is completely canceled in finite time and the states Q and R are directly connected by a Γ_- shock. This is the case when the point Q falls between R and P in Fig. 2.15(b). In any case the shock never disappears completely.

We note that the integral in the last term of Eq. (2.6.16) remains finite because we are concerned with finite feed data and the upperbound of the integration remains finite. As the interaction goes on, the factor $|b^r - b|$, which is a measure of the shock strength, becomes smaller and smaller. For large time, therefore, we see from Eq. (2.6.16) that

$$|b^r - b| \sim \frac{M}{\sqrt{x}} \qquad (2.6.17)$$

where M is a constant determined by the states P and R and the feed data. This implies that the shock strength decays like $1/\sqrt{x}$ for large time. This feature of the decay of shocks is consistent with the Glimm–Lax theory (1970; see Sec. 1.11), where the variable t takes the role of our x here, and is essentially the same as that we have observed with single equations (see Secs. 6.3, 6.6, and 7.9 of Vol. 1).

If the simple wave is centered at the point A, $\eta = \eta_1 = $ constant and $d\eta/db = 0$, so that we obtain a particularly simple expression

$$x(b) = x_0 \left(\frac{b^r - b^p}{b^r - b} \right)^2 = \frac{\eta_1(b^r - b^p)}{\rho a b^p (b^r - b)^2}, \qquad b^p \le b < b^r \qquad (2.6.18)$$

in which Eq. (2.6.11) has been introduced. This can be rearranged to give

$$b = b^r - \sqrt{\frac{\eta_1(b^r - b^p)}{\rho a b^p x}}$$

and substituting this into the first equation of Eq. (2.6.12) with $\eta = \eta_1$, we get

$$\sqrt{\tau - \eta_1 - x} = b^r \sqrt{\rho a x} - \sqrt{\eta_1 \left(\frac{b^r}{b^p} - 1 \right)} \qquad (2.6.19)$$

Hence the shock path S_+ in this case is given by a part of a parabola.

In other cases, we may take the point D as the new origin of the coordinate system to have $\eta_1 = 0$ and $x_0 = 0$, and thus

$$x(b) = \frac{1}{\rho a (b^r - b)^2} \int_{b^p}^{b} \left(\frac{b^r}{b} - 1 \right) \frac{d\bar{\eta}}{db} db$$

or, by applying integration by parts,

$$x(b) = \frac{1}{\rho a (b^r - b)} \left\{ \frac{\bar{\eta}(b)}{b} + \frac{b^r}{b^r - b} \int_{b^p}^{b} \frac{\bar{\eta}(b)}{b^2} db \right\}, \qquad b^p \le b < b^r \tag{2.6.20}$$

where $\bar{\eta}(b)$ represents the adjusted form of $\eta(b)$.

For the purpose of illustration we shall consider a linear profile for $c_2^{\xi}(\eta)$. Thus

$$\frac{c_2 - c_2^p}{c_2^q - c_2^p} = \frac{\eta - \eta_1}{\eta_2 - \eta_1}$$

or, by using Eq. (2.6.10), we have

$$\frac{1/b - 1/b^p}{1/b^q - 1/b^p} = \frac{\eta - \eta_1}{\eta_2 - \eta_1} \tag{2.6.21}$$

Substituting this in Eq. (2.6.16) and carrying out the integration, we obtain

$$x(b) = x_0 \left(\frac{b^r - b^p}{b^r - b}\right)^2 + \frac{(\eta_2 - \eta_1)b^q}{\rho a(b^p - b^q)} \frac{b^p - b}{b(b^r - b)^2}\left\{1 + \frac{b^r(b + b^p)}{2b^p b}\right\} \tag{2.6.22}$$

Let us consider another situation in which a Γ_+ simple wave is based on the x axis and a Γ_+ shock emanates from a point on the τ axis as depicted in Fig. 2.16. The initial and feed data may be expressed as

$$\text{at } \tau = 0: \quad c_2 = \begin{cases} c_2^0(\xi), & 0 \le \xi < \xi_1 \\ c_2^q = c_2^0(\xi_1), & \xi \ge \xi_1 \end{cases} \tag{2.6.23}$$

$$\text{at } x = 0: \quad c_2 = \begin{cases} c_2^q, & 0 \le \eta < \eta_1 \\ c_2^i, & \eta > \eta_1 \end{cases}$$

and

$$c_1 = \beta c_2 - \frac{k\beta}{\beta + h} \tag{2.6.24}$$

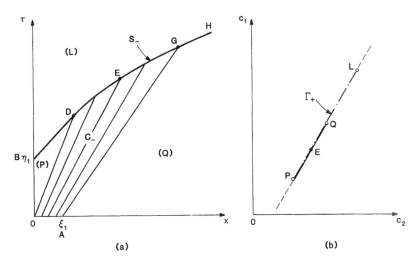

Figure 2.16

where β is fixed. For the arrangement shown in Fig. 2.16(b) we see that $\alpha^l < \alpha^q < \alpha^p$ and thus

$$a^l < a^q < a^p \qquad (2.6.25)$$

whereas b is now a constant.

For the C_- characteristic OD we have the equation

$$\tau = \sigma_-(a^p)x = \{1 + \rho b(a^p)^2\}x$$

and the shock path BD has the equation

$$\tau - \eta_1 = \tilde{\sigma}_-(a^l, a^p)x = (1 + \rho b a^l a^p)x$$

These equations are solved simultaneously to give the point of intersection $D(x_0, \tau_0)$, for which

$$\begin{aligned} x_0 &= \frac{\eta_1}{\rho b a^p (a^p - a^l)} \\ \tau_0 &= \frac{1 + \rho b(a^p)^2}{\rho b a^p (a^p - a^l)} \eta_1 \end{aligned} \qquad (2.6.26)$$

The two waves start to interact from D and they are gradually canceled. Since c_2^l increases, the Γ_+ shock accelerates during the interaction. If we pick an arbitrary point E on the shock path S_-, there exists a straight C_- characteristic connecting E to the initial line OA and carrying a constant value $a(\alpha)$ up to the right-hand side of the shock. Hence we can write two equations

$$\begin{aligned} \tau &= \sigma_-(a)\{x - \xi(a)\} \\ \frac{d\tau}{dx} &= \tilde{\sigma}_-(a) \end{aligned} \qquad (2.6.27)$$

where

$$\begin{aligned} \sigma_-(a) &= 1 + \rho b a^2 \\ \tilde{\sigma}_-(a) &= 1 + \rho b a^l a \end{aligned} \qquad (2.6.28)$$

and $\xi(a)$ is the inverse of the initial data

$$c_2 = c_2^0(\xi) = \frac{-1 + 1/a}{K_1 \gamma (\beta + h)} \qquad (2.6.29)$$

At this stage of interaction, the images of the interacting waves become $L \to E$ (shock) and $E \to Q$ (simple wave) as marked in Fig. 2.16(b).

We note that b as well as a^l remains constant in Eq. (2.6.27) and regard a as the parameter running along the shock path DG. We can then eliminate τ from Eq. (2.6.27) by applying the same procedure as in the previous case and obtain

$$\{\sigma_-(a) - \tilde{\sigma}_-(a)\}\frac{dx}{da} + \frac{d\sigma_-}{da}x = \frac{d}{da}\{\sigma_-(a)\xi(a)\} \qquad (2.6.30)$$

Substitution of Eq. (2.6.28) gives

$$(a - a')\frac{dx}{da} + 2x = \frac{1}{\rho ba}\frac{d}{da}\{(1 + \rho ba^2)\xi(a)\} \qquad (2.6.31)$$

which is subject to the initial condition

$$x = x_0 \quad \text{at} \quad a = a^p \qquad (2.6.32)$$

Putting Eq. (2.6.31) into the form of an exact differential and integrating, we obtain the solution in the form

$$x(a) = x_0\left(\frac{a^p - a'}{a - a'}\right)^2 + \frac{1}{\rho b(a - a')^2}\int_{a^p}^{a}\left(1 - \frac{a'}{a}\right)\frac{d}{d\xi}\{(1 + \rho ba^2)\xi(a)\}\,da \qquad (2.6.33)$$

or, by applying integration by parts,

$$x(a) = x_0\left(\frac{a^p - a'}{a - a'}\right)^2 + \left(a + \frac{1}{\rho ba}\right)\frac{\xi(a)}{a - a'}$$
$$- \frac{a'}{(a - a')^2}\int_{a^p}^{a}\left(1 + \frac{1}{\rho ba^2}\right)\xi(a)\,da, \qquad a^p \geq a \geq a^q \qquad (2.6.34)$$

We then use this in the first equation of Eq. (2.6.27) to determine τ as a function of a. Since $a' < a^q$, the interaction will be completed when $a = a^q$. This occurs at G in Fig. 2.16(a) and at this stage the simple wave is completely canceled. Beyond the point G the shock connects the states L and Q directly and so propagates with a constant speed along the path GH.

If we have $a' \geq a^q$ in Eq. (2.6.25), the shock path will never meet the C_- characteristic AG and the interaction will go on indefinitely. In this case we can also argue from Eq. (2.6.33) that $|a - a'|$ decreases like $1/\sqrt{x}$ for sufficiently large time. Now the shock decays from the bottom side and it does like $1/\sqrt{x}$ for large time.

If the simple wave is centered at the origin, we have $\xi \equiv 0$, so that

$$x(a) = x_0\left(\frac{a^p - a'}{a - a'}\right)^2 = \frac{\eta_1(a^p - a')}{\rho ba^p(a - a')^2}, \qquad a^p \geq a \geq a^q \qquad (2.6.35)$$

Otherwise, we can translate the origin of the coordinate system to the point D to have $x_0 \equiv 0$ in Eq. (2.6.34). With appropriate adjustment for $\xi(a)$ the last two terms of Eq. (2.6.34) determine x as a function of a.

In the region of a compression wave we expect to have a shock formed, which is obviously of the same family as the compression wave. Immediately after its birth the shock will interact with the compression wave. If the shock first forms on one of the boundary characteristics of the compression wave, the wave interaction can be analyzed by applying the procedure discussed above. Although the point of shock inception would be the singular point of the differential equation (2.16.14) or (2.6.31), the solution can be determined by direct integration since the differential equation can always be made exact. When the shock forms on one of the intermediate characteristics, it faces simple waves on both sides, so it interacts simultaneously with two simple waves. Since the states on both sides change as the shock propagates, the treatment above does not apply here. We shall discuss this situation in the following.

Simultaneous interactions of a shock with two simple waves

Consider a shock traveling between two simple waves. If the three waves are of the same family, the interaction may go on simultaneously on both sides of the shock. In such a case the state of concentrations varies on either side of the shock but the images of the interacting waves all fall on a single Γ characteristic. Hence one of the characteristic parameters α and β (or equivalently a and b) remains invariant throughout the process of interaction while the other parameter varies independently on each side of the shock. It then follows that the mathematical problem is essentially the same as that for the case of single equations treated in Sec. 6.4 of Vol. 1. We shall follow the same scheme with the variable parameter a or b in place of c.

Suppose now that a Γ_- shock interacts simultaneously with two Γ_- simple waves as illustrated in Fig. 2.17. In this case the parameter a is fixed and b varies independently on each side of the shock path S_+, its coordinates x and τ may be regarded as functions of b^l and b^r, so that, in accordance with Eq. (6.4.10) of Vol. 1, we can write

$$\frac{d\tau}{dx} = \tilde{\sigma}_+(b^l, b^r) = \frac{\dfrac{\partial \tau}{\partial b^l}\dfrac{db^l}{db^r} + \dfrac{\partial \tau}{\partial b^r}}{\dfrac{\partial x}{\partial b^l}\dfrac{db^l}{db^r} + \dfrac{\partial x}{\partial b^r}}$$

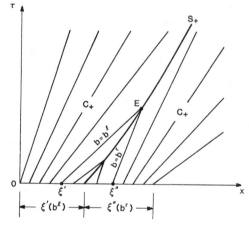

Figure 2.17

or, upon rearrangement,

$$\frac{db^l}{db^r} = -\frac{\tilde{\sigma}_+ \dfrac{\partial x}{\partial b^r} - \dfrac{\partial \tau}{\partial b^r}}{\tilde{\sigma}_+ \dfrac{\partial x}{\partial b^l} - \dfrac{\partial \tau}{\partial b^l}} = \psi(b^l, b^r) \tag{2.6.36}$$

The right-hand side is a function of b^l and b^r and the initial condition can be specified in the form

$$b^l = b_0^l \quad \text{when} \quad b^r = b_0^r \tag{2.6.37}$$

which corresponds to the point where simultaneous interactions start to take place. The solution of Eq. (2.6.36) subject to Eq. (2.6.37) provides the relationship between b^l and b^r along the shock path S_+.

If the two simple waves are both developed from the characteristic initial data $c_2^0(\xi)$, the layout of the C_+ characteristics and the shock path S_+ may be depicted as in Fig. 2.17. At an arbitrary point $E(x, \tau)$ on the shock path, we can always find two characteristics reaching point E from both sides, one from the point $(\xi', 0)$ bearing $b = b^l$ and the other from the point $(\xi'', 0)$ bearing $b = b^r$. Since E is an arbitrary point, we may consider ξ' or ξ'' as a function of b^l or b^r, respectively. We note that each of these functions is given by the inverse of the corresponding portion of the initial data. Thus we can write the equations of the two characteristics in the form

$$\begin{aligned} \tau &= \sigma_+(b')\{x - \xi'(b')\} \\ \tau &= \sigma_+(b'))\{x - \xi''(b')\} \end{aligned} \tag{2.6.38}$$

which may be solved to give

$$x = \frac{\sigma_+(b^l)\xi'(b^l) - \sigma_+(b^r)\xi''(b^r)}{\sigma_+(b^l) - \sigma_+(b^r)} \tag{2.6.39}$$

$$\tau = \sigma_+(b^l)\sigma_+(b^r)\frac{\xi'(b^l) - \xi''(b^r)}{\sigma_+(b^l) - \sigma_+(b^r)}$$

Substituting this into Eq. (2.6.36), we obtain

$$\frac{db^l}{db^r} = -\frac{\tilde{\sigma}_+ - \sigma_+^l}{\sigma_+^r - \tilde{\sigma}_+}\cdot\frac{\sigma_+^l(\xi'' - \xi')\dfrac{d\sigma_+^r}{db^r} - \sigma_+^r(\sigma_+^r - \sigma_+^l)\dfrac{d\xi''}{db^r}}{\sigma_+^r(\xi'' - \xi')\dfrac{d\sigma_+^l}{db^l} - \sigma_+^l(\sigma_+^r - \sigma_+^l)\dfrac{d\xi'}{db^l}} \tag{2.6.40}$$

in which we have used the abbreviations $\tilde{\sigma}_+ = \tilde{\sigma}_+(b^l, b^r)$, $\sigma_+^l = \sigma_+(b^l)$, and $\sigma_+^r = \sigma_+(b^r)$. If this equation is solved subject to the initial condition (2.6.37), we obtain the relationship between b^l and b^r and then these values can be used in Eq. (2.6.39) to determine the shock path S_+.

If a shock is formed in a compression wave region and interacts simultaneously on both sides with the compression wave, we have $b_0^l = b_0^r = b_0$ in Eq. (2.6.37), so that the initial point becomes the singular point of the differential equation (2.6.40). It is then required to determine the limit of the derivative db^l/db^r as $b^l \to b_0$ and $b^r \to b_0$ at the same time.

So far we have formulated the problem in a general context. If we had had Γ_+ waves interacting simultaneously clearly, we would have had $\tilde{\sigma}_-$, σ^l_-, σ^r_-, a^l, and a^r for $\tilde{\sigma}_+$, σ^l_+, σ^r_+, b^l, and b^r, respectively, in Eq. (2.6.40). In case of the Langmuir isotherm we have

$$\tilde{\sigma}_+ = \tilde{\sigma}_+(b^l, b^r) = 1 + \rho a b^l b^r$$

$$\sigma^l_+ = \sigma_+(b^l) = 1 + \rho a (b^l)^2 \tag{2.6.41}$$

$$\sigma^r_+ = \sigma_+(b^r) = 1 + \rho a (b^r)^2$$

and substituting these into Eq. (2.6.40) gives

$$\frac{db^l}{db^r} = -\frac{b^l}{b^r} \frac{2b^r \sigma^l_+ (\xi'' - \xi') - \sigma^r_+ \{(b^r)^2 - (b^l)^2\} \dfrac{d\xi''}{db^r}}{2b^l \sigma^r_+ (\xi'' - \xi') - \sigma^l_+ \{(b^r)^2 - (b^l)^2\} \dfrac{d\xi'}{db^l}} \tag{2.6.42}$$

which is subject to the initial condition (2.6.37). We note that the derivatives $d\xi''/db^r$ and $d\xi'/db^l$ are negative if the simple waves are expansive and positive if they are compressive.

To illustrate this we shall consider a portion of the sawtooth wave for the initial profile of c_2. This is shown in Fig. 2.18 and can be expressed as

$$c_2^0(\xi) = \begin{cases} c_2^t + (c_2^h - c_2^t)\xi, & 0 \le \xi < 1 \\ c_2^t + (c_2^h - c_2^t)(\xi - 1), & 1 < \xi < 2 \\ c_2^h, & \xi \ge 2 \end{cases} \tag{2.6.43}$$

$$c_2^f(\eta) = c_2^t, \qquad \eta > 0$$

and

$$c_1 = \alpha c_2 - \frac{k\alpha}{\alpha + h} \tag{2.6.44}$$

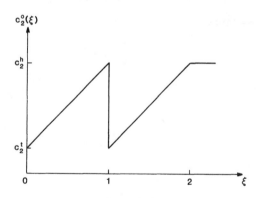

Figure 2.18

Two-Solute Chromatography with the Langmuir Isotherm   Chap. 2

where α is constant. The concentration c_2 is related to b by Eq. (2.6.10), so

$$c_2^0(\xi) = \frac{-1 + 1/b}{K_1 \gamma (\alpha + h)} \tag{2.6.45}$$

Combining this with Eq. (2.6.43), we obtain

$$\xi'(b^l) = \frac{1/b^l - 1/b^t}{1/b^h - 1/b^t} \tag{2.6.46}$$

and

$$\xi''(b^r) = 1 + \frac{1/b^r - 1/b^t}{1/b^h - 1/b^t} \tag{2.6.47}$$

Substituting these in Eq. (2.6.42) gives the equation

$$\frac{db^l}{db^r} = -\frac{2\{1 + \rho a (b^l)^2\}\left\{1 + \dfrac{b^l b^r (b^l - b^h)}{b^l b^h (b^l - b^r)}\right\} - \{1 + \rho a (b^r)^2\}\left(1 + \dfrac{b^l}{b^r}\right)\dfrac{b^l}{b^r}}{2\{1 + \rho a (b^r)^2\}\left\{1 + \dfrac{b^l b^r (b^l - b^h)}{b^l b^h (b^l - b^r)}\right\} - \{1 + \rho a (b^l)^2\}\left(1 + \dfrac{b^r}{b^l}\right)\dfrac{b^r}{b^l}} \tag{2.6.48}$$

and the initial condition is

$$b^l = b^h \quad \text{when} \quad b^r = b^t \tag{2.6.49}$$

As a numerical example let us take the case when

$$\epsilon = 0.4$$

$$N_1 = N_2 = 1.0 \text{ mol/liter,}$$

$$K_1 = 5 \text{ liters/mol}, \quad K_2 = 7.5 \text{ liters/mol}$$

$$c_2^t = 0.02 \text{ mol/liter}, \quad c_2^h = 0.12 \text{ mol/liter}$$

and

$$\alpha = -\frac{1}{2} = \text{fixed}$$

It follows that

$$\nu = 1.5, \quad h = 1, \quad \gamma = 1.5, \quad k = \frac{1}{15}, \quad \rho = 16.875$$

and

$$a(\alpha) = \frac{1}{2} = \text{fixed}$$

From Eq. (2.6.45) we have

$$b = \frac{1}{1 + K_1 \gamma (\alpha + h) c_2} = \frac{1}{1 + 3.75 c_2} \tag{2.6.50}$$

where c_2 is in mol/liter. Since the unit of the concentration is obvious, we shall just give the number in the following. By using Eq.(2.6.50) we obtain

$$b^l = \frac{40}{43} \quad \text{and} \quad b^h = \frac{20}{29}$$

Now Eq. (2.6.48) can be solved numerically to give the values of the pair (b^l, b^r) along the shock path. These values are then used in Eq. (2.6.39) with the aid of Eqs. (2.6.41), (2.6.46), and (2.6.47) to calculate the coordinates of the shock path S_+. The result is plotted in Fig. 2.19, in which the shock path tends to be linear almost from the beginning. In fact, it becomes virtually parallel to the C_+ characteristic for $c_2 = 0.07$ drawn by the dashed line as the shock propagates. The pair (b^l, b^r) is also used in Eq. (2.6.45) to obtain the pair (c_2^l, c_2^r), which in turn gives the values of c_1^l and c_1^r by the equation of the Γ_- characteristic

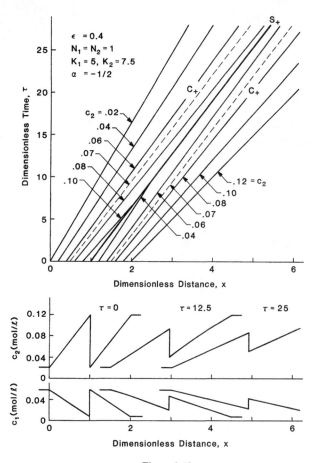

Figure 2.19

$$c_1 = \alpha c_2 - \frac{k\alpha}{\alpha + h} = -\frac{1}{2}c_2 + \frac{1}{15} \qquad (2.6.51)$$

The C_+ characteristics can be drawn backward on both sides of the shock path by using the slopes determined by Eq. (2.6.41). In Fig. 2.19 the values of c_2 are marked on each of the C_+ characteristics. With these values we can readily determine c_1 by Eq. (2.6.51), b by Eq. (2.6.50), σ_+ by Eq. (2.6.41), and ξ by Eq. (2.6.46) or Eq. (2.6.47). For example, with $c_2 = 0.02$ we obtain

$$c_1 = \frac{17}{300}, \qquad b = \frac{40}{43}, \qquad \sigma_+ = 8.3012, \qquad \xi' = 0$$

with $c_2 = 0.07$,

$$c_1 = \frac{19}{600}, \qquad b = \frac{80}{101}, \qquad \sigma_+ = 6.2936, \qquad \xi' = \frac{1}{2} \ \text{ or } \ \xi'' = \frac{3}{2}$$

and with $c_2 = 0.12$,

$$c_1 = \frac{1}{150}, \qquad b = \frac{20}{29}, \qquad \sigma_+ = 5.0131, \qquad \xi'' = 2$$

Thus the two simple waves can be constructed in the (x, τ)-plane.

The concentration profiles are determined from the (x, τ)-diagram at any time by simply reading the concentration values on the C_+ characteristics. These are shown in the lower part of Fig. 2.19 at two different times. The profiles are not quite linear at an earlier time ($\tau = 12.5$) but tend to become linear as time goes on (see those at $\tau = 25$). In the earlier period (at $\tau = 12.5$) it appears that the shock decays faster on the left-hand side than on the right, but later the shock strength tends to be balanced around the average concentrations, i.e., $c_2 = 0.07$ and $c_1 = 0.032$.

Since the shock path S_+ becomes parallel to the C_+ characteristic for $c_2 = 0.07$, we note that the solution in the region bounded by the two dashed lines is determined solely by the data specified over the interval $\frac{1}{2} < \xi < \frac{3}{2}$. This implies that if the initial data are specified by a sawtooth wave, i.e., a repetition of the data for $0 < \xi < 1$ shown in Fig. 2.18, then the solution is given by a successive train of the solution in the region bounded by the dashed lines in Fig. 2.19. Hence the periodic wave solution can be constructed without additional calculation.

Exercises

2.6.1. Consider a Γ_- simple wave developing from the characteristic initial data specified over the interval $0 \le \xi \le \xi_1$ and a Γ_- shock originating from the point $(\xi_2, 0)$ of the x-axis, where $\xi_1 < \xi_2$. Determine the shock path when the two waves are interacting and discuss.

2.6.2. For the case shown in Fig. 2.16, suppose that the simple wave is centered at the origin and find the equation relating τ and x along the shock path DG to show that it is part of a parabola.

2.6.3. If the data in reversed form of those shown in Fig. 2.18 are specified as the characteristic feed data, formulate the equation corresponding to Eq. (2.6.48). For the numerical example treated at the end of this section, solve the equation obtained here and construct the solution in the (x, τ)-plane as well as the concentration profiles.

2.7 Interactions between Waves of Different Families

Here the two interacting waves are of different families and are transmitted through each other during the course of interaction. The state of concentrations changes across each wave and neither of the characteristic parameters remains constant throughout the whole region. Since, however, only one of the characteristic parameters varies along the characteristics, it is possible to make an analytic treatment of the interaction analysis by using the characteristic parameters judiciously. We shall discuss three cases separately.

Transmission between two shocks

Suppose that a Γ_+ shock is traveling behind a Γ_- shock. The two shocks will meet and be transmitted through each other to produce a pair of transmitted shocks of different families as depicted in Fig. 2.20(a). The image of the solution is presented not only in the hodograph plane [Fig. 2.20(b)] but also in the (a, b)-plane [Fig. 2.20(c)]. The latter clearly shows which parameter varies and which remains invariant across each wave. We shall use the subscripts 0 and 1 on the characteristic parameters a and b to represent their lower and higher values, respectively. Then the characteristic parameters for the four constant states are

$$L(a_0, b_0), \qquad Q(a_0, b_1)$$
$$P(a_1, b_0), \qquad R(a_1, b_1) \tag{2.7.1}$$

Before the interaction the two shocks propagate along the path S_- and S_+ of slopes

$$\tilde{\sigma}_- = 1 + \rho a_0 a_1 b_0$$
$$\tilde{\sigma}_+ = 1 + \rho a_1 b_0 b_1 \tag{2.7.2}$$

respectively. Upon interaction the constant state P disappears and a new constant state Q appears. The interaction is instantaneous and the states L and Q are connected by the transmitted Γ_- shock, whereas Q and R are connected by the transmitted Γ_+ shock. These shocks propagate along the paths S'_+ and S'_-, respectively, and their slopes are given by

$$\tilde{\sigma}'_+ = 1 + \rho a_0 b_0 b_1$$
$$\tilde{\sigma}'_- = 1 + \rho a_0 a_1 b_1 \tag{2.7.3}$$

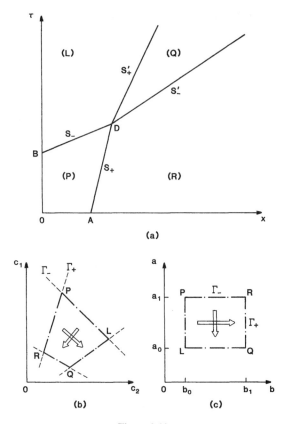

Figure 2.20

Clearly, we see that $\bar{\sigma}_+ > \bar{\sigma}'_+$ and $\bar{\sigma}_- < \bar{\sigma}'_-$, so that, upon interaction, the Γ_- shock is accelerated while the Γ_+ shock is decelerated. This observation is consistent with Rule. 2.6.

By definition [Eqs. (2.2.6) and (2.2.7)], we have

$$\alpha = -h\frac{1 - \gamma a}{1 - a}, \qquad \beta = h\frac{\gamma b - 1}{1 - b} \qquad (2.7.4)$$

and substituting these into Eqs. (2.1.33) and (2.1.34) gives

$$c_1 = \frac{(1 - \gamma a)(\gamma b - 1)}{K_1 \gamma(\gamma - 1)ab} \qquad (2.7.5)$$

$$c_2 = \frac{(1 - a)(1 - b)}{K_2(\gamma - 1)ab}$$

Hence the concentration values for the state Q are directly determined by putting $a = a_0$ and $b = b_1$ in Eq. (2.7.5).

Transmission between a simple wave and a shock

Let us now consider a Γ_+ simple wave traveling behind a Γ_- shock. The two waves will be transmitted through each other and produce the transmitted Γ_+ simple wave and the transmitted Γ_- shock. This is shown in Fig. 2.21(a) while the image of the solution is given in Fig. 2.21(b) and (c). Using the same scheme for notation as in the previous case, we have here the following arrangement:

$$L(a_1, b_0), \qquad Q(a_1, b_1)$$
$$P(a_0, b_0), \qquad R(a_0, b_1)$$

(2.7.6)

where $a_0 < a_1$ and $b_0 < b_1$.

As the interaction goes on, the image of the Γ_- shock gradually recedes from PR to $d \to e$ and then ultimately to LQ. Thus at one moment during the

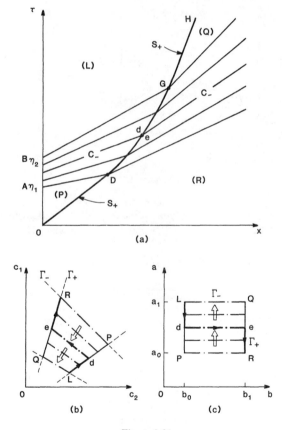

Figure 2.21

Two-Solute Chromatography with the Langmuir Isotherm   Chap. 2

interaction the image of the solution will be given by the path $L \to d \to e \to R$. Clearly, the parameter b remains constant on either side of the shock taking the values b_0 on the left and b_1 on the right-hand side, whereas the parameter a, being invariant across the shock, varies along the shock path S_+ from a_0 to a_1. Hence the shock path has a one-parameter representation in terms of a and the interaction starts when $a = a_0$ and terminates when $a = a_1$.

Again we shall assume that the Γ_+ simple wave is based on the characteristic feed data and the Γ_- shock originates from the origin. Thus the initial and feed data are prescribed as

$$
\begin{aligned}
\text{at } \tau = 0: \quad & c_2 = c_2^r && \text{for } \xi > 0 \\[4pt]
\text{at } x = 0: \quad & c_2 = \begin{cases} c_2^p = c_2^f(\eta_1) & \text{for } 0 < \eta \leq \eta_1 \\ c_2^f(\eta) & \text{for } \eta_1 < \eta < \eta_2 \\ c_2^l = c_2^f(\eta_2) & \text{for } \eta \geq \eta_2 \end{cases} &&
\end{aligned}
\tag{2.7.7}
$$

where

$$
c_1^p - c_1^r = \alpha_0(c_2^p - c_2^r)
\tag{2.7.8}
$$

and

$$
c_1^f(\eta) = \beta_0 c_2^f(\eta) - \frac{k\beta_0}{\beta_0 + h}
\tag{2.7.9}
$$

The characteristic feed data $c_2^f(\eta)$ will be assumed invertible so that we can express η in terms of a by using the equation

$$
c_2 = \frac{-1 + 1/a}{K_1 \gamma (\beta_0 + h)}
\tag{2.7.10}
$$

The C_- characteristic AD has the equation

$$
\tau - \eta_1 = \sigma_-(a_0)x = (1 + \rho a_0^2 b_0)x
$$

and the Γ_- shock first propagates along the straight line OD, for which we can write

$$
\tau = \tilde{\sigma}_+(b_0, b_1)x = (1 + \rho a_0 b_0 b_1)x
$$

The two meet at $D(x_0, \tau_0)$, where

$$
x_0 = \frac{\eta_1}{\tilde{\sigma}_+ - \sigma_-} = \frac{\eta_1}{\rho a_0 b_0 (b_1 - a_0)}
\tag{2.7.11}
$$

$$
\tau_0 = \frac{\tilde{\sigma}_+ \eta_1}{\tilde{\sigma}_+ - \sigma_-} = \frac{1 + \rho a_0 b_0 b_1}{\rho a_0 b_0 (b_1 - a_0)} \eta_1
$$

This is the point at which the interaction begins.

At an arbitrary point (x, τ) along the shock path DG we can locate the straight line C_- characteristic connecting the point to a point on AB and bearing the parameter value a. Thus we can write

$$\tau - \eta(a) = \sigma_-(a)x \tag{2.7.12}$$

$$\frac{d\tau}{dx} = \tilde{\sigma}_+(a)$$

in which

$$\sigma_-(a) = 1 + \rho b_0 a^2 \tag{2.7.13}$$

$$\tilde{\sigma}_+(a) = 1 + \rho b_0 b_1 a \tag{2.7.14}$$

Since a increases along the shock path DG, the shock continuously decelerates while interacting (see also Rule 2.6). Noting that a is the only variable parameter in Eq. (2.7.12), we differentiate the first equation with respect to a and combine with the second equation in the form $d\tau/da = \tilde{\sigma}_+(a) \, dx/da$ to obtain

$$\{\tilde{\sigma}_+(a) - \sigma_-(a)\}\frac{dx}{da} - \frac{d\sigma_-}{da}x = \frac{d\eta}{da} \tag{2.7.15}$$

or, by substituting Eqs. (2.7.13) and (2.7.14),

$$(b_1 - a)\frac{dx}{da} - 2x = \frac{1}{\rho b_0 a}\frac{d\eta}{da} \tag{2.7.16}$$

The initial condition is

$$x = x_0 \quad \text{at} \quad a = a_0 \tag{2.7.17}$$

The differential equation (2.7.16) takes the same form as Eq. (2.6.14) and can be solved in the same manner. Thus by multiplying throughout by $(b_1 - a)$, we put the equation into an exact differential form and integrate to have the solution in the form

$$x(a) = x_0\left(\frac{b_1 - a_0}{b_1 - a}\right)^2 + \frac{1}{\rho b_0(b_1 - a)^2}\int_{a_0}^{a}\left(\frac{b_1}{a} - 1\right)\frac{d\eta}{da}\,da \tag{2.7.18}$$

or, by applying integration by parts,

$$x(a) = x_0\left(\frac{b_1 - a_0}{b_1 - a}\right)^2 + \frac{1}{\rho b_0(b_1 - a)^2}\left\{\frac{b_1 - a}{a}\eta - \frac{b_1 - a_0}{a_0}\eta_1\right.$$
$$\left. + b_1\int_{a_0}^{a}\frac{\eta(a)}{a^2}da\right\}, \qquad a_0 \le a \le a_1 \tag{2.7.19}$$

This equation may be substituted into the first equation of Eq. (2.7.12) to determine τ as a function of a. The point G, at which the interaction terminates, is located by putting $a = a_1$ in Eq. (2.7.19) and beyond this point the transmitted Γ_- shock propagates with a constant speed along the path GH of slope

$$\tilde{\sigma}_+ = 1 + \rho a_1 b_0 b_1 \tag{2.7.20}$$

If the Γ_+ simple wave is centered, we may set $\eta \equiv \eta_1$ and $d\eta/da = 0$ to obtain

$$x(a) = x_0 \left(\frac{b_1 - a_0}{b_1 - a} \right)^2 = \frac{\eta_1(b_1 - a_0)}{\rho a_0 b_0 (b_1 - a)^2} \qquad (2.7.21)$$

This equation may be rewritten in the form

$$a = b_1 - \sqrt{\frac{\eta_1(b_1 - a_0)}{\rho a_0 b_0 x}}$$

which, upon substitution into Eqs. (2.7.12) and (2.7.13) with $\eta = \eta_1$, gives

$$\sqrt{\tau - \eta_1 - x} = b_1 \sqrt{\rho b_0 x} - \sqrt{\eta_1 \left(\frac{b_1}{a_0} - 1 \right)} \qquad (2.7.22)$$

In this case, therefore, the shock path DG is a portion of a parabola. In other cases we can translate the origin to the point D to make $x_0 \equiv 0$ and $\eta_1 \equiv 0$, so that

$$x(a) = \frac{1}{\rho b_0(b_1 - a)} \left\{ \frac{\bar{\eta}(a)}{a} + \frac{b_1}{b_1 - a} \int_{a_0}^{a} \frac{\bar{\eta}(a)}{a^2} da \right\} \qquad (2.7.23)$$

where $\bar{\eta}(a)$ represents the adjusted form of $\eta(a)$.

The transmitted Γ_+ simple wave is constructed by drawing the straight line C_- characteristics from various points on the shock path DG. Since b increases from b_0 to b_1 across the Γ_- shock, the slope of the C_- characteristic is given by

$$\sigma'_-(a) = 1 + \rho b_1 a^2, \qquad a_0 \le a \le a_1 \qquad (2.7.24)$$

Comparing this with Eq.(2.7.13), we see that $\sigma_- < \sigma'_-$ and so the C_- characteristics become steeper upon transmission through the Γ_- shock. The concentration values on each of these characteristics are determined by Eq. (2.7.5) with $b = b_1$ and the value of a borne on the particular characteristic. The concentrations for the new constant state Q appearing between the two transmitted waves are also given by Eq.(2.7.5) with $a = a_1$ and $b = b_1$.

Next we shall briefly discuss the case when a Γ_+ shock, originating from the point $(0, \eta_1)$ of the τ axis, interacts with a Γ_- simple wave developing from the portion $[0, \xi_1]$ of the x axis. This is shown in Fig. 2.22. The characteristic parameters (a, b) for the four constant states are

$$\begin{aligned} L(a_0, b_1), &\qquad Q(a_0, b_0) \\ P(a_1, b_1), &\qquad R(a_1, b_0) \end{aligned} \qquad (2.7.25)$$

where $a_0 < a_1$ and $b_0 < b_1$.

The shock path BD has the equation

$$\tau - \eta_1 = \tilde{\sigma}_-(a_0, a_1)x = (1 + \rho a_0 a_1 b_1)x$$

while the C_+ characteristic OD is the line

$$\tau = \sigma_+(b_1)x = (1 + \rho a_1 b_1^2)x$$

so that their intersection $D(x_0, \tau_0)$ has the coordinates

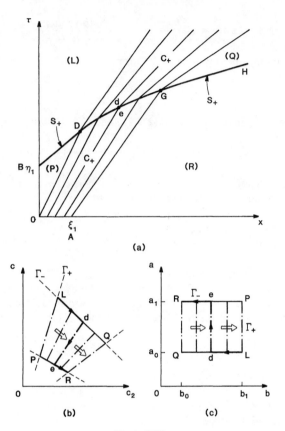

Figure 2.22

$$x_0 = \frac{\eta_1}{\rho a_1 b_1 (b_1 - a_0)} \tag{2.7.26}$$

$$\tau_0 = \frac{1 + \rho a_1 b_1^2}{\rho a_1 b_1 (b_1 - a_0)} \eta_1$$

At an arbitrary point (x, τ) on the shock path DG we can write two equations

$$\tau = \sigma_+(b)\{x - \xi(b)\} = (1 + \rho a_1 b^2)\{x - \xi(b)\}$$

$$\frac{d\tau}{dx} = \tilde{\sigma}_-(b) = 1 + \rho a_0 a_1 b \tag{2.7.27}$$

where $\xi(b)$ denotes the inverse of the characteristic initial data. We note that since b decreases from b_1 to b_0 along the path DG, the shock accelerates during the process of interaction.

Now eliminating τ between the two equations as in the previous case, we obtain the differential equation

$$(b - a_0)\frac{dx}{db} + 2x = \frac{1}{\rho a_1 b}\frac{d}{db}\{(1 + \rho a_1 b^2)\xi(b)\} \qquad (2.7.28)$$

which is subject to the initial condition

$$x = x_0 \quad \text{at} \quad b = b_1 \qquad (2.7.29)$$

Since Eq. (2.7.28) is essentially the same as Eq. (2.6.31), it can be solved by the same scheme to give

$$x(b) = x_0\left(\frac{b_1 - a_0}{b - a_0}\right)^2 + \left(b + \frac{1}{\rho a_1 b}\right)\frac{\xi(b)}{b - a_0}$$

$$+ \frac{a_0}{(b - a_0)^2}\int_{b_1}^{b}\left(1 + \frac{1}{\rho a_1 b^2}\right)\xi(b)\,db, \qquad b_1 \le b \le b_0 \qquad (2.7.30)$$

The interaction will be over at G, at which $x = x(b_0)$ and then the transmitted Γ_+ shock propagates along the straight line path GH of slope

$$\bar{\sigma}_- = 1 + \rho a_0 a_1 b_0 \qquad (2.7.31)$$

In the region of the transmitted Γ_- simple wave the refracted C_+ characteristics have slopes given by the equation

$$\sigma'_+(b) = 1 + \rho a_0 b^2 \qquad (2.7.32)$$

and, since $a_0 < a_1$, the characteristics become less steep. The concentration values borne on these characteristics are calculated by Eq. (2.7.5) with $a = a_0$ and the corresponding values of b. The new constant state Q is also determined by Eq. (2.7.5) with $a = a_0$ and $b = b_0$.

Transmission between two simple waves

Finally, let us take the case when two simple waves are interacting. A typical situation is depicted in Fig. 2.23 together with the image of the solution in the plane of the characteristic parameters a and b. The Γ_+ simple wave colliding with the Γ_- simple wave from behind, the characteristic parameters for the four constant states would be arranged as follows:

$$\begin{array}{ll} L(a_1, b_1), & Q(a_1, b_0) \\ P(a_0, b_1), & R(a_0, b_0) \end{array} \qquad (2.7.33)$$

where $a_0 < a_1$ and $b_0 < b_1$.

After first meeting at the point $d(x_0, \tau_0)$, the two simple waves, consisting of families of straight line C_- and C_+ characteristics, respectively, interact to form the penetration region $d-e-g-f$ bounded by the four characteristics. Upon entering this region the characteristics no longer remain straight because the concentrations vary along them. We note that along each of the C_+ characteristics, the parameter a increases while b remains constant, so the C_+ characteristics become steeper during the course of interaction. Similarly, the parameter b decreases while a remains invariant along each of the C_- characteristics so that

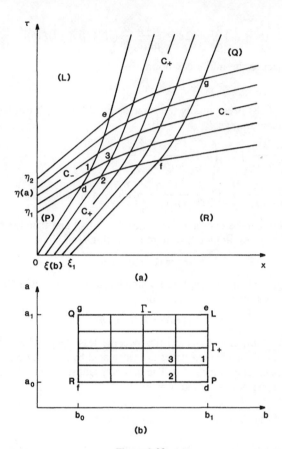

(a)

(b)

Figure 2.23

the C_- characteristics become less and less steep. Upon emerging from the penetration region the C_+ and C_- characteristics become again straight lines and compose the two transmitted simple waves.

The point $d(x_0, \tau_0)$ is located by the two characteristics intersecting there, the equations of which are

$$\tau = \sigma_+(b_1)x = (1 + \rho a_0 b_1^2)x$$

$$\tau - \eta_1 = \sigma_-(a_0)x = (1 + \rho a_0^2 b_1)x$$

and thus

$$x_0 = \frac{\eta_1}{\rho a_0 b_1 (b_1 - a_0)}$$

$$\tau_0 = \frac{1 + \rho a_0 b_1^2}{\rho a_0 b_1 (b_1 - a_0)} \eta_1 \qquad (2.7.34)$$

Two-Solute Chromatography with the Langmuir Isotherm   Chap. 2

For the portion $d-e$ of the C_+ characteristic the overall feature is compatible with the case shown in Fig. 2.21(a) except for the fact that the penetrating curve is a C_+ characteristic rather than the shock path S_+. Thus we have

$$\tau - \eta(a) = \sigma_-(a)x = (1 + \rho b_1 a^2)x$$

$$\frac{d\tau}{dx} = \sigma_+(a) = 1 + \rho b_1^2 a \qquad (2.7.35)$$

and by comparing these with Eqs. (2.7.12), (2.7.13), and (2.7.14) we find that the only change required is b_1 for b_0. Even the initial condition is the same as Eq. (2.7.17). Consequently, the curve $d-e$ is described by Eq. (2.7.19) with b_1 for b_0, where x_0 is given by Eq. (2.7.34). In particular, if the Γ_+ simple wave is centered at $(0, \eta_1)$, we have the equations

$$x(a, b_1) = x_0\left(\frac{b_1 - a_0}{b_1 - a}\right)^2, \qquad a_0 \le a \le a_1 \qquad (2.7.36)$$

and

$$\tau(a, b_1) = (1 + \rho a^2 b_1)x(a, b_1) + \eta_1 \qquad (2.7.37)$$

for the portion $d-e$.

Similarly, we notice that for the portion $d-f$ of the C_- characteristic, the situation is compatible with that presented in Fig. 2.22(a) but Eq. (2.7.27) must now read

$$\tau = \sigma_+(b)\{x - \xi(b)\} = (1 + \rho a_0 b^2)\{x - \xi(b)\}$$

$$\frac{d\tau}{dx} = \sigma_-(b) = 1 + \rho a^2 b \qquad (2.7.38)$$

Hence we need only to replace a_1 by a_0. This implies that the portion $d-f$ has the equation given by Eq. (2.7.30) with a_0 for a_1, where x_0 is given by Eq. (2.7.34). If the Γ_- simple wave is centered at the origin, we obtain the equations

$$x(a_0, b) = x_0\left(\frac{b_1 - a_0}{b - a_0}\right)^2, \qquad b_0 \le b \le b_1 \qquad (2.7.39)$$

and

$$\tau(a_0, b) = (1 + \rho a_0 b^2)x(a_0, b) \qquad (2.7.40)$$

for the portion $d-f$.

Coming to the penetration region $d-e-g-f$, we note that the characteristic parameters b and a remain constant along the C_+ and C_- characteristics, respectively. Thus b and a can be used as the Riemann invariants, J_+ and J_-, respectively (see Sec. 1.9). Now regarding both x and τ as functions of a and b, we can write

$$C_+: \quad \frac{\partial \tau}{\partial a} = \sigma_+(a, b)\frac{\partial x}{\partial a} = (1 + \rho ab^2)\frac{\partial x}{\partial a} \qquad (2.7.41)$$

and

$$C_-: \quad \frac{\partial \tau}{\partial b} = \sigma_-(a, b) \frac{\partial x}{\partial b} = (1 + \rho a^2 b) \frac{\partial x}{\partial b} \qquad (2.7.42)$$

Differentiating the first with respect to b and the second with respect to a and subtracting one from the other, we obtain the equation

$$\frac{\partial^2 x}{\partial a \, \partial b} + \frac{2}{b - a} \left(\frac{\partial x}{\partial a} - \frac{\partial x}{\partial b} \right) = 0 \qquad (2.7.43)$$

where $a < b$. This equation is to be solved in the rectangle $RQLP$ of Fig. 2.23(b). Since x is known along PL (or d–e) and PR (or d–f), the problem of determining the solution in the region d–e–g–f is equivalent to solving a characteristic initial value problem for Eq. (2.7.43). More specifically, Eq. (2.7.19) with b_1 for b_0 and Eq. (2.7.30) with a_0 for a_1 are the initial conditions prescribed at $b = b_1$ and at $a = a_0$, respectively. If the simple waves are both centered, we have Eqs. (2.7.36) and (2.7.39) for the initial data. Once x is determined, τ can be obtained from Eq. (2.7.41) or (2.7.42). Finally, for every set of a and b the concentrations c_1 and c_2 are given by Eq. (2.7.5).

Let us now determine the solution to the partial differential equation (2.7.43). First we notice that this equation is identical to Eq. (1.9.47) of isentropic, one-dimensional flow if we let $\gamma = 5/3$ and identify t, J_+, and J_- with x, b, and $-a$, respectively. The solution to the latter equation was first determined by Riemann and is found in the book by Courant and Friedrichs (1948). We shall derive the solution here by considering the equation

$$\frac{\partial^2 x}{\partial a \, \partial b} + \frac{\lambda}{b - a} \left(\frac{\partial x}{\partial a} - \frac{\partial x}{\partial b} \right) = 0 \qquad (2.7.44)$$

subject to the conditions

$$x(a_0, b) = x_0 \left(\frac{b_1 - a_0}{b - a_0} \right)^\lambda, \qquad b_0 \le b \le b_1 \qquad (2.7.45)$$

$$x(a, b_1) = x_0 \left(\frac{b_1 - a_0}{b_1 - a} \right)^\lambda, \qquad a_0 \le a \le a_1 \qquad (2.7.46)$$

where λ is a constant. If $\lambda = 2$, this reduces to our problem when the simple waves under interaction are both centered.

If we assume a solution of the form

$$x(a, b) = (b - a)^{-\lambda} u(a, b) \qquad (2.7.47)$$

then $u(a, b)$ must satisfy the equation

$$\frac{\partial^2 u}{\partial a \, \partial b} + \frac{\lambda(\lambda - 1)}{(b - a)^2} u = 0 \qquad (2.7.48)$$

We then combine the independent variables a and b into a single variable z as

$$z = \frac{(b - b_1)(a - a_0)}{(b_1 - a_0)(b - a)} \qquad (2.7.49)$$

which is motivated by the particular forms of the differential equation and the initial conditions. This transformation reduces Eq. (2.7.48) to the ordinary differential equation

$$z(1 - z)\frac{d^2u}{dz^2} + (1 - 2z)\frac{du}{dz} - \lambda(1 - \lambda)u = 0 \qquad (2.7.50)$$

This belongs to the class of hypergeometric differential equations and by identifying $\alpha = 1 - \lambda$, $\beta = \lambda$, and $\gamma = 1$, we see that the solution is given by

$$u(z) = KF(1 - \lambda, \lambda; 1; z)$$

where K is an arbitrary constant and F denotes the hypergeometric function defined as

$$F(\alpha, \beta; \gamma; z) = \frac{\Gamma(\gamma)}{\Gamma(\alpha)\Gamma(\beta)} \sum_{n=0}^{\infty} \frac{\Gamma(\alpha + n)\Gamma(\beta + n)}{\Gamma(\gamma + n)} \frac{z^n}{n!} \qquad (2.7.51)$$

where $\Gamma(\cdot)$ is the Gamma function. Hence we have

$$x(a, b) = K(b - a)^{-\lambda}F\left(1 - \lambda, \lambda; 1; \frac{(b - b_1)(a - a_0)}{(b_1 - a_0)(b - a)}\right)$$

and applying Eq. (2.7.45) gives $K = x_0(b_1 - a_0)^\lambda$ because $F(1 - \lambda, \lambda, 1; 0) = 1$, so that we obtain the solution

$$x(a, b) = x_0\left(\frac{b_1 - a_0}{b - a}\right)^\lambda F\left(1 - \lambda, \lambda; 1; \frac{(b - b_1)(a - a_0)}{(b_1 - a_0)(b - a)}\right) \qquad (2.7.52)$$

We note that this also satisfies Eq. (2.7.46).

Since we have $\lambda = 2$ in our problem, the solution of Eq.(2.7.43), satisfying the initial conditions (2.7.36) and (2.7.39), can be written as

$$x(a, b) = x_0\left(\frac{b_1 - a_0}{b - a}\right)^2 F\left(-1, 2; 1; \frac{(b - b_1)(a - a_0)}{(b_1 - a_0)(b - a)}\right)$$

But from the definition of the hypergeometric function it can be shown that $F(-1, 2, 1; z) = 1 - 2z$, so

$$x(a, b) = x_0\left(\frac{b_1 - a_0}{b - a}\right)^2\left\{1 - 2\frac{(b - b_1)(a - a_0)}{(b_1 - a_0)(b - a)}\right\} \qquad (2.7.53)$$

$$= x_0(b_1 - a_0)(b - a)^{-3}\{(b_1 - a)(b - a_0) + (b_1 - b)(a - a_0)\}$$

or by substituting Eq. (2.7.34) for x_0 we obtain

$$x(a, b) = \frac{\eta_1}{\rho} \frac{(b_1 - a)(b - a_0) + (b_1 - b)(a - a_0)}{a_0 b_1(b - a)^3} \qquad (2.7.54)$$

Now $\tau(a, b)$ can be determined by using either one of Eqs. (2.7.41) and

(2.7.42). Along each of the C_+ characteristics the parameter b remains fixed, so we can integrate Eq. (2.7.41) from $a = a_0$ along a C_+ characteristic to have

$$\tau(a, b) - \tau(a_0, b) = \int_{a_0}^{a} (1 + \rho a'b^2)x(a', b) \, da'$$

or, by applying integration by parts,

$$\tau(a, b) - \tau(a_0, b)$$
$$= (1 + \rho ab^2)x(a, b) - (1 + \rho a_0 b^2)x(a_0, b) - \rho b^2 \int_{a_0}^{a} x(a', b) \, da'$$

But the second term on the right-hand side cancels with $\tau(a_0, b)$ by Eq. (2.7.40), and after making the integration we obtain

$$\tau(a, b) = (1 + \rho a^2 b)x(a, b) - \rho b^2 \int_{a_0}^{a} x(a', b) \, da'$$

$$= (1 + \rho a^2 b)x(a, b) + \frac{\eta_1 b^2}{a_0 b_1}\left\{\frac{p}{b - a}(1 - \mu)\right. \qquad (2.7.55)$$

$$\left. - \frac{pb + q}{2(b - a)^2}(1 - \mu^2)\right\}$$

where

$$p = a_0 + b_1 - 2b$$
$$q = b_1(b - a_0) - a_0(b_1 - b) \qquad (2.7.56)$$
$$\mu = \frac{b - a}{b - a_0}$$

In particular, the point g at which the interaction terminates can be located by putting $a = a_1$ and $b = b_0$ in Eqs. (2.7.54) and (2.7.55), i.e.,

$$x = x(a_1, b_0) = \frac{\eta_1}{\rho}\frac{(b_1 - a_1)(b_0 - a_0) + (b_1 - b_0)(a_1 - a_0)}{a_0 b_1(b_0 - a_1)^3} \qquad (2.7.57)$$

and $\tau = \tau(a_1, b_0)$. The duration of the interaction is $\tau(a_1, b_0) - \tau(a_0, b_1) = \tau(a_1, b_0) - \tau_0$, where τ_0 is given by Eq. (2.7.34).

If the simple waves are not centered, we may construct the solution in the region d–e–g–f by applying the step-by-step mapping procedure discussed in Sec. 1.3. In this region we have

$$\sigma_+(a, b) = 1 + \rho ab^2$$
$$\sigma_-(a, b) = 1 + \rho a^2 b \qquad (2.7.58)$$

Thus knowing the values of a and b at each mesh point, the portion $1 \rightarrow 3$ of a C_- characteristic, for example, may be approximated by a straight line of slope

$$\sigma_- = \frac{1}{2}(\sigma_{-,1} + \sigma_{-,3}) \qquad (2.7.59)$$

and the portion $2 \to 3$ of a C_+ characteristic can be drawn to locate the point 3 at the intersection (see Fig. 2.23). Iteration of this procedure generates all the mesh points, so the solution in the penetration region is established. As the number of mesh points is increased, an increasingly accurate solution can be obtained. In fact, this scheme can be very effective even when the simple waves are centered.

The transmitted simple waves can be constructed by the same procedure as discussed above for transmission between a simple wave and a shock.

Exercises

2.7.1. Consider a Γ_+ simple wave based on the characteristic initial data specified over the interval $0 \leq \xi \leq \xi_1$ of the x axis and a Γ_- shock originating from the point $(\xi_2, 0)$ of the x axis, where $\xi_1 < \xi_2$. Determine the shock path when the two waves are interacting.

2.7.2. For the case depicted in Fig. 2.22, suppose that the simple wave is centered at the origin and show that the portion DG of the shock path S_+ is described by the equation

$$\sqrt{\tau - x} = a_0\sqrt{\rho a_1 x} + \sqrt{\eta_1\left(1 - \frac{a_0}{b_1}\right)}$$

so that it is given by a portion of a parabola.

2.7.3. Using the transformation formula

$$F(\alpha, \beta; \gamma; z) = (1 - z)^{-\beta}F\left(\beta, \gamma - \alpha; \gamma; \frac{z}{z - 1}\right)$$

show that Eq. (2.7.52) can be put into the form

$$x(a, b) = x_0\frac{(b_1 - a_0)^{2\lambda}}{(b - a_0)^\lambda(b_1 - a)^\lambda}F\left(\lambda, \lambda; 1; \frac{(b_1 - b)(a - a_0)}{(b - a_0)(b_1 - a)}\right)$$

2.7.4. If we identify t, J_+, and J_- in Eq. (1.9.47) with x, b, and $-a$, respectively, and further let $\gamma = \frac{5}{3}$, we obtain Eq. (2.7.43). This corresponds to the case $N = 1$ of Exercise 1.9.6 so that the solution may be written in the form

$$x = \frac{\partial}{\partial b}\frac{f(b)}{(b - a)^2} - \frac{\partial}{\partial a}\frac{g(a)}{(b - a)^2} + e$$

$$= -(b - a)^{-3}\{2f(b) - (b - a)f'(b) + 2g(a) + (b - a)g'(a)\} + e$$

Determine the arbitrary functions $f(b)$ and $g(a)$ as well as the arbitrary constant e by applying the initial conditions (2.7.36) and (2.7.39).

2.7.5. By using the characteristic parameter α and β instead of a and b in Eqs. (2.7.41) and (2.7.42), show that the differential equation corresponding to Eq. (2.7.43) is given by

$$\frac{\partial^2 x}{\partial \alpha' \partial \beta'} + \frac{2}{\alpha' - \beta'} \left\{ \frac{\alpha'}{\beta'} \frac{\partial x}{\partial \alpha'} - \frac{\beta'}{\alpha'} \frac{\partial x}{\partial \beta'} \right\} = 0$$

where $\alpha' = \alpha + h\gamma$ and $\beta' = \beta + h\gamma$. Consider a solution in the form

$$x(\alpha', \beta') = v(\alpha', \beta')u(\alpha', \beta')$$

and determine $v(\alpha', \beta')$ in such a way that the first-order derivative terms of $u(\alpha', \beta')$ vanish. With $v(\alpha', \beta')$ so determined, show that $u(\alpha', \beta')$ satisfies the equation

$$\frac{\partial^2 u}{\partial \alpha' \partial \beta'} + \frac{2}{(\beta' - \alpha')^2} u = 0$$

which is the same as Eq.(2.7.48).

2.7.6. Using Eqs. (2.7.37) and (2.7.54), integrate Eq. (2.7.42) from $b = b_1$ along a C_- characteristic and thus determine $\tau(a, b)$. Compare the result with Eq. (2.7.55).

2.8 Chromatographic Cycle for Two Solutes

It is well known that if two solutes are put on a chromatographic column as a mixture over a short period of time and then eluted with a stream of pure solvent, the two solutes can be separated. In fact, this is one of the standard operations in the practice of chromatography. We should expect the less strongly adsorbed solute A_1 to move ahead of the more strongly adsorbed one and would like to be able to determine when the separation is completed.

The mathematical problem here is to solve the pair of partial differential equations (2.1.10) subject to the following initial and feed conditions

$$\text{at } \tau = 0: \quad c_1 = 0, \quad\quad c_2 = 0 \quad\quad \text{for } x > 0$$

$$\text{at } x = 0: \begin{cases} c_1 = c_1^0, & c_2 = c_2^0 & \text{for } 0 < \tau < \tau_0 \\ c_1 = 0, & c_2 = 0 & \text{for } \tau > \tau_0 \end{cases} \quad (2.8.1)$$

Clearly, we are going to have a combination of Fig. 2.6 (the saturation problem) and Fig. 2.5 (the elution problem), each of which will be valid for small x. But after a short time the two sets of waves will meet each other and begin to interact. We shall try to follow the development step by step using the results of interaction analysis that are available from the previous sections.

The image of the solution in the hodograph plane is shown in Fig. 2.24 along with that in the (a, b)-plane, in which the Γ characteristics become parallel to the coordinate axes. Figure 2.25 shows the portrait of the solution in the (x, τ)-plane which will be elucidated in conjunction with Fig. 2.24. The concentration profiles are obtained from Figs. 2.24 and 2.25, as depicted in Fig. 2.26, at 10 different times marked on the τ axis of Fig. 2.25. These profiles clearly show characteristic features of the development of the two-solute chromatogram at various stages. Although they are only qualitative and not in scale, these will help understand the following discussion.

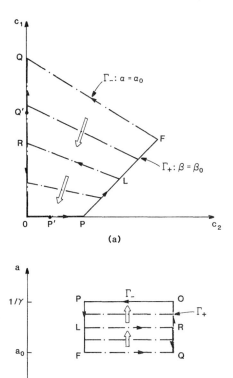

Figure 2.24

To begin with we shall use Eq. (2.1.31) to determine the characteristic parameters for the state (c_1^0, c_2^0), which maps into the point F in Fig. 2.24. Let these be α_0 and β_0. The corresponding values of a and b will be denoted by a_0 and b_0, respectively. The initial state has its image located at the point O and takes $\alpha = 0$ and $\beta = \infty$, so that $a = 1/\gamma$ and $b = 1$. The pairs of the Γ characteristics passing through F and O have intersections at P and Q as shown in Fig. 2.24. The characteristic parameters for these points are given in Table 2.8. Since the discussion will be made in terms of the characteristic parameters whenever convenient, we shall find this table very useful.

TABLE 2.8

State	α	β	a	b
O	0	∞	$1/\gamma$	1
P	0	β_0	$1/\gamma$	b_0
F	α_0	β_0	a_0	b_0
Q	α_0	∞	a_0	1

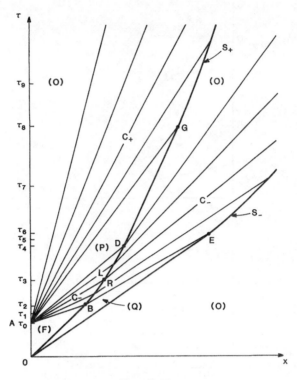

Figure 2.25

For $\tau < \tau_0$ we simply have a saturation problem. The steady inlet concentrations c_1^0 and c_2^0 are being fed to the column, which therefore becomes saturated with both solutes over the first part and with the less strongly adsorbed solute A_1 pushed out in the front. In Fig. 2.24 the image is given by the path $F \to Q \to O$ and to $F \to Q$ and $Q \to O$ there correspond a Γ_- shock and a Γ_+ shock, respectively. This problem has been treated in Sec. 2.3. The slopes of the two shock paths OB and OE in Fig. 2.25 are

$$\tilde{\sigma}_+(b_0, 1) = 1 + \rho a_0 b_0 \qquad (2.8.2)$$

and

$$\tilde{\sigma}_-\left(a_0, \frac{1}{\gamma}\right) = 1 + \frac{\rho a_0}{\gamma} \qquad (2.8.3)$$

respectively, while the state Q has the concentrations

$$c_1^q = c_1^0 - \alpha_0 c_2^0, \qquad c_2^q = 0 \qquad (2.8.4)$$

At $\tau = \tau_0$ the introduction of the sample of two solutes A_1 and A_2 is complete. The state $F(c_1^0, c_2^0)$ is found from the inlet to $x = \tau_0/\tilde{\sigma}_+(b_0, 1)$ and the

Two-Solute Chromatography with the Langmuir Isotherm   Chap. 2

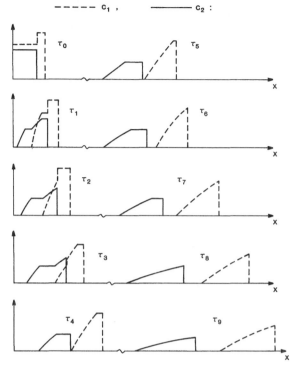

Figure 2.26

state Q of pure A_1 from this point to $x = \tau_0/\bar{\sigma}_-(a_0, 1/\gamma)$. This is shown in the first part of Fig. 2.26.

The feedstream is then switched to the flow of pure solvent which starts to elute the saturated part of the column. For $\tau_0 < \tau < \tau_2 = \tau_B$ the two shocks are still undisturbed and move with constant speed, but two simple waves develop in the region near point A. The solution in this region has also been treated in Sec. 2.3. It has its image along the path $O \rightarrow P \rightarrow F$ in Fig. 2.24. The portion $O \rightarrow P$ gives a Γ_- simple wave, the region of which is covered by a family of straight C_+ characteristics. These have slopes given by

$$\sigma_+(b) = 1 + \frac{\rho b^2}{\gamma} \qquad \text{for } b_0 \le b \le 1 \qquad (2.8.5)$$

and the concentration values borne on each of these characteristics are calculated by using $a = 1/\gamma$ and the corresponding value of b in Eq. (2.7.5). Here we obtain $c_1 = 0$, so that in the region of the Γ_- simple wave we have only A_2. Similarly, to the portion $P \rightarrow F$ there corresponds a Γ_+ simple wave which is represented by a family of straight C_- characteristics of slope

$$\sigma_-(a) = 1 + \rho b_0 a^2 \qquad \text{for } a_0 \le a \le \frac{1}{\gamma} \qquad (2.8.6)$$

Again the concentration values carried by each of these characteristics are determined by Eq. (2.7.5) with $b = b_0$ and the corresponding value of a. The new constant state P that separates the two simple wave regions has the concentrations

$$c_1^p = 0, \qquad c_2^p = c_2^0 - \frac{c_1^0}{\beta_0} \qquad (2.8.7)$$

so that it is a state of pure A_2. Typical profiles taken at $\tau = \tau_1$ are presented in the second part of Fig. 2.26.

Here we notice that the image of the solution makes a closed loop in the hodograph plane which may be expressed as $O \rightarrow P \rightarrow F \rightarrow Q \rightarrow O$, where the arrows indicate that the transition is in the positive x direction as usual. This loop will be referred to as the *chromatographic cycle* and we shall see that this cycle deforms continuously as the process goes on.

At $\tau = \tau_2 = \tau_B$ the head of the Γ_+ simple wave has just reached the Γ_- shock (see the third part of Fig. 2.26). The C_- characteristic AB has the equation

$$\tau - \tau_0 = \sigma_- x = (1 + \rho a_0^2 b_0)x$$

and the equation of the shock path OB is

$$\tau = \tilde{\sigma}_+(b_0, 1)x = (1 + \rho a_0 b_0)x$$

so that the point of intersection B has the coordinates

$$x_B = \frac{\tau_0}{\rho a_0 b_0 (1 - a_0)}$$

$$\tau_B = \left(1 + \frac{1}{\rho a_0 b_0}\right)\frac{\tau_0}{1 - a_0} \qquad (2.8.8)$$

From this point on, the Γ_+ simple wave and the Γ_- shock are involved in a transmissive interaction so that at a typical time $\tau = \tau_3$ during the interaction the chromatographic cycle will take the path $O \rightarrow P \rightarrow L \rightarrow R \rightarrow Q \rightarrow O$. From the concentration profiles at $\tau = \tau_3$ of Fig. 2.26 we can see the Γ_- shock weakening and also the transmitted Γ_+ simple wave developing on the right-hand side of the Γ_- shock. The situation here is the same as in Fig. 2.21, and since the simple wave in centered, we can use Eqs. (2.7.21) and (2.7.22) with $b_1 = 1$ and $\eta_1 = \tau_0$. Hence for the shock path BD we have the parametric representation

$$x = x_B \left(\frac{1 - a_0}{1 - a}\right)^2 = \frac{\tau_0(1 - a_0)}{\rho a_0 b_0 (1 - a)^2}$$

$$\tau = \tau_0 + (1 + \rho b_0 a^2)x \qquad (2.8.9)$$

for $a_0 \leq a \leq 1/\gamma$, or by eliminating a between the two equations, we obtain

$$\sqrt{\tau - \tau_0 - x} = \sqrt{\rho b_0 x} - \sqrt{\tau_0\left(\frac{1}{a_0 - 1}\right)} \qquad (2.8.10)$$

and this shows that the shock path BD is a part of a parabola.

To the right of the shock path BD the transmitted Γ_+ simple wave develops. Since $b = 1$ in this region and a remains unchanged across the Γ_- shock, the straight C_- characteristics have slopes given by

$$\sigma'_-(a) = 1 + \rho a^2 \qquad \text{for } a_0 \le a \le \frac{1}{\gamma} \qquad (2.8.11)$$

This is greater than $\sigma_-(a)$ of Eq. (2.8.6), so that the C_- characteristics become steeper upon transmission across the Γ_- shock. The concentration values borne on each of these characteristics are given by Eq. (2.7.5) with $b = 1$ and the corresponding value of a. We immediately find that $c_2 = 0$ and this implies that to the right of the shock path BD we have no A_2. It is interesting that the Γ_- shock plays the role of a sieve which passes the less strongly adsorbed solute A_1 but holds the more strongly adsorbed solute A_2.

At $\tau = \tau_4 = \tau_D$ the interaction is over and since the C_- characteristic AD bears $c_1 = 0$ (or $a = 1/\gamma$), we see on the right-hand side of the shock the state $c_1 = c_2 = 0$. Hence there develops a region of constant state $c_1 = c_2 = 0$ between the shock path and the region of the transmitted simple wave. At this stage the image of the Γ_- shock, which had been receding from $F \rightarrow Q$ toward $P \rightarrow O$, has just reached $P \rightarrow O$ so that the chromatographic cycle degenerates to the form $O \rightarrow P \rightarrow O \rightarrow Q \rightarrow O$. Clearly, we see that the separation is complete with the more strongly adsorbed solute A_2 on the left-hand side of the Γ_- shock and the less strongly adsorbed solute A_1 on the right as observed from the profiles at $\tau = \tau_4$ of Fig. 2.26. The point D of complete separation has the coordinates

$$x_D = \frac{\tau_0 \gamma^2 (1 - a_0)}{\rho (\gamma - 1)^2 a_0 b_0}$$
$$\tau_D = \tau_0 \left\{ 1 + \frac{1 - a_0}{(\gamma - 1)^2 a_0} + \frac{\gamma^2 (1 - a_0)}{\rho (\gamma - 1)^2 a_0 b_0} \right\} \qquad (2.8.12)$$

from Eq. (2.8.9) with $a = 1/\gamma$. These are the expressions for the length of column and the time required for a given separation and clearly both increase as $\gamma \rightarrow 1$, i.e., as the relative adsorptivity becomes smaller.

For $\tau_4 < \tau < \tau_8 = \tau_G$ the Γ_- shock propagates with a constant speed with $a = 1/\gamma$, $b^l = b_0$, and $b^r = 1$, so that the path DG has a slope

$$\bar{\sigma}'_+(b_0, 1) = 1 + \frac{\rho b_0}{\gamma} \qquad (2.8.13)$$

which is greater than $\bar{\sigma}_+(b_0, 1)$ of Eq. (2.8.2). Since the tail of the Γ_+ simple wave moves faster than the Γ_- shock, the two zones of pure component tend to separate more and more and a region of state O opens up between the two zones.

At $\tau = \tau_5$ the Γ_+ shock still remains undisturbed and propagates slightly ahead of the head of the Γ_+ simple wave. Thus the profile of c_1 shows a portion of the plateau of state Q, but at $\tau = \tau_6$ the Γ_+ shock is caught up by the head of the Γ_+ simple wave. These are observed from the profiles of c_1 at $\tau = \tau_5$ and $\tau = \tau_6$, respectively, of Fig. 2.26. The two waves begin to interact from the point E. Since the C_- characteristic BE has the equation

$$\tau - \tau_B = (1 + \rho a_0^2)(x - x_B)$$

and the shock path OE is the line

$$\tau = \left(1 + \frac{\rho a_0}{\gamma}\right)x$$

their intersection E has the coordinates

$$x_E = \frac{\tau_0 \gamma (b_0 - a_0)}{\rho a_0 b_0 (1 - a_0)(1 - \gamma a_0)}$$

$$\tau_E = \frac{\tau_0 \gamma (b_0 - a_0)(1 + \rho a_0/\gamma)}{\rho a_0 b_0 (1 - a_0)(1 - \gamma a_0)}$$

(2.8.14)

in which Eq. (2.8.8) has been introduced.

The situation here corresponds to the case shown in Fig. 2.15 and since the simple wave is not centered, we shall use Eq. (2.6.20). But the interacting waves are Γ_+ waves, so that we must interchange the roles of a and b. Now by measuring the vertical distance of each of the C_- characteristics above the point E we obtain the data $\bar{\eta}(a)$. The fixed parameters are identified as $b = 1$, $a^r = 1/\gamma$, and $a^p = a_0$. Consequently, the shock path S_- beyond the point E is described by

$$x(a) = x_E + \frac{\gamma}{\rho(1 - \gamma a)}\left\{\frac{\bar{\eta}(a)}{a} + \frac{1}{1 - \gamma a}\int_{a_0}^{a}\frac{\bar{\eta}(a)}{a^2}\,da\right\}$$

$$\tau(a) = \tau_E + \bar{\eta}(a) + (1 + \rho a^2)x(a)$$

(2.8.15)

for $a_0 \le a \le 1/\gamma$. Clearly, both x and τ diverge in the limit as $a \to 1/\gamma$, so the interaction goes on indefinitely. As the shock propagates, the two Γ_+ waves cancel each other and a typical picture of this phenomenon is depicted by the profile of c_1 at $\tau = \tau_7$ of Fig. 2.26. At this stage the chromatographic cycle becomes $O \to P \to O \to Q' \to O$, where the point Q' approaches the origin O asymptotically as the interaction proceeds.

At $\tau = \tau_8$, the head of the Γ_- simple wave catches up with the Γ_- shock (see the profile of c_2 at $\tau = \tau_8$ of Fig. 2.26) and the two waves start to interact at the point G. The C_+ characteristic AG is the line

$$\tau - \tau_0 = \left(1 + \frac{\rho b_0^2}{\gamma}\right)x$$

and the shock path DG has the equation

$$\tau - \tau_D = \left(1 + \frac{\rho b_0}{\gamma}\right)(x - x_D)$$

so that the point G has the coordinates

$$x_G = \frac{\tau_0 \gamma (1 - a_0)}{\rho(\gamma - 1)a_0 b_0 (1 - b_0)}$$

$$\tau_G = \tau_0 + \left(1 + \frac{\rho b_0^2}{\gamma}\right) x_G \tag{2.8.16}$$

in which Eq. (2.8.12) has been introduced.

The interaction here is identical to the case shown in Fig. 2.15. Although the shock path is not the straight line emanating from the origin, we can still use the first expression of Eq. (2.6.18) with $b^p = b_0$, $b^r = 1$, and $x_0 = x_G$. Hence beyond the point G the shock path S_+ has the parametric representation

$$x(b) = x_G \left(\frac{1 - b_0}{1 - b}\right)^2 = \frac{\tau_0 \gamma (1 - \gamma a_0)(1 - b_0)}{\rho(\gamma - 1)a_0 b_0 (1 - b)^2}$$

$$\tau(b) = \tau_0 + \left(1 + \frac{\rho b^2}{\gamma}\right) x(b) \tag{2.8.17}$$

for $b_0 \leq b < 1$. By eliminating b between the two equations we obtain

$$\sqrt{\tau - \tau_0 - x} = \sqrt{\frac{\rho}{\gamma}} x - \sqrt{\frac{\tau_0(1 - \gamma a_0)(1 - b_0)}{(\gamma - 1)a_0 b_0}} \tag{2.8.18}$$

so that the shock path is a portion of a parabola. Here again we notice that both x and τ diverge in the limit as $b \to 1$ and thus the interaction goes on indefinitely. By the process of interaction the two Γ_- waves gradually cancel each other, so the shock is continuously weakening, as shown in the last part of Fig. 2.26. The chromatographic cycle is now $O \to P' \to O \to Q' \to O$, where P', being an intermediate point on OP, approaches asymptotically the origin O.

It is obvious that continuance of the chromatography will lead only to the lengthening of each chromatogram of a pure component, which is not desirable although it gives a more distinct separation. Indeed, a column of length x_D will be good enough to obtain complete separation. In this case the outgoing stream contains no solute until $\tau = (1 + \rho a_0/\gamma)x_D$ when the Γ_+ shock reaches the exit and gives the breakthrough of the less strongly adsorbed solute A_1. We expect to see the solute A_1 emerging until $\tau = \tau_D$ [see Eq. (2.8.12)] when the Γ_- shock arrives at the exit and gives the breakthrough of the more strongly adsorbed solute A_2. The solute A_2 continues to emerge and, although the breakthrough curve may tail off, its rear end carrying zero concentration reaches the exit at $\tau = (1 + \rho/\gamma)x_D$, the time when the uppermost C_+ characteristic intersects the vertical line $x = x_D$ in Fig. 2.25. Since the sample was introduced during a period of time τ_0, the ratio of the volume of adsorbent required to that of sample treated is determined from Eq. (2.8.12) as

$$r = \frac{\text{volume of adsorbent required}}{\text{volume of sample treated}}$$

$$= \frac{(1 - \epsilon)z_D}{\epsilon V t_0} = \nu \frac{x_D}{\tau_0} = \frac{\nu(1 - a_0)}{\rho(1 - 1/\gamma)^2 a_0 b_0} \qquad (2.8.19)$$

Clearly, this ratio decreases as γ becomes larger and this implies that the more different the values of $N_1 K_1$ and $N_2 K_2$ are, the smaller a volume of adsorbent is required to treat a given volume of sample, which is just what we would expect qualitatively.

It is interesting to notice that this whole analysis uses only the fact that $\gamma > 1$; i.e., we distinguish between the two solutes on the grounds that one is more strongly adsorbed than the other. Then whatever the concentrations of the two in the sample may be, it is less strongly adsorbed solute that moves ahead and separates from the other.

To illustrate this further, we shall consider a numerical example for which we have

$$\epsilon = 0.4$$

$$N_1 = N_2 = 1 \text{ mol/liter}$$

$$K_1 = 5 \text{ liters/mol}, \quad K_2 = 7.5 \text{ liters/mol}$$

$$c_1^0 = \frac{2}{30} \text{ mol/liter}, \quad c_2^0 = \frac{1}{10} \text{ mol/liter}$$

$$\tau_0 = 5$$

The constant parameters are determined as

$$h = 1, \qquad \gamma = \frac{3}{2}, \qquad k = \frac{1}{15}, \qquad \rho = \frac{135}{8}$$

and from Eq. (2.1.31) we obtain

$$\alpha_0 = -\frac{2}{3}, \qquad \beta_0 = 1$$

or

$$a_0 = \frac{2}{5}, \qquad b_0 = \frac{4}{5}$$

Now the constant states P and Q are determined by Eqs. (2.8.4) and (2.8.7), i.e.,

$$P: \quad c_1 = 0, \qquad c_2 = \frac{1}{30} \text{ mol/liter}$$

$$Q: \quad c_1 = \frac{4}{30} \text{ mol/liter}, \qquad c_2 = 0$$

From Eqs. (2.8.8), (2.8.12), (2.8.14), and (2.8.16) we can directly calculate the coordinates of the points B, D, E, and G to obtain

$$B\left(\frac{125}{81}, \frac{800}{81}\right), \qquad D(5, 40)$$

$$E\left(\frac{125}{54}, \frac{1375}{108}\right), \qquad G\left(\frac{25}{3}, \frac{220}{3}\right)$$

The straight portions of the shock paths have the slopes

$$OB: \quad \bar{\sigma}_+(b_0, 1) = 6.4$$

$$OE: \quad \bar{\sigma}_-\left(a_0, \frac{1}{\gamma}\right) = 5.5$$

$$DG: \quad \bar{\sigma}_+\left(b_0, \frac{1}{\gamma}\right) = 10.0$$

and the curved portions are expressed as

$$BD: \quad \sqrt{\tau - x - 5} = \sqrt{13.5x} - \sqrt{7.5}$$

$$\text{beyond } G: \quad \sqrt{\tau - x - 5} = \sqrt{11.25x} - \sqrt{2.5}$$

while the shock path S_- beyond E involves numerical integration because Eq. (2.8.15) is used. It is now possible to locate the two shock paths S_+ and S_- as shown in Fig. 2.27.

The simple waves are constructed on the basis of the concentration values for the sake of convenience in determining the concentration profiles. The unit of concentrations is mol/liter everywhere. For the Γ_- simple wave we have $\alpha = 0$ (or $a = 1/\gamma$), so $c_1 = 0$. From Eq. (2.1.34) we can show that

$$b = \left\{1 + \frac{h(\gamma - 1)}{k} c_2\right\}^{-1} = \frac{1}{1 + K_2 c_2} = \frac{1}{1 + 7.5 c_2}$$

and the C_+ characteristics have the slope

$$\sigma_+(b) = 1 + \frac{\rho b^2}{\gamma} = 1 + \frac{45}{4} b^2$$

Thus for each value of c_2 we can draw a straight C_+ characteristic emanating from the point $(0, 5)$ and bearing that particular value of c_2 and $c_1 = 0$.

For the Γ_+ simple wave before interaction we have $\beta = \beta_0 = 1$, so c_1 is related to c_2 by the equation of the Γ_+ characteristic for $\beta = 1$, i.e.,

$$c_1 = \beta_0 c_2 - \frac{k\beta_0}{\beta_0 + h} = c_2 - \frac{1}{30}$$

It can be shown from Eq. (2.1.34) that

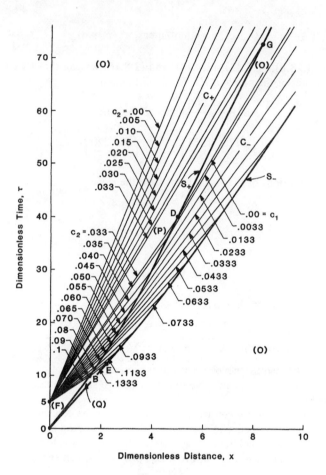

Figure 2.27

$$a = \{1 + \gamma(\beta_0 + 1)K_1 c_2\}^{-1} = \frac{1}{1 + 15c_2}$$

and the C_- characteristics have the slope

$$\sigma_-(a) = 1 + \rho b_0 a^2 = 1 + 13.5a^2$$

For each value of c_2, therefore, we obtain the values of c_1, a, and σ_- and construct the Γ_+ simple wave.

Finally, in the region of the transmitted Γ_+ simple wave we have $\beta = \infty$ (or $b = 1$), so that $c_2 = 0$. Since the Γ_- shock has its image on a Γ_- characteristic, we can write

$$c_1 - c_1' = \alpha(c_2 - c_2') = \alpha c_2$$

or

$$c_1' = c_1 - \alpha c_2, \qquad -\frac{2}{3} \leq \alpha \leq 0$$

where the prime denotes the values on the right-hand side of the shock. Thus c_1 jumps up across the shock. Since a remains invariant across the shock, the slope of the refracted C_- characteristic is given by

$$\sigma_-'(a) = 1 + \rho a^2 = 1 + 16.875a^2$$

Thus the Γ_+ simple wave can be constructed.

By using the layout of the characteristics and the shock paths given in Fig. 2.27 the concentration profiles are obtained at six different times as shown in Fig. 2.28. In the first part ($\tau = 5$) the sample has just been put on the column and the next two parts show the disappearance of the plateaus F and Q, respectively, and the beginning of two different interactions. At $\tau = 25$ the chromatogram is still being developed, and at $\tau = 40$ we clearly see that the two solutes A_1 and A_2 are separated. The position separating the two solutes is

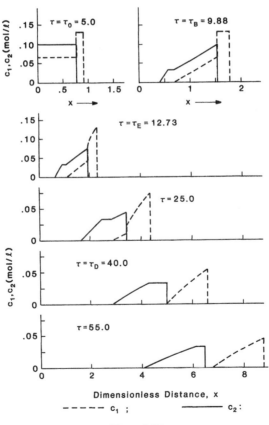

Figure 2.28

located at $x = x_D = 5$. By $\tau = 55$, the zone of pure A_1 has moved further than that of pure A_2, giving a more distinct separation.

The ratio of the volume of adsorbent required to that of the sample is equal to $r = \nu x_D / \tau_0 = 1.5$, which is remarkably low. This is the result of the particular values of $N_1 K_1$ and $N_2 K_2$ that were chosen for the example, but it does show that with strong adsorption and very different values of $N_i K_i$ the solutes can be separated remarkably easily.

Exercises

2.8.1. A mixture containing two solutes A_1 and A_2 is to be separated by dividing into equal portions such that a given column will just separate the two solutes in one sample. If these are fed to the column at as frequent an interval as possible, show that the period of elution is equal to or greater than

$$\frac{\tau_0 \gamma (1 - a_0)^2}{(\gamma - 1)^2 a_0 b_0}$$

where τ_0 is the period of sample introduction.

2.8.2. Consider the numerical example of this section with $K_2 = 10$ liters/mol while other data remain the same and construct the solution by following the procedure discussed in this section.

2.8.3. Determine the value of the ratio r in each case and discuss.
(a) $K_2 = 10$ liters/mol
(b) $K_2 = 6$ liters/mol
(c) $c_2 = \frac{4}{30}$ mol/liter
(d) $c_1 = \frac{1}{10}$ mol/liter
In each of the four cases above, other data remain the same as in the numerical example of this section.

2.9 Introduction to Displacement Development

Another important operation used in the practice of chromatography is displacement development. In this operation a mixture containing one or more adsorbable solutes is put on a column of adsorbent over a finite period of time and displaced through the column by adding a solution containing only one solute with higher adsorptivity than any of the solutes present. Then the solutes are gradually separated and eventually form bands of pure solute. These bands maintain sharp boundaries at both ends and propagate side by side with the same speed. The solute having the largest adsorptivity is called the *development agent* or simply the *developer*.

While the general problem is to be treated in Chapter 4, we should be able to analyze the development of a single-solute chromatogram since the overall system includes only two solutes. Although separation may not be of importance in this simple case, the process has been proven effective in the recovery of trace elements from mixtures. The treatment of such a simple case here will show all the characteristics of displacement development and certainly lay down a groundwork for the analysis of the general problem.

Suppose that we have a sample containing solute A_1 at concentration c_1^0 and this is put on a chromatographic column over a time period τ_0. At $\tau = \tau_0$ the feedstream is switched to the flow of mixture containing solute A_2 that is more strongly adsorbed than A_1. Here the solute A_2 plays the role of the developer. Thus the problem is now to determine the solution of Eq. (2.1.10) subject to the initial and feed conditions

at $\tau = 0$: $c_1 = 0$ and $c_2 = 0$ for $x > 0$

at $x = 0$: $\begin{cases} c_1 = c_1^0 & \text{and} \quad c_2 = 0 \quad \text{for } 0 < \tau < \tau_0 \quad (2.9.1) \\ c_1 = 0 & \text{and} \quad c_2 = c_2^* \quad \text{for } \tau > \tau_0 \end{cases}$

The situation is first examined in the hodograph plane as shown in Fig. 2.29. The state of the sample maps into F, a point on the c_1 axis, and the Γ_- characteristic passing through the point F has intercept on the c_2 axis at A. From Eq. (2.1.31) we immediately find that the characteristic parameters for the state $F(c_1^0, 0)$ are

$$\alpha = \alpha_0 = \frac{-hc_1^0}{k + c_1^0}, \qquad \beta = \infty \qquad (2.9.2)$$

and thus point A has

$$c_2 = -\frac{c_1^0}{\alpha_0} = \frac{k + c_1^0}{h} \qquad (2.9.3)$$

Also marked in Fig. 2.29 is the watershed point B on the c_2 axis at which we have

$$c_2 = \frac{k}{h} = \frac{\gamma - 1}{K_2} \qquad (2.9.4)$$

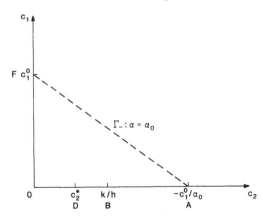

Figure 2.29

from Sec. 2.1. We recall that all the Γ_+ characteristics have their c_2 intercepts on the portion OB, whereas the Γ_- characteristics have their c_2 intercepts positioned to the right of B. The portion OB is the Γ_- characteristic with $\alpha = 0$ and the portion to the right of B is the Γ_+ characteristic with $\beta = \infty$. Now the state of the developer will have its image on a point D along the c_2 axis. It is then clear that the solution of the problem will have different structures depending on the level of the developer concentration c_2^*, i.e., depending on where point D is located on the c_2 axis. More specifically, there will be three different cases with point D falling on OB, BA, and to the right of A, respectively, and we shall discuss these cases one by one.

In all three cases, however, the solution for $\tau < \tau_0$ will be the same since the sample is simply fed to the column, which therefore becomes saturated with the solute A_1 over the first part. The solution is given by a Γ_+ shock that separates the states $F(c_1^0, 0)$ and $O(0, 0)$. The characteristic parameters for these states are as follows:

State	α	β	a	b
O	0	∞	$1/\gamma$	1
F	α	∞	a_0	1

Thus the shock path S_-, emanating from the origin of the (x, τ)-plane, has a slope given by

$$\tilde{\sigma}_-\left(a_0, \frac{1}{\gamma}\right) = 1 + \frac{\rho a_0}{\gamma} \qquad (2.9.5)$$

At $\tau = \tau_0$ the developer is introduced and the solution beyond this instant really depends on the value of c_2^*.

Case I: $c_2^* \le k/h = (\gamma - 1)/K_2$

Here the developer concentration c_2^* is fairly low and the point image D of the developer state falls on the portion OB of the c_2 axis, so that the image of the solution consists of the path $D \rightarrow O \rightarrow F \rightarrow O$ as we proceed in the x direction. Clearly, the portion $D \rightarrow O$ gives a Γ_- shock and $O \rightarrow F$ a Γ_+ simple wave while $F \rightarrow O$ represents the Γ_+ shock emanating from the origin. The layout of the shock paths and the straight C_- characteristics in the (x, τ)-plane is depicted in Fig. 2.30.

The characteristic parameters for the state D are determined from Eq. (2.1.31) as

$$\alpha = 0, \qquad \beta = \beta^* = \frac{k - hc_2^*}{c_2^*} \qquad (2.9.6)$$

and thus $a = 1/\gamma$ and $b = b^* = (\beta^* + h)/(\beta^* + h\gamma)$. The shock path $EK(S_+)$ has the slope

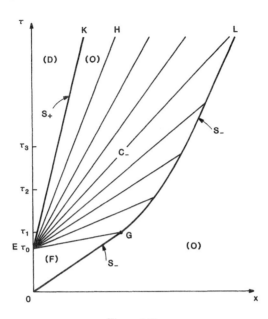

Figure 2.30

$$\bar{\sigma}_+(b^*, 1) = 1 + \frac{\rho b^*}{\gamma} \qquad (2.9.7)$$

The solute A_2 remains to the left of this shock while the state O of no solute develops on the right-hand side. It is immediately apparent that the developer A_2 has no influence over the behavior of the solute A_1 so that the solute A_1 is simply eluted by pure solvent. Since only one solute is present there, the situation to the right of the shock path EK is exactly the same as that of Fig. 6.12 of Vol. 1 and is fully discussed in Sec. 6.3 of Vol. 1. For completeness, however, we shall treat the analysis now in terms of the characteristic parameters.

The Γ_+ simple wave solution in the region GEH is given by a family of straight C_- characteristics which have slopes

$$\sigma_-(a) = 1 + \rho a^2, \qquad a_0 \le a \le \frac{1}{\gamma} \qquad (2.9.8)$$

since $b = 1$. The concentration c_1 borne on each of these characteristics is determined from Eq. (2.7.5) as

$$c_1 = \frac{1 - \gamma a}{K_1 \gamma a} \qquad (2.9.9)$$

Now the C_- characteristic EG has the equation

$$\tau - \tau_0 = (1 + \rho a_0^2)x$$

and the shock path OG is the line

$$\tau = \left(1 + \frac{\rho a_0}{\gamma}\right)x$$

so that the point of intersection G has the coordinates

$$x_G = \frac{\tau_0 \gamma}{\rho a_0 (1 - \gamma a_0)}$$

$$\tau_G = \frac{\tau_0 \gamma (1 + \rho a_0/\gamma)}{\rho a_0 (1 - \gamma a_0)} \qquad (2.9.10)$$

From this point on the two Γ_+ waves interact to cancel each other. The situation is the same as that of Fig. 2.15, but we have Γ_+ waves instead of Γ_- waves, so that the roles of a and b must be interchanged in Eq. (2.6.18). With the identification $x_0 = x_G$, $a^p = a_0$, and $a^r = 1/\gamma$, therefore, we obtain the parametric representation of the shock path beyond point G, i.e.,

$$x(a) = x_G\left(\frac{1 - \gamma a_0}{1 - \gamma a}\right)^2 = \frac{\tau_0 \gamma (1 - \gamma a_0)}{\rho a_0 (1 - \gamma a)^2}$$

$$\tau(a) = \tau_0 + (1 + \rho a^2)x(a) \qquad (2.9.11)$$

Eliminating a between the two equations, we find that

$$\sqrt{\tau - \tau_0 - x} = \frac{1}{\gamma}\sqrt{\rho x} - \sqrt{\tau_0\left(\frac{1}{\gamma a_0} - 1\right)} \qquad (2.9.12)$$

so that the shock path beyond G is a portion of a parabola.

In Fig. 2.31 the concentration profiles are shown at four different times marked in Fig. 2.30. The chromatogram of the solute A_1 is obtained at $\tau = \tau_0$ and this is to be developed by the developer A_2. But the two bands of pure solute remain apart from the beginning with the zone of no solute in between.

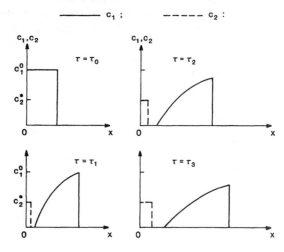

Figure 2.31

Two-Solute Chromatography with the Langmuir Isotherm   Chap. 2

Hence the sample is not developed by the development agent A_2 but simply eluted by the solvent so that the operation is not satisfactory in the sense that displacement development is not realized.

Case II: $k/h < c_2^* < (k + c_1^0)/h$

In this case the point D lies between B and A on the c_2 axis. The solution now has its image along the path $D \to P \to F \to O$, and to each of the three segments there corresponds a Γ_- shock, a Γ_+ simple wave, and a Γ_+ shock, respectively. Thus the structure of the solution is similar to that in the previous case, but the point P here represents a state of pure A_1. Figure 2.32 shows the portrait of the solution in the (x, τ)-plane.

The characteristic parameters for the state D are

$$\alpha = \alpha^* = -\frac{hc_2^* - k}{c_2^*}, \qquad \beta = 0 \qquad (2.9.13)$$

since $hc_2^* - k > 0$, so that $a = a^* = (\alpha^* + h)/(\alpha^* + h\gamma)$ and $b = 1/\gamma$. The state P has the concentrations

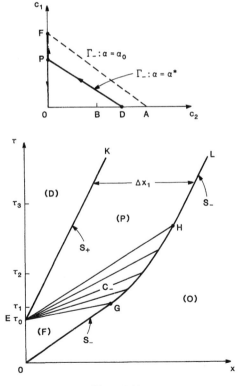

Figure 2.32

$$c_1 = c_1^* = -\alpha^* c_2^* = hc_2^* - k, \qquad c_2 = 0 \qquad (2.9.14)$$

and the characteristic parameters

$$\alpha = \alpha^*, \qquad \beta = \infty$$

or

$$\alpha = \alpha^*, \qquad b = 1$$

We note that c_1^*, the concentration of the state P, is less than c_1^0. Hence the shock path EK (S_+) has the slope

$$\tilde{\sigma}_+\left(\frac{1}{\gamma}, 1\right) = 1 + \frac{\rho a^*}{\gamma} \qquad (2.9.15)$$

which is smaller than $\tilde{\sigma}_+(b^*, 1)$ of Eq. (2.9.7). This implies that the Γ_- shock propagates faster here than in the previous case. Across this shock c_2 drops to zero but c_1 jumps up to c_1^*.

The straight C_- characteristics in the simple wave region GEH have slopes given by

$$\sigma_-(a) = 1 + \rho a^2, \qquad a_0 \leq a \leq a^* \qquad (2.9.16)$$

since b remains fixed at 1. Each of these characteristics carries a constant value of c_1 determined by Eq. (2.9.9). The C_- characteristic EG and the shock path OG are the same lines as those in the previous case, so the point of intersection G is located by Eq. (2.9.10). Furthermore, the portion GH of the shock path S_- is also described by Eq. (2.9.11) or Eq. (2.9.12). In this case, however, the interaction between the Γ_+ simple wave and the Γ_+ shock terminates when the C_- characteristic EH intersects the shock path. Since $a = a^*$ along the line EH, point H has the coordinates

$$x_H = \frac{\tau_0 \gamma (1 - \gamma a_0)}{\rho a_0 (1 - \gamma a^*)^2}$$
$$\tau_H = \tau_0 + \{1 + \rho (a^*)^2\} x_H \qquad (2.9.17)$$

Beyond point H the Γ_+ shock maintains constant strength c_1^* and the straight-line shock path HL (S_-) has the slope

$$\tilde{\sigma}_-\left(a^*, \frac{1}{\gamma}\right) = 1 + \frac{\rho a^*}{\gamma} \qquad (2.9.18)$$

This value of $\tilde{\sigma}_-(a^*, 1/\gamma)$ is the same as the value of $\tilde{\sigma}_+(1/\gamma, 1)$ given by Eq. (2.9.15), so that the two shocks are now propagating with the same speed and the two shock paths S_+ and S_- are parallel to each other. Between the two shocks we see that the band of pure A_1 is established, which has the concentration level

$$c_1^* = hc_2^* - k = hc_2^* - \frac{1}{K_1}\left(1 - \frac{1}{\gamma}\right) \qquad (2.9.19)$$

and the bandwidth

$$\Delta x_1 = x_H - \frac{\tau_H - \tau_0}{\bar{\sigma}_+(1/\gamma, 1)} = \frac{\tau_0}{1 + \rho a^*/\gamma} \left(\frac{a^*}{a_0} \cdot \frac{1 - \gamma a_0}{1 - \gamma a^*} \right)$$

or, by using Eq.(2.9.9),

$$\Delta x_1 = \frac{\tau_0 c_1^0 / c_1^*}{1 + \nu N_1 K_1 / (1 + K_1 c_1^*)} \tag{2.9.20}$$

Since the denominator is just the slope of the shock path, the vertical distance between the two shock paths is given by

$$\Delta \tau_1 = \frac{\tau_0 c_1^0}{c_1^*} \tag{2.9.21}$$

and this simply shows that the total amount of sample, $\tau_0 c_1^0$, initially put on is contained in the band.

Here the operation is successful and the column length required for a full development of the pure A_1 band is equal to x_H given by Eq. (2.9.17). With a column of this length the exit stream contains no solute for $\tau < \tau_H$, but at $\tau = \tau_H$ the Γ_+ shock reaches the exit to give the breakthrough of the solute A_1. Over the period of $\Delta \tau_1$ the solute A_1 continues to emerge at the concentration level c_1^*. Finally, the Γ_- shock arrives at the exit at $\tau = \tau_H + \Delta \tau_1$ and the developer A_2 makes the breakthrough as the concentration c_1 suddenly drops to zero.

The concentration profiles are shown in Fig. 2.33 at four different times. In the first part ($\tau = \tau_0$) we see the chromatogram of A_1 being formed and the next two parts show that while the solute A_1 is displaced by the developer A_2,

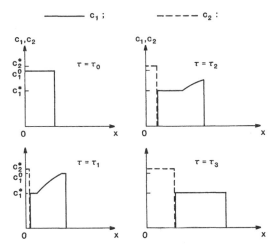

Figure 2.33

the band of pure A_1 is in the process of formation. The last part $(\tau = \tau_4 \geq \tau_H)$ shows the fully established band of pure A_1.

Case III: $c_2^* > (k + c_1^0)/h$

Here the developer concentration c_2^* is sufficiently high so that the point D falls to the right of A on the c_2 axis and the point P is now positioned above the point F on the c_1 axis as illustrated in Fig. 2.34. The image of the solution is given by the path $O \rightarrow P \rightarrow F \rightarrow O$, and to each of the portions $O \rightarrow P$ and $P \rightarrow F$ there corresponds a Γ_- shock and a Γ_+ shock, respectively, while the portion $F \rightarrow O$ is the image of the Γ_+ shock emanating from the origin. The three shock paths are shown in Fig. 2.35, which we are going to discuss.

The characteristic parameters for the state D are given by the same expressions as in Eq. (2.9.13) and the concentration c_1^* of the state P is also given by Eq. (2.9.14), but we note that $c_1^* > c_1^0$. Thus the characteristic parameters for the states D and P are as follows:

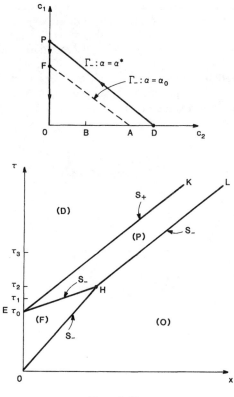

Figure 2.34

Two-Solute Chromatography with the Langmuir Isotherm   Chap. 2

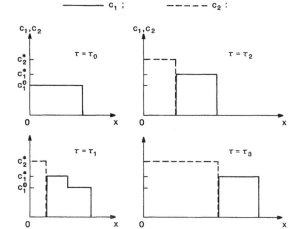

Figure 2.35

State	α	β	a	b
D	α^*	0	a^*	$1/\gamma$
P	α^*	∞	a^*	1

The shock path $EK(S_+)$ has the slope

$$\tilde{\sigma}_+\left(\frac{1}{\gamma},\, 1\right) = 1 + \frac{\rho a^*}{\gamma} \qquad (2.9.22)$$

which is the same as Eq. (2.9.15), but its value has decreased since a^* is smaller here than in the preceding case. Across the shock c_2 jumps down to zero while c_1 goes up to c_1^*. The shock path $EH(S_-)$ has the equation

$$\tau - \tau_0 = \tilde{\sigma}_-(a^*, a_0)x = (1 + \rho a^* a_0)x$$

and the shock path $OH(S_-)$ is the line

$$\tau = \tilde{\sigma}_-\left(a_0,\, \frac{1}{\gamma}\right)x = \left(1 + \frac{\rho a_0}{\gamma}\right)x$$

because $b = 1$ for the states P, F, and O. Hence the two shocks meet each other at H whose coordinates are

$$x_H = \frac{\tau_0 \gamma}{\rho a_0(1 - \gamma a^*)}$$

$$\tau_H = \left(1 + \frac{\rho a_0}{\gamma}\right)x_H \qquad (2.9.23)$$

Since the two shocks are of the same family, they are simply superposed upon interaction to give a new shock of the same family connecting the states P and O directly. The new Γ_+ shock propagates along the line HL which has the slope

$$\breve{\sigma}_-\left(a^*, \frac{1}{\gamma}\right) = 1 + \frac{\rho a^*}{\gamma} \tag{2.9.24}$$

and since this is the same as the slope of EK, the two shock paths now become parallel to each other. Between the two shocks we see the band of pure A_1 being fully developed, so displacement development is realized. The bandwidth Δx_1 and the vertical distance $\Delta \tau_1$ are given by Eqs. (2.9.20) and (2.9.21), respectively. Comparing Eq. (2.9.23) with Eq. (2.9.10), we find that $x_H < x_G$ because now $a_0 > a^*$. Thus the column length x_H required to obtain the band of pure A_1 is smaller here than in case II.

The concentration profiles are shown in Fig. 2.35. In the first part ($\tau = \tau_0$) the chromatogram of A_1 is just established and the second part ($\tau = \tau_1$) shows that the Γ_+ shocks are approaching each other. At $\tau = \tau_2 = \tau_H$ the two shocks are united and from this moment on ($\tau = \tau_3 > \tau_H$) the band of pure A_1 moves without change.

Incidentally, if $c_2^* = (k + c_1^*)/h$, the image of the developer state falls right on point A, so that the solution is given by a Γ_--shock ($D \rightarrow F$) and a Γ_+ shock ($F \rightarrow O$). The paths of these shocks are parallel from the beginning and thus a band of pure A_1 is established immediately after the developer is introduced.

In summary we have shown that there exists a critical value of the developer concentration

$$c_{2,cr}^* = \frac{k}{h} = \frac{1}{hK_1}\left(1 - \frac{1}{\gamma}\right) = \frac{\gamma - 1}{K_2} \tag{2.9.25}$$

such that for displacement development to be successful, the developer concentration c_2^* must be greater than this critical value. This critical value is given by the concentration c_2 at the watershed point and increases as the relative adsorptivity γ increases.

Before closing this section it is worthwhile to examine the effect of the developer concentration as well as that of the relative adsorptivity. First, we see from Eq. (2.9.19) that the concentration c_1^* of the pure A_1 band increases as the developer concentration c_2^* increases and also as the relative adsorptivity γ decreases. Hence as the developer concentration applied becomes higher, we obtain a narrower band of pure A_1 with a larger plateau value of concentration. Since $b = 1$ for the states F, P, and O, we can use Eq. (2.9.9) in the form

$$\gamma a = \frac{1}{1 + K_1 c_1}$$

and substituting this into Eq. (2.9.23) gives

$$x_H = \frac{\tau_0}{\rho} \gamma^2 (1 + Kc_1^0)\left(1 + \frac{1}{K_1 c_1^*}\right) \tag{2.9.26}$$

Now by applying the foregoing argument for c_1^*, we observe that x_H decreases as c_2^* increases or as γ decreases. This implies that the column length required to obtain the band of pure A_1 becomes shorter, as we use a higher value of the developer concentration or as the relative adsorptivity approaches its limiting value 1. The same argument can be proven valid in case II (see Exercise 2.9.2). Consequently, a higher value of the developer concentration and a lower value of the relative adsorptivity are favored in the application of displacement development. This observation with respect to the relative adsorptivity is somewhat unexpected.

Exercises

2.9.1. If we accept the idea that across the shock path EK of Fig. 2.32 or 2.34, c_2 drops to zero while c_1 jumps up to c_1^*, and if we apply the simple mass balance, $c_1^0 \tau_0 = c_1^* \Delta\tau_1$, of solute A_1, cases II and III can be treated by using the development for single-solute problems (see Chapter 6 of Vol. 1). Perform this analysis and show that all the formulas of this section can be derived.

2.9.2. By using Eq. (2.9.17), show that the column length required to obtain the band of pure A_1 decreases as the developer concentration increases and as the relative adsorptivity decreases.

2.9.3. A chromatographic column has been saturated with a sample of two solutes A_1 and A_2 over a period of time τ_0, where A_2 is more strongly adsorbed than A_1, and now this is to be treated by a mixture containing solute A_3 which is more strongly adsorbed than A_2. If we assume that entering the column, the developer A_3 first faces a shock across which the developer concentration drops to zero while the concentration of A_2 jumps up to a finite value c_2^*, the situation to the right of this shock path in the (x, τ)-plane poses a problem of two-solute chromatography. Depending on the value of c_2^* which would be a function of the developer concentration c_3^*, we can think of four different cases. Can you identify these cases? Discuss the structure of the solution in each case and proceed to carry out the analysis. In particular, show that in order to have a successful operation of displacement development, the value of c_2^* must be larger than $c_{2,cr}^*$ of Eq. (2.9.25), and this implies that we must have

$$c_3^* > c_{3,\,cr}^* = \frac{1}{K_3}\left(\frac{N_3 K_3}{N_1 K_1} - 1\right)$$

2.10 Shock Layer Analysis

Just as in the case of a single equation (see Sec. 7.10 of Vol. 1), we expect that axial dispersion in the fluid phase and interphase transfer resistance will smear out any sharp discontinuity in the concentration profiles. In actual systems, therefore, we would have *shock layers* (or constant-pattern profiles) rather than shocks. Here we would like to establish the existence of such shock layers and use them to examine the effect of axial dispersion and interphase mass transfer resistance in two-solute chromatography.

If we let E_i denote the effective axial dispersion coefficient of solute A_i and k_i the overall mass transfer coefficient of A_i based on the solid phase, the conservation equations can be written from Eq.(1.1.11) as

$$E_i \frac{\partial^2 c_i}{\partial z^2} = V \frac{\partial c_i}{\partial z} + \frac{\partial c_i}{\partial t} + \frac{1 - \epsilon}{\epsilon} \frac{\partial n_i}{\partial t} \qquad (2.10.1)$$

$$\frac{\partial n_i}{\partial t} = k_i(f_i(c_1, c_2) - n_i) \qquad (2.10.2)$$

for $i = 1, 2$, where $f_i(c_1, c_2)$ is now used to represent the adsorption equilibrium isotherm and other variables and parameters have their usual meanings. As before, we shall put

$$x = \frac{z}{Z}, \qquad \tau = \frac{Vt}{Z} \qquad (2.10.3)$$

$$\nu = \frac{1 - \epsilon}{\epsilon} \qquad (2.10.4)$$

$$Pe_i = \frac{VZ}{E_i} = \text{axial Peclet number for } A_i \qquad (2.10.5)$$

$$St_i = \frac{k_i Z}{V} = \text{Stanton number for } A_i \qquad (2.10.6)$$

where Z is a characteristic length of the system to be chosen arbitrarily, so that Eqs. (2.10.1) and (2.10.2) may be rewritten in the form

$$\frac{1}{Pe_i} \frac{\partial^2 c_i}{\partial x^2} = \frac{\partial c_i}{\partial x} + \frac{\partial c_i}{\partial \tau} + \nu \frac{\partial n_i}{\partial \tau} \qquad (2.10.7)$$

$$\frac{\partial n_i}{\partial \tau} = St_i(f_i(c_1, c_2) - n_i) \qquad (2.10.8)$$

for $i = 1, 2$. In order to exclude any end effects we shall assume that the chromatographic column is infinitely long and the boundary conditions are prescribed as

$$\begin{aligned} c_i &= c_i^l \text{(constant)} \quad \text{at } x = -\infty \\ c_i &= c_i^r \text{(constant)} \quad \text{at } x = +\infty \end{aligned} \qquad (2.10.9)$$

for $i = 1, 2$.

We now look for a moving coordinate ξ defined by

$$\xi = x - \tilde{\lambda}\tau \qquad (2.10.10)$$

with constant $\tilde{\lambda}$ in which the solution to Eqs. (2.10.7) and (2.10.8) can be expressed in the form

$$\begin{aligned} c_i(x, \tau) &= c_i(\xi) \\ n_i(x, \tau) &= n_i(\xi) \end{aligned} \qquad (2.10.11)$$

and satisfies the conditions

$$c_i = c_i^l, \qquad \frac{dc_i}{d\xi} = \frac{dn_i}{d\xi} = 0 \quad \text{at } \xi = -\infty$$

$$c_i = c_i^r, \qquad \frac{dc_i}{d\xi} = \frac{dn_i}{d\xi} = 0 \quad \text{at } \xi = +\infty$$

$$(2.10.12)$$

If there exists such a real value $\tilde{\lambda}$, Eq. (2.10.10) represents the shock layer of the system [Eqs. (2.10.7) and (2.10.8)], and $\tilde{\lambda}$ is the speed at which the shock layer propagates.

If we suppose that there exists a shock layer, the partial derivatives are transformed to ordinary derivatives with respect to ξ, so that Eqs.(2.10.7) and (2.10.8) are reduced to a system of ordinary differential equations

$$\frac{1}{\text{Pe}_i} \frac{d^2 c_i}{d\xi^2} - (1 - \tilde{\lambda}) \frac{dc_i}{d\xi} + \nu\tilde{\lambda} \frac{dn_i}{d\xi} = 0 \qquad (2.10.13)$$

$$-\tilde{\lambda} \frac{dn_i}{d\xi} = \text{St}_i(f_i(c_1, c_2) - n_i) \qquad (2.10.14)$$

Upon applying Eq. (2.10.12), we find the auxiliary conditions

$$\frac{d^2 c_i}{d\xi^2} = 0, \qquad n_i = f_i^l = f_i(c_1^l, c_2^l) \quad \text{at } x = -\infty$$

$$\frac{d^2 c_i}{d\xi^2} = 0, \qquad n_i = f_i^r = f_i(c_1^r, c_2^r) \quad \text{at } x = +\infty$$

$$(2.10.15)$$

that is, equilibrium must be established at both ends.

Substituting Eq. (2.10.14) into Eq. (2.10.13), differentiating the resulting equation with respect to ξ, and then eliminating $dn_i/d\xi$ by using Eq. (2.10.13), we obtain

$$\frac{\tilde{\lambda}}{\text{Pe}_i \text{St}_i} \frac{d^3 c_i}{d\xi^3} - \left\{ \frac{1}{\text{Pe}_i} + \frac{\tilde{\lambda}(1 - \tilde{\lambda})}{\text{St}_i} \right\} \frac{d^2 c_i}{d\xi^2} + (1 - \tilde{\lambda}) \frac{dc_i}{d\xi} - \nu\tilde{\lambda} \frac{df_i}{d\xi} = 0$$

which upon direct integration from $\xi = -\infty$, where both $dc_i/d\xi$ and $d^2 c_i/d\xi^2$ vanish, gives the equation

$$-\frac{\tilde{\lambda}}{\text{Pe}_i \text{St}_i} \frac{d^2 c_i}{d\xi^2} + \left\{ \frac{1}{\text{Pe}_i} + \frac{\tilde{\lambda}(1 - \tilde{\lambda})}{\text{St}_i} \right\} \frac{dc_i}{d\xi} = (1 - \tilde{\lambda})(c_i - c_i^l) - \nu\tilde{\lambda}(f_i - f_i^l)$$

$$(2.10.16)$$

for $i = 1, 2$. This is the basic equation to be solved subject to the second conditions of Eqs. (2.10.12) and (2.10.15) to determine the shock layer solution. Just as in the case of single solutes, dispersion $(1/\text{Pe}_i)$ and mass transfer resistance $(1/\text{St}_i)$ assume an equivalent role in Eq. (2.10.16), and in addition to their individual effects which are given by the first-order derivative term, there exists a coupled effect that appears as the second-order derivative term.

Now by applying the second conditions of Eqs. (2.10.12) and (2.10.15) to

Eq. (2.10.16), we obtain

$$\frac{1}{\tilde{\lambda}} = 1 + \nu \; \frac{f_i^l - f_i^r}{c_i^l - c_i^r} \qquad (2.10.17)$$

which contains the same expression as the *jump condition* (1.7.7) for the shock speed. Since $\tilde{\lambda}$ is independent of the subscript i, the following condition must be satisfied:

$$\frac{f_1^l - f_1^r}{c_1^l - c_1^r} = \frac{f_2^l - f_2^r}{c_2^l - c_2^r} \qquad (2.10.18)$$

but this is identical to the *compatibility relation* (1.7.6) which we have had in case of equilibrium chromatography. Consequently, if there exists a shock layer, the end states must satisfy the compatibility relation (2.10.18) and its propagation speed $\tilde{\lambda}$, being independent of Pe_i and St_i, is the same as that of the corresponding shock.

For physically relevant isotherms f_i [see Eqs. (1.1.8) and (1.1.9)], it can be shown that with the state at one end fixed, say (c_1^l, c_2^l), the compatibility relation (2.10.18) gives two separate curves emanating from the point (c_1^l, c_2^l) in the (c_1, c_2)-plane for the locus of the state at the other end, i.e., (c_1^r, c_2^r), or vice versa. These are the shock curves Σ_+ and Σ_- (see Sec. 1.7). Here the equilibrium relation is given by the Langmuir isotherm

$$f_i = \frac{N_i K_i c_i}{1 + K_1 c_1 + K_2 c_2}, \qquad i = 1, 2 \qquad (2.10.19)$$

and we know that in this case the shock curves become straight and coincide with the pair of Γ characteristics passing through the point (c_1^l, c_2^l). Hence if there exists a shock layer for a given state (c_1^l, c_2^l), the state at the other end must fall on either the Γ_+ or the Γ_- characteristic passing through the point $L(c_1^l, c_2^l)$ as shown in Fig. 2.36. In other words, the end states must satisfy the equation

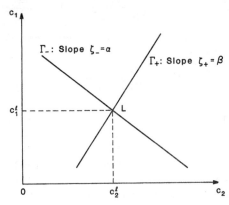

Figure 2.36

$$c_1^l - c_1^r = \zeta_{\pm}(c_2^l - c_2^r) \tag{2.10.20}$$

where $\zeta_+ = \beta$ and $\zeta_- = \alpha$. This implies that there can be two different kinds of shock layers, and for each kind we have to consider the two cases, $c_2^l > c_2^r$ and $c_2^l < c_2^r$.

Since Eq. (2.10.16) does not seem amenable to analytic treatment at the moment, we shall not analyze the simultaneous effect of dispersion and mass transfer resistance but will be concerned with the analysis of their individual effects. In the limiting case when all mass transfer resistances are negligible, we have $\mathrm{St}_i = \infty$ for $i = 1, 2$ and thus Eq. (2.10.16) becomes[†]

$$\frac{1}{\nu\lambda\,\mathrm{Pe}_1}\frac{dc_1}{d\xi} = \Omega(c_1 - c_1^l) - (f_1 - f_1^l) \equiv G(c_1, c_2)$$

$$\frac{1}{\nu\lambda\,\mathrm{Pe}_2}\frac{dc_2}{d\xi} = \Omega(c_2 - c_2^l) - (f_2 - f_2^l) \equiv H(c_1, c_2) \tag{2.10.21}$$

in which Ω is a constant parameter defined as

$$\Omega = \frac{1}{\nu}\left(\frac{1}{\lambda} - 1\right) = \frac{f_i^l - f_i^r}{c_i^l - c_i^r} \tag{2.10.22}$$

For the limiting case when axial dispersion has a negligible effect, we have $\mathrm{Pe}_i = \infty$ and so obtain the same pair of equations as Eq. (2.10.21) except for using $\mathrm{St}_i(1 - \lambda)$ instead of $\lambda\,\mathrm{Pe}_i$. Hence the discussion that we are going to establish for Eq. (2.10.21) will be equally applicable to the second case with proper adjustment.

We shall consider the situation in the (c_1, c_2)-plane. Suppose now that the states (c_1^l, c_2^l) and (c_1^r, c_2^r) fall on a Γ_--characteristic. The curves $G = 0$ and $H = 0$ intersect with each other twice: once at $L(c_1^l, c_2^l)$ and then at $R(c_1^r, c_2^r)$, as depicted in Fig. 2.37. We note that the locations of the two points L and R are interchangeable. Since there are no other singular points, the curves $G = 0$ and $H = 0$ separate the quadrant, $c_1 \geq 0$ and $c_2 \geq 0$, into five distinct regions and each region is simply connected. Both G and H assume a uniform sign in each region, and thus all the integral curves of Eq. (2.10.21) must satisfy these conditions. In particular, the curve $G = 0$ can be crossed only with a zero slope and the curve $H = 0$ only with an infinite slope.

The points L and R are the exclusive singular points of Eq. (2.10.21), and in order to investigate the direction field in the neighborhood of these points, we shall consider the characteristic equation of the system (2.10.21)

$$\begin{vmatrix} \dfrac{\partial G}{\partial c_1} - \dfrac{\mu}{\nu\lambda\,\mathrm{Pe}_1} & \dfrac{\partial G}{\partial c_2} \\[3mm] \dfrac{\partial H}{\partial c_1} & \dfrac{\partial H}{\partial c_2} - \dfrac{\mu}{\nu\lambda\,\mathrm{Pe}_2} \end{vmatrix} = 0$$

[†] Compare this pair of equations with Eq. (1.11.23) and see the discussion around the equation.

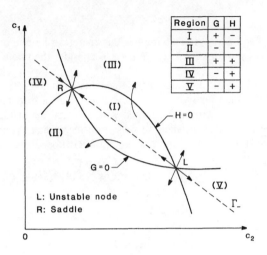

Figure 2.37

Region	G	H
I	+	-
II	-	-
III	+	+
IV	-	+
V	-	+

(In the figure:)
c_1

(III)

(IV)
R

(I)

H = 0

(II)

G = 0

L

(V)

Γ_-

L: Unstable node
R: Saddle

0

c_2

or

$$\mu^2 - \nu\tilde{\lambda}\left\{ Pe_1\left(\Omega - \frac{\partial f_1}{\partial c_1}\right) + Pe_2\left(\Omega - \frac{\partial f_2}{\partial c_2}\right)\right\}\mu$$
$$+ \nu^2\tilde{\lambda}^2\, Pe_1 Pe_2\left\{\left(\Omega - \frac{\partial f_1}{\partial c_1}\right)\left(\Omega - \frac{\partial f_2}{\partial c_2}\right) - \frac{\partial f_1}{\partial c_2}\frac{\partial f_2}{\partial c_1}\right\} = 0 \qquad (2.10.23)$$

Since both the derivatives $\partial f_1/\partial c_2$ and $\partial f_2/\partial c_1$ are negative, the discriminant is positive and there are two real, distinct values of μ which we shall call μ_1 and μ_2, respectively. Then we have

$$\mu_1 + \mu_2 = \nu\tilde{\lambda}\left\{ Pe_1\left(\Omega - \frac{\partial f_1}{\partial c_1}\right) + Pe_2\left(\Omega - \frac{\partial f_2}{\partial c_2}\right)\right\} \qquad (2.10.24)$$

and

$$\mu_1\mu_2 = \nu^2\tilde{\lambda}^2 Pe_1 Pe_2\left\{\Omega^2 - \left(\frac{\partial f_1}{\partial c_1} + \frac{\partial f_2}{\partial c_2}\right)\Omega + \frac{\partial f_1}{\partial c_1}\frac{\partial f_2}{\partial c_2} - \frac{\partial f_1}{\partial c_2}\frac{\partial f_2}{\partial c_1}\right\} \qquad (2.10.25)$$

Now if the points L and R fall on a Γ_- characteristic, we must put $\tilde{\lambda} = \tilde{\lambda}_+$, so

$$\nu\Omega = \frac{1}{\tilde{\lambda}_+} - 1 = \tilde{\sigma}_+ - 1 \qquad (2.10.26)$$

From Eq. (1.4.27) we have

$$\sigma_+ = 1 + \nu\frac{\partial f_1}{\partial c_1} + \frac{\nu}{\zeta_-}\frac{\partial f_1}{\partial c_2} = 1 + \nu\frac{\partial f_2}{\partial c_2} + \nu\zeta_-\frac{\partial f_2}{\partial c_1} \qquad (2.10.27)$$

and by substituting Eqs. (2.10.26) and (2.10.27), we can rewrite Eq. (2.10.24) as

$$\mu_1 + \mu_2 = \tilde{\lambda}_+ \left\{ \mathrm{Pe}_1 \left(\tilde{\sigma}_+ - \sigma_+ + \frac{\nu}{\zeta_-} \frac{\partial f_1}{\partial c_2} \right) + \mathrm{Pe}_2 \left(\tilde{\sigma}_+ - \sigma_+ + \nu \zeta_- \frac{\partial f_2}{\partial c_1} \right) \right\}$$

$$(2.10.28)$$

On the other hand, comparing Eq. (2.10.25) with Eqs. (1.4.18) and (1.4.19), we see that the right-hand side of Eq. (2.10.25) has two roots $(\sigma_\pm - 1)/\nu$ for Ω and applying this together with Eq. (2.10.26) to Eq. (2.10.25), we obtain

$$\mu_1 \mu_2 = \tilde{\lambda}_+^2 \, \mathrm{Pe}_1 \mathrm{Pe}_2 (\tilde{\sigma}_+ - \sigma_+)(\tilde{\sigma}_+ - \sigma_-) \qquad (2.10.29)$$

Here we note that Eqs. (2.10.28) and (2.10.29) hold for an arbitrary isotherm.

With the Langmuir isotherm we have Eqs. (2.2.8), (2.2.9), and (2.2.21) for σ_+, σ_-, and $\tilde{\sigma}_+$, respectively. In the present case the states L and R have a common value of the parameter a and $c_2^l > c_2^r$ implies $b^l < b^r$. Hence we obtain

$$\sigma_+^l < \tilde{\sigma}_+ < \sigma_+^r \qquad \text{if } c_2^l > c_2^r \qquad (2.10.30)$$

$$\sigma_+^l > \tilde{\sigma}_+ > \sigma_+^r \qquad \text{if } c_2^l < c_2^r \qquad (2.10.31)$$

and

$$\tilde{\sigma}_+ > \sigma_-^l, \sigma_-^r \qquad (2.10.32)$$

Let us first consider the case $c_2^l > c_2^r$. This is the case illustrated in Fig. 2.37. By applying Eqs. (2.10.30) and (2.10.32) to Eq. (2.10.29), we see that $\mu_1 \mu_2 > 0$ at L and $\mu_1 \mu_2 < 0$ at R, so that L is a node and R is a saddle point. On the other hand, it follows from Eqs. (2.10.28) and (2.10.30) that $\mu_1 + \mu_2 > 0$ at L since $\zeta_- = \alpha < 0$. Thus L is an unstable node.

From perturbation theory we know that there are exactly two integral curves of Eq. (2.10.21) which approach the saddle point R with negative slope μ_2 as $\xi \to +\infty$ and exactly two which approach it with positive slope μ_1 as $\xi \to -\infty$. Comparing this with the sign of the ratio G/H, we observe that one of the integral curves converging to R as $\xi \to +\infty$ approaches it from region I. Let this curve be designated by $\mathcal{G}(\xi)$.

We shall prove that $\mathcal{G}(\xi)$ is a shock layer. Consider the integral curves that pass through the points on the curves $G = 0$ and $H = 0$. Along the curve $G = 0$ we have $dc_2/d\xi = H < 0$, so that all the integral curves have horizontal tangent vectors and are directed outward from region I for increasing ξ. Similarly, along the curve $H = 0$ all the integral curves have vertical tangent vectors with $dc_1/d\xi = G > 0$ and are directed outwards from region I for increasing ξ. For decreasing ξ, therefore, all the integral curves crossing the curves $G = 0$ or $H = 0$ are directed into region I.

Let us now follow $\mathcal{G}(\xi)$ for decreasing ξ. Obviously, $\mathcal{G}(\xi)$ cannot intersect the curves $G = 0$ and $H = 0$ between L and R. Since there are no other singular points, it cannot terminate inside region I. Being monotonic with negative slope while in region I, $\mathcal{G}(\xi)$ must then approach the unstable node L as $\xi \to -\infty$. This establishes the existence of a shock layer.

Next we shall prove the uniqueness of the shock layer. Since the unstable node L can be approached by an integral curve only as $\xi \to -\infty$, the point R

should be approached as $\xi = +\infty$. The point R being a saddle point, we notice that there are two integral curves which approach R as $\xi \to +\infty$, one from region I and the other from region IV. Clearly, the former is identical to $\mathcal{S}(\xi)$ and the latter is the only integral curve other than $\mathcal{S}(\xi)$ that can connect the points L and R. Let this curve be denoted by $\mathcal{S}'(\xi)$. If $\mathcal{S}'(\xi)$ is also a shock layer, then $\mathcal{S}(\xi)$ and $\mathcal{S}'(\xi)$ compose a closed curve enclosing a simply connected region D in the $(c_1 c_2)$-plane. On the other hand, there are two integral curves that approach the saddle point R with positive slope μ_1 as $\xi \to -\infty$. One of these curves enters the region D as ξ increases. Since L and R are exclusive singular points, the integral curve cannot terminate or approach a limit cycle in the region D. Hence the integral curve must intersect $\mathcal{S}(\xi)$ or $\mathcal{S}'(\xi)$. This is a contradiction to the uniqueness of the integral curves and thus the uniqueness is established.

If $c_2^l < c_2^r$, the points L and R interchange their locations in Fig. 2.37 and we have Eqs. (2.10.31) and (2.10.32). Thus from Eqs. (2.10.28) and (2.10.29) we can show that R is an unstable node and so cannot be approached in any direction as $\xi \to +\infty$. Consequently, it is not possible to have a shock layer in this case.

We shall now briefly discuss the case when the states $L(c_1^l, c_2^l)$ and $R(c_1^r, c_2^r)$ fall on a Γ_+ characteristic as shown in Fig. 2.38. Following the same procedure as in the previous case, we obtain

$$\mu_1 + \mu_2 = \bar{\lambda}_- \left\{ \mathrm{Pe}_1 \left(\bar{\sigma}_- - \sigma_- + \frac{\nu}{\zeta_+} \frac{\partial f_1}{\partial c_2} \right) + \mathrm{Pe}_2 \left(\bar{\sigma}_- - \sigma_- + \nu \zeta_+ \frac{\partial f_2}{\partial c_1} \right) \right\}$$

(2.10.33)

and

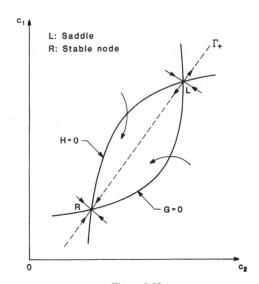

L: Saddle
R: Stable node

Figure 2.38

$$\mu_1\mu_2 = \tilde{\lambda}^2 \, \mathrm{Pe}_1\mathrm{Pe}_2(\tilde{\sigma}_- - \sigma_+)(\tilde{\sigma}_- - \sigma_-) \tag{2.10.34}$$

In case of the Langmuir isotherm we have

$$\sigma^l_- < \tilde{\sigma}_- < \sigma^r_- \quad \text{if} \quad c^l_2 > c^r_2 \tag{2.10.35}$$

$$\sigma^l_- > \tilde{\sigma}_- > \sigma^r_- \quad \text{if} \quad c^l_2 < c^r_2 \tag{2.10.36}$$

and

$$\tilde{\sigma}_- < \sigma^l_+ , \sigma^r_+ \tag{2.10.37}$$

If $c^l_2 > c^r_2$, it can be shown that L is a saddle point and R is a stable node. By applying an argument similar to that in the preceding case we can prove that there exists a unique shock layer. But if $c^l_2 < c^r_2$, L becomes a stable node and cannot be approached in any direction as $\xi \to -\infty$. Hence there is no shock layer.

Noting that Eqs. (2.10.30) and (2.10.32) or Eqs. (2.10.35) and (2.10.37) are exactly the shock conditions, we draw the following conclusions:

1. There exists a shock layer if and only if $c^l_2 > c^r_2$ or equivalently, if and only if the end states satisfy the shock condition.
2. For a given state (c^l_1, c^l_2) there can be two different shock layers, one having the end states on the Γ_+ characteristic and the other on the Γ_- characteristic.
3. If $\tilde{\lambda}$ or c^r_2 is further specified, the shock layer of either kind is unique.
4. A shock layer propagates with the same speed as the corresponding shock, which is independent of the dispersion coefficient and the mass transfer coefficient.

If the axial dispersion coefficients for both solutes are equal, so that $\mathrm{Pe}_1 = \mathrm{Pe}_2$, the mathematical treatment becomes much simpler and we can obtain an analytic expression for the shock layer. In this case Eq. (2.10.21) can be put in the form

$$\frac{dc_1}{dc_2} = \frac{\Omega(c_1 - c^l_1) - (f_1 - f^l_1)}{\Omega(c_2 - c^l_2) - (f_2 - f^l_2)} \tag{2.10.38}$$

As $c_1 \to c^l_1$ and $c_2 \to c^l_2$, the right-hand side becomes indeterminate and thus we apply L'Hospital's rule to obtain

$$\frac{\partial f_2}{\partial c_1}\left(\frac{dc_1}{dc_2}\right)^2 - \left(\frac{\partial f_1}{\partial c_1} - \frac{\partial f_2}{\partial c_2}\right)\frac{dc_1}{dc_2} - \frac{\partial f_1}{\partial c_2} = 0 \tag{2.10.39}$$

This is just the fundamental differential equation (2.1.12) and in case of the Langmuir isotherm this equation gives $dc_1/dc_2 = \zeta_\pm = $ constant, where $\zeta_+ = \beta$ and $\zeta_- = \alpha$. This implies that in the (c_1, c_2)-plane the shock layer curve starts from the singular point $L(c^l_1, c^l_2)$ in the direction of the $\Gamma(\Gamma_+$ or $\Gamma_-)$ characteristic. As we follow the Γ characteristic, however, Eq. (2.10.38) remains singular because in the expression

$$\frac{dc_1}{dc_2} = \frac{c_1 - c_1^l}{c_2 - c_2^l} \frac{\Omega - (f_1 - f_1^l)/(c_1 - c_1^l)}{\Omega - (f_2 - f_2^l)/(c_2 - c_2^l)}$$

the second ratio on the right-hand side becomes indeterminate. Hence it would be necessary to apply L'Hospital's rule for all points along the shock layer curve. It then follows that at every point on the shock layer curve we have $dc_1/dc_2 = \zeta_\pm$. In other words, the shock layer curve becomes coincident with the Γ characteristic connecting the two end points in the (c_1, c_2)-plane and thus satisfies the linear equation

$$c_1 - c_1^l = \zeta_\pm(c_2 - c_2^l) \tag{2.10.40}$$

with $\zeta_+ = \beta$ and $\zeta_- = \alpha$.

We shall use Eq. (2.10.40) in the form $c_1 = c_1^r + \zeta_\pm(c_2 - c_2^r)$ for $f_2(c_1, c_2)$ and in the form $c_1^l = c_1^r + \zeta_\pm(c_2^l - c_2^r)$ for $f_2^l = f_2(c_1^l, c_2^l)$ to obtain

$$f_2 = \frac{N_2' K_2' c_2}{1 + K_2' c_2}$$

$$f_2^l = \frac{N_2' K_2' c_2^l}{1 + K_2' c_2^l} \tag{2.10.41}$$

where

$$N_2' = \frac{N_2 K_2}{K_1 \zeta_\pm + K_2}$$

$$K_2' = \frac{K_1 \zeta_\pm + K_2}{1 + K_1(c_1^r - \zeta_\pm c_2^r)} \tag{2.10.42}$$

Since Eq. (2.10.41) is equivalent to the Langmuir isotherm for a single solute, it is obvious that the development in Sec. 7.10 of Vol. 1 may be applied here. If we put

$$p^l = 1 + K_2' c_2^l, \qquad p^r = 1 + K_2' c_2^r \tag{2.10.43}$$

Eq. (2.10.17) gives

$$\frac{1}{\bar{\lambda}} = 1 + \nu \frac{N_2' K_2'}{p^l p^r} \tag{2.10.44}$$

and the second equation of Eq.(2.10.21) can be rearranged to give

$$\left(1 + \nu \frac{N_2' K_2'}{p^l p^r}\right) \frac{1}{\nu \, Pe} \frac{dc_2}{d\xi} = \frac{N_2'(K_2')^2}{p^l p^r} \frac{(c_2 - c_2^l)(c_2 - c_2^r)}{1 + K_2' c_2}$$

Integrating this equation from a fixed point, say $c_2 = c_2^0$ at $\xi = 0$, we obtain an analytic expression for the shock layer $\mathscr{S}(\xi)$:

$$\left(\frac{c_2^l - c_2}{c_2^l - c_2^0}\right)^{p^l} \left(\frac{c_2^0 - c_2^r}{c_2 - c_2^r}\right)^{p^r} = \exp\left\{\frac{\nu N_2' K_2'(p^l - p^r)}{\nu N_2' K_2' + p^l p^r} Pe \; \xi\right\} \tag{2.10.45}$$

and $c_1(\xi)$ is then given by Eq. (2.10.40). We may specify two concentration bounds c_2^* and c_{2*} with a parameter ϵ of small value such as

$$\frac{c_2^l - c_2^*}{c_2^l - c_2^r} = \frac{c_{2*} - c_2^r}{c_2^l - c_2^r} = \epsilon$$

and define the shock layer thickness δ_ϵ as in the case of single solutes, i.e.,

$$\delta_\epsilon = \xi(c_{2*}) - \xi(c_2^*)$$

Applying Eq. (2.10.45) gives

$$\text{Pe } \delta_\epsilon = \left(1 + \frac{p^l p^r}{\nu N_2' K_2'}\right)\frac{p^l + p^r}{p^l - p^r}\ln\frac{1 - \epsilon}{\epsilon} \qquad (2.10.46)$$

which implies that δ_ϵ is inversely proportional to Pe or directly proportional to the axial dispersion coefficient.

To illustrate the foregoing development more clearly, we shall consider a numerical example for which we have

$$\epsilon = 0.4$$

$$N_1 = N_2 = 1 \text{ mol/liter}$$

$$K_1 = 5 \text{ liters/mol}, \quad K_2 = 10 \text{ liters/mol}$$

$$c_1^l = 0.05 \text{ mol/liter} \quad c_2^l = 0.10 \text{ mol/liter}$$

$$c_2^r = 0.06 \text{ mol/liter}$$

For the state (c_1^l, c_2^l), Eq. (2.1.31) gives $\zeta_+ = \beta = 1$ and $\zeta_- = \alpha = -\frac{1}{2}$.

When we take $\zeta_- = \alpha = -\frac{1}{2}$ and so the end states fall on the Γ_- characteristic, Eqs. (2.10.20) and (2.10.22) give $c_1^r = 0.07$ mol/liter and $\Omega = 400/117$, respectively. With the assumption of equal Peclet numbers, Eq. (2.10.21) is put in the form of the ratio $dc_1/dc_2 = G/H$ or $dc_2/dc_1 = H/G$. Integrating one of these equations numerically for various sets of data (c_1^0, c_2^0), we obtain the integral curves as shown in Fig. 2.39, in which the arrows denote the direction of decreasing ξ. It is clearly seen that L is an unstable node and R a saddle point. Consequently, the shock layer $\mathcal{S}(\xi)$ can be obtained only by a backward integration.

For $\zeta_+ = \beta = 1$ having the end states on the Γ_+ characteristic, we obtain $c_1^r = 0.01$ mol/liter and $\Omega = 200/199$. Assuming that $\text{Pe}_1 = 2\,\text{Pe}_2$ (or $E_1 = E_2/2$), we rearrange Eq. (2.10.21) in the form $dc_1/dc_2 = 2G/H$ or $dc_2/dc_1 = H/(2G)$ and perform numerical integration for various sets of data (c_1^0, c_2^0) to obtain the integral curves as presented in Fig. 2.40. The arrows here denote the direction of increasing ξ. Hence it is seen that L is a saddle point and R a stable node, and that the shock layer $\mathcal{S}(\xi)$ can be determined only by a forward integration.

To plot the shock layer $\mathcal{S}(\xi)$ we shall use the dimensionless concentrations defined by

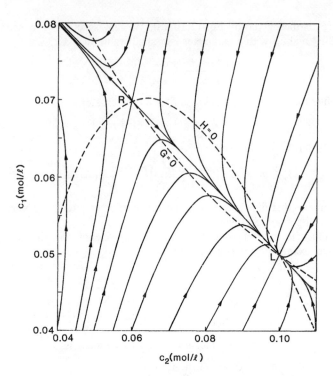

Figure 2.39

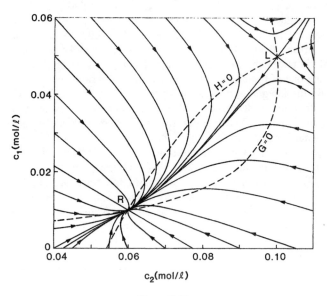

Figure 2.40

Two-Solute Chromatography with the Langmuir Isotherm   Chap. 2

$$\phi_i = \frac{c_i - \min(c_i^l, c_i^r)}{|c_i^l - c_i^r|}, \qquad i = 1, 2 \qquad (2.10.47)$$

In case of $Pe_1 = Pe_2 = Pe$, the shock layers are determined by Eq. (2.10.45) for various values of Pe and plotted by using the point for $\phi_2 = 0.5$ as a fixed point. While Fig. 2.41 shows the shock layers for $\zeta_- = \alpha = -\frac{1}{2}$, those for $\zeta_+ = \beta = 1$ are presented in Fig. 2.42. In both figures the sequence of shock layers (curves 1, 2, 3, and 4) converge to the corresponding shock (curve 5) as $Pe \to \infty$ and the shock layer thickness is inversely proportional to Pe, where $\delta(99\%)$ represents δ_ϵ for $\epsilon = 0.005$.

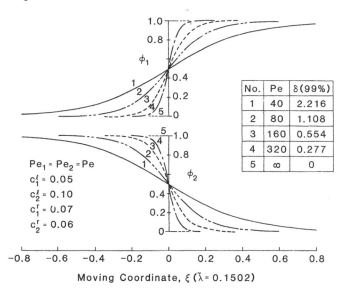

No.	Pe	$\delta(99\%)$
1	40	2.216
2	80	1.108
3	160	0.554
4	320	0.277
5	∞	0

$Pe_1 = Pe_2 = Pe$
$c_1^l = 0.05$
$c_2^l = 0.10$
$c_1^r = 0.07$
$c_2^r = 0.06$

Moving Coordinate, ξ ($\tilde{\lambda} = 0.1502$)

Figure 2.41

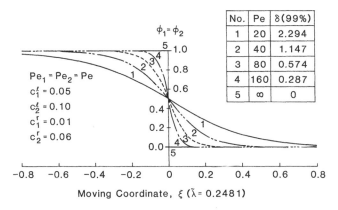

No.	Pe	$\delta(99\%)$
1	20	2.294
2	40	1.147
3	80	0.574
4	160	0.287
5	∞	0

$Pe_1 = Pe_2 = Pe$
$c_1^l = 0.05$
$c_2^l = 0.10$
$c_1^r = 0.01$
$c_2^r = 0.06$

Moving Coordinate, ξ ($\tilde{\lambda} = 0.2481$)

Figure 2.42

When the Peclet numbers are not the same, it is unlikely at the moment that Eq. (2.10.21) can be solved analytically but the numerical treatment should be straightforward since we have already investigated the direction field in the (c_1, c_2)-plane. We shall put Eq. (2.10.21) into the form

$$\frac{dc_1}{dc_2} = \frac{Pe_1}{Pe_2} \frac{G(c_1, c_2)}{H(c_1, c_2)} \qquad (2.10.48)$$

and integrate this equation numerically in the appropriate direction, i.e., backward if the end states fall on the Γ_- characteristic and forward if the end states fall on the Γ_+ characteristic. Since the initial point is singular, it is necessary to find the initial direction dc_1/dc_2 by applying L'Hospital's rule to Eq. (2.10.48), but the direction field is such that beginning from a point in the neighborhood of the saddle point, say $c_1 = c_1^r - 10^{-10}$ and $c_2 = c_2^r$ (see Fig. 2.37), integration generates the solution without recognizable discrepancy. Figure 2.43 presents various shock layer curves for the case when the end states fall on the Γ_- characteristic $(\alpha = -\frac{1}{2})$, whereas for the other case $(\beta = 1)$ we have Fig. 2.44. Both figures clearly show the effect of the ratio between two axial dispersion coefficients with the curves $G = 0$ and $H = 0$ representing the bounds of the effect.

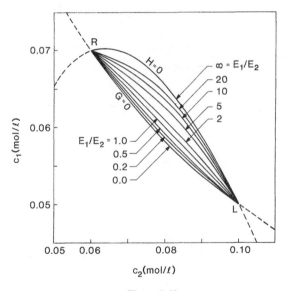

Figure 2.43

To study the effect of variation in one dispersion coefficient while the other remains fixed, the same example with $c_2^r = 0$ will be considered. In this situation the shock layer with the end states on the Γ_- characteristic $(\alpha = -\frac{1}{2})$ is the only kind and we find $c_1^r = 0.1$ mol/liter from Eq. (2.10.19). We select $c_2^r = 0$ here in order to compare the result later in this section with the transient solution of Eq. (2.10.7) for the saturation process.

Two-Solute Chromatography with the Langmuir Isotherm   Chap. 2

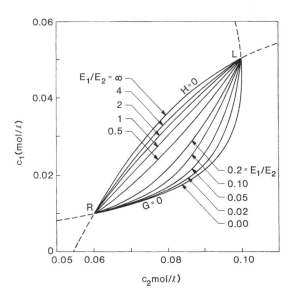

Figure 2.44

Equation (2.10.21) is integrated backward numerically and the concentrations in dimensionless form (2.10.47) are plotted against the moving coordinate ξ as shown in Figs. 2.45 and 2.46. The point for $\phi_1 = 0.5$ is used as a fixed point in Fig. 2.45 and the point $\phi_2 = 0.5$ in Fig. 2.46. In either case E_1 varying with E_2 fixed, or vice versa, the shock layer thickness is affected in a nonlinear fashion, and as the dispersion coefficient becomes larger, the influence becomes all the more significant. When the dispersion coefficient E_2 for the solute A_2, the more strongly adsorbed solute, varies while the other is fixed (see Fig. 2.46), the effect is seen to be much more pronounced than in the opposite situation (see Fig. 2.45). For example, when the dispersion coefficient under variation is quadrupled, the shock layer thickness grows nearly seven times as large in the former case, whereas it grows a little more than double in the latter. This implies that there exists a rather strong interaction between adsorption and axial dispersion or interphase mass transfer.

Finally, we would like to demonstrate the validity and usefulness of the shock layer solution. To accomplish this we shall solve numerically the transient equations, Eqs. (2.10.7) and (2.10.19), with the equilibrium assumption $n_i = f_i$ and make the comparison between the transient solution and the shock layer solution. This will illustrate how to make use of the shock layer analysis in practical problems.

As usual we employ here the central difference operator for the space derivatives and the forward difference operator for the time derivatives. All the variables and derivatives except for the time derivatives are averaged between two consecutive time steps. A simple iteration scheme is proven successful with rapid convergence.

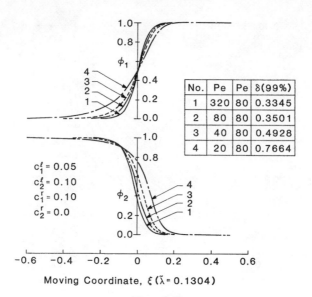

No.	Pe	Pe	δ(99%)
1	320	80	0.3345
2	80	80	0.3501
3	40	80	0.4928
4	20	80	0.7664

$c_1^\ell = 0.05$
$c_2^\ell = 0.10$
$c_1^r = 0.10$
$c_2^r = 0.0$

Moving Coordinate, ξ ($\bar\lambda = 0.1304$)

Figure 2.45

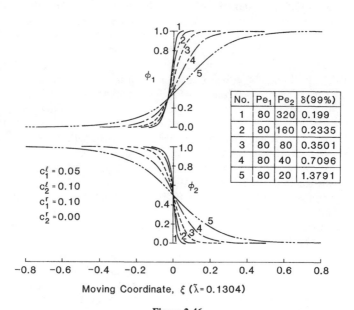

No.	Pe₁	Pe₂	δ(99%)
1	80	320	0.199
2	80	160	0.2335
3	80	80	0.3501
4	80	40	0.7096
5	80	20	1.3791

$c_1^\ell = 0.05$
$c_2^\ell = 0.10$
$c_1^r = 0.10$
$c_2^r = 0.00$

Moving Coordinate, ξ ($\bar\lambda = 0.1304$)

Figure 2.46

Since we are interested in the saturation process, the initial and boundary conditions are specified as follows:

at $\tau = 0$: $c_1 = 0$ and $c_2 = 0$

at $x = 0$: $c_1 = 0.05$ mol/liter   and $c_2 = 0.1$ mol/liter

at $x = X$: $\dfrac{dc_1}{dx} = 0$ and $\dfrac{dc_2}{dx} = 0$

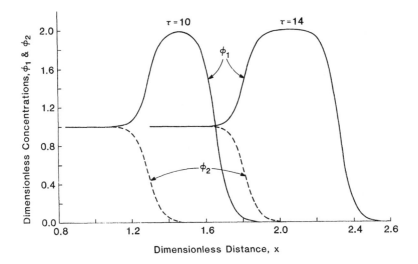

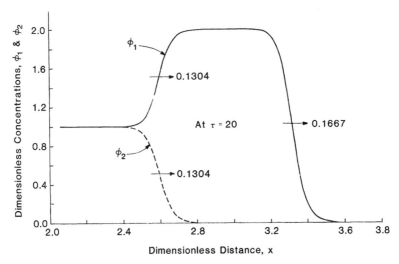

Figure 2.47

The fixed-state condition at $x = 0$ is considered reasonable from experience with chromatographic analysis. Since the adsorption fronts are to move downstream continuously and the plateau width increases with time, it is convenient in actual calculation to adjust the space variable in such a way that the point $x = X$ moves with the first adsorption front while the point $x = 0$ follows the second front.

For the case when $Pe_1 = Pe_2 = 80$, the concentration profiles at $\tau = 10$, 14, and 20 are presented in Fig. 2.47. At $\tau = 10$ the profiles for the second adsorption front are almost coincident with the shock layer solution (curve 3 of Fig. 2.46). Since only the solute A_1 is involved in the first front, the result of Sec. 7.10 of Vol. 1 can be applied there. At $\tau = 14$ we notice that the plateau value of ϕ_1 reaches the value 2.0 which would be predicted by Eq. (2.10.20) with $\zeta = -\frac{1}{2}$ and $c_2^\zeta = 0$ or by the analysis of the corresponding nondispersive system. At $\tau = 20$ the profiles become so similar to the shock layer that the discrepancy between the two cannot be shown in the figure. The numerical values in Fig. 2.47(b) represent the propagation speeds measured from the successive profiles and both values turn out to be the same as those predicted by Eq. (2.10.17). Moreover, the center point of each front can be located accurately by using the corresponding $\bar{\lambda}$ value.

For the cases of unequal Peclet numbers the transient solutions are also computed and presented in Fig. 2.48. The profiles for the second adsorption fronts are almost identical to the corresponding shock layers, curve 3 of Fig. 2.45 and curve 4 of Fig. 2.46. Other observations remain the same as in the above except for slight discrepancies in the location of center points.

For saturation processes, therefore, we can determine the concentration profiles for large time simply by applying the shock layer analysis rather than solving the transient equations.

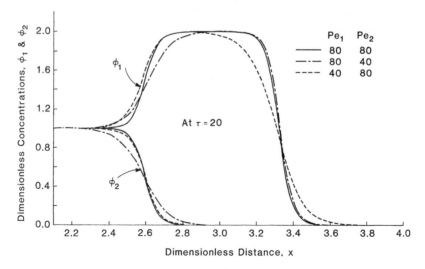

Figure 2.48

Two-Solute Chromatography with the Langmuir Isotherm   Chap. 2

Exercises

2.10.1. If the end states fall on a Γ_+ characteristic and $Pe_1 = Pe_2 = Pe$, show that Eq. (2.10.23) has two roots given by

$$\mu_1 = \bar{\lambda}_+ Pe(\bar{\sigma}_+ - \sigma_+)$$
$$\mu_2 = \bar{\lambda}_+ Pe\,(\bar{\sigma}_+ - \sigma_-)$$

and thus prove that if $c_2^l > c_2^r$, $L(c_1^l, c_2^l)$ is an unstable node, whereas $R(c_1^r, c_2^r)$ is a saddle point.

2.10.2. In the case of the Langmuir isotherm, express $(\Omega - \partial f_1/\partial c_1)$ and $(\Omega - \partial f_2/\partial c_2)$ in terms of the characteristic parameters α and β. By using the result in Eq. (2.10.24), show that $\mu_1 + \mu_2 > 0$ at L of Fig. 2.37 and $\mu_1 + \mu_2 < 0$ at R of Fig. 2.38.

2.10.3. Find from Eq. (2.10.48) the limiting value of dc_1/dc_2 as $c_1 \to c_1^r$ and $c_2 \to c_2^r$, and discuss.

REFERENCES

2.1–2.3. The theory of two-solute chromatography with the Langmuir isotherm was first given in this form by

E. GLUECKAUF, "Contributions to the theory of chromatography," *Proc. Roy. Soc. London* **A186**, 35–57 (1946).

E. GLUECKAUF, "Theory of chromatography: VII. The general theory of two solutes following non-linear isotherms," *Discuss. Faraday Soc.* **7**, 12–25 (1949).

The theory has been further elaborated with mathematical rigor by

H.-K. RHEE, "Studies on the theory of chromatography: Part II. Chromatography of two solutes," Ph.D. Thesis, University of Minnesota, 1968.

Three component ion exchange is equivalent to two-solute chromatography. For this treatment, see

G. KLEIN, D. TONDOUR, and T. VERMEULEN, "Multicomponent ion exchange in fixed beds: general properties of equilibrium systems," *Indust. Engrg. Chem. Fund.* **6**, 339–351 (1967).

D. TONDEUR and G. KLEIN, "Multicomponent ion exchange in fixed beds," *Indust. Engrg. Chem. Fund.* **6**, 351–361 (1967).

F. HELFFERICH and G. KLEIN, *Multicomponent Chromatography: Theory of Interference*, Marcel Dekker, New York, 1970.

2.4–2.8. The treatment here follows somewhat on the lines laid down in the thesis by Rhee (1968) cited above and

R. COURANT and K. O. FRIEDRICHS, *Supersonic Flow and Shock Waves*, Interscience, New York, 1948.

H.-K. RHEE, R. ARIS, and N. R. AMUNDSON, "On the theory of multicomponent chromatography," *Philos. Trans. Roy. Soc. London* **A267**, 419–455 (1970).

The Glimm–Lax theory on the decay of solutions mentioned in Sec. 2.6 is to be found in

J. Glimm and P. D. Lax, "Decay of solutions of systems of nonlinear hyperbolic conservation laws," *Mem. Amer. Math. Soc.* No. 101, 1–111 (1970).

The solution of Eq. (2.7.42) is given in the book by Courant and Friedrichs (1948) cited above.

An excellent lexicon for the hypergeometric functions is

M. Abramowitz and I. A. Stegun (eds.), *Handbook of Mathematical Functions,* National Bureau of Standards, Washington, D.C., 1964.

The chromatographic cycle (Sec. 2.8) was first discussed in the two papers by Glueckauf (1946, 1949) cited above.

2.9. The case treated here is the most elementary version of displacement development, of which the general theory is found in

H.-K. Rhee and N. R. Amundson, "Analysis of multicomponent separation by displacement development," *AIChE J.* **28,** 423–433 (1982).

The technique was first introduced by

A. Tiselius, "Displacement development in adsorption analysis," *Ark. Kemi Mineral. Geol.* **16A**(18), 1–11 (1943).

The theoretical features were discussed in the paper by Glueckauf (1946) cited above.

Experimental studies are to be found in the paper by Tiselius (1943) cited above and

E. Glueckauf and J. I. Coates, "Theory of chromatography: Part IV. The influence of incomplete equilibrium on the front boundary of chromatograms and on the effectiveness of separation," *J. Chem. Soc.* 1315–1321 (1947).

S. Fujine, K. Saito, and K. Shiba, "Transient behavior of lithium isotope separation by displacement chromatography," *Sep. Sci. Tech.* **17,** 1309–1325 (1982).

2.10. The development of this section is taken from

H.-K. Rhee and N. R. Amundson, "Shock layer in two solute chromatography: effect of axial dispersion and mass transfer," *Chem. Engrg. Sci.* **29,** 2049–2060 (1974).

The mathematical theory of the shock layer was first motivated by compressible fluid flow problems, and the development in this field is presented in

R. von Mises, "On the thickness of a steady shock wave," *J. Aeronaut. Sci.* **17,** 551–594 (1950).

D. Gilbarg, "The existence and limit of the one-dimensional shock layer," *Amer. J. Math.* **73,** 256–274 (1951).

H. Grad, "The profile of a steady plane shock waves," *Comm. Pure Appl. Math.* **5,** 257–300 (1952).

D. Gilbarg and D. Paolucci, "The structure of shock waves in the continuum theory of fluids," *J. Rational Mech. Anal.* **2,** 617–642 (1953).

J. Serrin, "Mathematical principles of classical fluid mechanics," in *Handbuch der physik,* Vol. 8, part 1, edited by S. Flügge, Springer-Verlag, Berlin, 1959, pp. 125–263.

L. Davison, "Shock wave structure in porous solids," *J. Appl. Phys.* **42,** 5503–5512 (1971).

See also

C. C. CONLEY and J. A. SMOLLER, "Viscosity matrices for two-dimensional nonlinear hyperbolic systems," *Comm. Pure Appl. Math* **23**, 867–884 (1970).

J. A. SMOLLER and C. C. CONLEY, "Viscosity matrices for two-dimensional nonlinear hyperbolic systems: II," *Amer. J. Math.* **94**, 631–650 (1972).

The concept of constant pattern profiles has been applied to analyze the effect of mass transfer resistance in fixed-bed adsorption process. For this treatment, see

D. O. COONEY and F. P. STRUSI, "Analytical description of fixed-bed sorption of two Langmuir solutes under nonequilibrium conditions," *Indust. Engrg. Chem. Fund.* **11**, 123–126 (1972).

W. G. BRADLEY and N. H. SWEED, "Rate controlled constant pattern fixed-bed sorption with axial dispersion and nonlinear multicomponent equilibria," *AIChE Symp. Ser.* **71**(152), 59–68 (1975).

For experimental evidence of the shock layer (or the constant pattern profile), see

W. J. THOMAS and J. L. LOMBARDI, "Binary adsorption of benzene-toluene mixtures," *Trans. Inst. Chem. Engrs. London* **49**, 240–250 (1971).

Numerical solution of the system (2.10.1) and (2.10.2) with $E_i = 0$ has been studied extensively in

J. S. C. HSIEH, R. M. TURIAN, and C. TIEN, "Multicomponent liquid phase adsorption in fixed bed," *AIChE J.* **23**, 263–275 (1977).

S.-C. WANG and C. TIEN, "Further work on multicomponent liquid phase adsorption in fixed beds," *AIChE J.* **28**, 565–572 (1982).

A more elaborate model taking into account the intraparticle multicomponent diffusion has been treated by using the orthogonal collocation method by

D. BARBA, G. DEL RE, and P. U. FOSCOLO, "Numerical simulation of multicomponent ion exchange equations," *Chem. Engrg. J.* **26**, 33–39 (1983).

Hyperbolic Systems of First-Order Quasilinear Equations and Multicomponent Chromatography

3

In this chapter we begin the development of the mathematical theory for quasilinear systems of more than two equations. For a general problem it is possible to formulate the characteristic directions exactly as in the case for pairs of equations and establish the existence and uniqueness theorem for a sufficiently small domain. However, we immediately notice that there exists a fundamental difference between the present problem and the problem with two equations in that Riemann invariants cannot exist in general. This obviously restricts our endeavors to cases for which the existence of Riemann invariants is ensured.

A typical example of such a case is found in the class of Riemann problems and it would seem only natural to treat this problem here in great detail, although in Chapter 4 we extend the treatment to problems with piecewise constant data. To be more specific, we will consider the problem of multicomponent chromatography with the Langmuir isotherm. In this case we can find explicit forms for the Riemann invariants and the characteristic parameters, so that the mathematical analysis can be done completely and affords a comprehensive picture of the solution. Clearly, this problem per se bears an important meaning and finds applications in a variety of fields. But also, we presume that this is the first step in obtaining a better understanding of the solution structure, and thus will definitely help us treat more general problems in Chapter 2.

In Sec. 3.1 we again introduce the chromatographic equations in a way to show how a system of first-order equations is realized in an actual situation. In Sec. 3.2 we introduce the method of characteristics and the notion of a totally hyperbolic system. Thus the system of m equations is transformed into a system of $2m$ ordinary differential equations, which may be integrated by a finite difference method. In Sec. 3.3 we treat the generalized Riemann invariants and the simple wave theory due to Lax (1957). We also argue that Riemann invariants strictly analogous to those for pairs of equations cannot exist except for some special cases. This argument naturally leads us to Sec. 3.4, in which we consider Riemann problems. Here we are assured that there exists a one-parameter representation of the solution and the Riemann invariants always exist. The system of m equations is now transformed to the $(m - 1)$ fundamental differential equations and one equation for the characteristic direction, which is proven identical to that defined in Sec. 3.2.

From this point on we restrict our discussion to multicomponent chromatography with the Langmuir isotherm. This isotherm is derived and discussed in Sec. 3.5. In Sec. 3.6 we solve the fundamental differential equations to find explicit expressions for the Riemann invariants and to show that there are m different kinds. In the course of this development we are led to introduce a continuous transformation of the dependent variables, which directly produces the set of characteristic parameters. It is also noted that there exists a linear relationship between any pair of dependent variables and thus the image of the solution falls on a sequence of straight lines (called Γ's) in the hodograph space. The transformation between the space of the dependent variables $\Phi(m)$ and the space of the characteristic parameters $\Omega(m)$ is investigated further in Sec. 3.7 to find the inverse transform and to prove that each of the Γ's is parallel to one of the coordinate axes in the space $\Omega(m)$. This observation turns out to be very instrumental in constructing the solutions to Riemann problems. In Sec. 3.8 we formulate the characteristic directions in terms of the characteristic parameters and establish the simple wave solution.

In Sec. 3.9 we take up the situation when the solution cannot be determined uniquely. As before, we introduce discontinuities in the solution and derive the jump condition as well as the compatibility relations. It is shown that the compatibility relations have the same solution as the fundamental differential equations and hence there are m different kinds of discontinuities. For each kind, the propagation direction is expressed in terms of the characteristic parameters and the pertinent entropy condition is given. It turns out that the discontinuities encountered in the present problem are all shocks in the sense of Lax (1957). In Sec. 3.10 we are concerned with the entropy change across a shock and prove that the thermodynamic entropy increases in the direction of increasing coverage of the adsorption sites.

The last two sections deal with the construction of solutions to Riemann problems. In Sec. 3.11 we establish the global existence and uniqueness of the solution and present a convenient scheme for the actual construction of solutions together with the relevant formulas in explicit forms. We also treat some special cases in which one or more solutes are absent from the feed state or from the initial state, for these are of potential interest in the practice of multicomponent chromatography. In Sec. 3.12 we present several numerical examples for the purpose of illustration. The solution scheme is explained in some detail and the results are given in the form of the physical plane portrait and the concentration profiles.

The reader may notice that Secs. 3.2 and 3.3 are concerned with the general theory for systems of quasilinear equations, whereas in other sections we confine overselves to Riemann problems and specifically deal with the theory of multicomponent chromatography with the Langmuir isotherm. Therefore, those readers with particular interest in application to chromatographic processes may as well skip Secs. 3.2 and 3.3 without loss of continuity. This will remain the same through Chapters 4 and 5, but we return to the general treatment in Chapter 6, although it will be still limited to Riemann problems.

3.1 Equations for the Equilibrium Chromatography of Many Solutes

Just as in the case of pairs of quasilinear equations of first order, we consider the equilibrium chromatography of many interacting solutes as a typical example here. Thus for each solute species A_i present in the chromatographic column with voidage ϵ and V as the interstitial velocity of the fluid phase, we can write a conservation equation in terms of the fluid-phase concentration $c_i(z, t)$ and the solid-phase concentration $n_i(z, t)$, both taken at distance z from the inlet and at time t. This was formulated in Secs. 1.2 and 1.3 of Vol. 1 to give the following equation:

$$V \frac{\partial c_i}{\partial z} + \frac{\partial c_i}{\partial t} + \frac{1 - \epsilon}{\epsilon} \frac{\partial n_i}{\partial t} = 0 \qquad (3.1.1)$$

If we put

$$\nu = \frac{1 - \epsilon}{\epsilon} \qquad (3.1.2)$$

and make the independent variables dimensionless; i.e.,

$$x = \frac{z}{Z}, \qquad \tau = \frac{\nu t}{Z} \qquad (3.1.3)$$

where Z is the characteristic length of the system. Eq. (3.1.1) becomes

$$\frac{\partial c_i}{\partial x} + \frac{\partial c_i}{\partial \tau} + v\frac{\partial n_i}{\partial \tau} = 0 \qquad (3.1.4)$$

Suppose that there are m different solutes competing for adsorption sites. Since we assume that adsorption equilibrium is established everywhere at any time, the solid-phase concentrations $n_1, n_2, \ldots, n_m$ are all functions of the fluid-phase concentrations $c_1, c_2, \ldots, c_m$. Hence we have

$$n_i = n_i(c_1, c_2, \ldots, c_m), \qquad i = 1, 2, \ldots, m \qquad (3.1.5)$$

These functions are essentially nonlinear due to the mutual influence among solutes and may be regarded, in general, as continuous, with as many derivatives as may be required. On physical grounds, we require the first-order derivatives to have uniform signs as

$$n_{i,j} = \frac{\partial n_i}{\partial c_j} \begin{cases} > 0 & \text{if } i = j \\ < 0 & \text{if } i \neq j \end{cases} \qquad (3.1.6)$$

The abbreviated symbol for the partial derivatives will be employed throughout this chapter and in the sequel. We also expect to have the conditions

$$|n_{i,i}| > |n_{j,i}| \qquad (3.1.7)$$

For the sake of convenience, we shall introduce the equilibrium column isotherm defined as

$$f_i = f_i(c_1, c_2, \ldots, c_m) = c_i + vn_i(c_1, c_2, \ldots, c_m) \qquad (3.1.8)$$

and rewrite Eq. (3.1.4) in the form

$$\frac{\partial c_i}{\partial x} + \frac{\partial f_i}{\partial \tau} = 0, \qquad i = 1, 2, \ldots, m \qquad (3.1.9)$$

Here it is evident that the column isotherm f_i satisfies the same inequality requirements as the adsorption isotherm n_i; that is,

$$f_{i,j} = \frac{\partial f_i}{\partial c_j} = \begin{cases} > 0 & \text{if } i = j \\ < 0 & \text{if } i \neq j \end{cases} \qquad (3.1.10)$$

and

$$|f_{i,i}| > |f_{j,i}| \qquad (3.1.11)$$

Substituting Eq. (3.1.8) into Eq. (3.1.9) gives

$$\frac{\partial c_i}{\partial x} + \sum_{j=1}^{m} f_{i,j}\frac{\partial c_j}{\partial \tau} = 0, \qquad i = 1, 2, \ldots, m \qquad (3.1.12)$$

so we obtain a system of m partial differential equations of first order. The natural initial data on these equations would be the specification of c_i at the inlet $x = 0$ and on the column at $\tau = 0$; that is,

$$\text{at } \tau = 0: \quad c_i = c_i^0(x) \quad \text{for } x > 0$$
$$\text{at } x = 0: \quad c_i = c_i^f(\tau) \quad \text{for } \tau > 0 \tag{3.1.13}$$

for $i = 1, 2, \ldots, m$, where the superscripts 0 and f represent the initial and feed conditions, respectively. These equations are entirely typical of the general system of equations that we shall study in Sec. 3.2 because the coefficients $f_{i,j}$ are functions of the dependent variables $c_1, c_2, \ldots, c_m$.

Let us now define column vectors as

$$\mathbf{c} = \begin{bmatrix} c_1 \\ c_2 \\ \cdot \\ \cdot \\ c_m \end{bmatrix}, \quad \mathbf{n} = \begin{bmatrix} n_1 \\ n_2 \\ \cdot \\ \cdot \\ n_m \end{bmatrix}, \quad \mathbf{f} = \begin{bmatrix} f_1 \\ f_2 \\ \cdot \\ \cdot \\ f_m \end{bmatrix} \tag{3.1.14}$$

so that we have

$$\mathbf{f}(\mathbf{c}) = \mathbf{c} + \nu\mathbf{n}(\mathbf{c}) \tag{3.1.15}$$

We can then write Eq. (3.1.9) in vector form as

$$\frac{\partial \mathbf{c}}{\partial x} + \frac{\partial}{\partial \tau}\mathbf{f}(\mathbf{c}) = \mathbf{0} \tag{3.1.16}$$

and Eq. (3.1.12) in matrix form as

$$\frac{\partial \mathbf{c}}{\partial x} + \mathbf{A}(\mathbf{c})\frac{\partial \mathbf{c}}{\partial \tau} = \mathbf{0} \tag{3.1.17}$$

with the $m \times m$ matrix $\mathbf{A}$ defined by

$$\mathbf{A} = [f_{i,j}] = \begin{bmatrix} f_{1,1} & f_{1,2} & \cdots & f_{1,m} \\ f_{2,1} & f_{2,2} & \cdots & f_{2,m} \\ \cdot & \cdot & \ddots & \cdot \\ \cdot & \cdot & & \cdot \\ f_{m,1} & f_{m,2} & \cdots & f_{m,m} \end{bmatrix} \tag{3.1.18}$$

It is immediately apparent that

$$\mathbf{A} = \nabla\mathbf{f} = \mathbf{I} + \nu\nabla\mathbf{n} \tag{3.1.19}$$

where ∇ denotes the gradient in the m-dimensional concentration space and $\mathbf{I}$ the identity matrix. In other words, the matrix $\mathbf{A}$ is given by the Jacobian matrix of the vector-valued function $\mathbf{f}$. This matrix is frequently referred to as the *Fréchet derivative* of $\mathbf{f}$ and denoted by d$\mathbf{f}$. Equation (3.1.17) is now subject to the initial and feed data prescribed as

$$\text{at } \tau = 0: \quad \mathbf{c} = \mathbf{c}^o(x) \quad \text{for } x < 0$$
$$\text{at } x = 0: \quad \mathbf{c} = \mathbf{c}^f(\tau) \quad \text{for } \tau > 0 \tag{3.1.20}$$

Exercise

3.1.1. Suppose that m different solutes are competing for adsorption sites in a continuous countercurrent moving-bed adsorber under equilibrium conditions, so that the conservation equations are given by Eq. (1.5.1) of Vol. 1. After introducing dimensionless parameters and variables defined by Eqs. (3.1.2) and (3.1.3), put the system of equations into matrix form.

3.2 Hyperbolic Systems of More than Two First-Order Equations

In this section we consider the general quasilinear system of first-order equations for m dependent variables $u_1, u_2, \ldots, u_m$ with two independent variables, x and y:

$$\mathcal{L}_i \equiv \sum_{j=1}^{m} a_{ij} \frac{\partial u_j}{\partial x} + \sum_{j=1}^{m} b_{ij} \frac{\partial u_j}{\partial y} + g_i = 0, \qquad i = 1, 2, \ldots, m \qquad (3.2.1)$$

where the coefficients a_{ij} and b_{ij} as well as the terms g_i are all continuous functions of x, y, $u_1, u_2, \ldots, u_m$. Again, as in Sec. 1.2 for pairs of equations, the system (3.2.1) is *quasilinear* since the partial derivatives all enter linearly. The system is called simply *linear* if the coefficients a_{ij} and b_{ij} are all dependent on x and y only. A linear system may be called *strictly linear* if the terms g_i are dependent on x and y only, *linear* if the terms g_i depend on $u_1, u_2, \ldots, u_m$ linearly, and *semilinear* if the terms g_i depend on $u_1, u_2, \ldots, u_m$ in some nonlinear manner.

In the differential operator $\mathcal{L}_i$ the m dependent variables $u_1, u_2, \ldots, u_m$ appear differentiated in different directions. We now ask for a curve $C: x(\omega)$, $y(\omega)$, such that a linear combination

$$\mathcal{L} = \sum_{i=1}^{m} \lambda_i \mathcal{L}_i = 0 \qquad (3.2.2)$$

can be formed in which all variables $u_1, u_2, \ldots, u_m$ appear differentiated in the direction of the curve C given by

$$\frac{dy}{dx} = \frac{\partial y / \partial \omega}{\partial x / \partial \omega} = \sigma \qquad (3.2.3)$$

In the operator $\mathcal{L}$, then, differentiations are to occur only with respect to the curve parameter ω. Such a direction, if it exists, is called a *characteristic direction* and the curve C a *characteristic*.

The condition that in $\mathcal{L}$ the variables $u_1, u_2, \ldots, u_m$ are differentiated in the direction of the tangent to the curve C is simply

$$\frac{\partial y}{\partial \omega} \sum_{i=1}^{m} \lambda_i a_{ij} = \frac{\partial x}{\partial \omega} \sum_{i=1}^{m} \lambda_i b_{ij}, \qquad j = 1, 2, \ldots, m \qquad (3.2.4)$$

for the coefficients of the derivatives $\partial u_j/\partial x$ and $\partial u_j/\partial y$ in $\mathscr{L}$ are given by $\Sigma \lambda_i a_{ij}$ and $\Sigma \lambda_i b_{ij}$, respectively. The differentiation of u_j in the direction of the curve C may be expressed as

$$\frac{\partial u_j}{\partial \omega} = \left(\frac{\partial u_j}{\partial x} + \sigma \frac{\partial u_j}{\partial y}\right)\frac{\partial x}{\partial \omega} = \left(\frac{1}{\sigma}\frac{\partial u_j}{\partial x} + \frac{\partial u_j}{\partial y}\right)\frac{\partial y}{\partial \omega} \qquad (3.2.5)$$

and thus the expression $\mathscr{L}$ can be written after multiplication with either $\partial x/\partial \omega$ or $\partial y/\partial \omega$ as

$$\mathscr{L}\frac{\partial x}{\partial \omega} = \sum_{i=1}^{m}\sum_{j=1}^{m} \lambda_i a_{ij}\frac{\partial u_j}{\partial \omega} + \sum_{i=1}^{m} \lambda_i g_i \frac{\partial x}{\partial \omega} = 0 \qquad (3.2.6)$$

or as

$$\mathscr{L}\frac{\partial y}{\partial \omega} = \sum_{i=1}^{m}\sum_{j=1}^{m} \lambda_i b_{ij}\frac{\partial u_j}{\partial \omega} + \sum_{i=1}^{m} \lambda_i g_i \frac{\partial y}{\partial \omega} = 0 \qquad (3.2.7)$$

Consequently, we find that a necessary and sufficient condition for the existence of a characteristic C is the compatibility of the following homogeneous linear equations for λ_i:

$$\sum_{i=1}^{m} \lambda_i\left(a_{ij}\frac{\partial y}{\partial \omega} - b_{ij}\frac{\partial x}{\partial \omega}\right) = 0,$$

$$\sum_{i=1}^{m} \lambda_i\left(\sum_{j=1}^{m} a_{ij}\frac{\partial u_j}{\partial \omega} + g_i\frac{\partial x}{\partial \omega}\right) = 0 \qquad j = 1, 2, \ldots, m \qquad (3.2.8)$$

$$\sum_{i=1}^{m} \lambda_i\left(\sum_{j=1}^{m} b_{ij}\frac{\partial u_j}{\partial \omega} + g_i\frac{\partial y}{\partial \omega}\right) = 0$$

This condition is tantamount to the vanishing of all the mth-order determinants taken from the matrix of the coefficients of λ_i. Thus a number of characteristic relations are expected to follow.

From the first m equations, in particular, we obtain the characteristic equation

$$\left| a_{ij}\frac{\partial y}{\partial \omega} - b_{ij}\frac{\partial x}{\partial \omega} \right| = 0 \qquad (3.2.9)$$

which is an algebraic equation of mth degree for the characteristic direction

$$\frac{\partial y/\partial \omega}{\partial x/\partial \omega} = \frac{dy}{dx} = \sigma \qquad (3.2.10)$$

If the characteristic equation (3.2.9) has m real roots $\sigma_1, \sigma_2, \ldots, \sigma_m$, the system is called *hyperbolic*. If the m roots are real and distinct, we say that the system is *totally hyperbolic,* and it is with this that we will be concerned. Thus there exist m different families of characteristics C^k satisfying the ordinary differential equation

$$\frac{dy}{dx} = \sigma_k, \qquad k = 1, 2, \ldots, m \qquad (3.2.11)$$

each family covering the domain of the (x, y)-plane under consideration. Clearly, σ_k is a function of x, y, u_1, u_2, . . . , u_m and it is referred to as the kth *characteristic field*. Similarly, the curve C^k is called the kth *characteristic*. Let the parameter running along a C^k be ω_k; then we can write

$$\frac{\partial y}{\partial \omega_k} - \sigma_k \frac{\partial x}{\partial \omega_k} = 0 \quad \text{along} \quad C_k \qquad (3.2.12)$$

for $k = 1, 2, \ldots, m$. The parameters ω_1, ω_2, . . . , ω_m are known as the *characteristic parameters*.

Next we take any m, except for the first m, equations from Eq. (3.2.8) and set the determinant of the matrix of the coefficients equal to zero to obtain the equation

$$\sum_{j=1}^{m} M_{kj} \frac{\partial u_j}{\partial \omega_k} + N_k \frac{\partial y}{\partial \omega_k} = 0 \quad \text{along} \quad C_k \qquad (3.2.13)$$

for $k = 1, 2, \ldots, m$. Here the coefficients M_{kj} and N_k are known functions of x, y, u_1, u_2, . . . , u_m with a nonvanishing determinant $|M_{kj}|$. If necessary, the last term on the left-hand side of Eq. (3.2.13) may be put in the form $N_k \sigma_k \partial x / \partial \omega_k$. Together with Eq. (3.2.12) these equations (3.2.13) constitute the characteristic differential equations for the original system of partial differential equations (3.2.1).

Considered along the characteristic C^k, Eq. (3.2.13) may be interpreted as an ordinary differential equation. For $m = 2$ this observation suggests introducing the two families of characteristics as new coordinate lines and the two parameters ω_1 and ω_2 as two new independent variables. Indeed, this is exactly what we have done in Chapters 1 and 2. For $m > 2$, however, the situation is not as simple, since the m characteristics imply m parameters, while we have only two independent variables. Nevertheless, the characteristic differential equations (3.2.12) and (3.2.13) lend themselves to a theoretical treatment which at the same time suggests various numerical schemes by finite difference methods.

We shall first assume that the original system of equations (3.2.1) is linear. Then the m families of characteristics are fixed curves in the (x, y)-plane given by the ordinary differential equations (3.2.11). Let us now consider the initial value problem: Given a curve I, nowhere characteristic, with prescribed initial values of u_1, u_2, . . . , u_m, find a solution at all points P with coordinates x, y in a suitably small domain R adjacent to the curve I. The domain is assumed to be such that the m characteristics through a point P intersect the curve I in distinct points Q_k and have different directions at P (see Fig. 3.1). Then the m equations in Eq. (3.2.13) can be written, after integration by parts, in the form

$$\sum_{j=1}^{m} M_{kj} u_j \big|_P = -N_k y \big|_P + N_k y \bigg|_{Q_k} + \sum_{j=1}^{m} M_{kj} u_j \bigg|_{Q_k} + \sum_{j=1}^{m} \int_{Q_k}^{P} u_j \, dM_{kj} \qquad (3.2.14)$$

for $k = 1, 2, \ldots, m$, in which the integration is extended over the arc $\overgroup{Q_k P}$ of C^k. We note that the first three members on the right-hand side are known

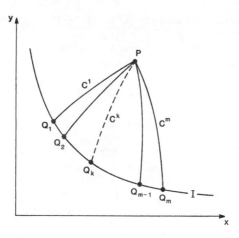

Figure 3.1

quantities. The m relations in Eq. (3.2.14) immediately suggest an iteration scheme: We insert on the right-hand side, for $u_1, u_2, \ldots, u_m$, arbitrary first estimates $u_1^{(1)}, u_2^{(1)}, \ldots, u_m^{(1)}$ satisfying the initial conditions. Now identifying $u_1, u_2, \ldots, u_m$ on the left-hand side with the next estimates $u_1^{(2)}, u_2^{(2)}, \ldots, u_m^{(2)}$, we obtain a system of m linear equations for the values of $u_1, u_2, \ldots, u_m$ at P. The determinant of the system is assumed nonvanishing. The resulting values $u_1^{(2)}, u_2^{(2)}, \ldots, u_m^{(2)}$, as functions of x, y are now used on the right-hand side again to produce expressions for the next approximation, and so on. Convergence of the procedure to a solution $u_1, u_2, \ldots, u_m$ of the original system of equations is not difficult to prove.

In the general case of a quasilinear system of equations (3.2.1) we start with the first estimates $u_1^{(1)}, u_2^{(1)}, \ldots, u_m^{(1)}$, substitute these values in the coefficients a_{ij} and b_{ij} of the original system of differential equations, and thus obtain a linear system. By applying the method discussed above, this linear system will produce as solution the second estimates $u_1^{(2)}, u_2^{(2)}, \ldots, u_m^{(2)}$. Iterating this procedure we obtain convergence to the desired solution in a suitably small domain.

Although this may be too schematic, the procedure given above shows how the solution can in principle be constructed. It shows, for instance, that to determine the solution at point P, we need only the initial data prescribed on the segment $\overset{\frown}{Q_1 Q_m}$ of the initial curve I cut out by the two outer characteristics through point P. This is illustrated in Fig. 3.2. For this reason the segment $\overset{\frown}{Q_1 Q_m}$ is called the *domain of dependence* of the solution at the point P. Similarly, the solution at every point of the region D, shown as the shaded area in Fig. 3.2, will be uniquely determined by the initial data specified on the segment $\overset{\frown}{Q_1 Q_m}$. Thus it is appropriate to call the region D the *domain of determinacy* of the solution with respect to the initial data prescribed on the segment $\overset{\frown}{Q_1 Q_m}$.

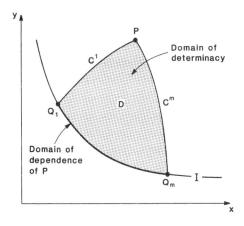

Figure 3.2

We also notice that the initial data at the point Q are needed in determining the solution at any point in the region R, which is contained between the two outer characteristics emanating from the point Q and shown as the shaded area in Fig. 3.3. In other words, the initial data at the point Q cannot influence the solution at any point outside the region R because all the characteristics C^k emanating from the point Q remain within the region R. It is in this sense that we call the region R the *range of influence* of the point Q. We may as well say that the range of influence of the point Q consists of all points whose domains of dependence contain the point Q.

A rigorous version of the procedure described above has been established by Friedrichs (1948), Courant and Lax (1949), and Lax (1953). Their conclusions may be embodied by the following theorem (see, e.g., Jeffrey, 1976):

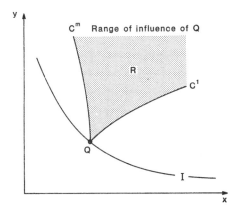

Figure 3.3

THEOREM 3.1: (Existence and Uniqueness): Let the quasilinear system of partial differential equations (3.2.1) be such that

1. It is totally hyperbolic throughout the (x, y)-plane.
2. The coefficients a_{ij} and b_{ij} as well as the terms g_i have Lipschitz continuous partial derivatives.
3. The initial data are prescribed on the curve I and the data have Lipschitz continuous derivatives.

Then a unique solution exists in a suitably small domain R adjacent to the initial curve I and, furthermore, within this domain the solution has Lipschitz continuous partial derivatives.

Now the existence of a unique solution is ensured only for a small region adjacent to the initial curve I. We have already observed in Secs. 6.1 and 7.8 of Vol. 1 and in Secs. 1.6, and 1.10 that Theorem 3.1 would not guarantee the existence of a global solution in the cases of single equations and pairs of equations. The question is, then, how the solution structure changes if nonuniqueness occurs and how one can extend the solution outside this region. These subjects will be investigated further in the latter part of this chapter as well as in Chapter 6.

Next we shall see how the characteristic directions can be determined for a system of equations that are written in matrix form. Let us start out with the system of chromatographic equations (3.1.12) and compare this to Eq. (3.2.1) with $y \equiv \tau$. We immediately find that

$$a_{ij} = \delta_{ij}, \qquad b_{ij} = f_{i,j}, \qquad g_i = 0$$

for $i, j = 1, 2, \ldots, m$, where δ_{ij} is the Kronecker delta. The characteristic equation (3.2.9) now reads

$$|\delta_{ij}\sigma - f_{i,j}| = 0$$

with

$$\sigma = \frac{d\tau}{dx} = \frac{\partial\tau/\partial\omega}{\partial x/\partial\omega} \tag{3.2.15}$$

or, in matrix form,

$$|\mathbf{A} - \sigma\mathbf{I}| = 0 \tag{3.2.16}$$

where the $m \times m$ matrix $\mathbf{A}$ represents the Jacobian matrix of $\mathbf{f}$ [see Eq. (3.1.18)], which is identical to the matrix of coefficients in Eq. (3.1.17). Therefore, the characteristic directions σ are given by the eigenvalues of the matrix of coefficients $\mathbf{A}$. This implies that if the matrix $\mathbf{A}$ has m real, distinct eigenvalues, the system of equations (3.1.17) is totally hyperbolic.

Indeed, the general system of equations (3.2.1) can be put in matrix form by defining the two column vectors

$$
\mathbf{u} = \begin{bmatrix} u_1 \\ u_2 \\ \cdot \\ \cdot \\ \cdot \\ u_m \end{bmatrix}, \qquad \mathbf{g}_0 = \begin{bmatrix} g_1 \\ g_2 \\ \cdot \\ \cdot \\ \cdot \\ g_m \end{bmatrix} \tag{3.2.17}
$$

and the two $m \times m$ matrices

$$
\mathbf{A}_0 = [a_{ij}], \qquad \mathbf{B}_0 = [b_{ij}] \tag{3.2.18}
$$

Thus from Eq. (3.2.1) we have the equation

$$
\mathbf{A}_0 \frac{\partial \mathbf{u}}{\partial x} + \mathbf{B}_0 \frac{\partial \mathbf{u}}{\partial y} + \mathbf{g}_0 = \mathbf{0} \tag{3.2.19}
$$

We shall assume that this equation is hyperbolic in the x-direction[†] so that the matrix $\mathbf{A}_0$ is positive definite and so has an inverse $\mathbf{A}_0^{-1}$. Premultiplying Eq. (3.2.19) by $\mathbf{A}_0^{-1}$ reduces the equation to the form

$$
\frac{\partial \mathbf{u}}{\partial x} + \mathbf{A} \frac{\partial \mathbf{u}}{\partial y} + \mathbf{g} = \mathbf{0} \tag{3.2.20}
$$

where

$$
\mathbf{A} = \mathbf{A}_0^{-1} \mathbf{B}_0 \tag{3.2.21}
$$

and

$$
\mathbf{g} = \mathbf{A}_0^{-1} \mathbf{g}_0 \tag{3.2.22}
$$

It is now clear that the characteristic directions σ are given by the eigenvalues of the matrix $\mathbf{A}$; i.e.,

$$
|\mathbf{A} - \sigma \mathbf{I}| = 0 \tag{3.2.23}
$$

where

$$
\sigma = \frac{dy}{dx} = \frac{\partial y / \partial \omega}{\partial x / \partial \omega} \tag{3.2.24}
$$

When an arbitrary system of quasilinear equations (3.2.1) can be reduced to the form of conservation laws (3.1.9), we say that the system is *completely conservative*. Although conservation equations can always be put into the form of a system of quasilinear equations, the converse is not necessarily true. Indeed, it has been shown by Rozdestvenskii (1957, 1959) that if the coefficients a_{ij} and b_{ij} as well as the term g_i are sufficiently smooth functions of x, y, u_1, u_2, ... , u_m, then the system (3.2.1) for $m \le 2$ is always conservative but for $m \ge 3$ the system is, in general, nonconservative.

[†] This is equivalent to saying that the hyperbolic system is *nondegenerate* and the free (noncharacteristic) direction is given by $x = $ const.

Exercises

3.2.1. Consider the system of equations (3.2.1) for $m = 3$ and show that the coefficients in Eq. (3.2.13) are defined as

$$M_{kj} = \sum_{i=1}^{3} \alpha_{ik} b_{ij}, \qquad N_k = \sum_{i=1}^{3} \alpha_{ik} g_i$$

for $k = 1, 2,$ and 3, where

$$\alpha_{1k} = \begin{vmatrix} a_{21}\sigma_k - b_{21} & a_{31}\sigma_k - b_{31} \\ a_{22}\sigma_k - b_{22} & a_{32}\sigma_k - b_{32} \end{vmatrix}$$

$$\alpha_{2k} = \begin{vmatrix} a_{11}\sigma_k - b_{11} & a_{31}\sigma_k - b_{31} \\ a_{12}\sigma_k - b_{12} & a_{32}\sigma_k - b_{32} \end{vmatrix}$$

$$\alpha_{3k} = \begin{vmatrix} a_{11}\sigma_k - b_{11} & a_{21}\sigma_k - b_{21} \\ a_{12}\sigma_k - b_{12} & a_{22}\sigma_k - b_{22} \end{vmatrix}$$

3.2.2 The conservation equations for a continuous countercurrent moving-bed adsorber were derived in matrix form in Exercise 3.1.1. Show that the characteristic direction σ' for this system is related to the characteristic direction for the corresponding fixed-bed system as

$$\sigma' = \frac{\sigma}{1 - \dfrac{V_s}{V_f}(\sigma - 1)}$$

where V_s and V_f are the velocities of the solid phase and the fluid phase, respectively. [*Hint:* Equation (3.2.16) may be rewritten in the form $|\nu\nabla n - (\lambda - 1)\mathbf{I}| = 0$.]

3.3 Generalized Riemann Invariants and Simple Waves

In Sec. 1.9 we observed that the Riemann invariants could play a key role in constructing the simple wave solutions for pairs of reducible equations. We are naturally tempted to ask whether this useful method of solution can be generalized and extended to systems of more than two equations. Here we confine our attention to homogeneous systems in which the coefficients are all dependent only on the dependent variables. Thus we have, in matrix form, the equation[†]

$$\frac{\partial \mathbf{u}}{\partial x} + \mathbf{A}(\mathbf{u})\frac{\partial \mathbf{u}}{\partial y} = \mathbf{0} \tag{3.3.1}$$

where $\mathbf{u}$ is a column vector with m components and $\mathbf{A}(\mathbf{u})$ is an $m \times m$ matrix

---

[†] In the mathematics literature it is common to have the matrix $\mathbf{A}(\mathbf{u})$ associated with the term $\partial\mathbf{u}/\partial x$, but it is really immaterial. This form of equation readily accommodates the system of chromatrographic equations (3.1.17).

with m real, distinct eigenvalues (i.e., the system is totally hyperbolic). The system (3.3.1) will be said to be reducible in the generalized sense.

Given by the eigenvalues of $\mathbf{A}(\mathbf{u})$

$$|\mathbf{A}(\mathbf{u}) - \sigma \mathbf{I}| = 0 \tag{3.3.2}$$

the characteristic directions depend only on $\mathbf{u}$; i.e., along a C^k

$$\frac{dy}{dx} = \sigma_k(\mathbf{u}), \qquad k = 1, 2, \ldots, m \tag{3.3.3}$$

Corresponding to these eigenvalues σ_k there will be m left eigenvectors $\boldsymbol{\ell}_k$ and m right eigenvectors $\mathbf{r}_k$, such that

$$\boldsymbol{\ell}_k^T \mathbf{A} = \sigma_k \boldsymbol{\ell}_k^T, \qquad k = 1, 2, \ldots, m \tag{3.3.4}$$

and

$$\mathbf{A}\mathbf{r}_k = \sigma_k \mathbf{r}_k, \qquad k = 1, 2, \ldots, m \tag{3.3.5}$$

Multiplying Eq. (3.3.1) on the left by $\boldsymbol{\ell}_k^T$ and using Eq. (3.3.4), we obtain

$$\boldsymbol{\ell}_k^T \frac{\partial \mathbf{u}}{\partial x} + \sigma_k \boldsymbol{\ell}_k^T \frac{\partial \mathbf{u}}{\partial y} = 0$$

or

$$\boldsymbol{\ell}_k^T \left(\frac{\partial \mathbf{u}}{\partial x} + \sigma_k \frac{\partial \mathbf{u}}{\partial y} \right) = 0 \tag{3.3.6}$$

If we let the symbol d/dk abbreviate differentiation in the kth characteristic direction, i.e.,

$$\frac{d}{dk} = \frac{\partial}{\partial x} + \sigma_k \frac{\partial}{\partial y} \tag{3.3.7}$$

Eq. (3.3.6) reduces to

$$\boldsymbol{\ell}_k^T \frac{d\mathbf{u}}{dk} = 0, \qquad k = 1, 2, \ldots, m \tag{3.3.8}$$

Suppose that there exists a function $J(\mathbf{u})$ which remains constant along a jth characteristic C^j. Then J must satisfy the condition

$$\frac{\partial J}{\partial u_1} du_1 + \frac{\partial J}{\partial u_2} du_2 + \cdots + \frac{\partial J}{\partial u_m} du_m = 0$$

or

$$\sum_{i=1}^{m} \frac{\partial J}{\partial u_i} du_i = 0 \quad \text{along} \quad \text{a } C^j$$

but because

$$du_i = \frac{\partial u_i}{\partial x} dx + \frac{\partial u_i}{\partial y} dy = \left(\frac{\partial u_i}{\partial x} + \sigma_j \frac{\partial u_i}{\partial y} \right) dx = \frac{du_i}{dj} dx$$

along a C^j, we have

$$\sum_{i=1}^{m} \frac{\partial J}{\partial u_i} \frac{du_i}{dj} = 0$$

which we can write in matrix form

$$(\nabla J)^T \frac{d\mathbf{u}}{dj} = 0 \tag{3.3.9}$$

where ∇ denotes the gradient in the m-dimensional space of dependent variables $\mathbf{u}$. When compared with Eq. (3.3.8), the Eq. (3.3.9) implies that the vector ∇J must be proportional to the left eigenvector ℓ_j. Since the left and right eigenvectors of a matrix are biorthogonal, i.e.,

$$\ell_j^T \mathbf{r}_k = 0 \qquad \text{for } j \neq k \tag{3.3.10}$$

we obtain

$$\mathbf{r}_k^T \nabla J = 0 \tag{3.3.11}$$

If this equation has a solution for J, it is called a k-Riemann invariant after Lax (1957), and this is the generalization of the Riemann invariants J_+ and J_- for pairs of equations (see Sec. 1.9). Now we see that there are $(m - 1)$ choices for ℓ_j, $j \neq k$, and thus there are $(m - 1)$ independent k-Riemann invariants, each of which remains constant along a C^j for $j \neq k$. We shall use the symbol J_i^k, $i = 1, 2, 3, \ldots, m - 1$, to denote the $(m - 1)$ k-Riemann invariants. In summary, we read:

DEFINITION 3.1:[†] A function $J_i^k(\mathbf{u})$ is a k-Riemann invariant of the system (3.3.1) if it satisfies the condition

$$\mathbf{r}_k^T \nabla J_i^k = 0 \tag{3.3.12}$$

for all values of $\mathbf{u}$, where r_k is the kth right eigenvector of $\mathbf{A}$.

THEOREM 3.2: The k-Riemann invariant J_i^k, if it exists, remains constant along one of the C^j's, $j \neq k$, and there exist exactly $(m - 1)$ independent k-Riemann invariants.

Next, we closely follow the development of the simple wave theory due to Lax (1957). For this we consider the solution in the (x, y)-plane and examine what meaning the Riemann invariants have with respect to the solution. For the moment, let us assume that the solution is continuous and consider a portion B of the boundary of a region of constant state. According to the uniqueness theorem (see Theorem 3.1), there is exactly one way of continuing the solution as a smooth solution across B. We also note that disturbances can propagate only

---

[†] This definition of Riemann invariants is compatible with that of J_+ and J_-, of Sec. 1.9. If the present notation were applied to J_+ and J_-, we would have J_1^1 and J_1^2 for J_+ and J_-, respectively.

along characteristics. Therefore, if B is not characteristic, the solution must be continued as a constant across B, and this is a contradiction to the fact that B is a portion of the boundary. This argument leads us to:

THEOREM 3.3: A region of constant state in the (x, y)-plane is bounded by characteristics which are necessarily straight lines.

Suppose that a portion B of the boundary of a region of constant state is given by a kth characteristic C^k and consider the solution on the other side of B. If we take the $(m - 1)$ independent k-Riemann invariants J_j^k, it follows from Eqs. (3.3.10) and (3.3.12) that both the left eigenvectors ℓ_j, $j \neq k$, and the gradients of J_i^k span the orthogonal complement of $\mathbf{r}_j$, and hence ℓ_j can be expressed as a linear combination of ∇J_i^k; i.e.,

$$\ell_j = \sum_{i=1}^{m-1} \beta_{ji} \nabla J_i^k, \qquad j \neq k \tag{3.3.13}$$

where the coefficients β_{ji} are functions of $\mathbf{u}$. Substituting this in Eq. (3.3.8) gives

$$\sum_{i=1}^{m-1} \beta_{ji} (\nabla J_i^k)^T \frac{d\mathbf{u}}{dj} = 0$$

which can be rewritten as

$$\sum_{i=1}^{m-1} \beta_{ji} \frac{dJ_i^k}{dj} = 0, \qquad j \neq k \tag{3.3.14}$$

If the solution $\mathbf{u}$ is given, the coefficients β_{ji} have known values. Thus Eq. (3.3.14) is a linear hyperbolic system of $(m - 1)$ equations for the $(m - 1)$ unknowns J_i^k. Clearly, the characteristic directions of this system are given by σ_j, $j \neq k$, so that C^k is not a characteristic of this system. By the same uniqueness theorem as before, the solution J_i^k of this system, which is constant on one side of B, must be continued as a constant across B. Consequently, we have established

THEOREM 3.4: If a portion B of the boundary of a region of constant state is a kth characteristic C^k, all k-Riemann invariants on the other side of B are constant.

We are now in a position to give a formal definition of a generalized simple wave by analogy with the case of systems of two equations.

DEFINITION 3.2: A solution in a region of the (x, y)-plane for which all k-Riemann invariants are constant is called a k-*simple wave*.

It is immediately apparent that a solution in a region adjacent to a region of constant state is a simple wave.

In a k-simple wave region the k-Riemann invariants J_i^k are constant and thus

$$\frac{dJ_i^k}{dk} = (\nabla J_i^k)^T \frac{d\mathbf{u}}{dk} = 0$$

Combining these $(m - 1)$ equations with Eqs. (3.3.8), we can write

$$\begin{bmatrix} (\nabla J_1^k)^T \\ (\nabla J_2^k)^T \\ \vdots \\ (\nabla J_{m-1}^k)^T \\ \boldsymbol{\ell}_k^T \end{bmatrix} \frac{d\mathbf{u}}{dk} = 0 \tag{3.3.15}$$

Since the $(m - 1)$ elements ∇J_i^k are linearly independent and orthogonal to $\mathbf{r}_k$ while $\boldsymbol{\ell}_k^T \mathbf{r}_k \neq 0$, the $m \times m$ matrix in Eq. (3.3.15) is nonsingular. It then follows that $d\mathbf{u}/dk = \mathbf{0}$, so $\mathbf{u}$ remains constant in the kth characteristic direction and the kth characteristics C^k are all straight lines. This gives the following theorem.

THEOREM 3.5: The kth characteristics C^k in a k-simple wave region are straight lines along each of which the solution remains constant.

Since the foregoing development is in complete analogy with that for pairs of equations, we might as well picture the image of a k-simple wave in the m-dimensional space of dependent variables which we shall call the *hodograph space*. The $(m - 1)$ k-Riemann invariants remain constant on this image and constitute $(m - 1)$ independent equations for $u_1, u_2, \ldots, u_m$. This implies that there is a functional relationship between any two of $u_1, u_2, \ldots, u_m$, so that we have, for example,

$$u_i = u_i(u_1), \qquad i = 2, 3, \ldots, m \tag{3.3.16}$$

It then follows that the image of a k-simple wave consists of a single curve in the hodograph space. We shall call this curve a Γ^k to make clear that it is the image of a k-simple wave and so the $(m - 1)$ k-Riemann invariants remain fixed along it.

Along a C^j, $j \neq k$, one of the k-Riemann invariants J_i^k remains unchanged and thus all the jth characteristics C^j in the k-simple wave region map onto a Γ^k in the hodograph space. This is true for all j except for $j = k$. Since a kth characteristic C^k carries a constant state in a k-simple wave region, its image falls on a single point along a Γ^k. This correspondence is in contrast to that for systems of two equations because in the latter case a Γ_+ is the image of a C_+, while a C_- has a point image along a Γ_+.

From Eq. (3.3.8) we have

$$\boldsymbol{\ell}_j^T \frac{d\mathbf{u}}{dj} = 0 \tag{3.3.17}$$

where $d\mathbf{u}/dj$ represents differentiation of $\mathbf{u}$ in the direction of a C^j. In case of a k-simple wave, a C^j, $j \neq k$, maps onto a Γ^k in the hodograph space and by use of Eq. (3.3.16) we can write

$$\frac{d\mathbf{u}}{dj} = \left(\frac{\partial u_1}{\partial x} + \sigma_j \frac{\partial u_1}{\partial y}\right)\frac{d\mathbf{u}}{du_1} = \frac{du_1}{dj}\frac{d\mathbf{u}}{du_1}, \qquad j \neq k \qquad (3.3.18)$$

Substitution of this into Eq. (3.3.17) gives

$$\ell_j^T \frac{d\mathbf{u}}{du_1} = 0, \qquad j \neq k \qquad (3.3.19)$$

which upon comparison with Eq. (3.3.10) establishes the fact that the differential $d\mathbf{u}$ taken in the direction of a Γ^k is proportional to the right eigenvector $\mathbf{r}_k$. An immediate consequence is that the curve Γ^k is everywhere tangent to the $\mathbf{r}_k$ and thus we are led to

THEOREM 3.6: A k-simple wave maps onto a curve Γ^k in the m-dimensional hodograph space of the dependent variables and the curve Γ^k is everywhere tangent to the right eigenvector $\mathbf{r}_k$.

This theorem gives rise to a set of m differential equations of first order,

$$\frac{du_1}{r_{1k}} = \frac{du_2}{r_{2k}} = \cdots = \frac{du_m}{r_{mk}} = d\omega \quad \text{along a } \Gamma^k \qquad (3.3.20)$$

where $r_{1k}, r_{2k}, \ldots, r_{mk}$ are the elements of the right eigenvector $\mathbf{r}_k$ and ω is a parameter running along a Γ^k. When integrated, Eq. (3.3.20) will give $(m - 1)$ linearly independent relations among m elements of $\mathbf{u}$ that would remain invariant along the Γ^k. This set of $(m - 1)$ invariants clearly determines the curve Γ^k and constitutes the set of $(m - 1)$ k-Riemann invariants J_i^k. In practice, we write

$$\frac{du_i}{d\omega} = r_{ik}, \qquad i = 1, 2, \ldots, m \qquad (3.3.21)$$

and determine u_i in terms of the parameter ω. Then elimination of ω among these m equations will give the $(m - 1)$ k-Riemann invariants.

According to Theorem 3.5, a k-simple wave region in the (x, y)-plane is covered by a family of straight-line characteristics C^k, and each of these characteristics has a point image, all falling on a single Γ^k, in the hodograph space. In picturing the layout of the characteristics C^k in a k-simple wave region just as in the case of pairs of equations we find it very important to know how the value of σ_k varies along a Γ^k. If σ_k varies monotonically, we would expect construction of the solution to be straightforward. Otherwise, we expect to encounter rather complicated situations. In the former case we say that the kth characteristic field is *genuinely nonlinear*. Thus we have:

DEFINITION 3.3 The kth characteristic field of a quasilinear hyperbolic system is called *genuinely nonlinear* if $\nabla \sigma_k$ is not orthogonal to the right eigenvector $\mathbf{r}_k$ for any value of $\mathbf{u}$, i.e., if

$$\mathbf{r}_k^T \nabla \sigma_k \neq 0 \qquad \text{for all } \mathbf{u} \qquad (3.3.22)$$

So far we have seen that the simple wave theory can be generalized and extended to systems of more than two equations. Now we raise a question: Does Eq. (3.3.12) always have a solution for the k-Riemann invariant J_i^k? We recall from Eqs. (3.3.8) and (3.3.9) that ∇J^k is proportional to $\boldsymbol{\ell}_j$. Since the left eigenvector is determined only up to an arbitrary multiplicative constant, the result $\nabla J^k \propto \boldsymbol{\ell}_j$ imposes $(m-1)$ conditions on the m elements of ∇J^k. In addition, we must require the equality of mixed derivatives

$$\frac{\partial}{\partial u_i}\left(\frac{\partial J^k}{\partial u_j}\right) = \frac{\partial}{\partial u_j}\left(\frac{\partial J^k}{\partial u_i}\right), \qquad i, j = 1, 2, \ldots, m \qquad (3.3.23)$$

which gives rise to m^2 constraints. Excluding from these the m identities corresponding to $i = j$, we have $(m^2 - m)$ constraints, and by the symmetry argument we see that there are only $\frac{1}{2}m(m-1)$ different constraints. Hence the m components of ∇J^k must satisfy a total of $(m-1) + \frac{1}{2}m(m-1) = \frac{1}{2}(m-1)(m+2)$ conditions. Apart from special cases, this is not possible in general unless $m = 2$, so it is obvious that Riemann invariants strictly analogous to those for systems of two equations cannot exist for $m > 2$.

One example for which there exist Riemann invariants is afforded by equations of one-dimensional, unsteady, nonisentropic flow. The system of equations for a polytropic gas is found in the book by Courant and Friedrichs (1948) (see also Sec. 1.12 of Vol. 1) and in matrix form, it is

$$\frac{\partial \mathbf{u}}{\partial t} + \mathbf{A}\frac{\partial \mathbf{u}}{\partial y} = \mathbf{0} \qquad (3.3.24)$$

with

$$\mathbf{u} = \begin{bmatrix} \rho \\ v \\ S \end{bmatrix}, \qquad \mathbf{A} = \begin{bmatrix} v & \rho & 0 \\ \dfrac{1}{\rho}\dfrac{\partial p}{\partial \rho} & v & \dfrac{1}{\rho}\dfrac{\partial p}{\partial S} \\ 0 & 0 & v \end{bmatrix} \qquad (3.3.25)$$

where ρ is the density, v the velocity in the y-direction, S the entropy, and p the pressure. The pressure is regarded as a function of ρ and S, and $c^2 = \partial p/\partial \rho$ is the square of the sound speed.

The characteristic directions, $dy/dt = \sigma$, represent some characteristic speeds of the system and are readily determined as

$$\sigma_1 = v - c, \qquad \sigma_2 = v, \qquad \sigma_3 = v + c \qquad (3.3.26)$$

along with the right eigenvectors

$$
\mathbf{r}^1 = \begin{bmatrix} \rho \\ -c \\ 0 \end{bmatrix}, \qquad \mathbf{r}^2 = \begin{bmatrix} \dfrac{\partial p}{\partial S} \\ 0 \\ -\dfrac{\partial p}{\partial \rho} \end{bmatrix}, \qquad \mathbf{r}^3 = \begin{bmatrix} \rho \\ c \\ 0 \end{bmatrix} \qquad (3.3.27)
$$

Since all three eigenvectors are in particularly simple form, it is not difficult to find the Riemann invariants from Eq. (3.3.13). They can be chosen as follows.

1-Riemann invariants:

$$
J_1^1 \equiv v + \int \frac{c}{\rho}\,d\rho = \text{const.}, \quad J_2^1 \equiv S = \text{const.} \qquad (3.3.28)
$$

2-Riemann invariants:

$$
J_1^2 \equiv p = \text{const.}, \quad J_2^2 \equiv v = \text{const.} \qquad (3.3.29)
$$

3-Riemann invariants:

$$
J_1^3 \equiv v - \int \frac{c}{\rho}\,d\rho = \text{const.}, \quad J_2^3 \equiv S = \text{const.} \qquad (3.3.30)
$$

If, however, there exists a functional relationship between any two of the dependent variables just as in the case of a simple wave, Eq. (3.3.23) is automatically satisfied, so we are assured that the Riemann invariants always exist. Indeed, Eq. (3.3.20) is proven valid in this case and gives rise to the k-Riemann invariants. Now the remaining question is: Under what circumstances do we have such functional relationships? Although many cases may occur, we can find an example in the Riemann problem and, since this is of practical importance, we shall treat this problem in detail in the forthcoming sections.

Exercises

3.3.1 Recall the treatment of Sec. 1.9 for pairs of equations. If we let $\sigma_1 = \sigma_-$ and $\sigma_2 = \sigma_+$, show that Γ_+ and Γ_- correspond to Γ^1 and Γ^2, respectively. How would you interpret J_+ and J_- in this case?

3.3.2. By substituting Eq. (3.3.27) in Eq. (3.3.21) and eliminating ω, show that the Riemann invariants are given by Eqs. (3.3.28), (3.3.29), and (3.3.30). From these results, argue that the flow must be isentropic whenever it takes place in a simple wave region, so the entropy equation is redundant.

3.4 Riemann Problem and Fundamental Differential Equations

In Secs. 3.2 and 3.3 we were concerned with the general system of first-order partial differential equations written in matrix form. Here we return to physical systems for which we have the equations in the form of conservation laws.

More specifically, we start with the system of chromatographic equations, which reads from Eq. (3.1.9)

$$\frac{\partial c_i}{\partial x} + \frac{\partial f_i}{\partial \tau} = 0 \qquad (3.4.1)$$

$$f_i = f_i(c_1, c_2, \ldots, c_m) = c_i + \nu n_i(c_1, c_2, \ldots, c_m) \qquad (3.4.2)$$

for $i = 1, 2, \ldots, m$, and consider Riemann problems for which the initial and boundary conditions are specified by two different constant states with a jump discontinuity at the origin. Thus we have

$$\text{at } \tau = 0: \quad c_i = c_i^0 \quad \text{(a constant)}$$
$$\text{at } x = 0: \quad c_i = c_i^f \quad \text{(a constant)} \qquad (3.4.3)$$

for $i = 1, 2, \ldots, m$ and

$$c_i^0 \neq c_i^f \qquad (3.4.4)$$

for some i, where the superscripts 0 and f represent the initial and feed data, respectively. As we have seen in Chapter 2, this class of problems is of special interest in the practice of chromatography, and treating this problem provides a basis for extension to problems with piecewise constant data.

As a preliminary observation we note that if $c_i(x, \tau)$ is a solution of the system (3.4.1) satisfying the conditions of Eq. (3.4.3), then the function $c_{i\alpha} = c_i(\alpha x, \alpha\tau)$ with any positive constant α is also a solution and takes on the same initial and boundary values. If we confine our attention to continuous solutions, Theorem 3.1 asserts the uniqueness of the solution, so we must have $c_{i\alpha} = c_i(x, \tau)$. This is true if and only if $c_i(x, \tau)$ is a function of x/τ only,[†] and this is exactly what we have observed in the case of single equations (see Sec. 5.2 of Vol. 1) as well as in the case of pairs of equations (see Secs. 1.5 and 2.3).

Let us suppose that the solution has been obtained. If this is denoted by

$$c_i = c_i\left(\frac{x}{\tau}\right), \qquad i = 1, 2, \ldots, m \qquad (3.4.5)$$

and I_i is the inverse function of c_i, so that

$$\frac{x}{\tau} = I_i(c_i) = I_j(c_j), \qquad i, j = 1, 2, \ldots, m$$

there must be a system of relations

$$c_j = c_j(I_i(c_i)) = g_{ji}(c_i), \qquad i, j = 1, 2, \ldots, m \qquad (3.4.6)$$

where $g_{ii}(c_i) = c_i$. In other words, all the concentrations $c_1, c_2, \ldots, c_m$ can be expressed in terms of a single concentration, say c_1, and this implies that there exists a unique one-parameter representation of the solution, i.e.,

[†] Note that $c_i(x, \tau)$ is a homogeneous function of degree zero.

$$c_i = c_i(\omega), \qquad i = 1, 2, \ldots, m \tag{3.4.7}$$

In the m-dimensional concentration space (i.e., the hodograph space) this equation represents a single curve with the parameter ω running along it. Such a curve is the image of the solution and will be called a Γ.

Both the initial and feed data have point images in the hodograph space, which will be denoted by I and F, respectively. Now we can picture a smooth, or piecewise smooth, curve connecting points F and I as the image of the solution. The situation is illustrated for $m = 3$ in Fig. 3.4. At this stage we note the possibility that the solution may be determined by simple waves.

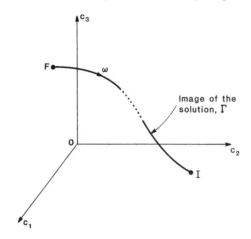

Figure 3.4

Let $d/d\omega$ represent differentiation in the direction of the curve Γ with ω as the independent variable. Then we have

$$\frac{\partial c_i}{\partial x} = \frac{dc_i}{d\omega}\frac{\partial \omega}{\partial x} \tag{3.4.8}$$

and since f_i can be expressed as a function of c_i along the curve Γ, we can write

$$\frac{\partial f_i}{\partial \tau} = \frac{\mathscr{D} f_i}{\mathscr{D} c_i}\frac{dc_i}{d\omega}\frac{\partial \omega}{\partial \tau} \tag{3.4.9}$$

The directional derivative $\mathscr{D} f_i/\mathscr{D} c_i$ can also be written in the form

$$\frac{\mathscr{D} f_i}{\mathscr{D} c_i} = \frac{df_i/d\omega}{dc_i/d\omega} = \sum_{j=1}^{m} f_{i,j}\frac{dc_j/d\omega}{dc_i/d\omega} \tag{3.4.10}$$

Substituting Eqs. (3.4.8) and (3.4.9) in Eq. (3.4.1), we obtain

$$\left(\frac{\partial \omega}{\partial x} + \frac{\mathscr{D} f_i}{\mathscr{D} c_i}\frac{\partial \omega}{\partial \tau}\right)\frac{dc_i}{d\omega} = 0, \qquad i = 1, 2, \ldots, m \tag{3.4.11}$$

Since c_i is not constant along the curve Γ, $dc_i/d\omega \neq 0$ and hence

$$\left(\frac{d\tau}{dx}\right)_\omega = -\frac{\partial\omega/\partial x}{\partial\omega/\partial\tau} = \frac{\mathcal{D}f_i}{\mathcal{D}c_i} = \sigma \qquad (3.4.12)$$

It follows from this equation that in the physical plane of x and τ there exists a specific direction along which the parameter ω is held constant. This direction is determined as a function of $c_1, c_2, \ldots, c_m$, but since ω is constant, all the concentrations remain fixed. Consequently, Eq. (3.4.12) gives rise to a straight line in the (x, τ)-plane which maps onto a single point on the curve Γ. We note that this straight line is equivalent to a kth characteristic C^k in a k-simple wave region and the curve Γ corresponds to a Γ^k.

On the other hand, Eq. (3.4.12) must be independent of the choice of i, so we have

$$\frac{\mathcal{D}f_1}{\mathcal{D}c_1} = \frac{\mathcal{D}f_2}{\mathcal{D}c_2} = \cdots = \frac{\mathcal{D}f_m}{\mathcal{D}c_m} \qquad (3.4.13)$$

Although motivated from a different point of view, Eq. (3.4.13) takes exactly the same form as the fundamental differential equation obtained for a system of two equations (see Sec. 1.4), but now it consists of $(m - 1)$ equations. Again we shall call these $(m - 1)$ equations the fundamental differential equations for Riemann problems.[†] It is evident that solution of the fundamental differential equations generates the one-parameter family [Eq. (3.4.6) or (3.4.7)], i.e., the curve Γ in the hodograph space.

It is yet to be shown how this formulation would be connected with the theoretical development in Secs. 3.2 and 3.3. For this purpose we expand Eq. (3.4.12) by using Eq. (3.4.10) to obtain

$$f_{i,1}\frac{dc_1}{d\omega} + f_{i,2}\frac{dc_2}{d\omega} + \cdots + (f_{i,i} - \sigma)\frac{dc_i}{d\omega} + \cdots + f_{i,m}\frac{dc_m}{d\omega} = 0$$

which is valid for all i. This system of m equations can be written in matrix form

$$(\mathbf{A} - \sigma\mathbf{I})\frac{d\mathbf{c}}{d\omega} = \mathbf{0} \qquad (3.4.14)$$

where

$$\mathbf{c} = \begin{bmatrix} c_1 \\ c_2 \\ \cdot \\ \cdot \\ \cdot \\ c_m \end{bmatrix}, \qquad \mathbf{f} = \begin{bmatrix} f_1 \\ f_2 \\ \cdot \\ \cdot \\ \cdot \\ f_m \end{bmatrix} \qquad (3.4.15)$$

and

$$\mathbf{A} = \nabla\mathbf{f} = [f_{i,j}] = d\mathbf{f} \qquad (3.4.16)$$

[†] Existence of the fundamental differential equations was recognized as early as in 1943 (see, e.g., DeVault, 1943 and Bayle and Klinkenberg, 1954).

Now Eq. (3.4.7) guarantees that there exists a nontrivial solution for $d\mathbf{c}/d\omega$, and this entails the condition that

$$|\mathbf{A} - \sigma\mathbf{I}| = 0 \qquad (3.4.17)$$

Compared with Eqs. (3.2.15) and (3.2.16), this equation shows that σ defined by Eq. (3.4.12) represents the characteristic direction of the system (3.4.1) in the (x, τ)-plane, and thus the curve Γ is equivalent to the image Γ^k of a k-simple wave in the hodograph space.

Incidentally, we can express the characteristic direction and the fundamental differential equations in terms of the adsorption isotherms n_i by using Eq. (3.4.2) in Eqs. (3.4.12) and (3.4.13). This gives

$$\left(\frac{d\tau}{dx}\right)_\omega = 1 + \nu\frac{\mathfrak{D}n_i}{\mathfrak{D}c_i} = \sigma \qquad (3.4.18)$$

and

$$\frac{\mathfrak{D}n_1}{\mathfrak{D}c_1} = \frac{\mathfrak{D}n_2}{\mathfrak{D}c_q} = \cdots = \frac{\mathfrak{D}n_m}{\mathfrak{D}c_m} \qquad (3.4.19)$$

Up to this point we have treated the problems with arbitrary nonlinear functions f_i, so that the development has general bearing. Starting from the next section, we are going to be concerned with a specific form of the nonlinear function given by the Langmuir isotherm. This will give a concrete picture of the solution and provide a solid basis for the treatment of more general problems, to be continued in Chapter 6.

3.5 Langmuir Isotherm for Multicomponent Adsorption

Retreating from the development of the general theory, we would like to treat a specific example so as not only to get a better feel for the solution but also to illustrate the application. For this purpose we consider the system of chromatographic equations with the Langmuir isotherm. The nicety of this problem is that we can find explicit expressions for the characteristic parameters and the Riemann invariants. By virtue of this fact, the analysis of equilibrium multicomponent chromatography can be done completely with mathematical rigor and affords a comprehensive picture of the solution. Apart from presenting a prototype example, the development to be presented henceforth will enjoy a whole variety of application in the practice of chromatography.

We start with a formulation of the Langmuir isotherm for multicomponent adsorption. Suppose that m different solutes $A_1, A_2, \ldots, A_m$ are competing for a limited number of adsorption sites on the solid phase. Let c_i and n_i denote the concentrations of A_i in the fluid and solid phases, respectively. Obviously, there exists a limiting value of n_i and, according to a thermodynamic argument, this limiting value must be the same for all solutes (see Sec. 1.3 of Vol.

1). However, this is not the case in reality, and thus we shall develop the formulation as such. If we let N_i be the limiting value of n_i, the fraction of vacant sites will be equal to $1 - \sum_{j=1}^{m} n_j/N_j$. The rate at which A_i is adsorbed might be expected to be proportional to the product of its concentration above the surface, c_i, and the availability of adsorption sites on the surface, so we have

$$r_{ai} = k_{ai} c_i \left(1 - \sum_{j=1}^{m} \frac{n_j}{N_j} \right)$$

We may take the rate of desorption to be proportional to the adsorbed concentration, n_i, i.e.,

$$r_{di} = k_{di} \frac{n_i}{N_i}$$

At equilibrium the two rates must be equal and thus we obtain

$$K_i c_i \left(1 - \sum_{j=1}^{m} \frac{n_j}{N_j} \right) = \frac{n_i}{N_i} \tag{3.5.1}$$

where the parameter K_i, being defined by

$$K_i = \frac{k_{ai}}{k_{di}} \tag{3.5.2}$$

represents the adsorption equilibrium constant for the solute A_i.

Since Eq. (3.5.1) holds for all i from 1 to m, we can sum these m equations to have

$$\sum_{j=1}^{m} K_j c_j = \frac{\sum_{j=1}^{m} \dfrac{n_j}{N_j}}{1 - \sum_{j=1}^{m} \dfrac{n_j}{N_j}} \tag{3.5.3}$$

or

$$\sum_{j=1}^{m} \frac{n_j}{N_j} = \frac{\sum_{j=1}^{m} K_j c_j}{1 + \sum_{j=1}^{m} K_j c_j} \tag{3.5.4}$$

and substitution of this into Eq. (3.5.1) gives

$$n_i = \frac{N_i K_i c_i}{1 + \sum_{j=1}^{m} K_j c_j}, \qquad i = 1, 2, \ldots, m \tag{3.5.5}$$

This is the Langmuir isotherm that describes the equilibrium of multicomponent adsorption, and its inverse is found directly from Eq. (3.5.1):

$$c_i = \frac{\dfrac{1}{K_i} \dfrac{n_i}{N_i}}{1 - \sum_{j=1}^{m} \dfrac{n_j}{N_j}}, \qquad i = 1, 2, \ldots, m \tag{3.5.6}$$

It can be shown in a straightforward manner that the Langmuir isotherm satisfies the elementary physical requirements given by Eqs. (3.1.6) and (3.1.7). The value of the equilibrium constant may be determined experimentally and is strongly dependent on the temperature:

$$K_i = K_i^0 T^{1/2} e^{-\Delta H_i/RT} \tag{3.5.7}$$

where ΔH_i is the heat of adsorption per mole of solute A_i. The limiting value N_i may be regarded as a constant intrinsic to the adsorbent and the solute A_i.

For convenience, we shall assume that the solutes are arranged in such a way that

$$\gamma_1 < \gamma_2 < \gamma_3 < \cdots < \gamma_{m-1} < \gamma_m \tag{3.5.8}$$

where γ_i is a dimensionless parameter defined by

$$\gamma_i = N_i K_i, \qquad i = 1, 2, \ldots, m \tag{3.5.9}$$

This is equivalent to numbering the solutes in the order of the adsorptivity from the smallest to the largest since the separation factor (or the relative adsorptivity) may be defined as

$$\beta_{ir} = \frac{n_i/c_i}{n_r/c_r} = \frac{N_i K_i}{N_r K_r} = \frac{\gamma_i}{\gamma_r} \tag{3.5.10}$$

where the subscript r denotes the reference species. Note that this separation factor can serve as a measure of the effectiveness of an adsorbent for separating the solute A_i from a mixture. For the Langmuir isotherm the separation factor turns out to be independent of concentrations.

The Langmuir isotherm carries an underlying assumption that the solvent or the carrier behaves as an inert component. If the solvent has a nonzero adsorptivity, but one that is smaller than any of the solutes, then Eq. (3.5.5) would become

$$n_i = \frac{N_i K_i c_i}{1 + \sum\limits_{j=0}^{m} K_j c_j}, \qquad i = 0, 1, 2, \ldots, m \tag{3.5.11}$$

in which the subscript 0 denotes the solvent. Under the conditions of constant temperature and pressure, we claim that

$$c_t = \sum_{j=1}^{m} c_j = \text{constant}$$

so Eq. (3.5.11) can be rearranged in the form

$$n_i = \frac{N_i' K_i' c_i}{1 + \sum\limits_{j=1}^{m} K_j' c_j}, \qquad i = 1, 2, \ldots, m \tag{3.5.12}$$

where

$$N_i' = \frac{N_i K_i}{K_i - K_0} \tag{3.5.13}$$

$$K_i' = \frac{K_i - K_0}{1 + K_0 c_t} \qquad (3.5.14)$$

for $i = 1, 2, \ldots, m$. Therefore, the Langmuir relation can be reduced to a form that excludes the solvent. Similarly, the constant-separation-factor equilibrium relations[†] introduced for ion-exchange systems can also be rewritten without difficulty in the form of Eq. (3.5.12), excluding the species with the smallest separation factor. This implies that the least adsorbable component (i.e., the solvent almost invariably) remains passive in the process of exchange and thus may be treated with the corresponding modifications given by Eqs. (3.5.13) and (3.5.14) as if it were an inert component.

Exercise

3.5.1. The constant-separation-factor equilibrium relation is commonly used for ion-exchange systems and takes the form

$$y_i = \frac{\alpha_k^i x_i}{\sum\limits_{j=1}^{m} \alpha_k^j x_j}, \qquad i = 1, 2, \ldots, m$$

where x_i and y_i denote the liquid-phase and solid-phase concentrations, respectively, and α_j^i the constant-separation-factor, defined by

$$\alpha_j^i = \frac{y_i/x_i}{y_j/x_j}$$

By using the conditions

$$\sum_{i=1}^{m} n_i = 1 \quad \text{and} \quad \sum_{i=1}^{m} y_i = 1$$

transform the equilibrium relation above into a Langmuir-type isotherm for $i = 1, 2, \ldots, m$.

3.6 Riemann Invariants for Multicomponent Chromatography with the Langmuir Isotherm

In the light of the form of the Langmuir isotherm (3.5.5), we shall introduce the following dimensionless variables:

$$\phi_i = K_i c_i, \qquad i = 1, 2, \ldots, m \qquad (3.6.1)$$

and

[†] See, for example, Tondeur and Klein (1967) and Helfferich and Klein (1970).

$$\delta = 1 + \sum_{i=1}^{m} K_i c_i = 1 + \sum_{i=1}^{m} \phi_i \qquad (3.6.2)$$

It then follows from Eq. (3.4.7) that δ may be regarded as a function of the parameter ω or ω as a function of δ and thus ϕ_i as well as n_i may be considered as a function of δ alone; i.e.,

$$\phi_i = \phi_i(\delta) \qquad (3.6.3)$$

and

$$n_i = N_i \frac{\phi_i(\delta)}{\delta} \qquad (3.6.4)$$

for $i = 1, 2, \ldots, m$. This implies that we choose δ as a parameter running along the curve Γ. We further remark that the total coverage Θ of adsorption sites can be expressed in terms of the parameter δ; i.e., from Eq. (3.5.4),

$$\Theta = \sum_{i=1}^{m} \frac{n_i}{N_i} = 1 - \frac{1}{\delta} \qquad (3.6.5)$$

Hence the coverage Θ varies in the same direction as the parameter δ.

By using δ in place of ω and introducing Eqs. (3.6.3) and (3.6.4), we can express the directional derivative $\mathcal{D}n_i/\mathcal{D}c_i$ in terms of ϕ_i:

$$\frac{\mathcal{D}n_i}{\mathcal{D}c_i} = \frac{\dfrac{dn_i}{d\delta}}{\dfrac{dc_i}{d\delta}} = \frac{N_i \dfrac{d}{d\delta}\left(\dfrac{\phi_i}{\delta}\right)}{\dfrac{1}{K_i}\dfrac{d\phi_i}{d\delta}}$$

$$= \gamma_i \frac{\dfrac{d}{d\delta}\left(\dfrac{\phi_i}{\delta}\right)}{\dfrac{d\phi_i}{d\delta}} \qquad (3.6.6)$$

$$= \frac{\gamma_i}{\delta^2}\left(\delta - \frac{\phi_i}{\dfrac{d\phi_i}{d\delta}}\right),$$

$$i = 1, 2, \ldots, m$$

Substituting this into the fundamental differential equations (3.4.19) and multiplying the resulting equations through by δ^2 (which never vanishes), we obtain

$$\gamma_1\left(\delta - \frac{\phi_1}{\dfrac{d\phi_1}{d\delta}}\right) = \gamma_2\left(\delta - \frac{\phi_2}{\dfrac{d\phi_2}{d\delta}}\right) = \cdots = \gamma_m\left(\delta - \frac{\phi_m}{\dfrac{d\phi_m}{d\delta}}\right) \qquad (3.6.7)$$

We shall differentiate these expressions once more with respect to δ and arrange each of the resulting expressions in the form of a ratio to get

$$\frac{\dfrac{d^2\phi_1}{d\delta^2}}{\dfrac{1}{\gamma_1\phi_1}\left(\dfrac{d\phi_1}{d\delta}\right)^2} = \frac{\dfrac{d^2\phi_2}{d\delta^2}}{\dfrac{1}{\gamma_2\phi_2}\left(\dfrac{d\phi_2}{d\delta}\right)^2} = \cdots = \frac{\dfrac{d^2\phi_m}{d\delta^2}}{\dfrac{1}{\gamma_m\phi_m}\left(\dfrac{d\phi_m}{d\delta}\right)^2}$$

$$= \frac{\displaystyle\sum_{i=1}^{m}\dfrac{d^2\phi_i}{d\delta^2}}{\displaystyle\sum_{i=1}^{m}\dfrac{1}{\gamma_i\phi_i}\left(\dfrac{d\phi_i}{d\delta}\right)^2} \tag{3.6.8}$$

in which the last expression immediately follows from the first m ratios.

Since we have

$$\sum_{i=1}^{m}\frac{d^2\phi_i}{d\delta^2} = 0$$

from Eq. (3.6.2), any solution to Eq. (3.6.8) must satisfy either

$$\sum_{i=1}^{m}\frac{1}{\gamma_i\phi_i}\left(\frac{d\phi_i}{d\delta}\right)^2 = 0 \tag{3.6.9}$$

or

$$\frac{d^2\phi_i}{d\delta^2} = 0, \qquad i = 1, 2, \ldots, m \tag{3.6.10}$$

Equation (3.6.9), however, cannot have any physical meaning, for it would require at least one of the concentrations $\phi_1, \phi_2, \ldots, \phi_m$ to be negative. Consequently, it is from Eq. (3.6.10) that the physically relevant solution may be determined. Direct integration of Eq. (3.6.10) gives

$$\phi_i - \phi_i^0 = J_i(\delta - \delta^0), \qquad i = 1, 2, \ldots, m \tag{3.6.11}$$

in which the superscript 0 denotes the fixed state of concentrations and J_i is the integration constant to be determined. Adding the m equations in the system (3.6.11), we find that

$$\sum_{i=1}^{m} J_i = 1 \tag{3.6.12}$$

The system (3.6.11), being a one-parameter form of the solution, represents a straight line[†] passing through the fixed point $\{\phi_i^0\}$ in the m-dimensional concentration space $\Phi(m)$. This straight line is the Γ and has a direction given by the invariant set $\{J_i\}$. Here and in the following we shall use braces, $\{\cdot\}$, to represent a collection of the m members of the quantity enclosed.

[†] It is understood that Eq. (3.6.9) will give the singular solution which represents the envelope of the family of these straight lines.

The integration constant J_i can be determined by substituting Eq. (3.6.11) back into the original differential equation (3.4.19). Since Eq. (3.6.6) yields

$$\frac{\mathcal{D}n_i}{\mathcal{D}c_i} = \frac{1}{\delta}\left(\gamma_i - \frac{K_i n_i}{J_i}\right), \qquad i = 1, 2, \ldots, m \qquad (3.6.13)$$

we obtain

$$\gamma_1 - \frac{K_1 n_1}{J_1} = \gamma_2 - \frac{K_2 n_2}{J_2} = \cdots = \gamma_m - \frac{K_m n_m}{J_m} = \omega \qquad (3.6.14)$$

in which the new parameter ω represents the common value of the directional derivative $\mathcal{D}f_i/\mathcal{D}c_i$ multiplied by δ. It follows from Eq. (3.6.14) that

$$J_i = \frac{K_i n_i}{\gamma_i - \omega}, \qquad i = 1, 2, \ldots, m \qquad (3.6.15)$$

which, upon substitution into Eq. (3.6.12), gives the equation[‡]

$$\sum_{i=1}^{m} \frac{K_i n_i}{\gamma_i - \omega} = 1 \qquad (3.6.16)$$

For a given state of concentrations $\{\phi_i\}$ or $\{n_i\}$, Eq. (3.6.16) presents an mth-order algebraic equation for ω. As we can see clearly in Fig. 3.5, there exist precisely m real, distinct, positive roots which can be arranged as

$$0 \leq \omega_1 \leq \gamma_1 \leq \omega_2 \leq \gamma_2 \leq \omega_3 \leq \cdots \leq \gamma_{m-1} \leq \omega_m \leq \gamma_m \qquad (3.6.17)$$

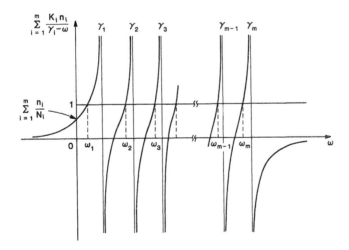

Figure 3.5

[‡] Note that Eq. (3.6.16) is equivalent to the h-transformation introduced by Helfferich and Klein (1970).

Consequently, there exist m different sets of $\{J_i\}$ and to each of these there corresponds a different Γ. We shall denote the set $\{J_i\}$ associated with ω_k as $\{J_i^k\}$ for $k = 1, 2, \ldots, m$; i.e.,

$$J_i^k = \frac{K_i n_i}{\gamma_i - \omega_k}, \qquad i = 1, 2, \ldots, m \qquad (3.6.18)$$

for $k = 1, 2, \ldots, m$, and the straight line determined by the set $\{J_i^k\}$ as Γ^k. Then it is obvious that the set $\{J_i^k\}$ remains invariant along a Γ^k, although n_i and ω_k change. Furthermore, Eq. (3.6.12) asserts that $(m - 1)$ members of the set $\{J_i^k\}$ are linearly independent. All these features now ensure that the invariant set $\{J_i^k\}$ corresponds to the set of the generalized k-Riemann invariants discussed in Sec. 3.3.

From Eqs. (3.6.17) and (3.6.18) one can readily observe that

$$0 < J_m^1 < J_m^2 < J_m^3 < \cdots < J_m^{m-1} < J_m^m \qquad (3.6.19)$$

and also that the sign of each Riemann invariant J_i^k is fixed as

$$J_i^k \begin{cases} < 0 & \text{for } i < k \\ > 0 & \text{for } i \geq k \end{cases} \qquad (3.6.20)$$

for $i, k = 1, 2, \ldots, m$. In particular, the 1-Riemann invariants J_i^1 are all positive and the m-Riemann invariants J_i^m are all negative except for J_m^m.

We also note from Eq. (3.6.14) that

$$\frac{\phi_i - \phi_i^0}{J_i^k} = \frac{\phi_j - \phi_j^0}{J_j^k} = \delta - \delta^0 \quad \text{along} \quad \text{a } \Gamma^k \qquad (3.6.21)$$

or

$$\frac{d\phi_i}{d\phi_j} = \frac{J_i^k}{J_j^k} \quad \text{along} \quad \text{a } \Gamma^k \qquad (3.6.22)$$

An immediate consequence is that in a k-simple wave region, the first $(k - 1)$ concentrations $c_1, c_2, \ldots, c_{k-1}$ all vary in one direction, while the next $(m - k + 1)$ concentrations $c_k, c_{k+1}, \ldots, c_m$ undergo variations in the opposite direction. In case of $m = 2$, we can write

$$\frac{dc_1}{dc_2} = \frac{K_2}{K_1} \frac{J_1^1}{J_2^1} = \beta \quad \text{along} \quad \text{a } \Gamma^1 \qquad (3.6.23)$$

and

$$\frac{dc_1}{dc_2} = \frac{K_2}{K_1} \frac{J_1^2}{J_2^2} = \alpha \quad \text{along} \quad \text{a } \Gamma^2 \qquad (3.6.24)$$

in which α and β are the characteristic parameters introduced in Sec. 2.1 [see, e.g., Eqs. (2.1.28) and (2.1.29)]. These equations show how the present approach is related to the formulation developed in Chapter 2.

Exercises

3.6.1 Show that the parameter ω satisfies the equation

$$\sum_{i=1}^{m} \frac{\phi_i}{\gamma_i/\omega - 1} = 1$$

3.6.2. By using Eq. (3.6.11), show that the product $\omega_k \delta$ remains invariant along a Γ^k.

3.6.3. Consider two solutes competing for adsorption sites on the surface of an adsorbent. The equilibrium relations follow the Langmuir isotherm with $N_1 = N_2 = 1$ mol/liter, $K_1 = 5$ liters/mol, and $K_2 = 7.5$ liters/mol. For the state of concentrations $c_1 = \frac{2}{30}$ mol/liter and $c_2 = \frac{1}{10}$ mol/liter, determine the Riemann invariants, J_1^1, J_2^1, J_1^2, and J_2^2. Comparing these with the information from Sec. 2.3, i.e., $\alpha = -\frac{2}{3}$ and $\beta = 1$, confirm that Eqs. (3.6.23) and (3.6.24) hold.

3.6.4. A three-solute adsorption system obeys the Langmuir isotherm with $N_1 = N_2 = N_3 = 1$ mol/liter, $K_1 = 5$ liters/mol, $K_2 = 2K_1$, and $K_3 = 3K_1$. For the state of concentrations $c_1 = c_2 = c_3 = 0.05$ mol/liter, solve Eq. (3.6.16) and thereby determine all the Riemann invariants.

3.7 Characteristic Parameters and the Space $\Omega(m)$

Another interpretation of Eq. (3.6.16) is that it defines a mapping of the m-dimensional concentration space $\Phi(m)$ onto the m-dimensional ω-space $\Omega(m)$. Since Eq. (3.6.15) always gives m real, distinct values for ω, the transformation is one-to-one.

To obtain the inverse transform we shall rearrange Eq. (3.6.16) in the form

$$\sum_{i=1}^{m} K_i n_i \prod_{j=1, i}^{m} (\gamma_j - \omega) = \prod_{j=1}^{m} (\gamma_j - \omega) \qquad (3.7.1)$$

and define a polynomial $F(\omega)$ of degree m as

$$F(\omega) = \prod_{j=1}^{m} (\gamma_j - \omega) - \sum_{i=1}^{m} K_i n_i \prod_{j=1, i}^{m} (\gamma_j - \omega) \qquad (3.7.2)$$

Since the equation $F(\omega) = 0$ has m roots $\omega_1, \omega_2, \ldots, \omega_m$, we can write

$$F(\omega) = \prod_{j=1}^{m} (\omega_j - \omega) \qquad (3.7.3)$$

By putting $\omega = \gamma_i$ in Eqs. (3.7.2) and (3.7.3) and making them equal to each other, we obtain

$$F(\gamma_i) = 0 - K_i n_i \prod_{j=1, i}^{m} (\gamma_j - \gamma_i) = \prod_{j=1}^{m} (\omega_j - \gamma_i)$$

or the equation

$$K_i n_i = (\gamma_i - \omega_i) \prod_{j=1,i}^{m} \frac{\gamma_i - \omega_j}{\gamma_i - \gamma_j}, \qquad i = 1, 2, \ldots, m \qquad (3.7.4)$$

which is the inverse of the transformation given by Eq. (3.6.16). It is obvious that the inverse transformation is also one-to-one. Consequently, the two spaces $\phi(m)$ and $\Omega(m)$ are homeomorphic with the homeomorphism given by Eqs. (3.6.16) and (3.7.4).

Alternatively, we can substitute Eq. (3.6.4) into Eq. (3.6.16) to obtain

$$\sum_{i=1}^{m} \frac{\gamma_i \phi_i}{\gamma_i - \omega} = \delta = 1 + \sum_{i=1}^{m} \phi_i$$

or, after rearrangement,

$$\omega \sum_{i=1}^{m} \phi_i \prod_{j=1,i}^{m} (\gamma_j - \omega) = \prod_{j=1}^{m} (\gamma_j - \omega) \qquad (3.7.5)$$

Let us now define a polynomial $G(\omega)$ of degree m as

$$G(\omega) = \prod_{j=1}^{m} (\gamma_j - \omega) - \omega \sum_{i=1}^{m} \phi_i \prod_{j=1,i}^{m} (\gamma_j - \omega) \qquad (3.7.6)$$

Noting that the coefficient of the term $(-\omega)^m$ is given by $1 + \Sigma \phi_i = \delta$ and that $G(\omega)$ has m zeros $\omega_1, \omega_2, \ldots, \omega_m$, we can express $G(\omega)$ in the form

$$G(\omega) = \delta \prod_{j=1}^{m} (\omega_j - \omega) \qquad (3.7.7)$$

For $\omega = 0$, we have

$$G(0) = \prod_{j=1}^{m} \gamma_j = \delta \prod_{j=1}^{m} \omega_j$$

and hence

$$\delta = \prod_{j=1}^{m} \frac{\gamma_j}{\omega_j} \qquad (3.7.8)$$

By putting $\omega = \gamma_i$, we find that

$$G(\gamma_i) = 0 - \gamma_i \phi_i \prod_{j=1,i}^{m} (\gamma_j - \gamma_i) = \delta \prod_{j=1}^{m} (\omega_j - \gamma_i)$$

or, upon using Eq. (3.7.8),

$$\phi_i = \delta \frac{\gamma_i - \omega_i}{\gamma_i} \prod_{j=1,i}^{m} \frac{\gamma_i - \omega_j}{\gamma_i - \gamma_j} = \frac{\gamma_i - \omega_i}{\omega_i} \prod_{j=1,i}^{m} \frac{\gamma_j(\gamma_i - \omega_j)}{\omega_j(\gamma_i - \gamma_j)}$$

$$= \left(\frac{\gamma_i}{\omega_i} - 1 \right) \prod_{j=1,i}^{m} \frac{\frac{\gamma_i}{\omega_j} - 1}{\frac{\gamma_i}{\gamma_j} - 1} \tag{3.7.9}$$

which is just another expression of the inverse transform.

To see how a solution to the quasilinear system (3.4.1) maps into the space $\Omega(m)$, we would like to find the image of a Γ^k in $\Omega(m)$. For this we shall consider Eq. (3.6.16) in the form

$$\sum_{i=1}^{m} \frac{K_i n_i}{\gamma_i - \omega_j} = 1 = \sum_{i=1}^{m} J_i^k, \qquad j \neq k \tag{3.7.10}$$

but since

$$K_i n_i = J_i^k (\gamma_i - \omega_k)$$

from Eq. (3.6.18), we obtain

$$\sum_{i=1}^{m} J_i^k \frac{\gamma_i - \omega_k}{\gamma_i - \omega_j} - \sum_{i=1}^{m} J_i^k = 0$$

or

$$(\omega_j - \omega_k) \sum_{i=1}^{m} \frac{J_i^k}{\gamma_i - \omega_j} = 0, \qquad j \neq k \tag{3.7.11}$$

Clearly, $\omega_j \neq \omega_k$, so

$$\sum_{i=1}^{m} \frac{J_i^k}{\gamma_i - \omega_j} = 0, \qquad j \neq k \tag{3.7.12}$$

Differentiating this equation with respect to δ in the direction of a Γ^k, along which the set $\{J_i^k\}$ remains invariant, we obtain

$$\frac{d\omega_j}{d\delta} \sum_{i=1}^{m} \frac{J_i^k}{(\gamma_i - \omega_j)^2} = 0, \qquad j \neq k \tag{3.7.13}$$

Since $(m - 1)$ members of the k-Riemann invariants $\{J_i^k\}$ are linearly independent, it can be shown that

$$\sum_{i=1}^{m} \frac{J_i^k}{(\gamma_i - \omega_j)^2} \neq 0$$

and hence we have

$$\frac{d\omega_j}{d\delta} = 0 \quad \text{along a } \Gamma^k, \quad j \neq k \tag{3.7.14}$$

An immediate consequence is that ω_j remains constant along a Γ^k if $j \neq k$. In other words, among the m parameters $\{\omega_j\}$ only ω_k varies along a Γ^k, and this implies that the image of a Γ^k falls on a straight line parallel to the ω_k-axis.

At this stage we notice that the ω-coordinate system is really equivalent to the characteristic coordinate system introduced for pairs of equations (see Sec. 1.3) and in this sense, we shall call ω_k the *generalized characteristic parameter*. The nature of the pertinent parameters is summarized in Table 3.1.

TABLE 3-1. INVARIANTS AND VARIABLES ALONG A Γ^k

Parameter	$j = k$	$j \neq k$
Riemann invariant, J_i^j	Invariant	Variable
Characteristic parameter, ω_j	Variable	Invariant

In the space $\Omega(m)$, the physically relevant portion is finite and bounded as follows:

$$\gamma_{k-1} \leq \omega_k \leq \gamma_k, \qquad k = 2, 3, \ldots, m$$
$$0 \leq \omega_1 \leq \gamma_1 \qquad\qquad (3.7.15)$$

If one of the concentrations is equal to zero; for example, $\phi_j = 0$, Eq. (3.6.16) requires that one of $\{\omega_k\}$ must be equal to γ_j and it immediately follows from Eq. (3.6.17) that either $\omega_j = \gamma_j$ or $\omega_{j+1} = \gamma_j$. Therefore, if we let π_j represent the $(m - 1)$-dimensional subspace in which $\phi_j = 0$, its image in $\Omega(m)$ consists of two planes $\omega_j = \gamma_j$ and $\omega_{j+1} = \gamma_j$. The only exception is the subspace π_m whose image is a single plane $\omega_m = \gamma_m$. The origin of the space $\Phi(m)$ represents a pure state (of no solute), $\{\phi_i = 0\}$, for which Eq. (3.6.16) generates m roots $\{\gamma_k\}$. Hence the image of the pure state in $\Phi(m)$ is given by the point $\omega_k = \gamma_k$, $k = 1, 2, \ldots, m$. The arguments above are summarized in Table 3.2 and in Fig. 3.6. The latter illustrates the region of $\Phi(3)$ that corresponds to the positive octant of $\Phi(3)$, with point O in $\Omega(3)$ corresponding to the origin of $\Phi(3)$. The subspace π_1 consists of the two planes $OABD$ on which $\omega_1 = \gamma_1$ and $ABGF$ on which $\omega_2 = \gamma_1$. Similarly, the subspace π_2 is composed of two planes: one is $ODHE$ on which $\omega_2 = \gamma_2$ and the other is $BDHG$ on which $\omega_3 = \gamma_2$. The subspace π_3, however, is given by the single plane $AOEF$ on which $\omega_3 = \gamma_3$.

TABLE 3-2. CORRESPONDENCE BETWEEN $\Phi(m)$ AND $\Omega(m)$

	Space	
	$\phi(m)$	$\Omega(m)$
State	$\{\phi_i\}$	$\{\omega_k\}$
Γ	Straight and slanted	Straight and parallel to axis
Pure state	$\{\phi_i = 0\}$	$\{\omega_k = \gamma_k\}$
Subspace, $\pi_j : \phi_j = 0$	Plane, $\phi_j = 0$	Two planes, $\omega_j = \gamma_j$ and $\omega_{j+1} = \gamma_j$

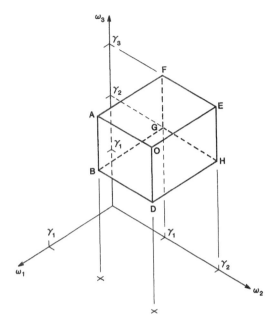

Figure 3.6

Exercises

3.7.1. Derive Eq. (3.7.9) from Eq. (3.7.4), or vice versa. Can you argue from Eq. (3.7.8) that $\omega_k \delta$ is invariant along a Γ^k?

3.7.2. For the case of $m = 2$, show the region of $\Omega(2)$ that corresponds to the positive quadrant of $\Phi(2)$. Identify the images of the subspaces π_1 and π_2 as well as the images of the origin of $\Phi(2)$ and the watershed point (i.e., point B in Fig. 2.1).

3.7.3 (a) By applying Cramer's rule to Eq. (3.6.16), derive Eq. (3.7.4).
   (b) Consider the resulting equation of Exercise 3.6.1 and apply Cramer's rule to obtain Eq. (3.7.9).

3.8 Characteristics and Simple Waves

Having established a thorough understanding of the structure of the image of a solution in the hodograph space $\Phi(m)$ or $\Omega(m)$, we are now in a position to construct the solution in the physical plane of x and τ. In principle, this can be done by mapping the image in $\Omega(m)$ back onto the (x, τ)-plane, and for this measure we need to use the characteristics C.

We have already observed in Sec. 3.4 that a characteristic direction is defined by Eq. (3.4.12) or equivalently, by Eq. (3.4.18). We also notice from Eqs. (3.6.13) and (3.6.14) that $\mathcal{D}n_i/\mathcal{D}c_i = \omega_k/\delta$ for $k = 1, 2, \ldots , m$ and hence there exists m different values for the characteristic direction. We shall put

$$\left(\frac{d\tau}{dx}\right)_{\omega_k} = 1 + \nu\left(\frac{\mathcal{D}n_i}{\mathcal{D}c_i}\right)_{\Gamma^k} = 1 + \nu\frac{\omega_k}{\delta} = \sigma_k, \qquad k = 1, 2, \ldots, m$$

(3.8.1)

Then by introducing Eq. (3.7.8) to this, we obtain

$$\left(\frac{d\tau}{dx}\right)_{\omega_k} = \sigma_k = 1 + \nu\omega_k \prod_{i=1}^{m} \frac{\omega_i}{\gamma_i}$$

(3.8.2)

$$= 1 + \nu p_k \omega_k^2$$

(3.8.3)

for $k = 1, 2, \ldots, m$, where p_k is defined as

$$p_k = \frac{1}{\gamma_k} \prod_{i=1,k}^{m} \frac{\omega_i}{\gamma_i}$$

(3.8.4)

which remains invariant along a Γ^k.

Consider a k-simple wave region in the (x, τ)-plane. Since its image lies on a single Γ^k, the parameter p_k is constant throughout the region. It is then obviously that a curve of slope σ_k is a straight line because ω_k is also held constant in that direction. This proves that a curve whose direction is given by σ_k is a kth characteristic C^k. Clearly, every C^k in a k-simple wave region maps on a single point along the same Γ^k and thus carries a constant state of concentrations. Since a Γ^k is parallel to the ω_k-axis in the space $\Omega(m)$, ω_k varies monotonically along a Γ^k and so does σ_k. We also recall that a Γ^k is tangent to the right eigenvector $\mathbf{r}_k$ of the matrix $\mathbf{A}$ and hence we can write

$$\mathbf{r}^T\boldsymbol{\nabla}\sigma_k \neq 0 \qquad \text{for all } \{c_i\}$$

(3.8.5)

This is true for every k from 1 to m and by Definition 3.3 we assert that the m characteristic fields of the system (3.4.1) are all genuinely nonlinear. Since $\mathbf{r}^T\boldsymbol{\nabla}\sigma_k$ represents the derivative of σ_k in the direction of a Γ^k, an alternative expression is directly obtained from Eq. (3.8.3) as

$$\frac{\mathcal{D}\sigma_k}{\mathcal{D}\omega_k} = 2\nu p_k \omega_k > 0, \qquad k = 1, 2, \ldots, m$$

(3.8.6)

which also implies that the kth characteristic field is genuinely nonlinear.

Let us now take a jth characteristic C^j in a k-simple wave region, where $j \neq k$. Since ω_j is held constant along the C^j, its image lies on the Γ^k, which is the image of the simple wave. This implies that ω_k varies along the C^j, so every C^j is a curve and traverses the k-simple wave region. In summary, a k-simple wave region is covered by a family of straight-line characteristics C^k and traversed by $(m - 1)$ families of curved characteristics C^j, $j \neq k$.

Applying Eq. (3.6.17) to Eq. (3.8.2), we find that

$$1 < \sigma_1 < \sigma_2 < \sigma_3 < \cdots < \sigma_{m-1} < \sigma_m$$

(3.8.7)

Note that the unity corresponds to the reciprocal of the fluid velocity in dimensionless form, whereas σ_k represents the reciprocal propagation speed of a disturbance. Hence, any disturbance in the concentration field is propagated at a lower speed than the fluid phase. In a k-simple wave region σ_k is constant and gives the direction of a C^k, along which the state of concentrations remains constant. Based on this argument, we may regard the reciprocal of σ_k as the propagation speed of a particular state of concentrations in a k-simple wave region.

From Theorem 3.3 we see that a region of constant state is bounded by pieces of straight characteristics. Since a constant state has a point image in $\Phi(m)$, the solution in a region adjacent to the region of constant state must have its image lain on a Γ emanating from the point image and thus one set of the Riemann invariants is constant. Let us assume that a portion B of the boundary of a region of constant state, say $\{\phi_i^0\}$, is a C^k and consider the solution $\{\phi_i\}$ on the other side of B. Suppose that the solution $\{\phi_i\}$ maps along the Γ^j, $j \neq k$, emanating from the point image of the state $\{\phi_i^0\}$. Then all the characteristic parameters except for ω_j remain constant on the other side of B. If we draw a C^j from a point P on B as shown in Fig. 3.7, we immediately find that this C^j must be extended as a straight line because ω_j is held constant along it. This C^j then carries the constant state $\{\phi_i^0\}$ along it and we encounter a contradiction that B is not the boundary. Therefore, the solution on the other side of B maps along the Γ^k issuing from the point image of the constant state. It then follows that all k-Riemann invariants are constant on the other side of B, so the solution is given by a k-simple wave. This argument clearly reconstitutes Theorem 3.4.

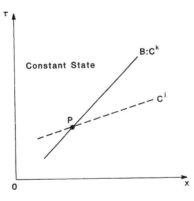

Figure 3.7

Now we have a clear picture about a k-simple wave in the (x, τ)-plane. It can be represented by a family of straight-line characteristics C^k as illustrated in Fig. 3.8. The $(m - 1)$ parameters ω_j, $j \neq k$, remain constant everywhere in the k-simple wave region, whereas ω_k is held constant only along each of the C^k's. Thus by choosing a value of ω_k we can evaluate σ_k from Eqs. (3.8.3) and (3.8.4) and locate the corresponding C^k starting from a point on the initial line. The state of concentrations carried along this C^k can be determined by

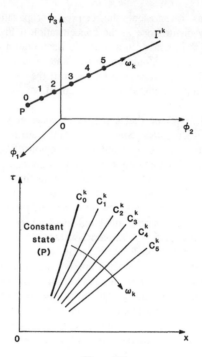

Figure 3.8

using Eq. (3.7.9). This establishes the k-simple wave solution, and the distribution of concentrations can be constructed at any time or at any fixed position simply by reading the concentrations on each of the kth characteristics C^k.

As we proceed in the x direction through a k-simple wave region, we expect the straight-line characteristics C^k to fan out clockwise as shown in Fig. 3.8, for otherwise the C^k's would eventually overlap and a physically impossible situation would be encountered. Just as in the case of single equations or pairs of equations, the latter situation can be resolved by introducing a discontinuity in the solution, and this is treated in Sec. 3.9. All we can say at this stage is that for a k-simple wave to be established as a continuous solution, we must have

$$\frac{\partial \sigma_k}{\partial x} = \frac{\mathcal{D}\sigma_k}{\mathcal{D}\omega_k} \frac{\partial \omega_k}{\partial x} < 0 \qquad (3.8.8)$$

or since $\mathcal{D}\sigma_k/\mathcal{D}\omega_k > 0$ from Eq. (3.8.6),

$$\frac{\partial \omega_k}{\partial x} < 0 \qquad (3.8.9)$$

Therefore, in a k-simple wave region the parameter ω_k is monotone decreasing in the x direction. We also find from Eqs. (3.7.8) and (3.6.21) that along a Γ^k,

$$\frac{d\delta}{d\omega_k} = -\frac{\gamma_k}{\omega_k^2} \prod_{j=1,k}^{m} \frac{\gamma_j}{\omega_j} < 0 \qquad (3.8.10)$$

and

$$\frac{d\phi_i}{d\delta} = J_i^k \begin{cases} <0 & \text{for } i < k \\ >0 & \text{for } i \geq k \end{cases} \qquad (3.8.11)$$

An immediate consequence is that in a k-simple wave region, δ increases in the x direction, while ϕ_i, $i < k$, decreases and ϕ_i, $i \geq k$, increases in the x direction. We summarize this to give:

RULE[†] 3.1: In a k-simple wave region, ω_k and ϕ_i, $i < k$, decrease in the x direction, while δ and ϕ_i, $i \geq k$, increase in the x direction.

The following deductions are immediately apparent from the signs of Riemann invariants:

COROLLARY:

1. Both the parameter δ and the concentration ϕ_m of the most strongly adsorbed component always increase in the x direction if the solution is given by a simple wave.
2. All the concentrations increase in the x direction in a 1-simple wave region.

We also deduce from Eq. (3.6.5) the following:

COROLLARY: In a simple wave region, the coverage Θ always increases in the x direction.

Exercise

3.8.1 According to Eqs. (3.1.19) and (3.2.16), the characteristic direction σ_k may be determined by solving the equation

$$|\nabla \mathbf{f} - \sigma \mathbf{I}| = 0$$

or

$$\left| \nabla \mathbf{n} - \frac{\sigma - 1}{\nu} \mathbf{I} \right| = 0$$

If the system is totally hyperbolic, this equation is equivalent to

$$\prod_{k=1}^{m} (\sigma_k - \sigma) = 0$$

and thus show by using Eq. (3.8.1) that for the Langmuir isotherm,

[†] The term "rule" is used here since this conclusion is valid specifically for the case of multi-component chromatography with the Langmuir isotherm.

$$|\nabla \mathbf{n}| = \delta^{-m} \prod_{i=1}^{m} \gamma_i$$

Also, derive this equation directly from Eq. (3.6.4).

3.9 Discontinuities: Shocks

If it happens that in a k-simple wave region,

$$\frac{\partial \sigma_k}{\partial x} > 0 \quad \text{or} \quad \frac{\partial \omega_k}{\partial x} < 0 \tag{3.9.1}$$

then the straight-line characteristics C^k would fan counterclockwise and so overlap. Under these circumstances the solution cannot be determined uniquely. Just as in the case of single equations or pairs of equations, this can be resolved by allowing discontinuities in the solution itself.

At a discontinuity Eq. (3.4.1) is no longer valid and must be replaced by conservation equations expressing the fact that the discontinuity propagates with such a speed that there is no accumulation of material or energy at the discontinuity. Since the solutes are mutually interacting and this is expressed by the equilibrium relations, it is evident that if one of the concentrations is discontinuous, then all other concentrations are also discontinuous. Let us suppose that all the concentrations undergo discontinuous changes at one position at the same time. Here again we shall use the superscripts l and r to denote, respectively, the left- and right-hand sides of a discontinuity, and brackets, $[\cdot]$, to represent the jump in the quantity enclosed across a discontinuity. Exactly as in Sec. 5.4 of Vol. 1 or as in Sec. 7.2 of Vol. 1, we set up the mass balance for solute A_i over the discontinuity to obtain the equation

$$\frac{d\tau}{dx} = \frac{f_i^l - f_i^r}{c_i^l - c_i^r} = \frac{[f_i]}{[c_i]} = \tilde{\sigma} \tag{3.9.2}$$

which is valid for any i.

As discussed in the preceding paragraph, the discontinuities in various concentrations are not separate but form an entity and must move as such. Therefore, the following equality condition must be satisfied across a discontinuity:

$$\frac{[f_1]}{[c_1]} = \frac{[f_2]}{[c_2]} = \cdots = \frac{[f_m]}{[c_m]} \tag{3.9.3}$$

This is a system of $(m - 1)$ algebraic equations, so if the state of concentrations on one side, say $\{c_i^l\}$, is given, we can determine the locus of the state $\{c_i^r\}$ in the concentration space. As the discontinuity propagates, the states on both sides must always satisfy Eq. (3.9.3). It is in this sense that the equations (3.9.3) are named the *generalized compatibility relations*.

Now Eq. (3.9.2) gives the reciprocal propagation speed of a discontinuity or directly its propagation direction, and this equation is frequently called the *jump condition*. If the state $\{c_i^l\}$ is given and $\tilde{\sigma}$ or one concentration of the state

$\{c_i^r\}$ is further specified, then the state $\{c_i^r\}$ can be determined by solving Eqs. (3.9.2) and (3.9.3) together. Equations (3.9.2) and (3.9.3) together constitute the *generalized Rankine–Hugoniot relations,* also called the jump conditions.

By substituting Eq. (3.4.2), we can express the jump condition as well as the compatibility relations in terms of the adsorption isotherm; i.e.,

$$\frac{d\tau}{dx} = 1 + \frac{[n_i]}{[c_i]} = \tilde{\sigma} \tag{3.9.4}$$

and

$$\frac{[n_1]}{[c_1]} = \frac{[n_2]}{[c_2]} = \cdots = \frac{[n_m]}{[c_m]} \tag{3.9.5}$$

We note that both the jump condition and the compatibility relations are symmetric with respect to the superscripts l and r. This implies that if two states $\{c_i^a\} = \{c_i^l\}$ and $\{c_i^b\} = \{c_i^r\}$ are connected by a discontinuity satisfying Eqs. (3.9.2) and (3.9.3), then the two states $\{c_i^b\} = \{c_i^l\}$ and $\{c_i^a\} = \{c_i^r\}$ can also be connected by a discontinuity because they satisfy Eqs. (3.9.2) and (3.9.3). Thus an ambiguity still exists concerning the direction of jumps and, in this sense, the jump condition and the compatibility relations are not sufficient to determine the physically relevant solution (i.e., the unique solution). As in the previous cases of single equations and pairs of equations, it is necessary to have an additional condition that regulates the direction of any jump discontinuity. Such a condition, which is yet to be formulated, will be called the *generalized entropy condition.* From the argument around Eq. (3.9.1) and Rule 3.1, we are tempted to say that the parameter δ (and thus the coverage Θ) decreases from the left to the right across a discontinuity. However, since we have not solved the compatibility relations (3.9.3), it could be incomplete or irrelevant if we deduce any conclusion at this stage. We will come back to this after the compatibility relations are solved.

In the limit as a discontinuity becomes very weak, the compatibility relations (3.9.3) tend to take the form of the fundamental differential equations (3.4.13). This implies that given the state $\{c_i^l\}$, the locus of the state $\{c_i^r\}$ in the concentration space becomes tangent to a curve Γ at the point $\{c_i^l\}$ and thus there may exist m different loci for the state $\{c_i^r\}$. We shall now introduce the Langmuir isotherm, and recalling the fact that the Γ's are all straight, we assert that the locus of the state $\{c_i^r\}$ entirely falls on a Γ emanating from the point $\{c_i^l\}$. This can be proven as follows. Along a Γ^k, for instance, we have from Eq. (3.6.21)

$$\frac{K_i[n_i]}{J_i^k} = \frac{K_j[c_j]}{J_j^k} \tag{3.9.6}$$

and

$$\mathcal{D}n_i = \frac{\mathcal{D}c_i}{\mathcal{D}c_j}\mathcal{D}n_j = \frac{K_j J_i^k}{K_i J_j^k}\mathcal{D}n_j$$

Integrating the latter relation along the Γ^k, we obtain

$$\frac{K_i[n_i]}{J_i^k} = \frac{K_j[n_j]}{J_j^k} \qquad (3.9.7)$$

and comparing this with Eq. (3.9.6), we find that the compatibility relations (3.9.5) are satisfied. This is true for any k and hence there exist m different solutions to Eq. (3.9.5).

To show that there are only m solutions and no more, we shall consider a discontinuity with the state $\{c_i^l\}$ given. Substituting the Langmuir isotherm in Eq. (3.9.4), we obtain

$$\frac{[n_i]}{[c_i]} = \gamma_i \frac{\phi_i^l/\delta^l - \phi_i^r/\delta^r}{\phi_i^l - \phi_i^r} = \frac{\bar{\sigma} - 1}{\nu} \equiv \mu, \qquad i = 1, 2, \ldots, m \qquad (3.9.8)$$

Rearranging this equation gives

$$\phi_i^r \left(\frac{\gamma_i}{\delta^r} - \mu \right) = \phi_i^l \left(\frac{\gamma_i}{\delta^l} - \mu \right)$$

or

$$\phi_i^r = \phi_i^l - (\delta^l - \delta^r) \frac{K_i n_i^l}{\gamma_i - \mu \delta^r}, \qquad i = 1, 2, \ldots, m \qquad (3.9.9)$$

We then add up the m equations in Eq. (3.9.9) together to obtain the equation

$$\sum_{i=1}^{m} \frac{K_i n_i^l}{\gamma_i - \mu \delta^r} = 1 \qquad (3.9.10)$$

By comparing this with Eq. (3.6.16) we find that for a given state $\{c_i^l\}$, there exist precisely m real, distinct values of $\mu\delta^r$. For each value of $\mu\delta^r$ so obtained, the state $\{c_i^r\}$ is uniquely determined if we further specify $\bar{\sigma}$ or one of c_i^r's. Indeed, if the value of $\bar{\sigma}$ (and thus the value of μ) is given, then δ^r is known, so Eq. (3.9.9) directly determines the state $\{c_i^r\}$. In case one of the c_i^r's (e.g., c_j^r) is further specified, Eq. (3.9.9) for $i = j$ gives δ^r, and with δ^r known, Eq. (3.9.9) for $i \neq j$ determines the state $\{c_i^r\}$. Consequently, we have established

RULE 3.2: The two states on both sides of a discontinuity necessarily have their images on a single Γ in the hodograph space and hence there are m different kinds of discontinuities.

DEFINITION 3.4: If the states on both sides of a discontinuity have their images on a Γ^k in the hodograph space, the discontinuity is called a *k-discontinuity*.

Let us now consider a k-discontinuity. Since the k-Riemann invariants remain constant across it, the states on both sides can be connected by a one-parameter family; that is,

$$[\phi_i] = J_i^k[\delta], \qquad i = 1, 2, \ldots, m \qquad (3.9.11)$$

To develop an expression for the propagation direction (to be denoted by $\bar{\sigma}_k$), we proceed from Eq. (3.9.8) as follows:

$$
\begin{aligned}
\frac{[n_i]}{[c_i]} &= \frac{\gamma_i}{\delta^l} \frac{\phi_i^l - \phi_i^r \delta^l/\delta^r}{\phi_i^l - \phi_i^r} = \frac{\gamma_i}{\delta^l}\left(1 - \frac{\phi_i^r}{\delta^r}\frac{[\delta]}{[\phi_i]}\right) \\
&= \frac{1}{\delta^l}\left(\gamma_i - \frac{K_i n_i^r}{J_i^k}\right)
\end{aligned}
\qquad (3.9.12)
$$

in which Eq. (3.9.11) has been substituted. But from the definition of J_i^k [see Eq. (3.6.18)], we have

$$\gamma_i - \frac{K_i n_i^r}{J_i^k} = \omega_k^r$$

and thus if δ^l is substituted from Eq. (3.7.8), we obtain

$$\frac{[n_i]}{[c_i]} = \omega_k^r \prod_{i=1}^{m} \frac{\omega_i^l}{\gamma_i} = \omega_k^l \omega_k^r \frac{1}{\gamma_k} \prod_{i=1,k}^{m} \frac{\omega_i^l}{\gamma_i} \qquad (3.9.13)$$

Across a k-discontinuity the $(m - 1)$ characteristic parameters ω_j, $j \neq k$, remain constant because the two states on both sides fall on a Γ^k, so we can put

$$p_k = \frac{1}{\gamma_k} \prod_{i=1,k}^{m} \frac{\omega_i^l}{\gamma_i} = \frac{1}{\gamma_k} \prod_{i=1,k}^{m} \frac{\omega_i^r}{\gamma_i} = \frac{1}{\gamma_k} \prod_{i=1,k}^{m} \frac{\omega_i}{\gamma_i} \qquad (3.9.14)$$

[see Eq. (3.8.4)] to have

$$\frac{[n_i]}{[c_i]} = p_k \omega_k^l \omega_k^r \qquad (3.9.15)$$

Finally, the propagation direction of a k-discontinuity is given by

$$\frac{d\tau}{dx} = \bar{\sigma}_k = 1 + \nu \omega_k^r \prod_{i=1}^{m} \frac{\omega_i^l}{\gamma_i} \qquad (3.9.16)$$

$$= 1 + \nu \omega_k^l \prod_{i=1}^{m} \frac{\omega_i^r}{\gamma_i} \qquad (3.9.17)$$

$$= 1 + \nu p_k \omega_k^l \omega_k^r \qquad (3.9.18)$$

We note that among the m characteristic parameters $\{\omega_j\}$, ω_k is the only parameter that varies across a k-discontinuity. However, we still have not determined whether ω_k increases or decreases across a k-discontinuity. If Eqs. (3.9.11) and (3.9.18) are satisfied with $\omega_k^l > \omega_k^r$, the same is true with $\omega_k^l < \omega_k^r$. According to Rule 3.1, however, we expect to obtain a k-simple wave (i.e., a continuous solution) if $\omega_k^l > \omega_k^r$. This argument leads us to the conclusion that a k-discontinuity is admissible if and only if $\omega_k^l < \omega_k^r$. This is precisely the entropy condition we discussed earlier in this section.

RULE 3.3 [Entropy Condition]: Across a k-discontinuity the characteristic parameter ω_k increases from the left to the right; that is, we have $\omega_k^l < \omega_k^r$.

Here the entropy condition is expressed in this particularly simple and direct form because the curves Γ are straight, so the compatibility relations have the same solution as the fundamental differential equations.[†] The entropy condition in a general context will be discussed in Chapter 6. Reflecting the entropy condition on Rule 3.1, we can easily deduce the following:

COROLLARY: Across a k-discontinuity:

1. The parameter δ, and therefore the coverage Θ, decrease from the left to the right.
2. ϕ_i decreases, if $i < k$, and increases, if $i \geq k$, from the left to the right.
3. The concentration ϕ_m of the most strongly adsorbed solute decreases from the left to the right.

Comparing Eq. (3.9.18) with Eq. (3.8.3) under the entropy condition, we immediately find that

$$\sigma_k^l < \tilde{\sigma}_k < \sigma_k^r \qquad (3.9.19)$$

From Eq. (3.8.2) we can write

$$\sigma_{k+1}^l = 1 + \nu\omega_{k+1}^l \prod_{i=1}^{m} \frac{\omega_i^l}{\gamma_i}$$

and

$$\sigma_{k-1}^r = 1 + \nu\omega_{k-1}^r \prod_{i=1}^{m} \frac{\omega_i^r}{\gamma_i}$$

Direct comparison of these equations with Eqs. (3.9.16) and (3.8.17) gives the inequalities

$$\sigma_{k+1}^l > \tilde{\sigma}_k > \sigma_{k-1}^r \qquad (3.9.20)$$

because $\omega_{k+1} > \omega_k > \omega_{k-1}$ under any circumstances [see Eq. (3.7.16)]. According to Lax (1957), the inequalities (3.9.19) and (3.9.20) are the conditions characterizing a *k-shock* and constitute what is known as the shock conditions. Thus we have

RULE 3.4: Every discontinuity appearing in multicomponent chromatography with the Langmuir isotherm is a shock.

The propagation path of a k-shock will be denoted by S^k and the concentration field along an S^k will be called a k-shock wave.

[†] Note also that here the kth characteristic field is genuinely nonlinear.

Physical interpretations of the inequalities (3.9.19) and (3.9.20) are worth examining here. Clearly, Eq. (3.9.19) implies that a shock propagates faster than the state on the right-hand side and with a lower speed than the state on the left-hand side. Therefore, a shock tends to remain sharp as usual and this is known as the self-sharpening tendency of a shock. Equation (3.9.20) tells us further that a k-shock will never catch up with the $(k - 1)$-simple wave traveling ahead and will never be caught up by the $(k + 1)$-simple wave traveling behind. We also note that the kth characteristics C^k impinge on the path S^k of a k-shock from both sides as the time increases, whereas a C^j passes through the S^k from the left to the right if $j < k$ and from the right to the left if $j > k$ as the time increases. These are illustrated in Fig. 3.9. From Eq. (3.8.2) we can write

$$\sigma_j = 1 + \nu\omega_j \prod_{i=1}^{m} \frac{\omega_i}{\gamma_i} = 1 + \nu\omega_j \frac{\omega_k}{\gamma_k} \prod_{i=1,k}^{m} \frac{\omega_i}{\gamma_i}, \qquad j \neq k$$

in which ω_k is the only variable across a k-shock. Since $\omega_k^l < \omega_k^r$, a C^j is refracted upward if $j < k$ and downward if $j > k$ as it passes through an S^k in the direction of increasing τ.

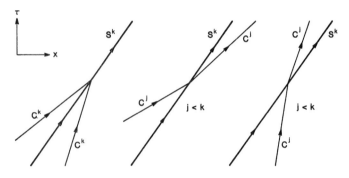

Figure 3.9

Suppose that we have a fixed state $\{c_i^l\}$ and this is to be connected by a shock to a different constant state $\{c_i^r\}$ on the right-hand side. If one concentration of the state $\{c_i^r\}$ is given, there exist, in general, m different candidates for the state $\{c_i^r\}$. Let these be denoted by $\{c_{ik}^r\}$, $k = 1, 2, \ldots, m$, respectively. To each of these states there corresponds a k-shock, which propagates in the direction given by Eq. (3.9.16); that is,

$$\tilde{\sigma}_k = 1 + \nu\omega_k^r \prod_{i=1}^{m} \frac{\omega_i^l}{\gamma_i}, \qquad k = 1, 2, \ldots, m$$

where ω_k^r is the kth characteristic parameter corresponding to the state $\{c_{ik}^r\}$. Although the states $\{c_{ik}^r\}$, $k = 1, 2, \ldots, m$, are all different, Eq. (3.6.17) shows that $\omega_k \leq \gamma_k \leq \omega_{k+1}$ under any circumstances and thus we have

$$1 < \tilde{\sigma}_1 < \tilde{\sigma}_2 < \tilde{\sigma}_3 < \cdots < \sigma_{m-1} < \tilde{\sigma}_m \qquad (3.9.21)$$

This is valid for the m shocks with distinct indices that have the state $\{c_i^l\}$ or $\{c_i^r\}$ in common.

Let us consider another situation in which two shocks of consecutive indices are traveling side by side. This is shown in Fig. 3.10 as a diagram of δ and x for a fixed τ. From Eqs. (3.9.16) and (3.9.17) we have

$$\tilde{\sigma}_k = 1 + \nu \omega_k^\gamma \prod_{i=1}^{m} \frac{\omega_i^\beta}{\gamma_i}$$

$$\tilde{\sigma}_{k+1} = 1 + \nu \omega_{k+1}^\alpha \prod_{i=1}^{m} \frac{\omega_i^\beta}{\gamma_i}$$

and since $\omega_k < \omega_{k+1}$, we obtain

$$\tilde{\sigma}_k < \tilde{\sigma}_{k+1}, \qquad k = 1, 2, \ldots, m - 1 \qquad (3.9.22)$$

This applies when the state on the right-hand side of the $(k + 1)$-shock is the same as the state on the left-hand side of the k-shock, or vice versa. In the latter case the k-shock will eventually catch up with the $(k + 1)$-shock, and this is one of the subjects treated in Chapter 4.

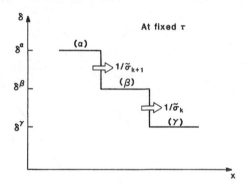

Figure 3.10

3.10 Entropy Change across a Shock

In this section we prove that the entropy condition given by Rule 3.3 does not violate the second law of thermodynamics. Let us consider a finite interval of the chromatographic column, $a(\tau) < x < b(\tau)$, where $a(\tau)$ and $b(\tau)$ denote the positions of the solute molecules that form the ends of the interval. Thus we have[†]

$$\frac{da}{d\tau} = \frac{db}{d\tau} = 1 \qquad (3.10.1)$$

[†] Note that, in terms of the dimensionless variables, the velocity of the fluid phase is equal to unity.

for the solute molecules in the fluid phase and

$$\frac{da}{d\tau} = \frac{db}{d\tau} = 0 \qquad (3.10.2)$$

for the solute molecules in the solid phase, since they are stationary. Suppose that a k-shock is contained in the interval at time τ. Denoting the position of the shock by $x = \bar{x}(\tau)$, we can write

$$\frac{d\bar{x}}{d\tau} = \frac{1}{\bar{\sigma}_k} = \tilde{\lambda}_k \qquad (3.10.3)$$

The equation of continuity for the solute A_i is given by

$$\frac{d}{d\tau} \int_a^b f_i(x, \tau)\, dx = 0, \qquad i = 1, 2, \ldots, m \qquad (3.10.4)$$

where

$$f_i = c_i + \nu n_i, \qquad i = 1, 2, \ldots, m \qquad (3.10.5)$$

The integrand being discontinuous, Eq. (3.10.4) can be rewritten by applying Reynold's transport theorem as

$$\frac{d}{d\tau}\left\{ \int_a^{\bar{x}} f_i(x, \tau)\, dx + \int_{\bar{x}}^b f_i(x, \tau)\, dx \right\} = 0$$

or

$$\int_a^{\bar{x}} \frac{\partial f_i}{\partial \tau}\, dx + \int_{\bar{x}}^b \frac{\partial f_i}{\partial \tau}\, dx + \{ f_i(\bar{x} - 0, \tau) - f_i(\bar{x} + 0, \tau) \}\frac{d\bar{x}}{d\tau}$$
$$- f_i(a, \tau)\frac{da}{d\tau} + f_i(b, \tau)\frac{db}{d\tau} = 0 \qquad (3.10.6)$$

We now take the limit as $a \to \bar{x}$ and $b \to \bar{x}$ simultaneously so that the first two integral terms vanish and the last two terms give, after Eqs. (3.10.1), (3.10.2), and (3.10.5) are substituted, $-(c_i^l - c_i^r)$. Hence we obtain

$$\{(c_i^l - c_i^r) + \nu(n_i^l - n_i^r)\}\tilde{\lambda}_k - (c_i^l - c_i^r) = 0$$

or, by using Eq. (3.9.18),

$$\nu\tilde{\lambda}_k\{p_k\omega_k^l\omega_k^r(c_i^r - c_i^l) - (n_i^r - n_i^l)\} = 0 \qquad (3.10.7)$$

It then follows that

$$\mathcal{M}_i^k \equiv \nu\tilde{\lambda}_k(p_k\omega_k^l\omega_k^r c_i - n_i) = \text{constant}, \qquad i = 1, 2, \ldots, m \qquad (3.10.8)$$

where $\mathcal{M}_i^k$ represents the molar flux of A_i through a k-shock.

From Eqs. (3.7.8) and (3.9.13) we find that

$$p_k\omega_k^l\omega_k^r = \frac{\omega_k^l}{\delta^r}$$

so

$$p_k \omega_k^l \omega_k^r c_i^r - n_i^r = \frac{\omega_k^l \gamma_i c_i^r}{\gamma_i \delta^r} - n_i^r = \frac{n_i^r}{\gamma_i}(\omega_k^l - \gamma_i)$$

Therefore, we can write

$$\mathcal{M}_i^k = \nu \tilde{\lambda}_k n_i^r (\omega_k^l - \gamma_i)/\gamma_i \begin{cases} >0 & \text{if } i < k \\ <0 & \text{if } i \geq k \end{cases} \qquad (3.10.9)$$

Since the coverage Θ of the adsorption sites decreases from the left to the right across a shock (see the corollary from Rule 3.3), we conclude that at a k-shock the molar flux of A_i is in the direction of increasing coverage Θ if $i \geq k$.

Here the process is isothermal and thus the entropy change across a k-shock can be expressed in the form

$$T[S]_r^l = [H]_r^l - [G]_r^l \qquad (3.10.10)$$

in which $[S]_r^l$, for instance, denotes the difference $S^l - S^r$. We shall assume that the adsorption mixture is ideal, so that

$$H = \sum_{i=1}^m (c_i h_i + \nu n_i H_i) \qquad (3.10.11)$$

$$G = \sum_{i=1}^m f_i \mu_i \qquad (3.10.12)$$

where h_i or H_i is the molar enthalpy of A_i in the fluid or solid phase, respectively, and μ_i is the chemical potential of A_i.

We note that both h_i and H_i remain unchanged across a shock, and thus by following the same procedure as above, we obtain

$$[H]_r^l = \lim_{b-a \to 0} \frac{d}{d\tau} \int_a^b H dx = \nu \tilde{\lambda}_k \sum_{i=1}^m \{p_k \omega_k^l \omega_k^r [c_i]_r^l h_i - [n_i]_r^l H_i\}$$

Substituting Eqs. (3.9.11) and (3.10.7) into this gives

$$[H]_r^l = \nu \tilde{\lambda}_k p_k \omega_k^l \omega_k^r [\delta]_r^l \sum_{i=1}^m (-\Delta H_i) \frac{J_i^k}{K_i} \qquad (3.10.13)$$

where ΔH_i is the enthalpy change of adsorption defined by

$$-\Delta H_i = h_i - H_i > 0 \qquad (3.10.14)$$

Since we are involved here with a thermodynamic consideration, we will accept the thermodynamic argument of Kemball, Rideal, and Guggenheim [*Trans. Faraday Soc.* **44**, 948 (1944)] and assume in this section that all the limiting values N_i are the same; i.e., $N_1 = N_2 = \cdots = N_m = N$. It then follows from Eqs. (3.5.7) and (3.5.8) that $(-\Delta H_i) > (-\Delta H_j)$ for $i > j$. Noting the fact that $J_i^k < 0$ for $i < k$ while $J_i^k > 0$ for $i \geq k$, we find that there exists a lower bound of the summation term in Eq. (3.10.13):

$$\sum_{i=1}^m (-\Delta H_i) \frac{J_i^k}{K_i} > (-\Delta H_{k-1}) \sum_{i=1}^m \frac{J_i^k}{K_i} \qquad (3.10.15)$$

Furthermore, each term J_i^k/K_i is invariant along the Γ^k on which the two states $\{c_i^l\}$ and $\{c_i^r\}$ fall. This Γ^k intersects at one end the subspace $\omega_k = \gamma_k$ in the space $\Omega(m)$ (see Fig. 3.6). Let this point of intersection be designated by $\{c_i^0: c_k \equiv 0\}$. Then we have

$$J_i^k = \frac{K_i n_i^0}{\gamma_i - \gamma_k} = \frac{K_i n_i^0}{N(K_i - K_k)}, \qquad i \neq k$$

$$J_k^k = 1 - \sum_{i=1,k}^{m} J_i^k$$

and thus

$$
\begin{aligned}
\sum_{i=1}^{k} \frac{J_i^k}{K_i} &= \frac{1}{K_k} + \sum_{i=1,k}^{m} \frac{J_i^k}{K_i}\left(1 - \frac{K_i}{K_k}\right) \\
&= \frac{1}{K_k} - \frac{1}{K_k} \sum_{i=1,k}^{m} \frac{n_i^0}{N} \\
&= \frac{1}{K_k}\left(1 - \sum_{i=1,k}^{m} \frac{n_i^0}{N}\right) \\
&> \frac{1}{K_k}\left(1 - \sum_{i=1}^{m} \frac{n_i^0}{N}\right) = \frac{1}{K_k \delta^0} > 0
\end{aligned}
\tag{3.10.16}
$$

Applying Eqs. (3.10.15) and (3.10.16) to Eq. (3.10.13) and noting that $\delta^l > \delta^r$ from the entropy condition, we find that

$$[H]_r^l = H^l - H^r > 0 \tag{3.10.17}$$

On the other hand, the chemical potential μ_i jumps from one side of the k-shock to the other, whereas it takes the same value for solutes A_i on both phases due to the adsorption equilibrium. Therefore, we obtain

$$[G]_r^l = \lim_{b-a\to 0} \frac{d}{d\tau} \int_a^b G \, dx = \nu\tilde{\lambda}_k \sum_{i=1}^{m} \{p_k\omega_k^l\omega_k^r[c_i\mu_i]_r^l - [n_i\mu_i]_r^l\}$$

which, upon introducing Eq. (3.10.8), can be reduced to

$$[G]_r^l = \sum_{i=1}^{m} \mathcal{M}_i^k [\mu_i]_r^l \tag{3.10.18}$$

Since $[\mu_i]_r^l = RT \ln (c_i^l/c_i^r)$, the sign of $[\mu_i]_r^l$ is the same as that of $[c_i]_r^l$. From Eq. (3.9.11) we have

$$[c_i]_r^l = \frac{J_i^k}{K_k}[\delta]_r^l \begin{cases} <0 & \text{for } i < k \\ >0 & \text{for } i \geq k \end{cases} \tag{3.10.19}$$

because $[\delta]_r^l = \delta^l - \delta^r > 0$. Applying Eqs. (3.10.9) and (3.10.19) to Eq. (3.10.18), we see that

$$[G]_r^l = G^l - G^r < 0 \tag{3.10.20}$$

Consequently, we obtain

$$[S]_r^l = S^l - S^r > 0 \qquad (3.10.21)$$

This implies that entropy increases in the direction of increasing total coverage Θ because $\Theta^l > \Theta^r$ across a shock. Thus the entropy condition is consistent with the second law of thermodynamics.

3.11 Solution of the Riemann Problem

We are now in a position to combine the results developed in previous sections to construct the solution of a Riemann problem. Given the constant initial data $\{\phi_i^0\}$ and the constant feed data $\{\phi_i^f\}$, it is only natural to begin with the homeomorphism, Eq. (3.6.16), to determine the two sets of characteristic parameters $\{\omega_k^0\}$ and $\{\omega_k^f\}$, which give the two point images of the data in the m-dimensional space $\Omega(m)$. From this we shall proceed to construct the image of the solution in $\Omega(m)$ and then to determine the map of this image in the physical plane of x and τ. In the course of this procedure we shall pay particular attention to the existence and uniqueness of the solution on top of the actual solution scheme.

Existence and uniqueness

In the space $\Omega(m)$ the images of the initial and feed data fall on two different points, I and F, respectively, so the image of a solution must be given by a continuous path that connects points I and F by a set of Γ's. A Γ^k being straight and parallel to the ω_k-axis, such a path can always be formed and hence there exists a solution image in $\Omega(m)$.

Before making any general observations, the situation for $m = 3$, as illustrated in Fig. 3.11, may serve to show the matter more clearly. Points F and I represent the feed and initial data and if we pass from F, the state when $x = 0$, to I, the state when $\tau = 0$, we shall be moving clockwise in the (x, τ)-plane. Now the inequalities on the σ_k and $\bar{\sigma}_k$ [i.e., Eqs. (3.8.7), (3.9.20), and (3.9.22)] require that we must use a sequence of Γ^k's in the order of $k = 3$, 2, 1 in passing from F to I[†]. In the space $\Omega(3)$ there is clearly a unique path $FABI$ consisting of segments of lines parallel to the axes of ω_3, ω_2, and ω_1 in that order, and this path has an image in the space $\Phi(3)$. If ω_3 decreases in going from F to A, then by Rule 3.1 we have a 3-simple wave. On the other hand, if ω_3 increases, then by Rule 3.3 it must be a 3-shock. Thus for each segment of the path we have a simple test of whether it represents a continuous or a discontinuous transition between states. We also expect to have two new constant states, corresponding to points A and B, respectively. Each of these constant states would intervene between two waves of consecutive indices in the (x, τ)-plane.

[†]Compare this argument with the discussion in the second half of Sec. 1.5; in particular, see Rule 1.3 there.

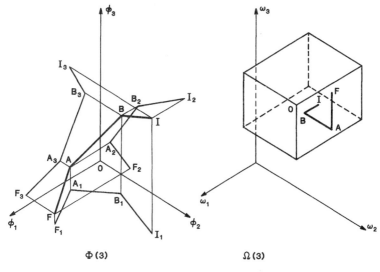

$\Phi(3)$ $\Omega(3)$

Figure 3.11

In the general situation, the path from F to I may consist of many segments of Γ's of various kinds. To a Γ^k there corresponds either a k-simple wave, which is represented by a family of straight-line characteristics C^k, or a k-shock wave, which is given by a straight shock path S^k, depending on the variation of the parameter ω_k. Between two waves of consecutive indices there appears a region of constant state. As in the previous case for $m = 3$, it can be shown by using Eqs. (3.8.7), (3.9.20), and (3.9.22) that the path must be composed of at most m segments each of which is part of a Γ of a distinct kind and which are arranged in the order of descending k as one passes from F to I. Since a Γ^k is straight and parallel to the ω_k-axis in $\Omega(m)$, such a path always exists and is unique. This establishes the existence of a unique solution.

Solution scheme

It is clear that the initial discontinuity appearing at the origin of the (x, τ)-plane has its range of influence centered at this point. This region is our main interest because outside we simply have two constant states corresponding to the initial and feed data, respectively. It is, therefore, convenient to construct the wave solutions in the (x, τ)-plane, from which the concentration profiles can be established without difficulty.

The general situation in the physical plane is depicted in Fig. 3.12, where it is seen that there is a succession of m waves which are arranged in the order of descending k as we proceed clockwise[†] from the feed state to the initial state. These waves are either centered simple waves (as represented by C^m, C^{k+1},

---

[†] Note that this is also in the direction of the fluid flow.

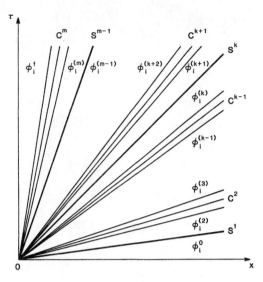

Figure 3.12

C^{k-1}, and C^2) or shock waves (as marked by S^{m-1}, S^k, and S^1), separating regions of constant state. The constant state appearing on the right-hand side of a k-wave will be called the *k-constant state* and denoted by a bracketed index, so that $\phi_i^{(k)}[\, = K_i c_i^{(k)}]$ is the dimensionless concentration of the ith solute A_i between the k- and $(k-1)$-wave regions. Clearly, the initial state $\{\phi_i^0\}$ corresponds to $\{\phi_i^{(1)}\}$, whereas the feed state $\{\phi_i^f\}$ corresponds to $\{\phi_i^{(m+1)}\}$. The angular region between the boundaries of the feed state and the initial state is the range of influence of the discontinuity at the origin.

Since ω_k varies only along a Γ^k, it is the only variable parameter across a k-wave, be it a simple wave or a shock wave, and thus we can readily identify the set of characteristic parameters for every constant state as illustrated in Table 3.3. We note that the parameters above the diagonal all carry the superscript f, whereas those below the diagonal bear the superscript 0. The arrows indicate the variation as we pass through the waves clockwise.

Now the k-constant state $\{\phi_i^{(k)}\}$ corresponds to the $(m-k+1)$th vertex from F, or the $(k-1)$th vertex from I, of the path in $\Omega(m)$ and so is characterized by the set of characteristic parameters

$$(\omega_1^f, \omega_2^f, \ldots, \omega_{k-1}^f, \omega_k^0, \ldots, \omega_{m-1}^0, \omega_m^0) \qquad (3.11.1)$$

These parameter values can be used in the inverse transform (3.7.9) to determine the concentrations of the k-constant state:

$$\phi_i^{(k)} = \left(\frac{\gamma_i}{\omega_i^f} - 1\right) \prod_{j=1,i}^{k-1} \frac{\gamma_i/\omega_j^f - 1}{\gamma_i/\gamma_j - 1} \prod_{j=k}^{m} \frac{\gamma_i/\omega_j^0 - 1}{\gamma_i/\gamma_j - 1} \qquad \text{for } i \leq k-1 \quad (3.11.2)$$

$$\phi_i^{(k)} = \left(\frac{\gamma_i}{\omega_i^0} - 1\right) \prod_{j=1}^{k-1} \frac{\gamma_i/\omega_j^f - 1}{\gamma_i/\gamma_j - 1} \prod_{j=k,i}^{m} \frac{\gamma_i/\omega_j^0 - 1}{\gamma_i/\gamma_j - 1} \qquad \text{for } i \geq k \quad (3.11.3)$$

Hyperbolic Systems and Multicomponent Chromatography   Chap. 3

TABLE 3-3. CONSTANT STATES AND CHARACTERISTIC PARAMETERS

Constant State	ω_m	ω_{m-1}	$\cdots$	ω_k	$\cdots$	ω_2	ω_1	Connecting Waves	Image
Feed state	ω_m^f	ω_{m-1}^f	$\cdots$	ω_k^f	$\cdots$	ω_2^f	ω_1^f	m-wave	Γ^m
m-constant state	ω_m^0	ω_{m-1}^f $\rightarrow$	$\cdots$	ω_k^f	$\cdots$	ω_2^f	ω_1^f	$(m-1)$-wave	Γ^{m-1}
$(m-1)$-constant state	ω_m^0	ω_{m-1}^0	$\cdots$	ω_k^f	$\cdots$	ω_2^f	ω_1^f		
.	.						.		
.	.		$\omega_{k+1}^f \rightarrow$				.		
$(k+1)$-constant state	ω_m^0	$\cdots$	ω_{k+1}^f	ω_k^f	$\cdots$	ω_2^f	ω_1^f		
k-constant state	ω_m^0	ω_{m-1}^0	$\cdots$	ω_k^0 $\omega_{k-1}^f \cdots$ $\rightarrow$	ω_2^f		ω_1^f	k-wave	Γ^k
$(k-1)$-constant state	ω_m^0	ω_{m-1}^0	$\cdots$	ω_k^0 $\omega_{k-1}^0 \cdots$	ω_2^f		ω_1^f	$(k-1)$-wave	Γ^{k-1}
.	.	.		.			.		
3-constant state	ω_m^0	ω_{m-1}^0	$\cdots$	ω_k^0	$\cdots$	$\omega_2^f \rightarrow$	ω_1^f	2-wave	Γ^2
2-constant state	ω_m^0	ω_{m-1}^0	$\cdots$	ω_k^0	$\cdots$	ω_2^0	$\omega_1^f \rightarrow$	1-wave	Γ^1
Initial state	ω_m^0	ω_{m-1}^0	$\cdots$	ω_k^0	$\cdots$	ω_2^0	ω_1^0		

Another approach is motivated by the recurrence formula

$$\phi_i^{(k)} - \phi_i^{(k+1)} = J_i^k(\delta^{(k)} - \delta^{(k+1)}) \qquad (3.11.4)$$

which is valid across a k-wave for $k = 1, 2, \ldots, m$. The Reimann invariant J_i^k remains unchanged across a k-wave and can be expressed as

$$J_i^k = \frac{\gamma_i \phi_i^{(k)}/\delta^{(k)}}{\gamma_i - \omega_k^0} = \frac{\gamma_i \phi_i^{(k+1)}/\delta^{(k+1)}}{\gamma_i - \omega_k^f}$$

From this we find that

$$J_i^k(\delta^{(k)} - \delta^{(k+1)}) = \frac{\gamma_i \phi_i^{(k)}}{\gamma_i - \omega_k^0} - \frac{\gamma_i \phi_i^{(k+1)}}{\gamma_i - \omega_k^f}$$

and substituting this into Eq. (3.11.4) gives

$$\frac{\phi_i^{(k)}}{1 - \gamma_i/\omega_k^0} = \frac{\phi_i^{(k+1)}}{1 - \gamma_i/\omega_k^f} \qquad k = 1, 2, \ldots, m \qquad (3.11.5)$$

This is a convenient recurrence formula which we can use successively to generate all the constant states. Applying Eq. (3.11.5) successively from $k = 1$ or reversely from $k = m$, we obtain

$$\phi_i^{(k)} = \phi_i^0 \prod_{j=1}^{k-1} \frac{1 - \gamma_i/\omega_j^f}{1 - \gamma_i/\omega_j^0} = \phi_i^f \prod_{j=k}^{m} \frac{1 - \gamma_i/\omega_j^0}{1 - \gamma_i/\omega_j^f} \qquad (3.11.6)$$

for $i = 1, 2, \ldots, m$ and $k = 2, 3, \ldots, m$.

Once every constant state is determined, it is straightforward to construct the wave solutions. For a k-wave, where $1 \leq k \leq m$, we proceed as follows:

1. If $\omega_k^f > \omega_k^0$, we have a k-simple wave and thus introduce Eq. (3.8.3); i.e.,

$$\sigma_k = 1 + \nu p_k \omega_k^2 \qquad \text{for } \omega_k^0 \leq \omega_k \leq \omega_k^f \qquad (3.11.7)$$

where

$$p_k = \frac{1}{\gamma_k} \prod_{i=1}^{k-1} \frac{\omega_i^f}{\gamma_i} \prod_{i=k+1}^{m} \frac{\omega_i^0}{\gamma_i} \qquad (3.11.8)$$

We then select as many values of ω_k as desired to calculate σ_k and draw straight line C^k's of slope σ_k from the origin to generate the centered k-simple wave. The constant values of concentrations $\{\phi_i\}$ borne along each C^k can be determined from Eq. (3.7.9) or more conveniently from an expression analogous to Eq. (3.11.5); that is, since in the region of a k-simple wave we can write

$$\phi_i = \phi_i^{(k)} + J_i^k(\delta - \delta^{(k)}) = \phi_i^{(k+1)} + J_i^k(\delta - \delta^{(k+1)})$$

instead of Eq. (3.11.4), it immediately follows that we obtain

$$\phi_i = \phi_i^{(k)} \frac{1 - \gamma_i/\omega_k}{1 - \gamma_i/\omega_k^0} = \phi_i^{(k+1)} \frac{1 - \gamma_i/\omega_k}{1 - \gamma_i/\omega_k^f}, \qquad i = 1, 2, \ldots, m \qquad (3.11.9)$$

in place of Eq. (3.11.5). If we use here the same value of ω_k as for σ_k, Eq. (3.11.9) determines the concentration state carried along that particular C^k.

2. If $\omega_k^f < \omega_k^0$, the k-wave is a shock. We then apply Eq. (3.9.18) in the form

$$\tilde{\sigma}_k = 1 + \nu p_k \omega_k^f \omega_k^0 \qquad (3.11.10)$$

where p_k is given by Eq. (3.11.8), and draw a straight line of slope $\tilde{\sigma}_k$ issuing from the origin. This is the shock path S^k and here the k-shock propagates with a constant speed $1/\tilde{\sigma}_k$ because the states on both sides remain constant.

3. If $\omega_k^f = \omega_k^0$, then by Eq. (3.11.5) we must have

$$\phi_i^{(k)} = \phi_i^{(k+1)} \qquad (3.11.11)$$

and thus neither a k-wave nor a k-constant state appears in the solution. For such a degenerate case, the image in $\Omega(m)$ will not contain a segment of Γ^k.

Special cases

If one or more solute species are absent from the feed mixture or from the initial bed of adsorbent, it is to be expected that some interesting features may appear. These are certainly of potential interest in the practice of multicomponent chromatography and hence worth examining in detail.

In Sec. 3.7 it was noticed that the subspace $\pi_j : \phi_j \equiv 0$ consists of two planes $\omega_j = \gamma_j$ and $\omega_{j+1} = \gamma_j$ in the space $\Omega(m)$ (see Table 3.2). The case $j = m$ is exceptional and π_m is the plane $\omega_m = \gamma_m$ in $\Omega(m)$. This implies that the subspace π_j has connections with the outside, $\Omega(m) - \pi_j$, only through Γ^j or Γ^{j+1}. Furthermore, it is observed that ω_j increases along Γ^j toward π_j while ω_{j+1} decreases along a Γ^{j+1} toward π_j. When associated with Rules 3.1 and 3.3, these arguments lead us to

RULE 3.5: As we pass in the x direction:

    1. A particular solute A_j can be exhausted across a j-shock or through a $(j + 1)$-simple wave.
    2. A particular solute A_j can emerge through a j-simple wave or across a $(j + 1)$-shock.
    3. The most strongly adsorbed solute A_m can be exhausted across an m-shock and emerge through an m-simple wave.

Let us now consider the conventional adsorption process (i.e., saturation of a fresh column) for which the initial bed contains no solute species. The feed mixture contains m solute species and each solute must vanish somewhere in the column. It follows directly from Rule 3.5, statements 1 and 3 that A_j must be exhausted across the j-shock for $1 \le j \le m$. In other words, we must have m shocks of different kinds and each solute disappears successively in the order of decreasing adsorptivity as we pass across shocks one by one in the x-direction.[†] This is illustrated for $m = 3$ in Table 3.4 and Fig. 3.13.

Clearly, we have $\omega_j^0 = \gamma_j$ for all j, so Eq. (3.11.8) gives

$$p_k = \frac{1}{\gamma_k} \prod_{i=1}^{k-1} \frac{\omega_i^f}{\gamma_i}$$

TABLE 3-4. CHARACTERISTIC PARAMETERS FOR THE SATURATION PROBLEM ($m = 3$)

Constant State	ω_3	ω_2	ω_1	Solute Present	Connecting Waves
Feed state	ω_3^f $\downarrow$	ω_2^f	ω_1^f	A_1, A_2, A_3	3-shock
3-constant state	γ_3	ω_2^f $\downarrow$	ω_1^f	A_1, A_2	2-shock
2-constant state	γ_3	γ_2	ω_1^f $\downarrow$	A_1	1-shock
Initial state	γ_3	γ_2	γ_1	None	

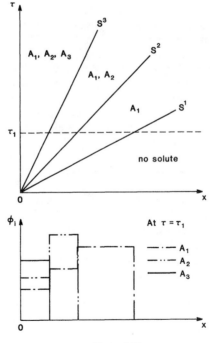

Figure 3.13

[†] This fact was noticed for an arbitrary adsorption isotherm by DeVault (1943).

Substituting this into Eq. (3.11.10) and letting $\omega_k^0 = \gamma_k$, we obtain

$$\bar{\sigma}_k = 1 + \nu \omega_k^f \prod_{i=1}^{k-1} \frac{\omega_i^f}{\gamma_i}, \qquad k = 1, 2, \ldots, m \qquad (3.11.12)$$

From Eq. (3.11.6) we can determine the k-constant state as

$$\phi_i^{(k)} = \phi_i^f \prod_{j=k}^{m} \frac{1 - \gamma_i/\gamma_j}{1 - \gamma_i/\omega_j^f}, \qquad i = 1, 2, \ldots, k-1$$

$$\phi_i^{(k)} = 0, \qquad i = k, k+1, \ldots, m \qquad (3.11.13)$$

We note that $\phi_i^{(k)} > \phi_i^{(k+1)}$ for $i = 1, 2, \ldots, k-1$, because $\gamma_j > \omega_j^f$ for all j. Therefore, as the solute A_k is exhausted across the k-shock, the solutes $A_1, A_2, \ldots, A_{k-1}$ (i.e., solutes that are less strongly adsorbed than A_k) all experience an increase in their concentrations. This is shown schematically in the lower part of Fig. 3.13.

Next we shall consider the conventional desorption process (i.e., elution of a uniformly presaturated column). Here we have the pure solvent flowing into the column, whereas the initial bed is a uniform chromatogram of m solute species. Therefore, each solute must emerge somewhere in the column. According to Rule 3.5, statements 2 and 3, the solute A_j must emerge through the j-simple wave for $1 \leq j \leq m$. This implies that the solution must contain m simple waves and each solute emerges successively in the order of decreasing adsorptivity as we pass through simple waves one by one in the x direction. This is shown for $m = 3$ in Table 3.5 and Fig. 3.14.

In this case we have $\omega_j^f = \gamma_j$ for all j, so that Eq. (3.11.6) gives, for the k-constant state, the expressions

$$\phi_i^{(k)} = 0, \qquad i = 1, 2, \ldots, k-1$$

$$\phi_i^{(k)} = \phi_i^0 \prod_{j=1}^{k-1} \frac{1 - \gamma_i/\gamma_j}{1 - \gamma_i/\omega_j^0}, \qquad i = k, k+1, \ldots, m \qquad (3.11.14)$$

We also find from Eq. (3.11.8) that

$$p_k = \frac{1}{\gamma_k} \prod_{i=k+1}^{m} \frac{\omega_i^0}{\gamma_i}$$

and, substituting this into Eq. (3.11.7), we obtain the characteristic direction

TABLE 3-5. CHARACTERISTIC PARAMETERS FOR THE ELUTION PROBLEM ($m = 3$)

Constant State	ω_3	ω_2	ω_1	Solute Present	Connecting Waves
Feed state	γ_3 $\downarrow$	γ_2	γ_1	None	3-simple wave
3-constant state	ω_3^0	γ_2 $\downarrow$	γ_1	A_3	2-simple wave
2-constant state	ω_3^0	ω_2^0	γ_1 $\downarrow$	A_2, A_3	1-simple wave
Initial state	ω_3^0	ω_2^0	ω_1^0	A_1, A_2, A_3	

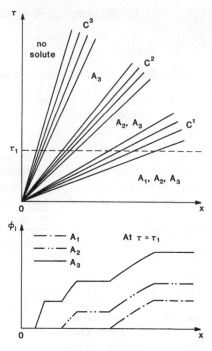

Figure 3.14

for the k-simple wave as

$$\sigma_k = 1 + \nu \frac{\omega_k^2}{\gamma_k} \prod_{i=k+1}^{m} \frac{\omega_i^0}{\gamma_i}, \qquad \omega_k^0 \le \omega_k \le \gamma_k \qquad (3.11.15)$$

Along each of these straight-line characteristics C^k the state of concentrations remain unchanged and this state is determined by Eq. (3.11.9) as

$$\phi_i = 0, \qquad i = 1, 2, \ldots, k-1$$
$$\phi_i = \phi_i^{(k)} \frac{1 - \gamma_i/\omega_k}{1 - \gamma_i/\omega_k^0}, \qquad i = k, k+1, \ldots, m \qquad (3.11.16)$$

Since $\gamma_j > \omega_j^0$ for all j, it follows from Eq. (3.11.14) that $\phi_i^{(k)} > \phi_i^{(k+1)}$ for $i = k+1, k+2, \ldots, m$. Hence as the solute A_k emerges through the k-simple wave, the solutes $A_{k+1}, A_{k+2}, \ldots, A_m$ (i.e., solutes that are more strongly adsorbed than A_k) all experience an increase in their concentrations. This is illustrated in the lower part of Fig. 3.14.

Another interesting example is produced by the alternating data, i.e.,

$$\phi_i^0 = 0 \qquad \text{if } i \text{ is even}$$
$$\phi_i^f = 0 \qquad \text{if } i \text{ is odd} \qquad (3.11.17)$$

or vice versa. It is clear that along the column, exhaustion of solute A_j alternates with emergence of solute A_{j-1}. According to Rule 3.5, however, solute A_{j-1} may emerge across the j-shock or be exhausted through the j-simple wave, while in these two kinds of wave solute A_j is, respectively, exhausted or emerges. A number of different cases can occur depending on the data specified. Some of these were suggested in the basic profile patterns for $m = 3$ by Klein, Tondeur and Vermeulen (1967) in relation to ion-exchange systems. In any case, once the sets $\{\omega_k^f\}$ and $\{\omega_k^0\}$ are determined, the construction of solution should be fairly straightforward and this will be illustrated in the next section.

Exercises

3.11.1. From Eq. (3.11.6), express the ratio ϕ_i^0/ϕ_i^f in terms of the characteristic parameters $\{\omega_j^f\}$ and $\{\omega_j^0\}$ and show that it is consistent with Eq. (3.7.9).

3.11.2. Consider the two situations shown in Fig. 3.15. In regions P and R all the solutes are present whereas only the solute A_j is absent from region Q. Noting that either $\omega_j = \gamma_j$ or $\omega_{j+1} = \gamma_j$ in region Q, show that the two regions P and Q can be connected either by a j-shock or by a ($j + 1$)-simple wave, whereas the two regions Q and R can be connected by a j-simple wave or a ($j + 1$)-shock.

3.11.3. Write Eqs. (3.11.12)–(3.11.16) for $m = 3$ and confirm all the discussions in the related paragraphs with the resulting equations.

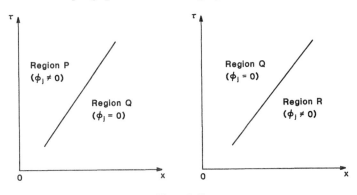

Figure 3.15

3.12 Illustrations

To illustrate the application, several numerical examples are treated in this section. Since each case was considered in a general fashion in Sec. 3.11, we shall add a brief discussion and present the solution in the form of graphs.

We assume in all examples that the limiting values N_i are all identical and given by

$$N_i = N = 1.0 \text{ mol/liter of adsorbent}$$

for $i = 1, 2, \ldots, m$. The bed voidage is given by

$$\epsilon = 0.4 \quad \text{or} \quad \nu = 1.5$$

Further information is given in the table that appears at the beginning of each example. The concentrations c_i are given in moles per liter of fluid phase and the concentrations n_i in moles per liter of adsorbent. The unit of K_i is, of course, the reciprocal of concentration. The characteristic parameters, determined by solving Eq. (3.6.16), are also presented in the table.

Example 3.1

General Case ($m = 3$)

	i		
	1	*2*	*3*
K_i	5.0	7.5	10.0
c_i^f	0.150	0.060	0.020
c_i^0	0.032	0.114	0.075
ω_i^f	2.445	6.667	9.586
ω_i^0	2.797	5.409	8.974

Since all three ω_k's vary from the feed state to the initial state, we expect to have three waves and two additional constant states. The 3- and 2-waves are centered simple waves because $\omega_3^f > w_3^0$ and $w_2^f > \omega_2^0$, whereas the 1-wave is a shock wave because $\omega_1^f < \omega_1^0$.

The two constant states are first identified by the set of characteristic parameters; that is, $(\omega_1^f, \omega_2^f, \omega_3^0)$ for the 3-constant state and $(\omega_1^f, \omega_2^0, \omega_3^0)$ for the 2-constant state, respectively. Thus the concentrations for each constant state can be readily determined by using Eq. (3.7.9), (3.11.5), or (3.11.6); for example, from Eq. (3.11.5) we obtain

$$\phi_i^{(3)} = \phi_i^f \frac{1 - \gamma_i/\omega_3^0}{1 - \gamma_i/\omega_3^f} = \phi_i^f \frac{1 - \gamma_i/8.974}{1 - \gamma_i/9.586}$$

$$\phi_i^{(2)} = \phi_i^{(3)} \frac{1 - \gamma_i/\omega_2^0}{1 - \gamma_i/\omega_2^f} = \phi_i^{(3)} \frac{1 - \gamma_i/5.409}{1 - \gamma_i/6.667}$$

These equations give

$$\phi_1^{(3)} = 0.694, \qquad \phi_2^{(3)} = 0.340, \qquad \phi_3^{(3)} = 0.529$$

for the 3-constant state and

$$\phi_1^{(2)} = 0.210, \qquad \phi_2^{(2)} = 1.052, \qquad \phi_3^{(2)} = 0.90$$

for the 2-constant state.

For the 3-simple wave we obtain

$$\sigma_3 = 1 + \frac{\nu \omega_1^f \omega_2^f \omega_3^2}{\gamma_1 \gamma_2 \gamma_3}$$

$$= 1 + \left(\frac{1.5}{10}\right)\left(\frac{2.445}{5}\right)\left(\frac{6.667}{7.5}\right)\omega_3^2$$

$$= 1 + 0.0652\omega_3^2$$

from Eqs. (3.11.7) and (3.11.8), and

$$\phi_i = \phi_i^f \frac{1 - \gamma_i/\omega_3}{1 - \gamma_i/\omega_3^f} = \phi_i^f \frac{1 - \gamma_i/\omega_3}{1 - \gamma_i/9.586}$$

from Eq. (3.11.9). These equations are to be used for various values of ω_3 between 8.974 and 9.586 to generate the 3-simple wave solution. With $\omega_3 = 9.35$, for example, we find

$$\sigma_3 = 6.70$$

$$\phi_1 = 0.729, \quad \phi_2 = 0.409, \quad \phi_3 = 0.322$$

and so this state of concentrations is carried along the straight line C^3 that starts from the origin of the (x, τ)-plane with a slope 6.70.

Similarly, for the 2-simple wave we obtain

$$\sigma_2 = 1 + \frac{\nu \omega_1^f \omega_3^0 \omega_2^2}{\gamma_1 \gamma_2 \gamma_3}$$

$$= 1 + \left(\frac{1.5}{7.5}\right)\left(\frac{2.445}{5}\right)\left(\frac{8.974}{10}\right)\omega_2^2$$

$$= 1 + 0.0878\omega_2^2$$

and

$$\phi_i = \phi_i^{(3)} \frac{1 - \gamma_i/\omega_2}{1 - \gamma_i/\omega_2^f} = \phi_i^{(3)} \frac{1 - \gamma_i/\omega_2}{1 - \gamma_i/6.667}$$

Use of these equations over the range $5.409 \le \omega_2 \le 6.667$ will produce the 2-simple wave solution.

The 1-shock simply connects the (2)-constant state to the initial state and propagates along a straight line path S^1, the slope of which is determined from Eqs. (3.11.8) and (3.11.10) as

$$\tilde{\sigma}_1 = 1 + \frac{\nu \omega_1^0 \omega_2^0 \omega_3^0 \omega_1^f}{\gamma_1 \gamma_2 \gamma_3}$$

$$= 1 + (1.5)\left(\frac{2.797}{5}\right)\left(\frac{5.409}{7.5}\right)\left(\frac{8.974}{10}\right)(2.445)$$

$$= 2.328$$

The solution is constructed in the physical plane as shown in the upper part of Fig. 3.16. The concentration profiles at $\tau = 4.0$ are determined from the physical plane portrait and given in the lower part of Fig. 3.16.

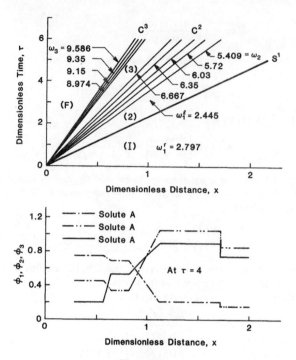

Figure 3.16

Example 3.2

Saturation of a fresh column ($m = 3$)

	1	2	3
	\multicolumn{3}{c}{i}		
K_i	5.0	10.0	15.0
c_i^f	0.05	0.05	0.05
c_i^0	0	0	0
ω_i^f	3.387	7.050	12.563
ω_i^0	5.0	10.0	15.0

This is the conventional process for the saturation of a clean bed of adsorbent. Since $\omega_k^f < \omega_k^0$ for all k, we have three shocks. The slope of each shock is determined directly from Eq. (3.11.12) to give

$$\tilde{\sigma}_1 = 1 + \nu\omega_1^f = 1 + (1.5)(3.387) = 6.081$$

$$\tilde{\sigma}_2 = 1 + \frac{\nu\omega_1^f\omega_2^f}{\gamma_1} = 1 + (1.5)(3.387)(\frac{7.05}{5}) = 8.164$$

$$\bar{\sigma}_3 = 1 + \frac{\nu \omega_1^f \omega_2^f \omega_3^f}{\gamma_1 \gamma_2}$$

$$= 1 + (1.5)(3.387)\left(\frac{7.05}{5}\right)\left(\frac{12.563}{10}\right) = 10.0$$

and the three shock paths, S^1, S^2, and S^3, of these slopes divide the physical plane in the manner shown in the upper part of Fig. 3.17.

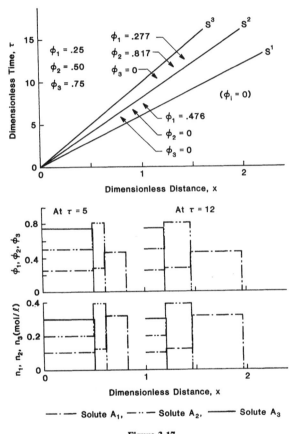

Figure 3.17

For the 3- and 2-constant states we use Eq. (3.11.13) to obtain

$$\phi_i^{(3)} = \phi_i^f \frac{1 - \gamma_i/\gamma_3}{1 - \gamma_i/\omega_3^f} = \phi_i^f \frac{1 - \gamma_i/15}{1 - \gamma_i/12.563}, \qquad i = 1, 2$$

with $\phi_3^{(3)} = 0$ and

$$\phi_i^{(2)} = \phi_i^f \frac{1 - \gamma_i/\gamma_2}{1 - \gamma_i/\omega_2^f} \frac{1 - \gamma_i/\gamma_3}{1 - \gamma_i/\omega_3^f}$$

$$= \phi_1^f \frac{1 - 5/10}{1 - 5/7.50} \left(\frac{1 - 5/10}{1 - 5/12.563} \right)$$

with $\phi_2^{(2)} = \phi_3^{(2)} = 0$. Thus we find that

$$\phi_1^{(3)} = 0.277, \qquad \phi_2^{(3)} = 0.817, \qquad \phi_3^{(3)} = 0$$

for the 3-constant state and

$$\phi_1^{(2)} = 0.476, \qquad \phi_2^{(2)} = 0, \qquad \phi_3^{(2)} = 0$$

for the 2-constant state. We note that both ϕ_1 and ϕ_2 jump up across the 3-shock and ϕ_1 again jumps up across the 2-shock. This can be seen in the lower part of Fig. 3.17, in which the profiles of ϕ_i as well as those of n_i are presented at $\tau = 5.0$ and $\tau = 12.0$.

Example 3.3

Elution of a uniformly presaturated column ($m = 3$)

		i	
	1	*2*	*3*
K_i	5.0	10.0	15.0
c_i^f	0	0	0
c_i^0	0.05	0.05	0.05
ω_i^f	5.0	10.0	15.0
ω_i^0	3.387	7.05	12.563

This is just the reverse of Example 3.2, representing the conventional process for the elution of a previously saturated bed of adsorbent. Here the feed state and the initial state are interchanged so that the arguments on the characteristic parameters are simply reversed from Example 3.2. Hence there appear three simple waves, one solute emerging through each of the waves. The simple wave regions are to be separated by two new constant states.

We first determine the new constant states by using Eq. (3.11.4). This gives

$$\phi_i^{(2)} = \phi_i^0 \frac{1 - \gamma_i/\gamma_1}{1 - \gamma_i/\omega_1^0} = \phi_i^0 \frac{1 - \gamma_i/5}{1 - \gamma_i/3.387}, \qquad i = 2, 3$$

with $\phi_1^{(2)} = 0$ and

$$\phi_3^{(3)} = \phi_3^0 \frac{1 - \gamma_3/\gamma_1}{1 - \gamma_3/\omega_1^0} \cdot \frac{1 - \gamma_3/\gamma_2}{1 - \gamma_3/\omega_2^0}$$

$$= \phi_3^0 \frac{1 - 15/5}{1 - 15/3.387} \left(\frac{1 - 15/10}{1 - 15/7.05} \right) = 0.2586\phi_3^0$$

with $\phi_1^{(3)} = \phi_2^{(3)} = 0$. Hence we obtain

$$\phi_1^{(2)} = 0, \qquad \phi_2^{(2)} = 0.256, \qquad \phi_3^{(2)} = 0.438$$

for the 2-constant state and

$$\phi_1^{(3)} = 0, \qquad \phi_2^{(3)} = 0, \qquad \phi_3^{(3)} = 0.194$$

for the 3-constant state.

For the 3-simple wave we use Eq. (3.11.15) to obtain

$$\sigma_3 = 1 + \frac{\nu \omega_3^2}{\gamma_3} = 1 + 0 \cdot 1 \omega_3^2, \qquad 12.563 \leq \omega_3 \leq 15$$

for the slope of the straight-line characteristics C^3. The state of concentrations carried along the C^3 is given by Eq. (3.11.16) as

$$\phi_1 = \phi_2 = 0$$

$$\phi_3 = \phi_3^{(3)} \frac{1 - 15/\omega_3}{1 - 15/\omega_3^0} = 0.194 \frac{1 - 15/\omega_3}{1 - 15/12.563}, \qquad 12.563 < \omega_3 < 15$$

Similarly, for the 2-simple wave we obtain

$$\sigma_2 = 1 + \frac{\nu \omega_3^0 \omega_2^2}{\gamma_2 \gamma_3} = 1 + 0.1256 \omega_2^2, \qquad 7.05 \leq \omega_2 \leq 10$$

and

$$\phi_1 = 0$$

$$\phi_i = \phi_i^{(2)} \frac{1 - \gamma_i/\omega_2}{1 - \gamma_i/\omega_2^0} = \phi_i^{(2)} \frac{1 - \gamma_i/\omega_2}{1 - \gamma_i/7.05}, \qquad i = 2, 3$$

We also find, for the 1-simple wave,

$$\sigma_1 = 1 + \frac{\nu \omega_2^0 \omega_3^0 \omega_1^2}{\gamma_1 \gamma_2 \gamma_3} = 1 + 0.1771 \omega_1^2, \qquad 3.387 \leq \omega_1 \leq 5$$

and

$$\phi_i = \phi_i^0 \frac{1 - \gamma_i/\omega_3}{1 - \gamma_i/\omega_3^0} = \phi_i^0 \frac{1 - \gamma_i/\omega_3}{1 - \gamma_i/12.563}, \qquad i = 1, 2, 3$$

The results are put together in Fig. 3.18, which shows the physical plane portrait with appropriate ω values and the profiles of ϕ_i and n_i at $\tau = 6.0$.

Example 3.4

Alternating data problem ($m = 3$)

	\multicolumn{3}{c}{i}		
	1	*2*	*3*
K_i	5.0	7.5	10.0
c_i^f	0	0.04	0
c_i^0	0.04	0	0.02
ω_i^f	5.0	5.769	10.0
ω_i^0	4.060	7.5	8.978

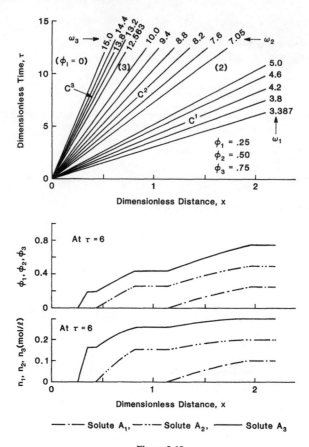

Figure 3.18

In this case a bed of adsorbent saturated uniformly with two solutes is being eluted by a stream carrying one solute of an intermediate adsorptivity. For the feed state, Eq. (3.6.16) is reduced to a linear equation for ω, whereas for the initial state it becomes a quadratic equation.

Comparing the values of ω_i^f and ω_i^0, we find that the solution includes a 3-simple wave, a 2-shock, and a 1-simple wave and these waves are intervened by new constant states. These constant states are determined by Eq. (3.11.6). Thus we obtain, for the 3-constant state,

$$c_1^{(3)} = c_1^0 \frac{1 - \gamma_1/\omega_1^f}{1 - \gamma_1/\omega_1^0} \frac{1 - \gamma_1/\omega_2^f}{1 - \gamma_1/\omega_2^0} = 0$$

$$c_2^{(3)} = c_2^f \frac{1 - \gamma_2/\omega_3^0}{1 - \gamma_2/\omega_3^f} = (0.04) \frac{1 - 7.5/8.978}{1 - 7.5/10} = 0.0263$$

$$c_3^{(3)} = c_3^0 \frac{1 - \gamma_3/\omega_1^f}{1 - \gamma_3/\omega_1^0} \frac{1 - \gamma_3/\omega_2^f}{1 - \gamma_3/\omega_2^0}$$

$$= (0.02)\frac{1 - 10/5}{1 - 10/4.06}\left(\frac{1 - 10/5.769}{1 - 10/7.5}\right) = 0.0301$$

and, for the 2-constant state,

$$c_1^{(2)} = c_1^0 \frac{1 - \gamma_1/\omega_1^f}{1 - \gamma_1/\omega_1^0} = 0$$

$$c_2^{(2)} = c_2^f \frac{1 - \gamma_2/\omega_2^0}{1 - \gamma_2/\omega_2^f} \frac{1 - \gamma_2/\omega_3^0}{1 - \gamma_2/\omega_3^f} = 0$$

$$c_3^{(2)} = c_3^0 \frac{1 - \gamma_3/\omega_1^f}{1 - \gamma_3/\omega_1^0} = (0.02)\frac{1 - 10/5}{1 - 10/4.06} = 0.0137$$

It is clearly seen that as we pass in the x-direction, the solute A_3 emerges through the 3-simple wave, the solute A_2 is exhausted across the 2-shock, and finally the solute A_1 emerges through the 1-simple wave.

For the 3-simple wave we use Eqs. (3.11.7) and (3.11.8) to obtain

$$\sigma_3 = 1 + \frac{\nu\omega_2^f \omega_3^2}{\gamma_2 \gamma_3} = 1 + 0.1153\omega_3^2, \qquad 8.978 \le \omega_3 \le 10$$

and Eq. (3.11.9) to have

$$c_1 = 0$$

$$c_i = c_i^{(3)}\frac{1 - \gamma_i/\omega_3}{1 - \gamma_i/\omega_3^0} = c_i^{(3)}\frac{1 - \gamma_i/\omega_3}{1 - \gamma_i/8.978}, \qquad i = 2, 3$$

for the concentrations borne on each of the characteristics C^3. Similarly, for the 1-simple wave we find

$$\sigma_1 = 1 + \frac{\nu\omega_3^0 \omega_1^2}{\gamma_1 \gamma_3} = 1 + 0.2693\omega_1^2, \qquad 4.06 \le \omega_1 \le 5$$

and

$$c_i = c_i^0 \frac{1 - \gamma_i/\omega_1}{1 - \gamma_i/\omega_1^0} = c_i^0 \frac{1 - \gamma_i/\omega_1}{1 - \gamma_i/4.06}, \qquad i = 1, 3$$

$$c_2 = 0$$

The 2-shock propagates in the direction given by Eqs. (3.11.8) and (3.11.10) as

$$\tilde{\sigma}_2 = 1 + \frac{\nu\omega_2^f \omega_3^0}{\gamma_3} = 1 + \frac{(1.5)(5.767)(8.978)}{10} = 8.766$$

The solution is constructed in the physical plane as shown in the upper part of Fig. 3.19, from which the profiles of c_i and n_i can be readily obtained. Those at $\tau = 10$ are presented in the lower part of Fig. 3.19. Clearly, emergence and exhaustion of solute species are alternating along the column.

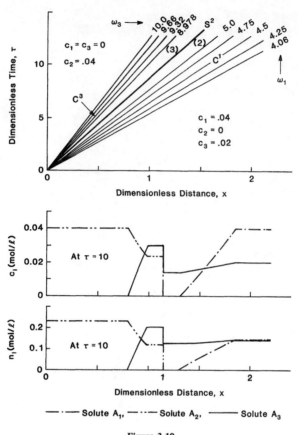

At τ = 10

At τ = 10

——·—— Solute A₁, ——··—— Solute A₂, ——— Solute A₃

Figure 3.19

Example 3.5

Alternating data problem ($m = 4$)

	i			
	1	*2*	*3*	*4*
K_i	5.0	10.0	12.5	20.0
c_i^f	0	0.06	0	0.04
c_i^0	0.04	0	0.02	0
ω_i^f	5.0	5.40	12.5	15.434
ω_i^0	4.075	10.0	10.575	20.0

The situation here is very similar to that in Example 3.4 except for the fact that the feed stream contains an additional solute which is more strongly adsorbed than any other solute. Here again Eq. (3.6.16) of Vol. 1 is reduced to a quadratic equation in ω for both the feed and initial states.

Based on the values of ω_i^f and ω_i^0, we immediately expect to have a 4-shock, a 3-simple wave, a 2-shock, and a 1-simple wave. In addition, the ω-values for the three new constant states are such that only A_2 would appear in the 4-constant state, whereas A_1 and A_4 would be absent from the 3-constant state, and finally the 1-constant state would contain A_3 alone. Hence, alternating along the column are not only the wave patterns but also the exhaustion and the emergence of solutes.

The nonvanishing concentrations in each of the three constant states can be determined from Eq. (3.11.5) or Eq. (3.11.6) to give

$$c_2^{(4)} = c_2^f \frac{1 - \gamma_2/\omega_4^0}{1 - \gamma_2/\omega_4^f} = (0.06)\frac{1 - 10/20}{1 - 10/15.434} = 0.0852$$

$$c_2^{(3)} = c_2^{(4)} \frac{1 - \gamma_2/\omega_3^0}{1 - \gamma_2/\omega_3^f} = (0.0852)\frac{1 - 10/10.575}{1 - 10/12.5} = 0.0232$$

$$c_3^{(2)} = c_3^0 \frac{1 - \gamma_3/\omega_1^f}{1 - \gamma_3/\omega_1^0} = (0.02)\frac{1 - 12.5/5}{1 - 12.5/4.075} = 0.0145$$

$$c_3^{(3)} = c_3^{(2)} \frac{1 - \gamma_3/\omega_2^f}{1 - \gamma_3/\omega_2^0} = (0.0145)\frac{1 - 12.5/5.4}{1 - 12.5/10} = 0.0763$$

The 4- and 2-shocks have their propagation directions given by Eqs. (3.11.8) and (3.11.10), i.e.,

$$\tilde{\sigma}_4 = 1 + \frac{\nu\omega_2^f\omega_4^f}{\gamma_2} = 1 + \frac{(1.5)(5.4)(15.434)}{10} = 13.502$$

$$\tilde{\sigma}_2 = 1 + \frac{\nu\omega_2^f\omega_3^0}{\gamma_3} = 1 + \frac{(1.5)(5.4)(10.575)}{12.5} = 7.853$$

For the 3-simple wave we have, from Eqs. (3.11.7) and (3.11.8),

$$\sigma_3 = 1 + \frac{\nu\omega_2^f\omega_3^2}{\gamma_2\gamma_3} = 1 + 0.0648\omega_3^2, \qquad 10.575 \le \omega_3 \le 12.5$$

and from Eq. (3.11.9),

$$c_i = c_i^{(3)} \frac{1 - \gamma_i/\omega_3}{1 - \gamma_i/\omega_3^0} = c_i^{(3)} \frac{1 - \gamma_i/\omega_3}{1 - \gamma_i/10.575}, \qquad i = 2, 3$$

with $c_1 = c_4 = 0$. For various values of ω_3, therefore, we can draw straight-line characteristics C^3 from the origin and calculate the concentrations borne on each of these C^3's. Similarly, for the 1-simple wave we obtain

$$\sigma_1 = 1 + \frac{\nu\omega_3^0\omega_1^2}{\gamma_1\gamma_3} = 1 + 0.2538\omega_1^2, \qquad 4.075 \le \omega_1 \le 5$$

and

$$c_i = c_i^0 \frac{1 - \gamma_i/\omega_1}{1 - \gamma_i/\omega_1^0} = c_i^0 \frac{1 - \gamma_i/\omega_1}{1 - \gamma_i/4.075}, \qquad i = 1, 3$$

with $c_2 = c_4 = 0$.

The physical plane portrait of the solution is shown in the upper part of Fig. 3.20, and from this diagram we construct the profiles of c_i and n_i at $\tau = 10$ as presented in the lower part of Fig. 3.20.

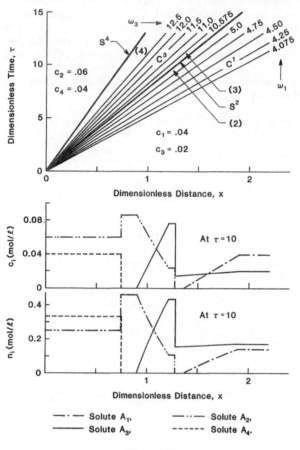

Figure 3.20

Exercises

3.12.1. For the numerical examples treated in Sec. 2.3, apply the formulae developed in Sec. 3.11 and construct the solutions as illustrated in this section. Compare your results with Figs. 2.5, 2.6, and 2.8.

3.12.2. Add another solute A_4 with $K_4 = 20$ liters/mol to the system associated with Examples 3.2 and 3.3. If $c_4^f = 0.05$ mol/liter and $c_4^0 = 0$ in Example 3.2, analyze the saturation problem as illustrated in this section. If $c_4^f = 0$ and $c_4^0 = 0.05$ mol/liter in Example 3.3, carry out the analysis of the elution problem as much in detail as you can and present your result in the form of the (x, τ)-diagram and (c_i, x)-diagram.

3.12.3. Suppose that the feed data and the initial data are interchanged in Example 3.4 and also in Example 3.5. For these two problems, follow the line of treatment illustrated in this section to construct the solutions and then compare your solutions with those of Examples 3.4 and 3.5.

REFERENCES

3.2. The treatment here follows on the line laid down in

R. COURANT and K. O. FRIEDRICHS, *Supersonic Flow and Shock Waves,* Interscience, New York, 1948.

For the proof of the existence and uniqueness theorem, see the development in

K. O. FRIEDRICHS, "Nonlinear hyperbolic differential equations for functions of two independent variables," *Amer. J. Math.* **70,** 555–589 (1948).

R. COURANT and P. D. LAX, "On nonlinear partial differential equations with two independent variables," *Comm. Pure Appl. Math.* **2,** 255–273 (1949).

P. D. LAX, "Nonlinear hyperbolic equations," *Comm. Pure Appl. Math.* **6,** 231–258 (1953).

A. JEFFREY, *Quasilinear Hyperbolic Systems and Waves,* Pitman, London, 1976.

Conservativeness of an arbitrary system of quasilinear equations is discussed in

B. L. Rozdestvenskii, "On systems of quasilinear equations," *Dokl. Akad. Nauk SSSR* **115,** 454–457 (1957); English translation in *Amer. Math. Soc. Transl.,* Ser 2, **42,** 13–17 (1964).

B. L. ROZDESTVENSKII, "Conservativeness of systems of quasilinear equations," *Uspehi Mat. Nauk (N.S.)* **14,** 217–218 (1959); English translation in *Amer. Math. Soc. Transl.,* Ser. 2, **42,** 37–39 (1964).

3.3. Generalization of Riemann invariants and simple wave theory was first established in the celebrated paper of

P. D. LAX, "Hyperbolic systems of conservation laws: II," *Comm. Pure Appl. Math.* **10,** 537–566 (1957).

For further discussions, see

A. JEFFREY and T. TANIUTI, *Nonlinear Wave Propagation,* Academic Press, New York, 1964.

J. SMOLLER, *Shock Waves and Reaction-Diffusion Equations,* Springer-Verlag, New York, 1983.

See also Jeffrey's book (1976) listed above for Sec. 3.2.

The equations of nonisentropic flow are found, for example, in the book by Courant and Friedrichs (1948) cited above for Sec. 3.2.

The generalized Riemann invariants for a system with n independent variables are treated in

A. M. GRUNDLAND, "Riemann invariants," in *Wave Phenomenon: Modern Theory and Applications,* edited by C. Rogers and T. B. Moodie, North-Holland, Amsterdam, 1984, pp. 123–152.

3.4–3.12. The basic references for most of these nine sections are

H.-K. RHEE, "Studies on the theory of chromatography: Part III. Chromatography of many solutes," Ph.D. thesis, University of Minnesota, 1968.

H.-K.RHEE, R. ARIS, and N. R. AMUNDSON, "On the theory of multicomponent chromatography," *Philos. Trans. Roy. Soc. London,* **A267,** 419–455 (1970).

H.-K. RHEE, "Equilibrium theory of multicomponent chromatography," in *Percolation Processes: Theory and Applications,* edited by A. E. Rodrigues and D. Tondeur, Sijthoff and Noordhoff, Alphen aan den Rijn, The Netherlands, 1981, pp. 285–328.

A similar approach, virtually parallel to this line, has been applied to ion-exchange systems by

F. HELFFERICH and G. KLEIN, *Multicomponent Chromatrography: Theory of Interference,* Marcel Dekker, New York, 1970.

3.4. Earlier works recognizing the fundamental differential equations are found in

D. DEVAULT, "The theory of chromatography," *J. Amer. Chem. Soc.* **65,** 532–540 (1943).

G. G. BAYLE and A. KLINKENBERG, "On the theory of adsorption chromatography for liquid mixtures," *Recl. Trav. Chim. Pays-Bas Belg.* **73,** 1037–1057 (1954).

3.5. For the constant-separation-factor equilibrium relations, see

D. TONDEUR and G. KLEIN, "Multicomponent ion exchange in fixed beds: constant-separation-factor equilibrium," *Indust. Engrg. Chem. Fund.* **6,** 351–361 (1967).
See also the book by Helfferich and Klein (1970) cited above.

3.6. The *h*-transformation takes the role of the starting point in the book by Helfferich and Klein (1970) cited above.

3.9. This definition of a k-shock is due to Lax (1957). See his paper cited above for Sec. 3.3.

3.10. For the formulation of the law of mass conservation, see Sec. 54 of the book by Courant and Friedrichs (1948) cited above for Sec. 3.2.

3.11. The saturation problem with an arbitrary isotherm is discussed to some extent in the paper by De Vault (1943) cited above.

3.12. The basic profile-patterns in ion exchange systems with alternating data are to be found in

G. KLEIN, D. TONDEUR, and T VERMEULEN, "Multicomponent ion exchange: General properties of equilibrium systems," *Ind. Eng. Chem. Fundam.* **6,** 339-351 (1967).

Wave Interactions in Multicomponent Chromatography

4

The theoretical development of multicomponent chromatography with the Langmuir isotherm is extended here to treat problems with piecewise constant initial and/or feed data. Since we are concerned with hyperbolic systems of equations, the solution may be composed of analytically different portions in different domains of the physical plane. Thus it can be constructed by pieces, starting independently from each point of discontinuity. As time goes on, however, the wave solutions developing from different points of discontinuity are bound to interact with each other. On the other hand, in the hodograph space the image of the solution as a whole is readily established by virtue of the linearity of the curve Γ^k.

By combining the image in the hodograph space and the physical plane portrait of the solution, we can thoroughly examine the fundamental features of wave interaction. Furthermore, the existence of explicit expressions for the Riemann invariants and the characteristic parameters enables us to analyze completely all the interactions that can take place. Application is illustrated by treating the problem of multicomponent separation by the chromatographic cycle and that by displacement development.

In Sec. 4.1 we introduce the piecewise constant data problem and wave interactions. We then continue to identify the pairs of interacting waves and establish the fundamental principles of wave interaction. Interactions between waves of the same family are treated in Sec. 4.2. We consider two cases: one with two shocks and the other with a simple wave and a shock, and develop analytic expressions describing the interactions. The decay of shocks is also discussed. In Sec. 4.3 we take up the interactions between waves of different families and present analytical descriptions of three different interactions: interactions between two shocks, between a simple wave and a shock, and between two simple waves.

The following three sections are concerned with the application of the theoretical development of the previous sections. For this we consider two common chromatographic processes that are practiced to separate multicomponent mixtures into pure components. In Sec. 4.4 the chromatographic cycle for a three-solute mixture is analyzed in a general context and illustrated with a numerical example. The treatment can readily be extended to a system with more than three solutes. In Sec. 4.5 we deal with displacement development and present explicit formulas for the critical value of the developer concentration, the concentration plateau of each pure solute band, and the bandwidth. This treatment is illustrated in Sec 4.6 with a numerical example involving three solutes plus the developer. For different levels of the developer concentration the process is fully analyzed to demonstrate the validity of the formulas and to examine the effect of the developer concentration on the column length required for complete separation.

4.1 Piecewise Constant Data and Patterns of Interaction

As an extension of the development for multicomponent chromatography in Chapter 3, we consider a class of problems associated with piecewise constant initial and/or feed data with an arbitrary number of discontinuities. This class of problems is of central importance in the practice of chromatographic processes, as we shall see while treating applications in the latter part of this chapter.

We recall that both the fundamental differential equations (3.4.19) and the compatibility relations (3.9.5) are expressed in terms of the dependent variables and hence the Riemann invariants as well as the characteristic parameters are determined in terms of the dependent variables only. Thus the image of the solution to a Riemann problem is completely determined by the two constant states and remains the same no matter where the initial discontinuity may be located. In other words, the fundamental structure of the solution is entirely dependent on the two constant states and remains unaffected under the translation of the point of discontinuity. Furthermore, we note that the solution to a hyperbolic system of equations is not necessarily analytic and may be constructed piece by piece (see the latter part of Sec. 1.2). Therefore, the simple wave theory as well as the theory of shocks can be applied to determine the solution of a problem with piecewise constant data at least in the neighborhood of the initial curve. Indeed, we can start out by constructing separately the wave solutions centered at each point of discontinuity.

As the waves propagate, we expect to see pairs of waves confronting and eventually interfering with each other. The two waves under mutual interference originate obviously from two different points of discontinuity, and hence the solution at this stage is influenced by two different sets of data. As before in Sec. 2.5, such a phenomenon will be called an *interaction between waves*. Thus in determining the solution to a piecewise constant data problem it is essential to treat the wave interactions.

Here we are concerned with identifying the pairs of interacting waves and with establishing the fundamental principles of wave interaction. Since these are very similar to those for the case of pairs of equations (see Sec. 2.5), we will discuss these principles only briefly, and the actual analysis of wave interactions will be treated in the next two sections, followed by three more sections dealing with applications.

First, let us consider pairs of waves belonging to the same family and traveling side by side. In the case of two k-simple waves the pair will never collide because the boundary characteristics C^k confronting each other are necessarily parallel. However, if one or both waves are shocks, we must have Eq. (3.9.19) satisfied, and this implies that the confronting waves will eventually meet each other and interact.

When the two waves, traveling side by side, are of different families, we use the inequalities (3.8.7), (3.9.20), and (3.9.21) to discern various cases. Thus if a j-wave (either simple wave or shock) travels behind a k-simple wave or a k-shock, where $j < k$, the two will interact with each other. Similarly, we see that if a j-wave travels ahead of a k-simple wave or a k-shock, where $k < j$, the two waves are bound to meet each other and interact. Otherwise, the two waves would travel away from each other. The argument above then leads us to

RULE 4.1   [Pairs of Interacting Waves]:

1. Two k-shocks or a k-simple wave and a k-shock necessarily interact with each other.
2. A k-wave (either a simple wave or a shock) necessarily interacts with a j-wave, traveling behind it if $j < k$, or traveling ahead of it if $j > k$.

Corollary [Noninteracting Waves]:

1. Two simple waves of the same family do not interact with each other.
2. A k-wave does not interact with a j-wave, traveling behind it if $j > k$, or traveling ahead of it if $j < k$.

It seems appropriate to make a remark at this point that Rule 4.1 has general significance, but the discussions to follow are valid only for the case of multicomponent chromatography with the Langmuir isotherm, for which the Γ's are all straight lines. Wave interactions in a general context present a much more complex problem which is yet to be resolved.

Suppose now that two waves are interacting and picture the situation in the hodograph space. If the two waves belong to the same family, say the kth one, the images fall on a single Γ^k along which ω_k is the only variable. This is illustrated in Fig. 4.1(a) and (b). In Fig. 4.1(a) we have two shocks $L \to P$ and $P \to R$ and when they meet each other, we are taking up a new Riemann problem with two constant states L and R. But these states are directly connected by a single k-shock. This implies that two shocks of the same family are superposed upon interaction and the constant state P in between disappears. This is known as the *confluence* or *union* of two shocks. In Fig. 4.1(b) we see the k-simple wave $Q \to P$ catching up with the k-shock $P \to R$ from behind. Throughout the course of interaction all the characteristic parameters except for ω_k remain fixed, so the image always lies on the Γ^k. Hence if the constant state P disappears at the beginning of the interaction, the state on the left-hand side of the shock will recede along the Γ^k toward R. When this state is given by the point E, the solution would consist of a continuous variation of state $Q \to E$ and a discontinuous change $E \to R$. The portion $E \to P$ of the k-simple wave has been canceled by the portion $P \to E$ of the k-shock. This is what we call the *cancellation* between a simple wave and a shock or the *decay* of a shock. In the limit as the point E approaches to the point R; that is, as the shock becomes weak, the shock speed $1/\bar{\sigma}_k$ tends to approach the characteristic speed $1/\sigma_k$ of the state R. It follows that the interaction will go on indefinitely. If the state R falls to the right of the state Q, the interaction will terminate when E reaches Q. The simple wave is then completely canceled and the resulting shock $Q \to R$ resumes a constant speed.

If the two waves in interaction belong to different families, the situation ap-

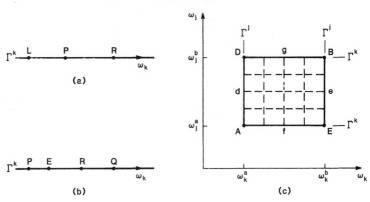

Figure 4.1

pears to be much more complicated. Let us suppose that a j-wave is interacting with a k-wave traveling ahead of it. This implies that we have $j < k$. We note that ω_j varies through the j-wave and ω_k through the k-wave while the other $(m - 2)$ parameters remain fixed throughout. Thus the images of the two waves will fall on a plane parallel to the plane composed of the ω_j-axis and the ω_k-axis in the space $\Omega(m)$. In case of two shocks, therefore, we would have $A \rightarrow D$ for the j-shock and $D \rightarrow B$ for the k-shock as shown in Fig. 4.1(c). When the two meet, the constant state D in between disappears and we are just looking at a new Riemann problem with constant states A and B. Clearly, the solution to this problem consists of the k-shock $A \rightarrow E$ trailing the j-shock $E \rightarrow B$ and the new constant state E in between. Now it is immediately apparent that the two shocks are transmitted through each other to generate a pair of new shocks of their respective families. This phenomenon is known as the *transmission* between waves.

In case we have a j-simple wave catching up to a k-shock, their images will be given by $D \rightarrow A$ and $A \rightarrow E$, respectively. At first, the head of the j-simple wave led by the state A will pass through the k-shock and we know that a new simple wave will develop to the right of the k-shock because this is the region adjacent to the region of the constant state E. But ω_k is the only variable across the k-shock and thus the new simple wave must have its image fall on segment EB of Fig. 4.1(c), which is a Γ^j. This implies that the new simple wave is a j-simple wave. At a particular stage of the interaction the solution would consist of a continuous variation $D \rightarrow d$ (rear part of the original simple wave), a discontinuous change $d \rightarrow e$ (k-shock), and another continuous variation $e \rightarrow E$ (leading portion of the new simple wave). As the interaction goes on, therefore, the image of the shock recedes continuously from $A \rightarrow E$, to $d \rightarrow e$, and ultimately to $D \rightarrow B$. When the interaction is over, the solution is given by the k-shock $D \rightarrow B$ trailing the j-simple wave $B \rightarrow E$ and the new constant state B in between. Here again the two waves are transmitted through each other. We shall distinguish the simple waves before and after transmission by calling them, respectively, the original and the transmitted simple waves.

If we have a j-simple wave catching up to a k-simple wave, their images will fall on $B \rightarrow E$ (a Γ^j) and $E \rightarrow A$ (a Γ^k) as marked in Fig. 4.1(c). By the same argument as above, the two waves are transmitted through each other while interacting and emerge as two new simple waves of their respective families. Here the interaction takes place in a finite region in the (x, τ)-plane called the *penetration region*. In this region we have a non-simple wave, but still a continuous, solution. A particular C^j-passing through the penetration region will have its image on the segment $e \rightarrow d$ (a Γ^k) and a particular C^k on the segment $f \rightarrow g$ (a Γ^j) as labeled in Fig. 4.1(c). It is clear that the transmitted simple waves will have their images fall on $B \rightarrow D$ (k-simple wave) and $D \rightarrow A$ (j-simple wave), so the two simple waves will be separated by a region of new constant state D.

These descriptions on the nature of wave interactions may be summarized as

RULE 4.2 [Fundamental Principles of Interaction]:

1. *Confluence (or union) of two shocks:* Two shocks of the same family are instantly superposed upon interaction.
2. *Cancellation between a simple wave and a shock:* A simple wave and a shock of the same family are gradually canceled by each other while interacting.
3. *Transmission between two waves:* Two waves of different families are transmitted through each other if they interact.

These are in complete analogy with those for the case of two solute chromatography discussed in Sec. 2.5 and hence we naturally expect that the interaction analysis may be accomplished by following the track laid down in Secs. 2.6 and 2.7.

When two waves are engaged in an interaction, the state of concentrations may vary along a C^k or on either side of an S^k and thus the characteristics and/or the shock paths may no longer remain straight, or they may be refracted across a shock path. These phenomena are worth examining in more detail before indulging in the interaction analysis.

For a C^k Eq. (3.8.2) gives

$$\sigma_k = 1 + \nu\omega_k \frac{\omega_j}{\gamma_j} \prod_{i=1,\,j}^{m} \frac{\omega_i}{\gamma_i} \qquad (4.1.1)$$

If this C^k is to pass through an S^j, $k < j$, we see that ω_j must jump to a higher value while the other parameters are fixed. Hence the C^k will be refracted upward in the (x, τ)-plane. If the C^k is to pass through a j-simple wave region, $k < j$, ω_j continuously decreases along the C^k and so does the slope σ_k.

For an S^k we have, from Eq. (3.9.18),

$$\tilde{\sigma}_k = 1 + \nu p_k \omega_k^l \omega_k^r \qquad (4.1.2)$$

$$= 1 + \nu\omega_k^r \frac{\omega_j^l}{\gamma_j} \prod_{i=1,\,j}^{m} \frac{\omega_i^l}{\gamma_i} \qquad (4.1.3)$$

If this k-shock is interacting with another k-wave (either simple wave or shock), only one of ω_k^l and ω_k^r changes while P_k remains invariable and thus $\tilde{\sigma}_k$ changes accordingly. If the k-shock is in interaction with a j-wave, $j < k$, then ω_j^l is the only variable in Eq. (4.1.3), so that the slope $\tilde{\sigma}_k$ changes in the same direction as ω_j^l.

Now we recall from Rules 3.1 and 3.3 that the concentration c_m of the most strongly adsorbed solute increases in the x direction through a k-simple wave and decreases from the left to the right across a k-shock. This is true for any k from one to m. We also note that this variation of c_m through a k-wave is in the opposite direction to the change in the parameter ω_k. Consequently, by interpreting σ_k and $\tilde{\sigma}_k$ as the reciprocal propagation speeds of a disturbance and a shock, respectively, we can summarize the foregoing findings as

RULE 4.3:

1. A disturbance in the concentration field accelerates (or decelerates) if c_m increases (or decreases) as it propagates along a characteristic.

2. A discontinuity in the concentration field accelerates (or decelerates) if c_m^l and/or c_m^r increase (or decrease) as it propagates along a shock path.

It is not difficult to see that Rule 4.3 will read exactly the same with δ or Θ in place of c_m.

Treating the problems with piecewise constant initial data is certainly one way to approach the general problem with arbitrary initial data. Indeed, it can be shown that arbitrary initial data may be approximated by a sequence of piecewise constant initial data (see e.g., Zhang and Guo, 1965). Considering the large-time behavior of solutions, we can imagine that at a certain stage all the interactions between waves of different families (i.e., transmissions) are completed, and afterward the interactions are all between waves of the same family. Thus the waves tend to cancel or unite and attain a simple form as time goes to infinity. It is not difficult to imagine that the asymptotic behavior of these waves will be determined by the solution of the Riemann problem with two constant states corresponding to the states specified at $x = -\infty$ and at $x = +\infty$, respectively.

For genuinely nonlinear (or linearly degenerate) systems of equations, it was shown by DiPerna (1977) that if the initial data are constant outside a bounded interval (i.e., if the initial data have compact support), the solution decays to zero in the total variation norm as $t \to \infty$. In several steps Liu (1977a, b, 1981) has discussed the large-time behavior of solutions of initial value problems. He starts out with the system of equations

$$\frac{\partial \mathbf{u}}{\partial t} + \frac{\partial}{\partial x} \mathbf{f}(\mathbf{u}) = \mathbf{0} \qquad (4.1.4)$$

and the initial data of the form

$$\mathbf{u}(x, 0) = \begin{cases} \mathbf{u}^l & \text{for } x < -N \\ \mathbf{u}_0(x) & \text{for } -N \le x \le N \\ \mathbf{u}^r & \text{for } x > N \end{cases} \qquad (4.1.5)$$

where $\mathbf{u}_0(x)$ is a measurable function, $\mathbf{u}^l$ and $\mathbf{u}^r$ are two constant states, and N is a positive constant. For this problem it is shown that the solution converges to that of the corresponding Riemann problem with the initial data

$$\mathbf{u}(x, 0) = \begin{cases} \mathbf{u}^l & \text{for } x < 0 \\ u^r & \text{for } x > 0 \end{cases} \qquad (4.1.6)$$

at algebraic rates in the norm of total variation. Next he considers more general initial data

$$\mathbf{u}(x, 0) = \mathbf{u}_0(x), \qquad -\infty < x < \infty \qquad (4.1.7)$$

where $\mathbf{u}_0(x)$ is a bounded function so that the limiting values of $\mathbf{u}_0$ at $x = \pm\infty$ exist:

$$\mathbf{u}^l = \mathbf{u}_0(-\infty), \qquad \mathbf{u}^r = \mathbf{u}_0(+\infty) \qquad (4.1.8)$$

and establishes that the solution tends to elementary waves (i.e., simple waves and shocks in case of a genuinely nonlinear system) determined by the values $\mathbf{u}^l$ and $\mathbf{u}^r$. This implies that the solution of Eqs. (4.1.4) and (4.1.7) tends to the solution of the corresponding Riemann problem with the constant states $\mathbf{u}^l$ and $\mathbf{u}^r$.

4.2 Interactions between Waves of the Same Family

Here we have two different cases: one with two shocks and the other with one simple wave and one shock. In either case the waves belong to the same family and their images fall on a single Γ in the space $\Omega(m)$ as shown in Fig. 4.1(a) and (b). It then follows that there exists a one-parameter representation of the solution and thus the analysis should be straightforward.

Confluence of two shocks

Suppose that we have two k-shocks, one originating from the origin O and the other from the point $A(0, \eta_1)$. The situation is depicted in Fig. 4.2 and its image would be just as shown in Fig. 4.1(a). Clearly, ω_k is the only variable parameter, so from Eq. (3.9.8) we can write

$$\tilde{\sigma}_{k1} = 1 + \nu p_k \omega_k^{\varrho} \omega_k^r \qquad (4.2.1)$$

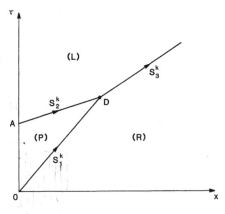

Figure 4.2

$$\bar{\sigma}_{k2} = 1 + vp_k \omega_k^l \omega_k^p \qquad (4.2.2)$$

where the lowercase superscript is used to indicate that the value is for the state designated by the corresponding capital letter.

Since $\omega_k^l < \omega_k^p < \omega_k^r$ by the entropy condition, the two shocks will collide at point $D(x_0, \tau_0)$ given by

$$x_0 = \frac{\eta_1}{\bar{\sigma}_{k1} - \bar{\sigma}_{k2}} = \frac{\eta_1}{vp_k \omega_k^p (\omega_k^r - \omega_k^l)}$$

$$\tau_0 = \bar{\sigma}_{k1} x_0 = \frac{1 + vp_k \omega_k^p \omega_k^r}{vp_k \omega_k^p (\omega_k^r - \omega_k^l)} \eta_1 \qquad (4.2.3)$$

Upon interaction, the two shocks are simply united to give a new shock with the states L and R on both sides. Therefore, the new shock propagates in the direction

$$\bar{\sigma}_{k3} = 1 + vp_k \omega_k^l \omega_k^r \qquad (4.2.4)$$

which takes an intermediate value between $\bar{\sigma}_{k1}$ and $\bar{\sigma}_{k2}$. In this case the interaction is an instantaneous process.

Cancellation between a simple wave and a shock

We can think of two situations, a k-simple wave trailing a k-shock and a k-shock overtaking a k-simple wave. We shall consider the former and assume that the simple wave is not centered but based on a finite interval of the τ-axis. This is equivalent to saying that the feed concentrations are distributed over that interval but they are subject to a one-parameter representation; that is, if we have $c_i = c_i^f(\eta)$ at $x = 0$, the parameter η runs along a Γ^k and thus we have $\eta = \eta(\omega_k)$. This is illustrated in Fig. 4.3 and the image in $\Omega(m)$ can be found in Fig. 4.1(b).

Now the initial and feed data may be expressed as

$$\text{at } \tau = 0: \quad \omega_k = \omega_k^r$$

$$\text{at } x = 0: \quad \omega_k = \begin{cases} \omega_k^p = \omega_k^f(\eta_1) & \text{for } 0 < \eta \leq \eta_1 \\ \omega_k^f(\eta) & \text{for } \eta_1 < \eta < \eta_2 \\ \omega_k^q = \omega_k^f(\eta_2) & \text{for } \eta \geq \eta_2 \end{cases} \qquad (4.2.5)$$

with the inequality relations

$$\omega_k^p < \omega_k^r < \omega_k^q \qquad (4.2.6)$$

The concentration values can be directly reproduced by using the values of ω_k in Eq. (3.7.9).

The k-simple wave is represented by a family of straight-line characteristics C^k emanating from points on AB of the τ axis, and to each of these there corresponds the equation

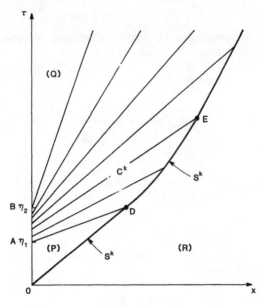

Figure 4.3

$$\tau - \eta(\omega_k) = \sigma_k(\omega_k)x = (1 + \nu p_k \omega_k^2)x \tag{4.2.7}$$

where $\eta(\omega_k)$ is the inverse function of $\omega_k = \omega_k^f(\eta)$.

The leading boundary of the simple wave region is the line AD which is a C^k carrying the state P and thus we have the equation

$$\tau - \eta_1 = \{1 + \nu p_k(\omega_k^p)^2\}x$$

Since the shock path OD has the equation

$$\tau = (1 + \nu p_k \omega_k^p \omega_k^r)x$$

we find that the two waves begin to interact from the point $D(x_0, \tau_0)$, where

$$x_0 = \frac{\eta_1}{\nu p_k \omega_k^p (\omega_k^r - \omega_k^p)}$$

$$\tau_0 = \frac{1 + \eta p_k \omega_k^p \omega_k^r}{\nu p_k \omega_k^p (\omega_k^r - \omega_k^p)} \eta_1 \tag{4.2.8}$$

As the interaction goes on, the two waves gradually cancel each other and hence ω_k^l increases (or c_m^l decreases) with time while ω_k^r remains fixed. This implies that the shock decelerates continuously as depicted in Fig. 4.3. Taking an arbitrary point $E(x, \tau)$ on the shock path and assigning ω_k for the state on the left-hand side of the shock, we can draw a straight line C^k of slope $1 + \nu p_k \omega_k \omega_k^r$ from E to reach a point on AB. The intercept is given by $\eta(\omega_k)$ and thus, at point E, we can write the two equations

$$\tau - \eta(\omega_k) = \sigma_k x = (1 + \nu p_k \omega_k^2)x \qquad (4.2.9)$$

$$\frac{d\tau}{dx} = \tilde{\sigma}_k = 1 + \nu p_k \omega_k \omega_k^r \qquad (4.2.10)$$

We note that both p_k and ω_k^r remain constant and regard the pair of equations as a parametric representation of the shock path beyond the point D. Considering both x and τ as functions of ω_k, we differentiate Eq. (4.2.9) with respect to ω_k, rewrite Eq. (4.2.10) in the form $d\tau/d\omega_k = \tilde{\sigma}_k \, dx/d\omega_k$, and then combine the resulting equations to eliminate $d\tau/d\omega_k$. This gives

$$(\tilde{\sigma}_k - \sigma_k)\frac{dx}{d\omega_k} - \frac{d\sigma_k}{d\omega_k}x = \frac{d\eta}{d\omega_k} \qquad (4.2.11)$$

or

$$(\omega_k^r - \omega_k)\frac{dx}{d\omega_k} - 2x = \frac{1}{\nu p_k \omega_k}\frac{d\eta}{d\omega_k} \qquad (4.2.12)$$

The initial condition is

$$x = x_0 \quad \text{at} \quad \omega_k = \omega_k^p \qquad (4.2.13)$$

Multiplying throughout by the factor $(\omega_k^r - \omega_k)$, we can put Eq. (4.2.12) into an exact differential form

$$\frac{d}{d\omega_k}\{(\omega_k^r - \omega_k)^2 x\} = \frac{1}{\nu p_k}\left(\frac{\omega_k^r}{\omega_k} - 1\right)\frac{d\eta}{d\omega_k}$$

and integrate this with the initial condition to have

$$x(\omega_k) = x_0\left(\frac{\omega_k^r - \omega_k^p}{\omega_k^r - \omega_k}\right)^2 + \frac{1}{\nu p_k(\omega_k^r - \omega_k)^2}\int_{\omega_k^p}^{\omega_k}\left(\frac{\omega_k^r}{\omega_k} - 1\right)\frac{d\eta}{d\omega_k}d\omega_k \qquad (4.2.14)$$

This may be substituted into Eq. (4.2.9) to give τ as a function of ω_k. The set (x, τ) so determined is the parametric representation of the shock path beyond the point D.

Since we are concerned here with bounded, smooth feed data, the integral in the last term of Eq. (4.2.14) remains finite, so $x(\omega_k) \to \infty$ as $\omega_k \to \omega_k^r$. This implies that the interaction will go on indefinitely, as we observed in the preceding section. If $\omega_k^q < \omega_k^r$, however, the interaction will terminate when $\omega_k = \omega_k^q$ and the k-simple wave will be completely canceled to give a direct connection by a k-shock between the states Q and R.

In cases when the k-simple wave is centered at the point A, we have $\eta = \eta_1 = $ constant and $d\eta/d\omega_k = 0$, and thus Eq. (4.2.14) yields

$$x(\omega_k) = x_0\left(\frac{\omega_k^r - \omega_k^p}{\omega_k^r - \omega_k}\right)^2 = \frac{\eta_1(\omega_k^r/\omega_k^p - 1)}{\nu p_k(\omega_k^r - \omega_k)^2} \qquad (4.2.15)$$

Rearranging this into the form

$$\omega_k = \omega_k^r - \sqrt{\frac{\eta_1(\omega_k^r/\omega_k^p - 1)}{\nu p_k x}}$$

and substituting into Eq. (4.2.9) with $\eta = \eta_1$, we obtain

$$\sqrt{\tau - \eta_1 - x} = \omega_k^r \sqrt{\nu p_k x} - \sqrt{\eta_1\left(\frac{\omega_k^r}{\omega_k^\ell} - 1\right)} \qquad (4.2.16)$$

Thus the shock path beyond the point D is a part of a parabola.

If the k-simple wave is not centered, we can translate the origin to the point D to make $x_0 \equiv 0$ and $\eta_1 \equiv 0$. Introducing this in Eq. (4.2.14) and applying integration by parts, we obtain the equation

$$x(\omega_k) = \frac{1}{\nu p_k(\omega_k^r - \omega_k)}\left\{\frac{\overline{\eta}}{\omega_k} + \frac{1}{1 - \omega_k/\omega_k^r}\int_{\omega_k^\ell}^{\omega_k}\frac{\overline{\eta}(\omega_k)}{\omega_k^2}d\omega_k\right\} \qquad (4.2.17)$$

in which $\overline{\eta}$ represents the adjusted form of η; i.e.,

$$\overline{\eta}(\omega_k) + \tau_0 - \eta(\omega_k) = \sigma_k(\omega_k)x_0 \qquad (4.2.18)$$

If the function $\overline{\eta}(\omega_k)$ is specified by numerical data, calculation of the integral of Eq. (4.2.27) should be straightforward. We shall see this in the latter part of this chapter.

We also notice that $|\omega_k^r - \omega_k|$ is a measure of the strength of the k-shock and decreases continuously as the two waves cancel each other (i.e., as the interaction goes on). Since the integral in the last term of Eq. (4.2.14) remains finite, it is immediately apparent from Eq. (4.2.14) that, for large time, we have

$$|\omega_k^r - \omega_k| \sim \frac{M}{\sqrt{x}} \qquad (4.2.19)$$

where M is a constant determined by the constant states P and R and the feed data. Therefore, the shock decays like $1/\sqrt{x}$ for large values of x. This feature of the large-time behavior of solutions was first conjectured by Lax (1963) and proved later by Liu (1977c). In fact, these authors are concerned with the system (4.1.4) and the initial data that take a common constant value $\mathbf{u}_0$ outside a finite interval:

$$\mathbf{u}(x, 0) = \mathbf{u}_0 \qquad \text{for } |x| > x_0 \qquad (4.2.20)$$

and show that the asymptotic form of the solution consists of m distinct N-waves each propagating at one of the characteristic speeds $(1/\sigma_k)$ of the state $\mathbf{u}_0$. It is further shown that the solution for large time approaches the constant state $\mathbf{u}_0$ like $1/\sqrt{t}$. We note that the variable t here takes the role of our x. The description above of the asymptotic behavior of solutions is entirely analogous to those for single equations (see Secs. 6.3, 6.6, and 7.9 of Vol. 1) and pairs of equations (see Sec. 2.6).

Exercises

4.2.1. Suppose that a k-simple wave is based on the interval $0 \leq \xi \leq \xi_1$, of the x-axis so that we have a relationship $\xi = \xi(\omega_k)$ for $0 \leq \xi \leq \xi_1$. If a k-shock starts from the point $(0, \eta_1)$ on the τ-axis and trails the k-simple wave with a constant state in between, determine the shock path when the two waves are interacting (see Fig. 2.16). If the k-simple wave is centered at the origin, find an equation relating τ and x along the shock path to show that it is a part of a parabola. Also, discuss the decay of the shock for large time.

4.2.2. If the k-simple wave is centered at the point A in Fig. 4.3 and if $\omega_k^\ell < \omega_k^q < \omega_k^r$, sketch the solution in the (x, τ)-plane and the profiles of the concentration c_m of the most strongly adsorbed solute at successive times. In particular, find the time when the interaction is completed and compare the propagation speeds of the k-shock before and after the interaction. Repeat the same for the case of Exercise 4.2.1.

4.3 Interactions between Waves of Different Families

Under this category we expect to have three different cases; the first case is with two shocks, the second with one simple wave and one shock, and finally, the third with two simple waves. Since there is a unique characteristic parameter that varies across each wave of distinct family, we have now a two-parameter system as a whole. We note, however, that as we follow one wave as it passes through the other, only one of the two parameters varies along the passage and this feature makes the interactions amenable to analytic treatment.

We shall standardize the situation in such a way that a j-wave catches up to a k-wave and these interact with each other, where $j < k$, and the ranges of the two parameters ω_j and ω_k are given by

$$\omega_j^a \le \omega_j \le \omega_j^b < \omega_k^a \le \omega_k \le \omega_k^b \tag{4.3.1}$$

Thus the four constant states [e.g., the states A, B, D, and E of Fig. 4.1(c)] are characterized by combinations of ω_j and ω_k.

Transmission between two shocks

Here a j-shock collides with a k-shock from behind and the two shocks are instantly transmitted across each other to produce two transmitted shocks of their respective families. This is illustrated in Fig. 4.4, in which we also see the image in the space $\Omega(m)$. In this case the characteristic parameters for each constant state are given by

$$\begin{aligned} L(\omega_j^a, \omega_k^a), \qquad Q(\omega_j^a, \omega_k^b) \\ P(\omega_j^b, \omega_k^a), \qquad R(\omega_j^b, \omega_k^b) \end{aligned} \tag{4.3.2}$$

Since ω_j and ω_k are the only variable parameters, we see that the quantity

$$q_{jk} = \frac{1}{\gamma_j \gamma_k} \prod_{i=1, j, k}^{m} \frac{\omega_i}{\gamma_i} = \frac{\gamma_j}{\omega_j} p_k \tag{4.3.3}$$

is an invariant here.

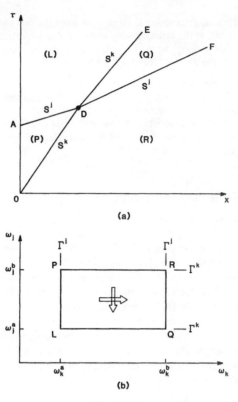

Figure 4.4

Up until the point of collision D, the two shocks propagate along the paths $AD(S^j)$ and $OD(S^k)$ having slopes

$$\tilde{\sigma}_j = 1 + \nu q_{jk} \omega_j^a \omega_j^b \omega_k^a \qquad (4.3.4)$$

and

$$\tilde{\sigma}_k = 1 + \nu q_{jk} \omega_j^b \omega_k^a \omega_k^b \qquad (4.3.5)$$

respectively. But since $\omega_j < \omega_k$, the two lines are to intersect. Upon interaction the constant state P disappears and the two shocks are transmitted across each other. The interaction is instantaneous and generates a new constant state Q, which is connected to the constant state L by the transmitted k-shock and to the constant state R by the transmitted j-shock.

The transmitted shocks propagate along the straight paths $DE(S^k)$ and $DF(S^j)$, whose slopes are given by

$$\tilde{\sigma}_k' = 1 + \nu q_{jk} \omega_j^a \omega_k^a \omega_k^b \qquad (4.3.6)$$

and

$$\bar{\sigma}_j' = 1 + \nu q_{jk} \omega_j^a \omega_j^b \omega_k^b \qquad (4.3.7)$$

respectively. Direct comparison shows that $\bar{\sigma}_j < \bar{\sigma}_j'$ and $\bar{\sigma}_k > \bar{\sigma}_k'$. Hence the j-shock decelerates while the k-shock accelerates upon interaction. The constant state Q can be determined by using Eq. (3.7.9).

Transmission between a simple wave and a shock

Just as in Sec. 4.2, we shall assume that the j-simple wave is based on the feed data distributed over a finite interval of the τ-axis. This implies that over the interval the ordinate η can be expressed as a function of ω_j. Although it appears somewhat artificial, this arrangement gives a wider range of application and, in fact, we shall find it necessary when we deal with applications in the following sections.

If a k-shock starts from the origin, where $j < k$, the head of the simple wave will collide with this shock and begin an interaction. From the discussion of Sec. 4.1 we understand that the two waves will be transmitted through each other to produce two transmitted waves of their respective families and a new constant state separating the two waves. The situation is illustrated in Fig. 4.5. Here again, the quantity q_{jk} defined by Eq. (4.3.3) is an invariant and the four constant states are now characterized by pairs of characteristic parameters ω_j and ω_k as follows:

$$\begin{aligned} L(\omega_j^b, \omega_k^a), \qquad Q(\omega_j^b, \omega_k^b) \\ P(\omega_j^a, \omega_k^a), \qquad R(\omega_j^a, \omega_k^b) \end{aligned} \qquad (4.3.8)$$

The image of the solution can be seen in Fig. 4.5(b). Before the interaction it consists of $L \rightarrow P$ for the j-simple wave and $P \rightarrow R$ for the k-shock. During the process of interaction the image of the k-shock recedes continuously from $P \rightarrow R$ to $d \rightarrow e$ and ultimately to $L \rightarrow Q$. When the image of the shock lies on $d \rightarrow e$, the leading portion $P \rightarrow d$ of the j-simple wave has been transmitted across the k-shock to give $R \rightarrow e$ as the image of the leading portion of the transmitted j-simple wave. At the final stage, therefore, the image of the solution will be composed of $L \rightarrow Q$ for the transmitted k-shock, Q for the new constant state appearing between the two waves, and $Q \rightarrow R$ for the transmitted j-simple wave.

For the j-simple wave before transmission we have a family of straight-line characteristics C^j issuing from points on AB as shown in Fig. 4.5(a), and each of these has the slope

$$\sigma_j(\omega_j) = 1 + \nu q_{jk} \omega_j^2 \omega_k^a \qquad (4.3.9)$$

Hence the lower boundary characteristic AD has the equation

$$\tau - \eta_l = \{1 + \nu q_{jk}(\omega_j^a)^2 \omega_k^a\} x$$

The equation of the shock path OD is

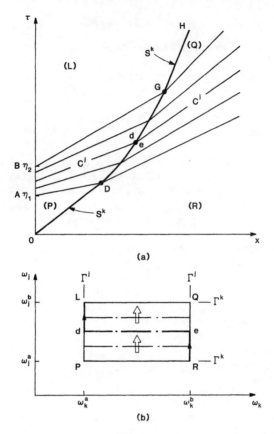

Figure 4.5

$$\tau = \tilde{\sigma}_k x = (1 + \nu q_{jk} \omega_j^a \omega_k^a \omega_k^b) x$$

and solving these equations simultaneously, we can find the point of intersection $D(x_0, \tau_0)$; i.e.,

$$x_0 = \frac{\eta_1}{\nu q_{jk} \omega_j^a \omega_k^a (\omega_k^b - \omega_j^a)}$$

$$\tau_0 = \frac{1 + \nu q_{jk} \omega_j^a \omega_k^a \omega_k^b}{\nu q_{jk} \omega_j^a \omega_k^a (\omega_k^b - \omega_j^a)} \eta_1 \qquad (4.3.10)$$

In the course of interaction the parameter ω_k remains unchanged on either side of the k-shock, ω_k^a on the left-hand side and ω_k^b on the right, but the parameter ω_j continuously increases from ω_j^a to ω_j^b. It follows that the k-shock gradually decelerates as the interaction goes on. To an arbitrary point (x, τ) along the shock path DG there corresponds a straight line C^j carrying the parameter value ω_j up to the left-hand side of the k-shock and hence we can

Wave Interactions in Multicomponent Chromatography   Chap. 4

write two equations for x and τ as

$$\tau - \eta(\omega_j) = \sigma_j x = (1 + \nu q_{jk}\omega_j^2 \omega_k^a)x \qquad (4.3.11)$$

$$\frac{d\tau}{dx} = \tilde{\sigma}_k = 1 + \nu q_{jk}\omega_j \omega_k^a \omega_k^b \qquad (4.3.12)$$

Since ω_j is the only variable parameter, this pair of equations gives a parametric representation of the shock path DG.

Now regarding both x and τ as functions of ω_j, we differentiate Eq. (4.3.11) with respect to ω_j to obtain

$$\frac{d\tau}{d\omega_j} - \frac{d\eta}{d\omega_j} = \sigma_j \frac{dx}{d\omega_j} + \frac{d\sigma_j}{d\omega_j}x$$

and rewrite Eq. (4.3.12) in the form

$$\frac{d\tau}{d\omega_j} = \tilde{\sigma}_k \frac{dx}{d\omega_j}$$

These equations can be combined to give

$$(\tilde{\sigma}_k - \sigma_j)\frac{dx}{d\omega_j} - \frac{d\sigma_j}{d\omega_j}x = \frac{d\eta}{d\omega_j} \qquad (4.3.13)$$

or, by substituting the expressions for σ_j and $\tilde{\sigma}_k$,

$$(\omega_k^b - \omega_j)\frac{dx}{d\omega_j} - 2x = \frac{1}{\nu q_{jk}\omega_k^a \omega_j}\frac{d\eta}{d\omega_j} \qquad (4.3.14)$$

This equation is subject to the initial condition

$$x = x_0 \quad \text{at} \quad \omega_j = \omega_j^a \qquad (4.3.15)$$

Here again Eq. (4.3.14) can be made exact by multiplying throughout by $(\omega_k^b - \omega_j)$. so that we have

$$\frac{d}{d\omega}\{(\omega_k^b - \omega_j)^2 x\} = \frac{1}{\nu q_{jk}\omega_k^a}\left(\frac{\omega_k^b}{\omega_j} - 1\right)\frac{d\eta}{d\omega_j}$$

or, upon direct integration,

$$x(\omega_j) = x_0\left(\frac{\omega_k^b - \omega_j^a}{\omega_k^b - \omega_j}\right)^2 + \frac{1}{\nu q_{jk}\omega_k^a(\omega_k^b - \omega_j)^2}\int_{\omega_j^a}^{\omega_j}\left(\frac{\omega_k^b}{\omega_j} - 1\right)\frac{d\eta}{d\omega_j}d\omega_j \qquad (4.3.16)$$

This equation is then substituted into Eq. (4.3.11) to determine τ as a function of ω_j. The point G, at which the interaction is completed, is found by putting $\omega_j = \omega_j^b$ in Eq. (4.4.16). Beyond the point G the k-shock resumes a constant speed and propagates along the path GH of slope

$$\tilde{\sigma}_k' = 1 + \nu q_{jk}\omega_j^b \omega_k^a \omega_k^b \qquad (4.3.17)$$

which is greater than the slope of the segment OD.

If the j-simple wave is centered at point A, we put $\eta \equiv \eta_1$ and

$d\eta/d\omega_j = 0$, and thus obtain a particularly simple expression

$$x(\omega_j) = x_0 \left(\frac{\omega_k^b - \omega_j^a}{\omega_k^b - \omega_j}\right)^2 = \frac{\eta_1(\omega_k^b - \omega_j^a)}{\nu q_{jk}\omega_j^a \omega_k^a (\omega_k^b - \omega_j)^2} \qquad (4.3.18)$$

in which Eq. (4.3.10) has been substituted. This equation may be rearranged in the form

$$\omega_j = \omega_k^b - \sqrt{\frac{\eta_1(\omega_k^b - \omega_j^a)}{\nu q_{jk}\omega_j^a \omega_k^a x}}$$

and if it is substituted into Eq. (4.3.11) with $\eta = \eta_1$, we obtain the equation

$$\sqrt{\tau - \eta_1 - x} = \omega_k^b \sqrt{\nu q_{jk}\omega_k^a x} - \sqrt{\eta_1 \left(\frac{\omega_k^b}{\omega_j^a} - 1\right)} \qquad (4.3.19)$$

for the shock path DG. Hence the shock path is given by a portion of a parabola.

In other cases, we may take the point D as the new origin of the coordinate system to have $\eta_1 = 0$ and $x_0 = 0$, so that Eq. (4.3.16) gives

$$\begin{aligned} x(\omega_k) &= \frac{1}{\nu q_{jk}\omega_k^a (\omega_k^b - \omega_j)^2} \int_{\omega_j^a}^{\omega_j} \left(\frac{\omega_k^b}{\omega_j} - 1\right) \frac{d\bar{\eta}}{d\omega_j} d\omega_j \\ &= \frac{1}{\nu q_{jk}\omega_k^a (\omega_k^b - \omega_j)} \left\{ \frac{\bar{\eta}}{\omega_j} + \frac{1}{1 - \omega_j/\omega_k^b} \int_{\omega_j^a}^{\omega_j} \frac{\bar{\eta}(\omega_j)}{\omega_j^2} d\omega_j \right\} \end{aligned} \qquad (4.3.20)$$

in which $\bar{\eta}$ represents the adjusted form of η. The two are related by the equation

$$\bar{\eta}(\omega_j) + \tau_0 - \eta(\omega_j) = \sigma_j(\omega_j)x_0 \qquad (4.3.21)$$

The transmitted j-simple wave can be constructed by locating straight-line characteristics C^j emanating from various points on the shock path DG. If we cross the k-shock along a C^j, we see ω_j remaining fixed but ω_k increasing from ω_k^a to ω_k^b and thus the C^j is refracted upward. The refracted C^j has a slope

$$\sigma_j' = 1 + \nu q_{jk}\omega_j^2 \omega_k^b \qquad (4.3.22)$$

and carries the state of concentrations characterized by the pair, ω_j and ω_k^b. In fact, the concentration c_i borne on a particular characteristic C^j changes discontinuously across the shock path S^k and the concentration values before and after the jump are related by

$$c_i^r = c_i^l \frac{1 - \gamma_i/\omega_k^b}{1 - \gamma_i/\omega_k^a}, \qquad i = 1, 2, \ldots, m \qquad (4.3.23)$$

Next we consider the opposite situation in which a j-shock collides with a k-simple wave based on a finite interval of the x-axis as depicted in Fig. 4.6(a). Thus we are assuming that over the interval OA the abscissa is given as a function of the parameter ω_k. The image of the solution lies on a plane parallel to the plane of the ω_j and ω_k axes as shown in Fig. 4.6(b) and the four constant states are characterized by the parameters ω_j and ω_k as

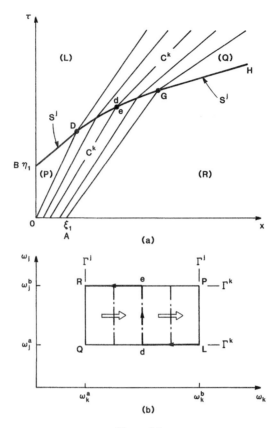

Figure 4.6

$$L(\omega_j^a,\ \omega_k^b), \qquad Q(\omega_j^a,\ \omega_k^a)$$
$$P(\omega_j^b,\ \omega_k^b), \qquad R(\omega_j^b,\ \omega_k^a) \tag{4.3.24}$$

Since the line of analysis would be very similar to the previous one, we will simply present the formulation with a brief discussion.

The k-simple wave is represented by a family of straight-line characteristics C^k, whose slopes are given by

$$\sigma_k(\omega_k) = 1 + \nu q_{jk}\omega_j^b\,\omega_k^2 \tag{4.3.25}$$

Hence the uppermost characteristic OD has the equation

$$\tau = \{1 + \nu q_{jk}\omega_j^b\,(\omega_k^b)^2\}x$$

whereas the shock path AD is the line

$$\tau - \eta_1 = \tilde{\sigma}_j x = (1 + \nu q_{jk}\omega_j^a\,\omega_j^b\,\omega_k^b)x$$

The two lines meet at the point $D(x_0,\ \tau_0)$, where

$$x_0 = \frac{\eta_1}{\nu q_{jk} \omega_j^b \omega_k^b (\omega_k^b - \omega_j^a)}$$

$$\tau_0 = \frac{1 + \nu q_{jk} \omega_j^b (\omega_k^b)^2}{\nu q_{jk} \omega_j^b \omega_k^b (\omega_k^b - \omega_j^a)} \eta_1 \tag{4.3.26}$$

and the two waves begin to interact with each other.

As before, we can take an arbitrary point (x, τ) along the shock path DG and write two equations

$$\tau = \sigma_k(\omega_k)\{x - \xi(\omega_k)\} = (1 + \eta q_{jk} \omega_j^b \omega_k^2)\{x - \xi(\omega_k)\} \tag{4.3.27}$$

$$\frac{d\tau}{dx} = \tilde{\sigma}_j = 1 + \nu q_{jk} \omega_j^a \omega_j^b \omega_k \tag{4.3.28}$$

where $\xi(\omega_k)$, $0 < \xi < \xi_1$, represents the nature of the initial data. During the process of interaction, ω_j remains fixed on either side of the j-shock with ω_j^a on the left-hand side and ω_j^b on the right, but ω_k decreases continuously from ω_k^a to ω_k^a. It follows from Eq. (4.3.28) that the j-shock accelerates as depicted in Fig. 4.6(a).

We note that both x and τ may be regarded as functions of ω_k and then eliminate τ from Eqs. (4.3.27) and (4.3.28) to obtain

$$(\sigma_k - \tilde{\sigma}_j)\frac{dx}{d\omega_k} + \frac{d\sigma_k}{d\omega_k}x = \frac{d}{d\omega_k}(\sigma_k \xi) \tag{4.3.29}$$

or, by substituting the expressions for $\tilde{\sigma}_j$ and σ_k,

$$(\omega_k - \omega_j^a)\frac{dx}{d\omega_k} + 2x = \frac{1}{\nu q_{jk} \omega_j^b \omega_k} \frac{d}{d\omega_k}\{(1 + \nu q_{jk} \omega_j^b \omega_k^2)\xi(\omega_k)\} \tag{4.3.30}$$

which is subject to the initial condition

$$x = x_0 \quad \text{at} \quad \omega_k = \omega_k^b \tag{4.3.31}$$

Equation (4.3.30) can be put into the form of an exact differential and then direct integration gives

$$x(\omega_k) = x_0 \left(\frac{\omega_k^b - \omega_j^a}{\omega_k - \omega_j^a}\right)^2 + \left(\omega_k + \frac{1}{\nu q_{jk} \omega_j^b \omega_k}\right)\frac{\xi}{\omega_k - \omega_j^a}$$

$$- \frac{\omega_j^a}{(\omega_k - \omega_j^a)^2} \int_{\omega_k^b}^{\omega_k} \left(1 + \frac{1}{\nu q_{jk} \omega_j^b \omega_k^2}\right)\xi(\omega_k)\, d\omega_k \tag{4.3.32}$$

in which we have applied integration by parts. With this expression τ can be determined from Eq. (4.3.27). If the k-simple wave is centered at the origin, we have $\xi \equiv 0$ and thus

$$x(\omega_k) = x_0 \left(\frac{\omega_k^b - \omega_j^a}{\omega_k - \omega_j^a}\right)^2 = \frac{\eta_1(\omega_k^b - \omega_j^a)}{\nu q_{jk} \omega_j^b \omega_k^b (\omega_k - \omega_j^a)^2} \tag{4.3.33}$$

from Eqs. (4.3.32) and (4.3.26).

The interaction terminates at the point G, which can be located simply by putting $\omega_k = \omega_k^a$ in Eq. (4.3.32) or Eq. (4.3.33). From this point on the j-shock propagates again with a constant speed along the straight-line path GH of slope

$$\tilde{\sigma}_j' = 1 + v q_{jk} \omega_j^a \omega_j^b \omega_k^a \qquad (4.3.34)$$

which is less steep than the line OD. The transmitted k-simple wave may be established by tracing the characteristics C^k across the shock path DG. When we cross the path DG (an S^j) in the direction of increasing τ, we see that ω_k remains unchanged whereas ω_j drops from ω_j^b to ω_j^a. Hence each of the new characteristics C^k will have a slope given by

$$\tilde{\sigma}_k' = 1 + v q_{jk} \omega_j^a \omega_k^2 \qquad (4.3.35)$$

which is less than σ_k of Eq. (4.3.25). This implies that a C^k is refracted forward upon crossing the shock path DG as shown in Fig. 4.6(a). The state of concentrations borne on a C^k is determined by the relations

$$c_i^l = c_i^r \frac{1 - \gamma_i / \omega_j^b}{1 - \gamma_i / \omega_j^a}, \qquad i = 1, 2, \ldots, m \qquad (4.3.36)$$

where c_i^r and c_i^l denote, respectively, the values of c_i before and after crossing the shock path DG (an S^j).

Transmission between two simple waves

We shall finally take up the case when a j-simple wave overtakes a k-simple wave traveling ahead of it. The two waves are transmitted through each other and emerge as two transmitted simple waves of their respective families. This is illustrated in Fig. 4.7 along with the image of the solution in the plane of the characteristic parameters ω_j and ω_k. In the penetration region d–e–g–f, bounded by the four characteristics, we have a nonsimple wave but still a continuous solution which may be determined by establishing the network of the characteristics C^j and C^k. The characteristic parameters for each of the four constant states are

$$L(\omega_j^b, \omega_k^b), \qquad Q(\omega_j^b, \omega_k^a)$$
$$P(\omega_j^a, \omega_k^b), \qquad R(\omega_j^a, \omega_k^a) \qquad (4.3.37)$$

The j-simple wave collides with the k-simple wave at the point d and the two are engaged in an interaction, which takes place over the penetration region d–e–g–f. If we trace a particular C^j, it remains straight up to the boundary d–e, but upon crossing the boundary it becomes less and less steep because ω_k decreases although ω_j is fixed. Similarly, a C^k remains straight up to the boundary d–f and then gradually takes larger slopes since ω_j increases along the C^k while ω_k remains constant. When both characteristics C^j and C^k emerge through the boundaries f–g and e–g, respectively, they become straight lines again and constitute the two transmitted simple waves.

The characteristics O–d and A–d have the equations

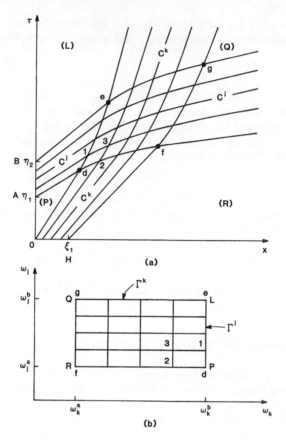

Figure 4.7

$$\tau = \sigma_k x = \{1 + \nu q_{jk}\omega_j^a(\omega_k^b)^2\}x$$
$$\tau - \eta_1 = \sigma_j x = \{1 + \nu q_{jk}(\omega_j^a)^2\omega_k^b\}x \tag{4.3.38}$$

respectively, and the two meet at the point $d(x_0, \tau_0)$, where

$$x_0 = \frac{\eta_1}{\nu q_{jk}\omega_j^a\omega_k^b(\omega_k^b - \omega_j^a)}$$
$$\tau_0 = \frac{1 + \nu q_{jk}\omega_j^a(\omega_k^b)^2}{\nu q_{jk}\omega_j^a\omega_k^b(\omega_k^b - \omega_j^a)}\eta_1 \tag{4.3.39}$$

Comparing the boundary d–e with the shock path DG of Fig. 4.5(a), we immediately find that the two situations are entirely analogous except for the fact that we have a C^k here instead of an S^k. At an arbitrary point (x, τ) on d–e, therefore, we can write

$$\tau - \eta(\omega_j) = \sigma_j x = (1 + \nu q_{jk}\omega_j^2\omega_k^b)x$$

$$\frac{d\tau}{dx} = \sigma_k = 1 + \nu q_{jk}\omega_j(\omega_k^b)^2 \qquad (4.3.40)$$

but these equations would be identical to Eqs. (4.3.11) and (4.3.12) if we replaced ω_k^a in the latter by ω_k^b. Even the initial conditions would become the same. This implies the equation for the boundary d–e is given by Eq. (4.3.16) with ω_k^b substituted for ω_k^a and x_0 introduced from Eq. (4.3.39). If the k-simple wave is centered at point A, we obtain

$$x(\omega_j,\ \omega_k^b) = x_0\left(\frac{\omega_k^b - \omega_j^a}{\omega_k^b - \omega_j}\right)^2$$

$$= \frac{\eta_1(\omega_k^b - \omega_j^a)}{\nu q_{jk}\omega_j^a \omega_k^b (\omega_k^b - \omega_j)^2}, \qquad \omega_j^a \le \omega_j \le \omega_j^b \qquad (4.3.41)$$

and

$$\tau(\omega_j,\ \omega_k^b) = \eta_1 + (1 + \nu q_{jk}\omega_j^2 \omega_k^b)x(\omega_j,\ \omega_k^b) \qquad (4.3.42)$$

for the boundary d–e.

Similarly, for the boundary d–f we observe that the overall feature is compatible with the situation shown in Fig. 4.6(a) except for the fact that d–f is a C^j, whereas DG is an S^j. It is obvious that, at an arbitrary point (x, τ) on d–f, we can write the two equations

$$\tau = \sigma_k\{x - \xi(\omega_k)\} = (1 + \nu q_{jk}\omega_j^a \omega_k^2)\{x - \xi(\omega_k)\}$$

$$\frac{d\tau}{dx} = \sigma_j = 1 + \nu q_{jk}(\omega_j^a)^2\omega_k \qquad (4.3.43)$$

We would have obtained exactly the same equations if we had substituted ω_j^a for ω_j^b in Eqs. (4.3.27) and (4.3.28). With this substitution the initial condition would also have been the same. It then follows that the boundary d–f has the equation given by Eq. (4.3.32) with ω_j^b replaced by ω_j^a and x_0 introduced from Eq. (4.3.39). In particular, if the k-simple wave is centered at the origin, we have the equations

$$x(\omega_j^a,\ \omega_k) = x_0\left(\frac{\omega_k^b - \omega_j^a}{\omega_k - \omega_j^a}\right)^2$$

$$= \frac{\eta_1(\omega_k^b - \omega_j^a)}{\nu q_{jk}\omega_j^a \omega_k^b (\omega_k - \omega_j^a)^2}, \qquad \omega_k^a \le \omega_k \le \omega_k^b \qquad (4.3.44)$$

and

$$\tau(\omega_j^a,\ \omega_k) = (1 + \nu q_{jk}\omega_j^a \omega_k^2)x(\omega_j^a,\ \omega_k) \qquad (4.3.45)$$

for the portion d–f.

Within the penetration region d–e–g–f, the parameter ω_j remains fixed along a C^j, while ω_k is kept constant along a C^k, and thus there exists a two-parameter representation of the solution. Since the solution would be continuous everywhere, it is allowed to regard both x and τ as functions of ω_j and ω_k, and this enables us to write

$$C^j \quad : \quad \frac{\partial \tau}{\partial \omega_k} = \sigma_j(\omega_j, \, \omega_k)\frac{\partial x}{\partial \omega_k} = (1 \, + \, \nu q_{jk}\omega_j^2 \, \omega_k)\frac{\partial x}{\partial \omega_k} \qquad (4.3.46)$$

and

$$C^k \quad : \quad \frac{\partial \tau}{\partial \omega_j} = \sigma_k(\omega_j, \, \omega_k)\frac{\partial x}{\partial \omega_j} = (1 \, + \, \nu q_{jk}\omega_j\omega_k^2)\frac{\partial x}{\partial \omega_j} \qquad (4.3.47)$$

To eliminate τ we differentiate Eq. (4.3.46) with respect to ω_j and Eq. (4.3.47) with respect to ω_k and subtract one of the resulting equations from the other. This gives the second-order differential equation

$$\frac{\partial^2 x}{\partial \omega_j \, \partial \omega_k} + \frac{2}{\omega_k - \omega_j}\left(\frac{\partial x}{\partial \omega_j} - \frac{\partial x}{\partial \omega_k}\right) = 0 \qquad (4.3.48)$$

where $\omega_j < \omega_k$. Since this equation is to be solved for the penetration region d–e–g–f, the appropriate boundary conditions are given by Eq. (4.3.16) with ω_k^b for ω_k^a and Eq. (4.3.32) with ω_j^a for ω_j^b specified at $\omega_k = \omega_k^b$ and $\omega_j = \omega_j^a$, respectively.

If the j-simple wave is centered at A and the k-simple wave at the origin, the boundary conditions are specified as given by Eqs. (4.3.41) and (4.3.44). The differential equation (4.3.48) subject to these boundary conditions has been treated in Sec. 2.7 and from the development there we can write the solution in the form

$$x(\omega_j, \, \omega_k) = x_0\left(\frac{\omega_k^b - \omega_j^a}{\omega_k - \omega_j}\right)^2\left\{1 \, - \, 2\frac{(\omega_k - \omega_k^b)(\omega_j - \omega_j^a)}{(\omega_k^b - \omega_j^a)(\omega_k - \omega_j)}\right\}$$

$$= x_0(\omega_k^b - \omega_j^a)(\omega_k - \omega_j)^{-3}\{(\omega_k^b - \omega_j)(\omega_k - \omega_j^a) \qquad (4.3.49)$$

$$+ (\omega_k^b - \omega_k)(\omega_j - \omega_j^a)\}$$

which is equivalent to Eq. (2.7.53). Substituting x_0 from Eq. (4.3.39), we obtain

$$x(\omega_j, \, \omega_k) = \frac{\eta_1}{\nu q_{jk}}\frac{(\omega_k^b - \omega_j)(\omega_k - \omega_j^a) \, + \, (\omega_k^b - \omega_k)(\omega_j - \omega_j^a)}{\omega_j^a \omega_k^b(\omega_k - \omega_j)^3} \qquad (4.3.50)$$

Once x is determined, $\tau(\omega_j, \, \omega_k)$ can be obtained from Eqs. (4.3.50) and (4.3.47). Along each of the characteristics C^k the parameter ω_k remains fixed, and thus we can integrate Eq. (4.3.47) from $\omega_j = \omega_j^a$ along a C^k. This gives

$$\tau(\omega_j, \, \omega_k) \, - \, \tau(\omega_j^a, \, \omega_k) = \int_{\omega_j^a}^{\omega_j} (1 \, + \, \nu q_{jk}\omega_j' \, \omega_k^2)\frac{\partial x}{\partial \omega_j'} \, d\omega_j'$$

or, after integration by parts,

$$\tau(\omega_j, \, \omega_k) \, - \, \tau(\omega_j^a, \, \omega_k) = (1 \, + \, \nu q_{jk}\omega_j\omega_k^2)x(\omega_j, \, \omega_k)$$

$$-(1 \, + \, \nu q_{jk}\omega_j^a \omega_k^2)x(\omega_j^a, \, \omega_k) \, - \, \nu q_{jk}\omega_k^2\int_{\omega_j^a}^{\omega_j} x(\omega_j', \, \omega_k) \, d\omega_j'$$

According to Eq. (4.3.45), the term $\tau(\omega_j^a, \omega_k)$ cancels with the second term on the right-hand side. Carrying out the integration and making some rearrangements, we obtain

$$\tau(\omega_j, \omega_k) = (1 + \nu q_{jk}\omega_j\omega_k^2)x(\omega_j, \omega_k) + \frac{\eta_1\omega_k^2}{\omega_j^a\omega_k^b}\left\{\frac{p}{\omega_k - \omega_j}(1 - \mu)\right.$$
$$\left. - \frac{p\omega_k + q}{2(\omega_k - \omega_j)^2}(1 - \mu^2)\right\} \tag{4.3.51}$$

in which

$$p = \omega_j^a + \omega_k^b - 2\omega_k$$
$$q = \omega_k^b(\omega_k - \omega_j^a) - \omega_j^a(\omega_k^b - \omega_k) \tag{4.3.52}$$
$$\mu = \frac{\omega_k - \omega_j}{\omega_k - \omega_j^a}$$

The interaction is to be completed at the point g, which we can locate by putting $\omega_j = \omega_j^b$ and $\omega_k = \omega_k^a$ in Eqs. (4.3.50) and (4.3.51). Thus the point g has the coordinates

$$x = x(\omega_j^b, \omega_k^a) = \frac{\eta_1}{\nu q_{jk}}\frac{(\omega_k^b - \omega_j^b)(\omega_k^a - \omega_j^a) + (\omega_k^b - \omega_k^a)(\omega_j^b - \omega_j^a)}{\omega_j^a\omega_k^b(\omega_k - \omega_j)^3}$$
$$\tau = \tau(\omega_j^b, \omega_k^a) \tag{4.3.53}$$

The interaction goes on for a period $\tau(\omega_j^b, \omega_k^a) - \tau(\omega_j^a, \omega_k^b) = \tau(\omega_j^b, \omega_k^a) - \tau_0$ and the penetration region covers the range of $x(\omega_j^b, \omega_k^a) - x(\omega_j^a, \omega_k^b)$ $= x(\omega_j^b, \omega_k^a) - x_0$, where x_0 and τ_0 are given by Eq. (4.3.39). For every pair of ω_j and ω_k values in the penetration region, the state of concentrations can be determined by using Eq. (3.7.9).

The transmitted simple waves are constructed by drawing straight-line characteristics C^j and C^k from points on f–g and e–g, respectively, and again they carry constant states of concentrations determined by the pair of ω_j and ω_k values borne on the lines.

In case the simple waves are not centered, the solution in the penetration region may be constructed by applying the step-by-step mapping procedure. At every point in this region we can write

$$C^j \; : \; \frac{d\tau}{dx} = \sigma_j(\omega_j, \omega_k) = 1 + \nu q_{jk}\omega_j^2\omega_k \tag{4.3.54}$$

$$C^k \; : \; \frac{d\tau}{dx} = \sigma_k(\omega_j, \omega_k) = 1 + \nu q_{jk}\omega_j\omega_k^2 \tag{4.3.55}$$

Since we know the values of ω_j and ω_k at every mesh point, both σ_j and σ_k can be evaluated. Hence the portion $1 \rightarrow 3$ of a C^j, for instance, may be approximated by a straight line that starts from point 1 with a slope

$$\sigma_j = \frac{1}{2}(\sigma_{j1} + \sigma_{j3}) \tag{4.3.56}$$

Similarly, the portion $2 \to 3$ of a C^k may be approximated by another straight line emanating from point 2 with a slope

$$\sigma_k = \frac{1}{2}(\sigma_{k2} + \sigma_{k3}) \qquad (4.3.57)$$

and point 3 can be located at the intersection of the two lines. By iterating this procedure we can locate all the mesh points and determine the concentrations at each point by using Eq. (3.7.9). This gives an approximate of the solution in the penetration region and, as we take more mesh points, an increasingly accurate solution can be obtained.

Exercises

4.3.1 For the case shown in Fig. 4.6, suppose that the k-simple wave is centered at the origin and show that the portion DG of the shock path S^j is given by a part of a parabola. Sketch the profiles of the concentration c_m of the most strongly adsorbed solute before, during, and after the interaction.

4.3.2. By using Eqs. (4.3.42) and (4.3.50), integrate Eq. (4.3.46) from $\omega_k = \omega_k^b$ along a C^j and thus determine $\tau(\omega_j, \omega_k)$. Compare the result with Eq. (4.3.51).

4.4 Chromatographic Cycle for m Solutes

We have already seen in Sec. 2.8 that two solutes can be separated by putting them on a chromatographic column as a mixture over a short period of time and then eluting with a stream of solvent. This has been treated almost exhaustively and illustrated with a numerical example. Here we would like to extend the treatment to the case of more than two solutes by applying the theoretical developments of the previous and present chapters.

Let us suppose that a fresh bed of adsorbent is first irrigated with a fluid mixture containing m solutes A_i, $i = 1, 2, \ldots, m$, laying down the sample in the form of a chromatogram. After a finite period of time, τ_0, the feedstream will be changed from the mixture to the pure solvent so that the chromatogram starts to be eluted. Since the initial and feed data include two discontinuities, one at the origin and the other at the point $(0, \tau_0)$ of the (x, τ)-plane, we have a problem with piecewise constant data and expect to see wave interactions.

We shall discuss the problem for $m = 3$, but the same procedure can be applied equally well for any m greater than 3. Noting that the initial state and the ultimate feed state (for $\tau > \tau_0$) would map on the same point of pure state, we can readily picture the image of the solution in the hodograph space as illustrated in Fig. 4.8, where the upper part shows the image in the concentration space $\Phi(3)$ and the lower part the image in the space $\Omega(3)$. The point image of the pure state is marked as O, whereas the point corresponding to the feed mixture is denoted by F. Clearly, the image in $\Phi(3)$ or $\Omega(3)$ of the state along the column at any moment is given by a closed loop. Such a loop we shall call a

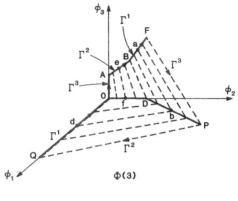

──── Simple wave,   ─ ─ ─ ─ Shock

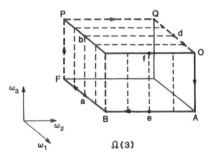

Figure 4.8

chromatographic cycle. While interactions occur between pairs of waves, this cycle will deform continuously. By pursuing the course of deformation of the chromatographic cycle, we shall see how different solutes are separated. However, we would like to point out that although this is certainly a useful visualization, it is not necessary to use the image in the hodograph space for the construction of a solution. This is important since picturing the image in the hodograph space would be difficult for $m \geq 4$.

By applying the results of the previous sections, a complete analysis can be achieved, giving the physical plane portrait of the solution as shown in Fig. 4.9. In the remainder of this section, we shall discuss in detail how a solution can be constructed. The set of the characteristic parameters corresponding to the state F of the feed mixture will be denoted as $(\omega_1^f, \omega_2^f, \omega_3^f)$. The set for the pure state O is obviously given by $(\gamma_1, \gamma_2, \gamma_3)$, and thus each ω_k is bounded as

$$\omega_k^f \leq \omega_k \leq \gamma_k, \qquad k = 1, 2, 3 \tag{4.4.1}$$

Up until the moment $\tau = \tau_1$, when the first interaction starts, the solution will be given by a combination of Fig. 3.13 and Fig. 3.14, each of which will be valid as long as the waves are not engaged in interactions. Thus the four

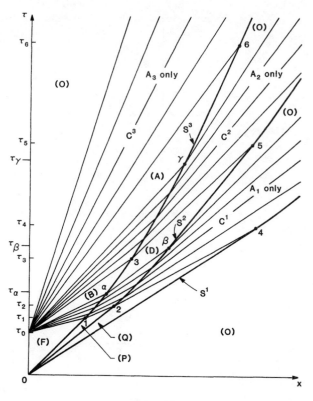

Figure 4.9

new constant states A, B, P and Q can readily be identified in terms of the characteristic parameters as listed in Table 4.1. The distribution of solutes is represented by the chromatographic cycle

$$O \longrightarrow A \longrightarrow B \longrightarrow F \longrightarrow P \longrightarrow Q \longrightarrow O$$

in which the arrows denote the directions of transition as we move from the feed state to the initial state.

At $\tau = \tau_1$, the 1-simple wave starts to interact with the 3-shock, so the constant state F disappears and we see the image of the 3-shock receding from FP to BD in Fig. 4.8. While interacting, the two waves are transmitted through each other and thus the shock path S^3 is located by Eq. (4.3.18) with $j = 1$ and $k = 3$, whereas the transmitted 1-simple wave can be constructed by using Eqs. (4.3.22) and (4.3.23). In the meantime the chromatographic cycle becomes, for instance,

$$O \longrightarrow A \longrightarrow B \longrightarrow a \longrightarrow b \longrightarrow P \longrightarrow Q \longrightarrow O$$

where a and b denote some intermediate points on BF and DP, respectively.

TABLE 4-1. CHARACTERISTIC PARAMETERS FOR CONSTANT STATES

Constant State	ω_1	ω_2	ω_3
O	γ_1	γ_2	γ_3
A	γ_1	γ_2	ω_3^f
B	γ_1	ω_2^f	ω_3^f
F	ω_1^f	ω_2^f	ω_3^f
P	ω_1^f	ω_2^f	γ_3
Q	ω_1^f	γ_2	γ_3
D	γ_1	ω_2^f	γ_3

When the interaction is completed at $\tau = \tau_\alpha$, the image of the 3-shock has receded to BD. Hence the shock path S^3 separates a mixture containing A_2 and A_3 from that containing A_1 and A_2 from the left to the right.

Since ω_1 and ω_3 are bounded according to Eq. (4.4.1), we can locate the point (x_1, τ_1), at which the interaction begins, by using Eq. (4.3.10) with $j = 1$ and $k = 3$; that is,

$$x_1 = \frac{\tau_0\,\gamma_1\,\gamma_2\,\gamma_3}{\nu\omega_1^f\,\omega_2^f\,\omega_3^f(\gamma_3 - \omega_1^f)}$$

$$\tau_1 = \left(1 + \frac{\nu\omega_1^f\,\omega_2^f\,\omega_3^f}{\gamma_1\,\gamma_2}\right)x_1 = \frac{\tau_0\,\gamma_3}{\gamma_3 - \omega_1^f}\left(1 + \frac{\gamma_1\,\gamma_2}{\nu\omega_1^f\,\omega_2^f\,\omega_3^f}\right)$$
(4.4.2)

Similarly, the point (x_α, τ_α) at which the interaction terminates can be determined from Eqs. (4.3.11) and (4.3.18):

$$x_\alpha = \frac{\tau_0\,\gamma_1\,\gamma_2\,\gamma_3(\gamma_3 - \omega_1^f)}{\nu\omega_1^f\,\omega_2^f\,\omega_3^f(\gamma_3 - \gamma_1)^2}$$

$$\tau_\alpha = \tau_0 + \left(1 + \frac{\nu\gamma_1\,\omega_2^f\,\omega_3^f}{\gamma_2\,\gamma_3}\right)x_\alpha$$
(4.4.3)

It is obvious that both x_α and τ_α decrease as the ratio γ_3/γ_1 increases.

At $\tau = \tau_2$ (possibly $\tau_2 \leq \tau_\alpha$), the once transmitted 1-simple wave is again engaged in an interaction with the 2-shock. This is the point of intersection between the lower boundary characteristic C^1 and the shock path S^2 emanating from the origin and thus has the coordinates in the (x, τ)-plane

$$x_2 = \frac{x_1\,\gamma_1\,\gamma_2}{\nu\omega_1^f\,\omega_2^f(\gamma_2 - \omega_1^f)}\left\{\frac{\tau_1}{x_1} - 1 - \frac{\nu\,(\omega_1^f)^2\,\omega_2^f}{\gamma_1\,\gamma_2}\right\} = x_1\,\frac{\omega_3^f - \omega_1^f}{\gamma_2 - \omega_1^f}$$

$$\tau_2 = \left(1 + \frac{\nu\omega_1^f\,\omega_2^f}{\gamma_1}\right)x_2$$
(4.4.4)

For the curved shock path S^2 we can use Eq. (4.3.20) with $j = 1$ and $k = 2$ and with properly adjusted $\bar\eta(\omega_1)$. The twice transmitted 1-simple wave is determined by Eqs. (4.3.22) and (4.3.23). The chromatographic cycle is now reduced to

$$O \longrightarrow A \longrightarrow B \longrightarrow D \longrightarrow b \longrightarrow d \longrightarrow Q \longrightarrow O$$

where d is an intermediate point on OQ.

After the interaction is over at $\tau = \tau_\beta$, the image of the 2-shock falls on DO and thus we observe only A_1 to the right of the shock path S^2, whereas on the left-hand side we have a region of constant state D containing the solute A_2 alone (see the characteristic parameters for D in Table 4.1). This implies that chromatograms of pure A_1 and of pure A_2 are obtained at $\tau = \tau_\beta$. The point (x_β, τ_β) at which the interaction terminates can be located by using Eq. (4.3.20) and the equation of the upper boundary characteristic C^2, for which we can write

$$\bar{\eta}(\gamma_1) = (\tau_\alpha - \tau_2) + \left(1 + \frac{\nu\gamma_1\omega_2^f}{\gamma_2}\right)(x_2 - x_\alpha) \tag{4.4.5}$$

Using this in Eq. (4.3.20) we obtain

$$x_\beta = x_2 + \frac{\gamma_1\gamma_2}{\nu\omega_2^f(\gamma_2 - \gamma_1)}\left\{\frac{\bar{\eta}(\gamma_1)}{\gamma_1} + \frac{\gamma_2}{\gamma_2 - \gamma_1}\int_{\omega_1^f}^{\gamma_1}\frac{\bar{\eta}(\omega_1)}{\omega_1^2}\,d\omega_1\right\}$$

$$\tau_\beta = \tau_\alpha + \left(1 + \frac{\nu\gamma_1\omega_2^f}{\gamma_2}\right)(x_\beta - x_\alpha) \tag{4.4.6}$$

and clearly we see that both x_β and τ_β decrease as the ratio γ_2/γ_1 increases.

The 2-simple wave, on the other hand, also starts an interaction with the 3-shock at $\tau = \tau_3$. During the course of interaction the image of the 3-shock recedes from BD to e–f, and ultimately to AO. At an intermediate stage, therefore, the chromatogram cycle would become

$$O \longrightarrow A \longrightarrow e \longrightarrow f \longrightarrow D \longrightarrow O \longrightarrow Q \longrightarrow O$$

where e and f denote some intermediate points on AB and OD, respectively. When this interaction is over at $\tau = \tau_\gamma$, the image of the 3-shock falls on AO, so the chromatographic cycle is now reduced to

$$O \longrightarrow A \longrightarrow O \longrightarrow D \longrightarrow O \longrightarrow Q \longrightarrow O$$

This implies that the shock path S^3 separates a mixture containing A_3 only from that containing A_2 alone from the left to the right. Hence a complete separation of three solutes, A_1, A_2, and A_3, is accomplished at $\tau = \tau_\gamma$.

The point (x_3, τ_3), at which the interaction begins, is located at the intersection between the lower boundary characteristic C^2 and the shock path S^3, so that we obtain

$$x_3 = \frac{x_\alpha\gamma_2\gamma_3}{\nu\omega_2^f\omega_3^f(\gamma_3 - \omega_2^f)}\left(1 + \frac{\nu\omega_2^f\omega_3^f}{\gamma_2} - \frac{\tau_\alpha - \tau_0}{x_\alpha}\right) = x_\alpha\frac{\gamma_3 - \gamma_1}{\gamma_3 - \omega_2^f}$$

$$\tau_3 = \tau_0 + \left\{1 + \frac{\nu(\omega_2^f)^2\omega_3^f}{\gamma_2\gamma_3}\right\}x_3 \tag{4.4.7}$$

The curved portion of the S^3 is given by Eq. (4.3.18) with x_0 replaced by x_3, and the transmitted 2-simple wave can be constructed by Eqs. (4.3.22) and

(4.3.23) with $j = 2$ and $k = 3$. Therefore, the point (x_γ, τ_γ) where the interaction terminates has the coordinates given by

$$x_\gamma = x_3 \left(\frac{\gamma_3 - \omega_2^f}{\gamma_3 - \gamma_2} \right)^2 = \frac{\tau_0 \gamma_1 \gamma_2 \gamma_3 (\gamma_3 - \omega_1^f)(\gamma_3 - \omega_2^f)}{\nu \omega_1^f \omega_2^f \omega_3^f (\gamma_3 - \gamma_1)(\gamma_3 - \gamma_2)^2}$$

$$\tau_\gamma = \tau_0 + \left(1 + \frac{\nu \gamma_2 \omega_3^f}{\gamma_3} \right) x_\gamma \tag{4.4.8}$$

It is immediately apparent that as the ratios γ_3/γ_2 and γ_3/γ_1 increase, both x_γ and τ_γ decrease.

Meanwhile, the twice-transmitted 1-simple wave starts an interaction with the 1-shock at $\tau = \tau_4$. As time goes on, the once-transmitted 2-simple wave also begins to overtake the 2-shock at $\tau = \tau_5$. The shock paths S^1 and S^2 at these stages can be located by applying Eq. (4.2.17) with $k = 1$ and $k = 2$, respectively, and with properly adjusted $\bar{\eta}$. Also, the 3-simple wave will eventually start to overtake the 3-shock at $\tau = \tau_6$ and this interaction is described by Eq. (4.2.15) with $k = 3$ and x_6 for x_0. Each of these interactions is accompanied with the disappearance of the constant state Q, D, or A, respectively, and thus the chromatographic cycle will take the ultimate form

$$O \longrightarrow A' \longrightarrow O \longrightarrow D' \longrightarrow O \longrightarrow Q' \longrightarrow O$$

where A', for example, denotes an intermediate point on OA. These interactions will go on indefinitely while the waves involved cancel each other, making the bands of pure solute tend to expand in width and decay in height.

Finally, the physical plane portrait of the solution is completed as depicted in Fig. 4.9, from which the distribution of solutes can be read at any moment. This we shall illustrate in the examples that follow. Now we are assured that exactly the same procedure can be applied to a system involved with more than three solutes.

Incidentally, we have noticed in the course of the analysis above that the 3-shock path S^3 takes the role of a sieve which holds only the solute A_3 and passes other solutes. Similarly, the 2-shock path S^2 acts like a sieve that holds the solute A_2 while passing the solute A_1. The 1-shock path S^1 holds the solute A_1 that has passed through the two sieves. It is obvious that this argument can be extended to systems with more than three solutes.

If we use a column of length x_γ, which is expected to be good enough for a complete separation, we shall see indeed the bands of pure solute emerging one by one from the exit of the column. At first, the outgoing stream would contain no solute up until $\tau = (1 + \nu \omega_1^f) x_\gamma$, the time when the 1-shock reaches the exit to give the breakthrough of the solute A_1. After the tail of the A_1 band passes the exit, we expect to see again no solute for a short period of time and then the breakthrough of the solute A_2 follows. The solute A_2 continues to emerge until $\tau = \tau_\gamma$, the time when the solute A_3 makes its breakthrough. The solute A_3 keeps emerging and, although the A_3 band tails off, its rear end bearing zero concentration reaches the exit at $\tau = (1 + \nu \gamma_3) x_\gamma$, the time when the uppermost characteristic C^3 intersects the line $x = x_\gamma$.

Just as in the case of two solutes, we can compare the volume of sample treated with that of adsorbent required for a complete separation. The ratio between the two can be determined by using Eq. (4.4.8) as

$$r = \frac{\text{volume of adsorbent required}}{\text{volume of sample treated}} = \frac{(1 - \epsilon)z_\gamma}{\epsilon V t_0} = \frac{x_\gamma}{\tau_0}$$

$$= \frac{\gamma_1 \gamma_2 \gamma_3 (\gamma_3 - \omega_1^f)(\gamma_3 - \omega_2^f)}{\omega_1^f \omega_2^f \omega_3^f (\gamma_3 - \gamma_1)(\gamma_3 - \gamma_2)^2} \quad (4.4.9)$$

It is immediately apparent that this ratio decreases as the ratios γ_3/γ_2 and γ_3/γ_1 increase. Consequently, the more the adsorptivities γ_1, γ_2, and γ_3 differ from one another, the smaller a volume of adsorbent is required to separate a given volume of sample into pure solutes. It is noted, however, that the effect of the ratio γ_3/γ_2 is of second order, whereas the ratio γ_3/γ_1 has a first-order effect.

In case the ratio γ_2/γ_1 is very small compared to the ratio γ_3/γ_2, the value of x_β can become quite large so that we may have $x_\beta > x_\gamma$. If so, we shall need a column of length x_β and, according to Eq. (4.4.6), this length becomes larger as the ratio γ_2/γ_1 decreases. In this case the ratio r will obtain another expression based on Eq. (4.4.6), but it certainly increases as γ_2/γ_1 decreases. With a column of length x_β, we shall see the bands of pure solute emerging one by one from the exit as discussed in the above, the only difference being that the short interval of no solute now appears between the A_2 and A_3 bands rather than between the A_1 and A_2 bands.

Next, we shall take up a numerical example and illustrate in more detail the development discussed above. The system to be examined is the same as those for Examples 3.2 and 3.3. Thus we have

$$\epsilon = 0.4 \quad \text{or} \quad \nu = 1.5$$

and

$$N_i = N = 1.0 \text{ mol/liter of adsorbent}$$

for $i = 1, 2, 3$. The values of the adsorption equilibrium constants K_i in liter/mol and the feed and initial concentrations in mol/liter are listed in Table 4.2 along with the characteristic parameters for the feed and initial states. We note that the sample is introduced over a period of $\tau_0 = 10$.

TABLE 4-2. PARAMETERS AND CONDITIONS FOR THE EXAMPLE

	i			Remark
	1	2	3	
K_i	5.0	10.0	15.0	
c_i^f	0	0	0	For $\tau > 10$
c_i^f	0.05	0.05	0.05	For $0 < \tau < 10$
c_i^0	0	0	0	For $x > 0$
ω_i^f	5.0	10.0	15.0	For $\tau > 10$
ω_i^f	3.387	7.05	12.563	For $0 < \tau < 10$
ω_i^0	5.0	10.0	15.0	For $x > 0$

Clearly, this numerical example corresponds to the superposition and interaction of Examples 3.2 and 3.3. Presented in Fig. 4.10 is the physical plane portrait of the solution showing the interactions between the simple waves and the shocks. To the left of the shock path S^3 the situation is the same as that of Example 3.3, with three simple waves separated by two new constant states A and B whereas the three shock paths and the two new constant states P and Q are identical to those of Example 3.2. The characteristic parameters for each of the constant states are readily identified as given in Table 4.3, in which we also present the concentration values.

The simple waves are constructed in such a way that, for the j-simple wave, each of the straight-line characteristic C^j bears the ω_j-value as given in Table 4.4. We note that this particular value of ω_j remains unchanged along the C^j even after it is refracted across a shock path. Other parameters ω_k, $j \neq k$, are

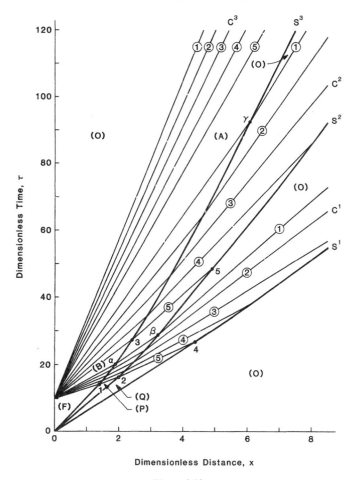

Figure 4.10

TABLE 4-3. PARAMETERS AND CONCENTRATIONS FOR CONSTANT STATES

Constant State	ω_1	ω_2	ω_3	c_1	c_2	c_3
O	5.0	10.0	15.0	0	0	0
A	5.0	10.0	12.563	0	0	0.0129
B	5.0	7.05	12.563	0	0.0256	0.0292
F	3.387	7.05	12.563	0.05	0.05	0.05
P	3.387	7.05	15.0	0.0554	0.0817	0
Q	3.387	10.0	15.0	0.0953	0	0
D	5.0	7.05	15.0	0	0.0418	0

TABLE 4-4. CHARACTERISTIC PARAMETER ω_k BORNE ON EACH OF THE STRAIGHT–LINE CHARACTERISTICS C^j

Simple Wave	Straight–Line Characteristic	Characteristic Parameter	Number on Straight–Line Characteristic				
			1	2	3	4	5
3-simple wave	C^3	ω_3	15.0	14.4	13.8	13.2	12.563
2-simple wave	C^2	ω_2	10.0	9.25	8.5	7.75	7.05
1-simple wave	C^1	ω_1	5.0	4.6	4.2	3.8	3.387

invariant in the j-simple wave region, but they will take new values if the j-simple wave is transmitted across a shock. These values can be read directly from Table 4.3 by looking at the two constant states separated by the j-simple wave. Given the set of the characteristic parameters, we can use Eqs. (4.3.9) and (4.3.22) to determine the slope of the straight-line characteristics and then apply Eqs. (3.11.5) and (4.3.23) to calculate the concentration values carried along each of the straight-line characteristics.

The slopes of the shock paths, when they are straight, can be determined by using any one of Eqs. (3.9.16), (3.9.17), and (3.9.18), and they are listed in Table 4.5. Now we can locate the points of intersection between a boundary characteristic C^j and the shock paths S^k by making use of Eqs. (4.4.2)–(4.4.8). These are presented in Table 4.6. For x_β the integration in Eq. (4.4.6) may be carried out numerically after the once-transmitted 1-simple wave is established, and experience tells that the convergence is very good.

All the interactions involved in this example can be analyzed without difficulty by following the procedure discussed in Secs. 4.2 and 4.3. The interaction between the 1-simple wave and the 3-shock starts from the point (x_1, τ_1) and terminates at the point (x_α, τ_α). The shock path S^3 between these points is described by Eq. (4.3.18), which gives

$$x(\omega_1) = \frac{\tau_0 \gamma_1 \gamma_2 \gamma_3 (\gamma_3 - \omega_1^f)}{\nu \omega_1^f \omega_2^f \omega_3^f (\gamma_3 - \omega_1)^2} = \frac{193.561}{(15 - \omega_1)^2} \qquad (4.4.10)$$

$$\tau(\omega_1) = 10 + (1 + 0.1771\omega_1^2)x(\omega_1)$$

TABLE 4-5. SLOPES OF VARIOUS SHOCK PATHS

State on the Left-Hand Side	State on the Right-Hand Side	Family of the Shock, k	Slope of the Shock Path, $\bar{\sigma}_k$
F	P	3	10.0
P	Q	2	8.164
Q	O	1	6.081
B	D	3	14.285
D	O	2	11.575
A	O	3	19.845

TABLE 4-6. COORDINATES OF THE POINT OF INTERSECTION

Point	Coordinates	
	x	τ
1	1.435	14.352
2	1.991	16.255
3	2.435	27.639
4	4.521	27.492
5	5.025	49.537
6	12.627	222.919
α	1.936	20.509
β	3.208	28.507
γ	6.155	93.483

for $3.387 \leq \omega_1 \leq 5.0$. These equations may be combined to give the equation [see also Eq. (4.3.19)]

$$\sqrt{\tau - x - 10} = 6.3132 \sqrt{x} - 5.8555 \qquad (4.4.11)$$

which represents a parabola.

The once-transmitted 1-simple wave again starts an interaction with the 2-shock from the point (x_2, τ_2). Here we may take the origin temporarily at this point and read the vertical intercepts $\bar{\eta}$ of all the characteristics C^2. This information can be used in Eq. (4.3.20), which reads

$$
\begin{aligned}
\bar{x}(\omega_1) &= \frac{\gamma_1 \gamma_2}{\nu \omega_2^f (\gamma_2 - \gamma_1)} \left\{ \frac{\bar{\eta}(\omega_1)}{\omega_1} + \frac{\gamma_2}{\gamma_2 - \gamma_1} \int_{\omega_1^f}^{\omega_1} \frac{\bar{\eta}(\omega_1)}{\omega_1^2} d\omega_1 \right\} \\
&= 0.9456 \left\{ \frac{\bar{\eta}(\omega_1)}{\omega_1} + 2 \int_{3.387}^{\omega_1} \frac{\bar{\eta}(\omega_1)}{\omega_1^2} d\omega_1 \right\}
\end{aligned}
\qquad (4.4.12)
$$

Then the curved portion of the shock path S^2 is given by

$$
\begin{aligned}
x(\omega_1) &= x_2 + \bar{x}(\omega_1) \\
\tau(\omega_1) &= \bar{\eta}(\omega_1) + (1 + 0.2115\omega_1^2)x(\omega_1)
\end{aligned}
\qquad (4.4.13)
$$

for $3.387 \leq \omega_1 \leq 5.0$. This interaction is over at the point (x_β, τ_β) and yields the separation between the solutes A_1 and A_2.

On the other hand, the 2-simple wave starts to interact with the once-transmitted 3-shock from the point (x_3, τ_3) and the interaction continues to the point (x_γ, τ_γ), at which we see the solutes A_2 and A_3 being separated completely. This portion of the shock path S^3 is also described by Eq. (4.3.19) with x_3 for x_0; i.e.,

$$x(\omega_2) = \frac{\tau_0 \gamma_1 \gamma_2 \gamma_3 (\gamma_3 - \omega_1^f)(\gamma_3 - \omega_2^f)}{\nu \omega_1^f \omega_2^f \omega_3^f (\gamma_3 - \gamma_1)(\gamma_3 - \omega_2)^2} = \frac{153.8806}{(15 - \omega_2)^2}$$

$$\tau(\omega_2) = 10 + (1 + 0.1256\omega_2^2)x(\omega_2)$$

$$(4.4.14)$$

for $7.05 \leq \omega_2 \leq 10.0$ or, by eliminating ω_2 from these equations, we obtain

$$\sqrt{\tau - x - 10} = 5.3160\sqrt{x} - 4.3963 \qquad (4.4.15)$$

and the curved portion of the shock path S^3 is again given by a part of a parabola.

The interactions between the twice-transmitted 1-simple wave and the 1-shock begins at the point (x_4, τ_4) and goes on indefinitely while the two waves cancel each other. The shock path S^1 at this stage can be determined by using Eq. (4.2.17) with properly adjusted $\overline{\eta}(\omega_1)$. Similarly, the once-transmitted 2-simple wave starts to overtake the once-transmitted 2-shock at the point (x_5, τ_5). The two waves are gradually canceled by each other when they meet along the shock path S^2, which we can locate by applying Eq. (4.2.17) with $k = 2$. It is obvious that the 3-simple wave will eventually overtake the twice-transmitted 3-shock. This will take place from the point (x_6, τ_6), not shown in Fig. 4.10, and the shock path S^3 beyond this point can be described by Eq. (4.2.15) with $k = 3$ and x_6 for x_0. Hence this is also given by a portion of a parabola.

From the physical plane portrait of the solution we can construct the concentration profiles inside the column at successive times. The profiles are taken at various times, that are representative of distinct stages of the process, and presented in Fig. 4.11. In the first part of the figure ($\tau = 10$) the sample has just been put on the column to give the three solute chromatogram. The next three parts ($\tau = 13$, $\tau = 15.5$, and $\tau = 23$) clearly show how the development of the chromatogram proceeds. By $\tau = 35$, the solute A_1 has been separated from A_2 and A_3. The chromatographic mixture of these two solutes continues to develop ($\tau = 55$), and finally, the two have also separated as shown in the last part ($\tau = 95$). In the last three parts of Fig. 4.11 we can see the tailing off of the pure solute bands and the decay of shocks with time.

The ratio r of the adsorbent volume required to the sample volume is equal to $\nu x_\gamma / \tau_0 = 0.9233$ from Eq. (4.4.9), and this value is even lower than that for the two-solute problems treated in Sec. 2.8. But this is not an unexpected matter because we had $\gamma_2/\gamma_1 = 1.5$ there in comparison to $\gamma_2/\gamma_1 = 2$ and $\gamma_3/\gamma_1 = 3$ here. If we treat a problem with $\gamma_1 = 5$, $\gamma_2 = 7.5$, and $\gamma_3 = 15$, we would expect to have $x_\beta > x_\gamma$, and the ratio r may become larger than 1.5 (see Example 4.4.3). Such a small value of r is attributed to very different val-

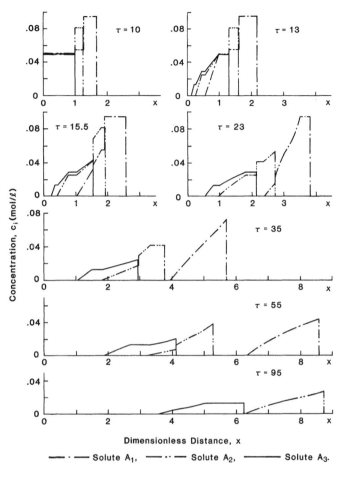

Figure 4.11

ues of γ_i chosen, and this example demonstrates that the chromatographic cycle can be very effective in separating solutes with very different values of γ_i.

Exercises

4.4.1 Derive equations for the shock path S^3 between the points (x_1, τ_1) and (x_α, τ_α) and also between points (x_3, τ_3) and (x_γ, τ_γ) in Fig. 4.10, and show that each part of the S^3 is a portion of a parabola. By using the resulting equations, confirm Eqs. (4.4.3) and (4.4.8).

4.4.2. A mixture containing three solutes A_1, A_2, and A_3 is to be separated by dividing into equal portions such that a given chromatographic column will just separate the three solutes contained in one sample. Assume that the values of γ_i and the concentrations are such that $x_\beta < x_\gamma$. If the samples are fed to the column at intervals as frequently as possible, show that one period would consist of an interval τ_0 for sample introduction and another interval

$$\tau_1 = \nu(\gamma_3 - \omega_1^f)x_\gamma = \frac{\tau_0 \gamma_1 \gamma_2 \gamma_3(\gamma_3 - \omega_1^f)(\gamma_3 - \omega_2^f)}{\omega_1^f \omega_2^f \omega_2^f(\gamma_3 - \gamma_1)(\gamma_3 - \gamma_2)^2}$$

for elution.

4.4.3. For the example treated in this section, determine the shock path S^2 between the points (x_2, τ_2) and (x_β, τ_β) by applying Eq. (4.4.6) and confirm the coordinates (x_β, τ_β) and (x_5, τ_5) given in Table 4.6.

4.4.4. Consider a chromatographic system with $\gamma_2 = 7.5$ and otherwise the same as that of the numerical example treated in this section and repeat the analysis. In particular, check if x_β is greater than x_γ and determine the ratio r of the adsorbent volume required to the sample volume to be treated.

4.5 Multicomponent Separation by Displacement Development

In this and following sections we shall take up the subject of displacement development for which an introductory treatment was already given in Sec. 2.9. Here we have a mixture containing $(m - 1)$ solutes and, as before, put this on a column of adsorbent over a finite period of time to form a chromatogram. This is then displaced through the column by a solution of a solute that is more strongly adsorbed than any solute present in the chromatogram. This solute is known as the development agent or the developer. When the operation is successful, the solutes are gradually separated and form pure solute bands. At the ultimate stage these bands maintain sharp boundaries at both ends and propagate side by side in sequence all with the same speed.

Before treating the general problem it is well to analyze in some detail the process applied to a two-solute mixture and elucidate various features that are of importance. Although we treated the development of a single-solute chromatogram in Sec. 2.9, we shall see that the development here is enough, just adequate for a generalization.

Suppose that we have a fresh bed of adsorbent and saturate it with a mixture containing two solutes A_1 and A_2 over a finite time interval τ_0. The feed is then changed to a stream containing only A_3, which is the developer. Thus we have a three-solute system. If the feed mixture and the developer solution maintain constant concentrations, we again have a piecewise constant data problem with two points of discontinuity, one at the origin and the other at the point $(0, \tau_0)$. Two sets of wave solutions will appear in the (x, τ)-plane and the waves will be engaged in interactions as illustrated in Fig. 4.12 for a particular case.

Here and in the following section we shall use capital letters without

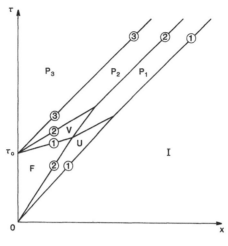

Figure 4.12

parentheses in the figure to represent the constant state. Shock paths and characteristics will be denoted by heavier lines and lighter lines, respectively, labeled with encircled numbers corresponding to their families. For brevity we also use the notation $D[\omega_1, \omega_2, \ldots, \omega_k, \ldots, \omega_m]$ to represent the set of the characteristic parameters $\{\omega_k\}$ corresponding to the state D.

For the state I of the initial bed we have $I[\gamma_1, \gamma_2, \gamma_3]$, whereas for the state F of the mixture Eq. (13.6.16) gives one root γ_3 and a quadratic equation from which we obtain two more roots ω_1^f and ω_2^f such that $0 < \omega_1^f \le \gamma_1 \le \omega_2^f \le \gamma_2$, so we have $F[\omega_1^f, \omega_2^f, \gamma_3]$. It is immediately apparent that in the (x, τ)-plane, two shocks will develop from the origin with the 3-wave missing because $\omega_3 = \gamma_3$ for both states I and F. Thus the solution in the neighborhood of the origin is identical to that of the two-solute saturation problem for which the solute A_2 is exhausted across the 2-shock and the solute A_1 across the 1-shock. Between the two shocks a new constant state U will appear, and clearly we have $U[\omega_1^f, \gamma_2, \gamma_3]$. With the ω-values known for F, U, and I, the propagation directions of the two shocks can readily be determined from Eq. (3.9.16) or (3.9.17) to give

$$\bar{\sigma}_2 = 1 + \frac{\nu \omega_1^f \omega_2^f}{\gamma_1}$$

$$\bar{\sigma}_1 = 1 + \nu \omega_1^f \tag{4.5.1}$$

For the state P_3 of the developer stream which contains A_3 only, Eq. (3.3.16) yields two roots, γ_1 and γ_2, and another root ω^* determined by the developer concentration c_3^* as

$$\frac{K_3 n_3^*}{\gamma_3 - \omega_3^*} = 1$$

or

$$\omega^* = \frac{\gamma_3}{1 + \phi_3^*} \tag{4.5.2}$$

since the value of ω^* can vary from zero to γ_3 depending on the value of ϕ_3^*, the ordering of the parameters ω^*, γ_1, and γ_2 is not fixed. If $\omega^* < \gamma_1 < \gamma_2$, for example, we have $P_3[\omega^*, \gamma_1, \gamma_2]$, and thus the wave solution from the point $(0, \tau_0)$ consists of a 3-shock, a 2-shock, and a 1-shock if $\omega^* < \omega_1^f$ or a 1-simple wave if $\omega^* > \omega_1^f$. Across the 3-shock ω_3 increases from γ_2 to γ_3 so that A_3 is exhausted while A_2 emerges. The constant state P_2 which will appear between the 3- and 2-shocks has the characteristic parameters $P_2[\omega^*, \gamma_1, \gamma_3]$. This implies that the state P_2 contains A_2 only and thus represents the pure solute band for A_2. Between the 2-shock and the 1-wave we have the constant state $V[\omega^*, \omega_2^f, \gamma_3]$. The three waves developing from the point $(0, \tau_0)$ can be established by using Eq. (3.8.2) or (3.9.16):

$$\tilde{\sigma}_3 = 1 + \nu\omega^*$$

$$\tilde{\sigma}_2 = 1 + \frac{\nu\omega^*\omega_2^f}{\gamma_2}$$

$$\tilde{\sigma}_1 = 1 + \frac{\nu\omega^*\omega_1^f\omega_2^f}{\gamma_1\gamma_2} \qquad \text{if } \omega^* < \omega_1^f \tag{4.5.3}$$

$$\sigma_1 = 1 + \frac{\nu\omega_1^2\omega_2^f}{\gamma_1\gamma_2} \quad \text{for } \omega_1^f \leq \omega_1 \leq \omega^* \qquad \text{if } \omega^* > \omega_1^f$$

All the interactions can be analyzed without difficulty by applying the results of Secs. 4.2 and 4.3. The new constant state will turn out to have the characteristic parameters $P_1[\omega^*, \gamma_2, \gamma_3]$, so a pure solute band for A_1 is formed.

If the displacement development is to be successful, the process should eventually reach a stationary state at which the constant state P_i contains A_i only for $i = 1, 2, 3$, respectively, and the four constant states P_3, P_2, P_1, and I must be partitioned by three shocks as shown in Fig. 4.12. Obviously, two of the characteristic parameters for each P_i will be given by γ_j, $j \neq i$, while for the state I we have $I[\gamma_1, \gamma_2, \gamma_3]$.

We shall now start from the initial state $I[\gamma_1, \gamma_2, \gamma_3]$ and cross the 1-shock from the right to the left. Here the parameter ω_1 is the only variable and must decrease from γ_1 to a lower value ω_1^* to give $P_1[\omega_1^*, \gamma_2, \gamma_3]$. Next when we pass through the 2-shock in the same direction, the only variable is the parameter ω_2, which is to decrease from γ_2 to γ_1. For the state P_2, therefore, we have $P_2[\omega_1^*, \gamma_1, \gamma_3]$. Finally, across the 3-shock only the parameter ω_3 is allowed to vary and actually decreases from γ_3 to γ_2 so that we have $P_3[\omega_1^*, \gamma_1, \gamma_2]$. This argument is summarized in Table 4.7, in which the arrow indicates the variation when the shock is crossed from right to left.

It is clear from the argument above that ω_1^* must be equal to ω^* given by Eq. (4.5.2) and thus $\omega^* < \gamma_1$ or

TABLE 4-7.   CONSTANT STATES AND CHARACTERISTIC PARAMETERS

Constant State	Solute Present	Parameters $\{\omega_k\}$			Connecting Wave
		ω_1	ω_2	ω_3	
I	None	γ_1	γ_2	γ_3	1-shock
P_1	A_1	ω_1^*	γ_2	γ_3	2-shock
P_2	A_2	ω_1^*	γ_1	γ_3	3-shock
P_3	A_3	ω_1^*	γ_1	γ_2	

$$\phi_3^* > \frac{\gamma_3}{\gamma_1} - 1 \equiv \phi_{3,\mathrm{cr}}^* \tag{4.5.4}$$

For a successful operation, therefore, the developer concentration ϕ_3^* must be higher than the critical value $\phi_{3,\mathrm{cr}}^*$ defined by Eq. (4.5.4). We note that this critical value depends only on the adsorption characteristics of the developer A_3 and the least strongly adsorbed solute A_1 and independent of that of the solute A_2 as well as the state of the feed mixture to be separated.

Since the value ω_1^* is common to the states P_1, P_2, and P_3 in Table 4.7, we can write

$$\omega^* = \omega_1^* = \frac{\gamma_3}{1 + \phi_3^*} = \frac{\gamma_2}{1 + \phi_2^*} = \frac{\gamma_1}{1 + \phi_1^*} \tag{4.5.5}$$

in which ϕ_1^* and ϕ_2^* denote the concentrations of the pure solute in their respective pure solute bands. These values are directly determined from Eq. (4.5.5) to give

$$\phi_i^* = \frac{\gamma_i}{\gamma_3}(1 + \phi_3^*) - 1, \qquad i = 1, 2 \tag{4.5.6}$$

which is independent of the state of the original mixture to be treated.[†]

The propagation directions of shocks are readily determined by using the characteristic parameters listed in Table 4.7. We immediately find that the three directions are given by the same expression

$$\tilde{\sigma} = \tilde{\sigma}_1 = \tilde{\sigma}_2 = \tilde{\sigma}_3 = 1 + \nu\omega^*$$
$$= 1 + \frac{\nu\gamma_i}{1 + \phi_i^*} \tag{4.5.7}$$

which is compatible with Eq. (4.5.6). This implies that the three shock paths are all parallel to one another, and thus the widths of pure solute bands, P_1 and P_2, remain constant as they propagate. This is just what we would expect from conservation of mass.

[†] This was observed experimentally by Tiselius as early as in 1943 and also theoretically by Glueckauf (1946) and Claesson (1949).

To illustrate the development further we shall treat various cases classified in terms of the developer concentration ϕ_3^*. Let us first consider the case when $\phi_3^* > \phi_{3,\text{cr}}^*$. We have $\omega^* < \gamma_1$ and hence the process will reach ultimately the state shown in Fig. 4.12 with the characteristic parameters given in Table 4.7. However, two different situations are expected to occur in the intermediate stage, depending on the comparison between ω^* and ω_1^f.

Suppose now that the developer concentration ϕ_3^* is sufficiently high to give $\omega^* < \omega_1^f < \gamma_1$. Then the connection between the state $P_3[\omega^*, \gamma_1, \gamma_2]$ and the state $F[\omega_1^f, \omega_2^f, \omega_3^f]$ are made by three shocks because all three parameters ω_1, ω_2, and ω_3 increase from P_3 to F. The physical plane portrait is exactly the same as illustrated in Fig. 4.12.

If ϕ_3^* takes a somewhat lower value so that $\omega_1^f < \omega^* < \gamma_1$, the overall feature remains the same except for the fact that the 1-wave developing from the point $(0, \tau_0)$ is a centered simple wave as shown in Fig. 4.13. This 1-simple wave meets the 2-shock originating from the origin and the two are transmitted through each other while interacting. The characteristic parameters for the constant states U and V are given by $U[\omega_1^f, \gamma_2, \gamma_3]$ and $V[\omega^*, \gamma_1, \gamma_3]$, respectively, and since ω_k is the only variable across a k-wave, for the state P_1 we have $P_1[\omega^*, \gamma_2, \gamma_3]$. The transmitted 1-simple wave will interact with the 1-shock originating from the origin, and the two cancel each other until the 1-simple wave is completely canceled because $\omega^* < \gamma_1$. Consequently, pure solute bands are established in a satisfactory manner. However, the length of the column required will become larger than that in the previous case.

Next we shall consider the case when $\phi_3^* < \phi_{3,\text{cr}}^*$. Here we have $\omega^* > \gamma_1$, so the process is expected to become unsatisfactory in the sense that pure sol-

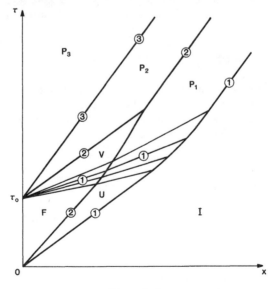

Figure 4.13

ute bands are not partitioned by sharp boundaries. We shall discuss two different cases to be encountered depending on the value of ω^*.

If the developer concentration ϕ_3^* is slightly lower than the critical value $\phi_{3,cr}^*$, then we have $\gamma_1 < \omega^* < \gamma_2$ and $P_3[\gamma_1, \omega^*, \gamma_2]$. Since ω_3 increases from γ_2 to γ_3 in the x-direction, the 3-wave developing from the point $(0, \tau_0)$ is a shock across which the developer A_3 is exhausted. For the state P_2 we have $P_2[\gamma_1, \omega^*, \gamma_3]$. The 2-wave developing from the point $(0, \tau_0)$ will be a shock if $\omega^* < \omega_2^f$ or a simple wave if $\omega^* > \omega_2^f$. In either case we have $V[\gamma_1, \omega_2^f, \gamma_3]$. The 1-wave, however, is necessarily a simple wave because ω_1 is to decrease from γ_1 to ω_1^f in the x-direction.

Since we have $U[\omega_1^f, \gamma_2, \gamma_3]$ and $V[\gamma_1, \omega_2^f, \gamma_2]$, the interaction between the 1-simple wave centered at the point $(0, \tau_0)$ and the 2-shock originating from the origin will generate a constant state $O[\gamma_1, \gamma_2, \gamma_3]$ which represents the state of no solute as shown in Fig. 4.14. Both states O and I have $\omega_1 = \gamma_1$, so the interaction between the transmitted 1-simple wave and the 1-shock originating from the origin will go on indefinitely. An immediate consequence is that the pure solute band for A_1 does not propagate side by side with that for A_2 and it tails off continuously in the rear while the height of the front decreases. This is precisely what we would expect to obtain when a chromatogram of A_1 is eluted by pure solvent. The interaction between the 2-waves will ultimately give rise to a 2-shock, which connects states $P_2[\gamma_1, \omega^*, \omega_3]$ and $O[\gamma_1, \gamma_2, \gamma_3]$.

Consequently, the displacement development becomes only partially successful in the sense that the pure solute band for A_2 maintains sharp boundaries, whereas that for A_1 not only stays away from the A_2-band, but also its rear boundary is diffuse. Here we may say that the presence of the devel

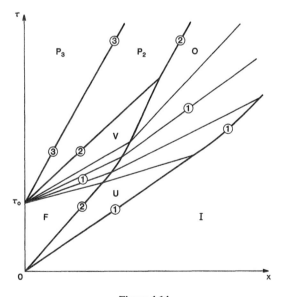

Figure 4.14

oper A_3 has no influence over the behavior of solute A_1. We note, however, that this situation can be rather advantageous if A_1 is an impurity to be removed from the mixture completely or if it is desired to recover A_1 in high purity.

Let us now suppose that the developer concentration ϕ_3^* is extremely low so that we have $\gamma_2 < \omega^* < \gamma_3$ and $P_3[\gamma_1, \gamma_2, \omega^*]$. Since ω_3 increases from ω^* to γ_3 in the x-direction, the 3-wave developing from the point $(0, \tau_0)$ is a shock, but on the right hand side we have the state of no solute $O_2[\gamma_1, \gamma_2, \gamma_3]$, as marked in Fig. 4.15. It is immediately apparent that the situation to the right of the 3-shock path is exactly the same as that which we would obtain in the conventional process of elution of a two-solute chromatogram by pure solvent (see Figs. 2.25 and 2.27). Clearly, the developer A_3 has no influence over the behavior of both solutes A_1 and A_2, so that the two solutes are simply eluted by pure solvent.

Since both ω_2 and ω_1 decrease from the state P_3 to the state F, the 2- and 1-waves developing from the point $(0, \tau_0)$ are simple waves and for the constant state V we have $V[\gamma_1, \omega_2^f, \gamma_3]$. The 1-simple wave collides with the 2-shock originating from the origin and the two are transmitted through each other. With $U[\omega_1^f, \gamma_2, \gamma_3]$ and $V[\gamma_1, \omega_2^f, \gamma_3]$ the interaction will give a new constant state $O_1[\gamma_1, \gamma_2, \gamma_3]$ as shown in Fig. 4.15. This implies that the two solutes are separated into pure solute bands with the state O_1 of no solute in between. Both bands tend to tail off in the rear and be lessened in height. Consequently, the process is unsuccessful.

In summary, we conclude that displacement development becomes successful only if $\phi_3^* > \phi_{3,cr}^*$. Otherwise, the developer has no influence over both

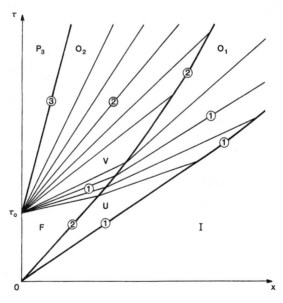

Figure 4.15

Wave Interactions in Multicomponent Chromatography   Chap. 4

solutes A_1 and A_2 or over the least strongly adsorbed solute A_1, which are then simple eluted by pure solvent.

Finally, we would like to extend the development in such a way that it can be applied directly to multisolute systems. Thus we shall consider a mixture containing $(m - 1)$ different solutes, $A_1, A_2, \ldots, A_{m-1}$, and a solution containing the developer A_m only, which is more strongly adsorbed than any solute in the original mixture. As before, an initially clean column (I) is first saturated with the original mixture (F) for a finite period of time τ_0 and then fed with the developer solution (P_m).

The characteristic parameters for the initial state I of the column are given as

$$I[\gamma_1, \gamma_2, \ldots, \gamma_k, \ldots, \gamma_m]$$

and for the state F of the original mixture we obtain, from Eq. (3.6.16),

$$F[\omega_1^f, \omega_2^f, \ldots, \omega_k^f, \ldots, \omega_{m-1}^f, \gamma_m]$$

where $\omega_1^f \leq \gamma_1 \leq \omega_2^f \leq \gamma_2 \leq \cdots \leq \omega_{m-1}^f \leq \gamma_{m-1} < \gamma_m$. Since ω_m remains the same between the states F and I, the set of waves that will develop from the origin consists of $(m - 1)$ waves with the m-wave missing as illustrated in Fig. 4.16. These are all shocks because ω_k increases from ω_k to γ_k when we pass through the k-wave from left to right, and this is true for all k from 1 to $m - 1$.

For the state P_m of the developer solution Eq. (3.6.16) gives $(m - 1)$ roots $\gamma_1, \gamma_2, \ldots, \gamma_{m-1}$ directly and another root ω^* in terms of the developer con-

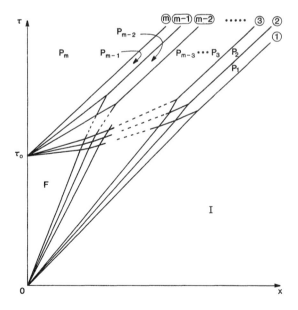

Figure 4.16

centration ϕ_m^* as

$$\omega^* = \frac{\gamma_m}{1 + \phi_m^*} \qquad (4.5.8)$$

Now ω^* can take any value between zero and γ_m depending on the value of ϕ_m^*, and thus the ordering of the parameters $\{\omega_k\}$ is not fixed. If $\gamma_{j-1} < \omega^* < \gamma_j$, for instance, ω^* is the jth root of Eq. (3.6.16), so that we have

$$P_m[\gamma_1, \gamma_2, \ldots, \gamma_{j-1}, \omega^*, \gamma_j, \ldots, \gamma_{m-1}]$$

In this case the set of waves that will develop from point $(0, \tau_0)$ is composed of $(m - j)$ shocks on the left-hand side and $(j - 1)$ waves on the right-hand side. The j-wave will be a shock if $\omega^* < \omega_j^f$ or a simple wave if $\omega^* > \omega_j^f$.

As time goes on, the two sets of waves developing from the origin and the point $(0, \tau_0)$ will be engaged in a series of interactions, but these can be analyzed by applying the results of Secs. 4.2 and 4.3. If the displacement development is to be successful, we must have m shocks when all the interactions are completed. These shocks should propagate along parallel paths as shown in Fig. 4.16.

Suppose that we start from the state I and cross the 1-shock from right to left. Being the only variable parameter, ω_1 decreases from γ_1 to a lower value ω_1^*; i.e., $\omega_1^* < \gamma_1$. Since ω_1 varies only across the 1-wave, such a value ω_1^* remains unchanged as we cross successively the 2-shock through the m-shock. This implies that the value ω_1^* is the smallest root of Eq. (3.6.16) for every constant state P_i, $i = 1, 2, \ldots, m$.

We now recall that for the state F Eq. (3.6.16) gives $(m - 1)$ roots $\gamma_1, \gamma_2, \ldots, \gamma_{m-1}$ directly and another root ω^* determined by Eq. (4.5.8). It then follows that

$$\omega^* = \omega_1^* = \frac{\gamma_m}{1 + \phi_m^*} < \gamma_1$$

or

$$\phi_m^* > \frac{\gamma_m}{\gamma_1} - 1 \equiv \phi_{m,\,\mathrm{cr}}^* \qquad (4.5.9)$$

It is noticed that the critical value of the developer concentration is determined by the adsorption characteristics of itself and the least strongly adsorbed solute in the system. It is independent of the properties of intermediate solutes as well as of the state of the original mixture. When the developer concentration ϕ_m^* is greater than $\phi_{m,\,\mathrm{cr}}^*$, all the waves at the ultimate stage will be shocks that propagate with the same speed and form boundaries of $(m - 1)$ pure solute bands traveling side by side without interval. Hence the displacement development becomes successful.

At the ultimate stage we have m constant states P_j, $j = 1, 2, \ldots, m$, representing m pure solute bands. We have already observed that for every state

P_i, $i = 1, 2, \ldots, m$, the smallest characteristic parameter ω_1 is equal to ω^*, while the other $(m - 1)$ parameters are given by γ_k, $k \neq i$. Therefore, we can write

$$\omega^* = \frac{\gamma_1}{1 + \phi_1^*} = \frac{\gamma_2}{1 + \phi_2^*} = \cdots = \frac{\gamma_i}{1 + \phi_i^*} = \cdots = \frac{\gamma_m}{1 + \phi_m^*} \qquad (4.5.10)$$

where ϕ_i^* represents the concentration of A_i in the pure solute band for A_i. These equations can be rearranged to express ϕ_i^* in terms of the developer concentration ϕ_m^*; i.e.,

$$\phi_i^* = \frac{\gamma_i}{\gamma_m}(\phi_m^* - 1) - 1, \qquad i = 1, 2, \ldots, m-1 \qquad (4.5.11)$$

This particularly simple form of equations immediately suggests the use of displacement development not only to identify an unknown mixture as noticed by Claesson (1949) but also to determine the isotherm parameters γ_i and K_i.

Since the characteristic parameters are given for all the constant states P_i, it is not difficult to find the propagation directions of shocks and it turns out that they are given by

$$\tilde{\sigma} = \tilde{\sigma}_1 = \tilde{\sigma}_2 = \cdots = \tilde{\sigma}_m = 1 + \nu\omega^*$$
$$= 1 + \frac{\nu\gamma_i}{1 + \phi_i^*} \qquad (4.5.12)$$

Thus all the shock paths are parallel in the (x, τ)-plane.

If we let Δx_i denote the horizontal width of the A_i band, a simple material balance yields

$$(c_i^* + \nu n_i^*)\, \Delta x_i = \tau_0 c_i^f$$

or, after substituting the Langmuir isotherm for n_i^*,

$$\Delta x_i = \frac{\tau_0 c_i^f}{\tilde{\sigma} c_i^*} \qquad (4.5.13)$$

The vertical distance $\Delta\tau_i$ between the two shock paths S^{i+1} and S^i is then given by

$$\Delta\tau_i = \tau_0 \frac{c_i^f}{c_i^*} \qquad (4.5.14)$$

and this is the width of the pure solute b and for A_j which we would expect to observe in the breakthrough curve. Since the path of the m-shock is given a priori, we can locate all the shock paths at the ultimate stage without going through the analysis of interactions.

However, the column length required for complete separation can be determined only after all the interactions are analyzed. All the waves being shocks, the analysis should be straightforward, although it may become tedious as the number of components involved increases. If $\omega^* > \omega_1^f$, the 1-wave from the point $(0, \tau_0)$ is a simple wave and thus the successive interactions between this 1-simple wave and each of the shocks originating from the origin must be analyzed. This case will be illustrated for $m = 4$ in the next section.

4.6 Example: Three-Solute Separation by Displacement Development

To illustrate the application of the theoretical development of the preceding section, we shall consider here a specific example with numericial values for all the parameters and discuss the analysis in some detail. Suppose that we have

$$\epsilon = 0.4 \quad \text{or} \quad \nu = 1.5$$

$$N_i = N = 1.0 \text{ mol/liter} \quad \text{for} \quad i = 1, 2, 3, 4$$

$$\tau_0 = 10$$

while other parameters and conditions are specified as given in Table. 4.8.

TABLE 4-8.   NUMERICAL VALUES FOR THE PARAMETERS AND CONDITIONS

	i			
	1	2	3	4
K_i (liters/mol)	5	10	15	20
c_i^0 (mol/liter)	0	0	0	0
c_i^f (mol/liter)	0.05	0.05	0.05	0

The critical value of the developer concentration is

$$c_{4,\text{cr}}^* = \frac{1}{20}\left(\frac{20}{5} - 1\right) = 0.15 \text{ mol/liter}$$

which is very high when compared to the feed concentrations. As before we shall use the symbol $D[\omega_1, \omega_2, \omega_3, \omega_4]$ to represent the set of characteristic parameters $\{\omega_k\}$ corresponding to the state D. For the initial state I of the column we have $I[5, 10, 15, 20]$. For the state F Eq. (3.6.16) becomes a cubic equation for ω which is solved to give the first three ω-values: $\omega_1^f = 3.387$, $\omega_2^f = 7.050$, and $\omega_3^f = 12.563$. The largest root ω_4^f is equal to γ_4 and hence we have $F[3.387, 7.050, 12.563, 20]$.

It is obvious that three shocks will develop from the origin as shown in the neighborhood of the origin in Fig. 4.17. For the two constant states U_1 and U_2 we have the characteristic parameters given by $U_1[3.387, 10, 15, 20]$ and $U_2[3.387, 7.050, 15, 20]$. The slopes of the three shock paths are determined by applying Eq. (3.9.17):

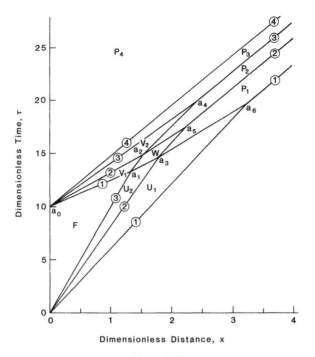

Figure 4.17

$$\tilde{\sigma}_3 = 1 + (1.5)(12.563)\left(\frac{3.387}{5}\right)\left(\frac{7.050}{10}\right)\left(\frac{15}{15}\right)\left(\frac{20}{20}\right) = 10.0$$

$$\tilde{\sigma}_2 = 1 + (1.5)(7.050)\left(\frac{3.387}{5}\right)\left(\frac{10}{10}\right)\left(\frac{15}{15}\right)\left(\frac{20}{20}\right) = 8.164$$

$$\tilde{\sigma}_1 = 1 + (1.5)(3.387)\left(\frac{5}{5}\right)\left(\frac{10}{10}\right)\left(\frac{15}{15}\right)\left(\frac{20}{20}\right) = 6.081$$

The discussion above is valid for any value of the developer concentration c_4^*.

Case for which $c_4^* > c_{4,cr}^*$ and $\omega^* < \omega_1^f$

Let us first consider a case for which the developer concentration is much greater than its critical value 0.15 mol/liter. We shall take $c_4^* = 0.35$ mol/liter, so that from Eq. (4.5.8) we have

$$\omega^* = \frac{20}{1 + (20)(0.35)} = 2.5 \ < \ \omega_1^f$$

Thus the state P_4 is characterized as $P_4[2.5, 5, 10, 15]$.

We expect to obtain pure solute bands at the ultimate stage as depicted in Fig. 4.17 and the concentration plateaus are determined by using Eq. (4.5.11); i.e.,

$$c_1^* = \frac{1 + (20)(0.35)}{20} - \frac{1}{5} = 0.20 \text{ mol/liter}$$

$$c_2^* = \frac{1 + (20)(0.35)}{20} - \frac{1}{10} = 0.30 \text{ mol/liter}$$

$$c_3^* = \frac{1 + (20)(0.35)}{20} - \frac{1}{15} = \frac{1}{3} \text{ mol/liter}$$

The four shocks, forming the boundaries of the pure solute bands, will propagate along parallel lines in the (x, τ)-plane, whose slopes are given by Eq. (4.5.12):

$$\tilde{\sigma} = 1 + \frac{(1.5)(20)}{1 + (20)(0.35)} = 4.75$$

The band widths are then determined by Eqs. (4.5.13) and (4.5.14) to give

$$\Delta x_3 = \frac{(10)(0.05)}{(4.75)(1/3)} = 0.316, \qquad \Delta \tau_3 = 1.5$$

$$\Delta x_2 = \frac{(10)(0.05)}{(4.75)(0.3)} = 0.351, \qquad \Delta \tau_2 = 1.667$$

$$\Delta x_1 = \frac{(10)(0.05)}{(4.75)(0.2)} = 0.526, \qquad \Delta \tau_1 = 2.5$$

The path S^4 of the 4-shock originating from the point $a_0(0, 10)$ can be located by using the slope $\tilde{\sigma}$ and then the paths of the other three shocks by using the band widths Δx_j or $\Delta \tau_j$ with the same slope. This procedure will complete the solution attained at the ultimate stage. In this manner we can obtain all the information needed for practical purposes except for the length of column required for complete separation. To determine the length of the column, it is necessary to go through the analysis of interactions, which will be discussed here in detail.

Since we have $P_4[2.5, 5, 10, 15]$ in the upstream side and $F[3.387, 7.050, 12.568, 20]$ in the downstream side, four shocks are expected to develop from the point $a_0(0, 10)$ as shown in Fig. 4.17. The three intermediate constant states will have the characteristic parameters given by $P_3[2.5, 5, 10, 20]$, $V_2[2.5, 5, 12.563, 20]$, and $V_1[2.5, 7.050, 12.563, 20]$. We note that state P_3 contains A_3 only and thus the A_3-band is established from the beginning. The slopes of the shock paths are determined by using Eq. (3.9.16) to give

$$\tilde{\sigma}_4 = 1 + (1.5)(20)\left(\frac{2.5}{5}\right)\left(\frac{5}{10}\right)\left(\frac{10}{15}\right)\left(\frac{15}{20}\right) = 4.75$$

$$\bar{\sigma}_3 = 1 + (1.5)(12.563)\left(\frac{2.5}{5}\right)\left(\frac{5}{10}\right)\left(\frac{10}{15}\right)\left(\frac{20}{20}\right) = 4.141$$

$$\bar{\sigma}_2 = 1 + (1.5)(7.050)\left(\frac{2.5}{5}\right)\left(\frac{5}{10}\right)\left(\frac{12.563}{15}\right)\left(\frac{20}{20}\right) = 3.214$$

$$\bar{\sigma}_1 = 1 + (1.5)(3.387)\left(\frac{2.5}{5}\right)\left(\frac{7.050}{10}\right)\left(\frac{12.563}{15}\right)\left(\frac{20}{20}\right) = 2.50$$

The shocks developed from the origin are engaged in a sequence of interactions with the shocks originating from the point $a_0(0, 10)$. When interactions between shocks of different families are over, new constant states W, P_2, and P_1 will appear as depicted in Fig. 4.17. The characteristic parameters for each of these constant states are easily identified by looking at the characteristic parameters for the constant states U_1, U_2, V_1, V_2, P_3, and I. Hence we have $W[2.5, 7.050, 15, 20]$, $P_2[2.5, 5, 15, 20]$, and $P_1[2.5, 10, 15, 20]$. It follows that the states P_1 and P_2, respectively, contain A_1 and A_2 only. After the interactions between shocks of the same family that take place at points a_4, a_5, and a_6, respectively, the four shock paths become parallel and the pure solute bands for A_1, A_2, and A_3 are fully established. Beyond the point a_6 the pure solute bands are simply propagating without changing their shape and size.

The slopes of the shock paths after each interaction can be calculated by using the characteristic parameters on both sides in Eq. (3.9.16) or (3.9.17). Once these slopes are known, the points of interaction are determined by computation or by a graphical method. The slopes of the shock paths and the coordinates of the points of interaction are listed in Tables 4.9 and 4.10, respectively.

We now have all the information needed to complete the solution in the (x, τ)-plane as shown in Fig. 4.17. At the ultimate stage the shock paths become parallel as expected and the band widths are found to be the same as those predicted a priori. Since the solution at the ultimate stage is obtained at the point a_6, the length of the column required for a full development of pure solute

TABLE 4-9.   SLOPES OF VARIOUS SHOCK PATHS

State on the Left-Hand Side	State on the Right-Hand Side	Index of the Shock, k	Slope of the Shock Path, $\bar{\sigma}_k$
V_1	W	3	7.643
V_2	P_2	3	5.711
P_3	P_2	3	4.750
P_2	W	2	3.644
W	P_1	2	6.288
P_2	P_1	2	4.750
W	U_2	1	2.791
P_1	U_1	1	3.540
P_1	I	1	4.750

TABLE 4-10. COORDINATES OF THE POINT OF INTERACTION

Point of Interaction	Coordinates		Mode of Interaction
	x	τ	
a_1	1.333	13.333	Transmission
a_2	1.549	14.978	Transmission
a_3	1.789	14.605	Transmission
a_4	2.464	20.203	Confluence
a_5	2.261	17.573	Confluence
a_6	3.255	19.794	Confluence

bands is 3.255. It should be noted, however, that the state U_1 also contains A_1 alone and thus complete separation actually obtains at the point a_4. Hence we may proceed with a column of length 2.464 as well. In this case the state U_1 will break through first and then the state P_1 follows. The ratio of the adsorbent volume required to the sample volume treated is determined by Eq. (4.4.9), i.e.,

$$r = \nu \frac{x_6}{\tau_0} = \frac{(1.5)(2.464)}{10} = 0.3696$$

Such a low value for this ratio is attained at the expense of the high developer concentration $c_4^* = 0.35$ mol/liter. If the same volume of sample mixture is to be separated by using pure solvent ($c_4^* = 0$), the ratio will be 0.9233, as we have seen in Sec. 4.4. This implies that with $c_4^* = 0.35$ mol/liter we can reduce the adsorbent volume required by 60%.

So far the constant states are represented in terms of the characteristic parameters. The concentrations can be determined by using Eq. (3.7.9) or (3.11.5). Only nonvanishing concentrations for various constant states are given below, with units in mol/liter.

U_1: $\quad c_1 = 0.0952$

U_2: $\quad c_1 = 0.0554, c_2 = 0.0817$

V_1: $\quad c_1 = 0.1050, c_2 = 0.0768, c_3 = 0.0729$

V_2: $\quad c_2 = 0.1836, c_3 = 0.1293$

W: $\quad c_1 = 0.1163, c_2 = 0.1255$

P_1: $\quad c_1 = 0.20$

P_2: $\quad c_2 = 0.30$

P_3: $\quad c_3 = 1/3$

The concentrations for the states P_1, P_2, and P_3 are the same as the values obtained previously by applying Eq. (4.5.11).

We are now in a position to construct the concentration profiles at various stages in the course of the development. These are presented in Fig. 4.18, from which we gain better understanding of how the separation is realized. At

$\tau = 10$ [Fig. 4.18(a)], the feed mixture has just been loaded on the column and Fig. 4.18(b) (at $\tau = 12$) shows the development of the pure solute band for the solute A_3. At $\tau = 14$ [Fig. 4.18(c)] the constant state W appears after an interaction and the solute A_1 is almost separated from the solute A_3. In Fig. 4.18(d) at $\tau = 16$ we clearly see that A_1 is completely separated from A_3 and A_2 is being separated from both A_1 and A_3. The next two parts at $\tau = 18$ and at $\tau = 21$ show the separation of A_1 from A_2 and that of A_2 from A_3, respectively. Afterward the pure solute bands for A_1, A_2, and A_3 do not change in shape and simply propagate side by side with the same speed.

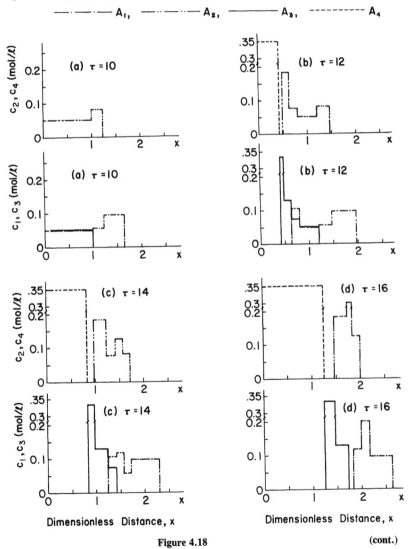

$$\text{—}\cdot\text{—}\cdot\text{— } A_1, \quad \text{—}\cdot\cdot\text{—}\cdot\cdot\text{— } A_2, \quad \text{———— } A_3, \quad \text{------ } A_4$$

Figure 4.18 (cont.)

Figure 4.18 (cont.)

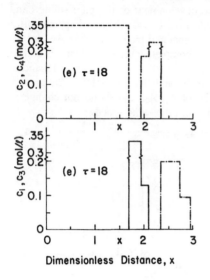

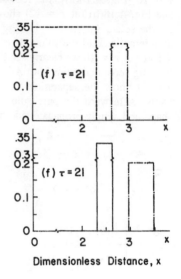

Dimensionless Distance, x

Dimensionless Distance, x

Case for which $c_4^* > c_{4,cr}^*$ and $\omega_1^f < \omega^* < \gamma_1$

Here we shall consider a case in which the developer concentration is slightly higher than its critical value, so $\omega^* > \omega_1^f$. Since $\omega_1^f = 3.387$, we find from Eq. (4.5.11) that $\omega^* = \omega_1^f = 3.387$ if $c_4^* = 0.245$ mol/liter. We shall take $c_4^* = 0.20$ mol/liter to give

$$\omega^* = \frac{20}{1 + (20)(0.2)} = 4.0$$

so we have $P_4[4, 5, 10, 15]$.

Once again we expect to observe pure solute bands forming in a satisfactory fashion at the ultimate stage and the concentration plateau in each band will be given by

$$c_1^* = 0.050 \text{ mol/liter}$$

$$c_2^* = 0.150 \text{ mol/liter}$$

$$c_3^* = 0.1833 \text{ mol/liter}$$

The boundaries will propagate in the same direction in the (x, τ)-plane, which is

$$\bar{\sigma} = 1 + \frac{(1.5)(20)}{1 + (20)(0.2)} = 7.0$$

The bandwidths are determined as

$$\Delta x_3 = 0.3897, \qquad \Delta \tau_3 = 2.7278$$

$$\Delta x_2 = 0.4762, \qquad \Delta \tau_2 = 3.3333$$
$$\Delta x_1 = 1.4286, \qquad \Delta \tau_1 = 10.0$$

Since $\omega^* > \omega_1^f$, the 1-wave developing from point $a_0(0, 10)$ is a simple wave, but otherwise, the physical plane portrait of the solution will be similar to that in the previous case. The 1-simple wave will interact successively with the three shocks originating from the origin, and these interactions can be fully analyzed by applying the results of Secs. 4.2 and 4.3.

The solution is first constructed in the physical plane as shown in Fig. 4.19. The 3- and 2-shocks originating from the origin are engaged in interactions with the 1-simple wave but, after interactions with the 2- and 3-shocks originating from point a_0, the shock paths S^3 and S^2 ultimately become parallel to the path of the 4-shock emanating from point a_0. The 1-shock starts an interaction with the 1-simple wave from point a_8 and this interaction is over at point a_9, beyond which the shock path S^1 also beomes parallel to other shock

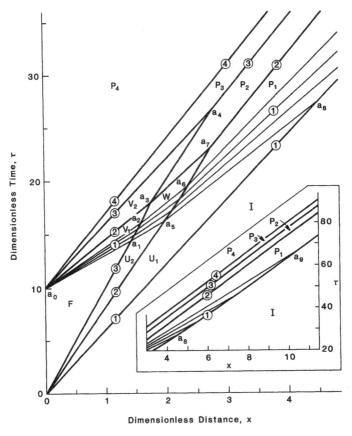

Figure 4.19

paths. Hence the pure solute bands for A_1, A_2, and A_3 are fully established. The bandwidths observed from Fig. 4.19 match very well the values determined in the above. The coordinates of those points marked as a_i, $i = 1, 2, \ldots, 9$, in Fig. 4.19 are listed in Table 4.11.

TABLE 4-11.   COORDINATES OF THE POINTS a_i IN FIG. 4.19

| Point | Coordinates | | Mode of Interaction |
	x	τ	
a_1	1.435	14.351	Transmission
a_2	1.599	16.131	Transmission
a_3	1.759	17.991	Transmission
a_4	2.796	26.846	Confluence
a_5	1.991	16.255	Transmission
a_6	2.349	19.419	Transmission
a_7	2.741	23.126	Confluence
a_8	4.522	27.496	Cancellation
a_9	10.576	67.134	Cancellation

The solution at the ultimate stage is obtained at the point $a_9(10.576, 67.134)$ due to the slow interaction between the 1-simple wave and the 1-shock. The complete separation of solutes, however, is actually accomplished at the point $a_4(2.796, 26.846)$. If a column of length 2.796 is used, the pure solute band for A_1 in the breakthrough curve will consist of the plateau U_1, a gradual change from U_1 to P_1, and the plateau P_1, whereas for A_2 and A_3 we obtain uniform bands. The ratio of the adsorbent volume required to the sample volume treated is 0.4194, and this implies that the column length becomes 13.5% greater than that in the previous case. Still, it is fairly low in comparison to the value 0.9233 for the case $c_4^* = 0$.

The concentrations for each constant state as well as those borne on each of the characteristics C^1 can be determined in the same way as before. Combining these values with the portrait in Fig. 4.19, we can construct the concentration profiles at various times. These are presented in Fig. 4.20. In the first part at $\tau = 13$, we see how the influence of the developer propagates through the column, while the next two parts [Fig. 4.20(b) and (c)] show how the solute A_1 is separated from the solute A_3. At $\tau = 21$ [Fig. 4.20(d)], the solute A_2 is being separated from both A_1 and A_3. In Fig. 4.20(e) at $\tau = 24$ we clearly see that A_1 is completely separated from A_2 but A_2 is still being separated from A_3. At $\tau = 28$ [Fig. 4.20(f)], we see the complete separation of the three solutes, although the A_1-band still undergoes a change in itself due to the interaction between the 1-simple wave and the 1-shock. If the column is long enough, the fully developed A_1-band will be eventually obtained as shown in the last part [Fig. 4.20(g) [$\tau = 70$].

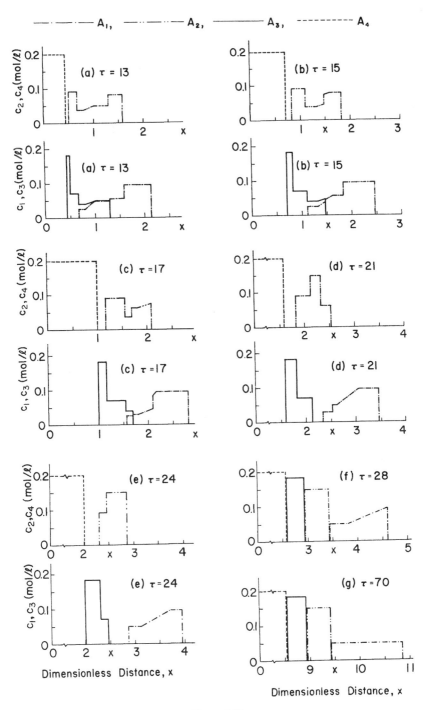

Figure 4.20

Cases for which $c_4^* < c_{4,cr}^*$

We expect here the development to become unsatisfactory in the sense that not all the three pure solute bands are bounded by sharp boundaries. We would like to comment briefly on three different cases.

If $c_4^* = 0.1$ mol/liter, we obtain $\omega^* = 6.667$. This is a case with $\gamma_1 < \omega^* < \gamma_2$, and thus we have $P_4[5, 6.667, 10, 15]$. It follows that the 1-wave developing from point $a_0(0, 10)$ is a simple wave and hence the overall feature will be the same as that of Fig. 4.19. However, since we have $P_3[5, 6.667, 10, 20]$ and $P_2[5, 6.667, 15, 20]$, it is obvious that the constant state marked as P_1 in Fig. 4.19 is characterized by the set $[5, 10, 15, 20]$, which represents the state of no solute. The value of ω_1 here is the same as that of the initial state I, and thus the interaction between the 1-simple wave and the 1-shock will go on indefinitely. This implies that although we would obtain pure solute bands for A_2 and A_3 propagating side by side with sharp boundaries, the A_1-band will have a diffuse boundary in the rear and departs from the A_2-band. In other words, the solute A_1 is not aware of the presence of the developer and is simply being eluted by the solvent (see Fig. 4.14).

In case $c_4^* = 0.04$ mol/liter, we find $\omega^* = 11.111$, which falls between γ_2 and γ_3. Thus we have $P_4[5, 10, 11.111, 15]$, and comparing this with $F[3.387, 7.050, 12.563, 20]$, we notice that the 1- and 2-waves developing from point $a_0(0, 10)$ are both simple waves. Furthermore, with $P_3[5, 10, 11.111, 20]$ the constant state marked as P_2 in Fig. 4.19 is characterized by the set $[5, 10, 15, 20]$, so we see the state of no solute developing in front of the A_3-band. In other words, the situation to the right of the shock path S^3 is similar to that shown in Fig. 4.15. We expect, therefore, that at the ultimate stage both the A_1- and A_2-bands have diffuse rear boundaries, while the A_3-band has sharp boundaries at both ends and remains adjacent to the A_4-band without interval. The A_3- and A_2-bands are intervened by the state of pure solvent and so are the A_2- and A_1-bands. Here the developer is supplied in such a low concentration that it exercises no influence over the behavior of solutes A_2 and A_1. As far as these two solutes are concerned, the process is equivalent to the conventional elution using pure solvent.

Finally, let us suppose that the developer concentration is very low, say $c_4^* = 0.01$ mol/liter. Now we find $\omega^* = 16.667$, which is even larger than γ_3, and thus we have $P_4[5, 10, 15, 16.667]$. It then follows that the state on the right-hand side of the 4-shock is characterized by the set $[5, 10, 15, 20]$, and thus a region of pure state (solvent only) develops to the right of the shock path S^4 from the beginning. This implies that the feed state F remains indifferent to the developer and the feed mixture is not developed by the displacement agent A_4 but by the solvent. Consequently, the physical plane portrait of the solution would be identical to that shown in Fig. 4.10 except for the additional path for the 4-shock (slope = 26) emanating from the point $a_0(0, 10)$ to remain on the left-hand side of the 3-simple wave. Although a complete separation can be attained, all three pure solute bands will have diffuse boundaries in the rear and propagate independently from one another. We recall from Sec. 4.4 that the ratio of the adsorbent volume required to the sample volume treated is 0.9233, and it is clear from the development in this section that this ratio continuously increases to reach the value of 0.9233 as the developer concentration becomes lower.

Exercises

4.6.1 Analyze the displacement development of a two-solute chromatogram for which we have the prameter values

$$\epsilon = 0.4, \quad N_1 = N_2 = N_3 = 1 \text{ mol/liter}$$

$$K_1 = 5 \text{ liters/mol}, \quad K_2 = 10 \text{ liters/mol}, \quad K_3 = 15 \text{ liters/mol}$$

and the initial and boundary conditions

at $\tau = 0$: $\quad c_1 = c_2 = c_3 = 0$

at $x = 0$: $\quad c_1 = c_2 = 0.1$ mol/liter $\quad$ and $\quad c_3 = 0 \quad$ for $0 \leq \tau \leq 10$

$\quad\quad\quad\quad c_1 = c_2 = 0$ $\quad$ and $\quad c_3 = 0.6$ mol/liter $\quad$ for $\tau > 10$

Determine the position in the (x, τ)-plane at which complete separation is accomplished and the concentration plateau and the width of each pure solute band obtained at the ultimate stage. Discuss how the column performance may change if the developer concentration c_3^* is given by 0.3, 0.1, and 0.03 mol/liter, respectively.

4.6.2 For the numerical example treated in this section, consider the cases when $c_4^* = 0.1$ mol/liter and 0.04 mol/liter, respectively, and complete the physical plane portrait of the solution and the concentration profiles at various times for each case.

REFERENCES

The basic references for much of this chapter are

H.-K. RHEE, "Studies on the theory of chromatography: Part III. Chromatography of many solutes," Ph.D. thesis, University of Minnesota, 1968.

H.-K. RHEE, R. ARIS, and N. R. AMUNDSON, "On the theory of chromatography," *Philos Trans. Roy. Soc. London* **A267**, 419–455 (1970).

H.-K. RHEE, "Equilibrium theory of multicomponent chromatography," in *Percolation Processes: Theory and Applications,* edited by A. E. Rodrigues and D. Tondeur, Sijthoff & Noordhoff, Alphen aan den Rijn, The Netherlands, 1981, pp. 285–328.

See also

F. HELFFERICH and G. KLEIN, *Multicomponent Chromatography: Theory of Interference,* Marcel Dekker, New York, 1970.

4.1. Approximation of arbitrary initial data by a sequence of piecewise constant initial data is discussed in

T. ZHANG and Y.-H. GUO, "A class of initial value problems for systems of aerodynamic equations," *Acta Math. Sinica* **15**, 386–396 (1965); English translation in *Chinese Math.* **7**, 90–101 (1965).

For the large-time behavior of solutions of initial value problems, see the treatment in

R. J. DiPERNA, "Decay of solutions of hyperbolic systems of conservation laws with a convex extension," *Arch. Rational Mech. Anal.* **64**, 1–46 (1977).

T.-P. LIU, "Linear and nonlinear large-time behavior of solutions of general systems of hyperbolic conservation laws," *Comm. Pure Appl. Math.* **30**, 767–796 (1977a).

T.-P. LIU, "Large-time behavior of solutions of initial and initial-boundary value problems of a general system of hyperbolic conservation laws," *Comm. Math. Phys.* **55**, 163–177 (1977b).

T.-P. LIU, "Admissible solutions of hyperbolic conservation laws," *Mem. Amer. Math. Soc.* **30**, No. 240, 1–78 (1981).

4.2. The asymptotic form of the solution is discussed in

P. D. LAX, "Nonlinear hyperbolic systems of conservation laws," in *Nonlinear Problems*, edited by R. L. Langer, University of Wisconsin Press, Madison, Wis. 1963, pp. 3–12.

T.-P. LIU, "Decay to N-waves of solutions of general systems of nonlinear hyperbolic conservation laws," *Comm. Pure Appl. Math.* **30**, 585–610 (1977c).

4.3. For the solution of Eq.(4.3.47), see Sec. 2.8 and the references cited there.

4.5. and **4.6.** The development here is taken from

H.-K. RHEE and N. R. AMUNDSON, "Analysis of multicomponent separation by displacement development," *AIChE J.* **28**, 423–433 (1982).

Earlier works on displacement development are to be found in

A. TISELIUS, "Displacement development in adsorption analysis," *Ark. Kemi Mineral. Geol.* **16A**(18), 1–11 (1943).

E. GLUECKAUF, "Contributions to the theory of chromatography," *Proc. Roy. Soc. London* **A186**, 35–57 (1946).

E. GLUECKAUF, "Theory of chromatography: VII. The general theory of two solutes following non-linear isotherms," *Discuss. Faraday Soc.* **7**, 12–25 (1949).

S. CLAESSON, "Theory of frontal analysis and displacement development," *Discuss. Faraday Soc.* **7**, 35–38 (1949).

L. G. SILLÉN, "On filtration through a sorbent layer: Part IV. The ψ-condition, a simple approach to the theory of sorption columns," *Ark. Kemi* **2**, 477–498 (1950).

Displacement development applied to ion-exchange systems are treated in

J. E. POWELL, H. R. BURKHOLDER, and D. B. JAMES, "Elution requirements for the resolution of ternary rare-earth mixtures," *J. Chromatogr.* **32**, 559–566 (1968).

D. B. JAMES, J. E. POWELL, and H. R. BURKHOLDER, "Displacement ion-exchange separation of ternary rare-earth mixtures with chelating agents," *J. Chromatogr.* **35**, 423–429 (1968).

F. HELFFERICH and D. B. JAMES, "An equilibrium theory for rare-earth separation by displacement development," *J. Chromatogr.* **46**, 1–28 (1970).

The last paper presents a general theory than can be applied to a system of any number of components and illustrates application to the separation of a 15-component rare-earth mineral euxenite.

Experimental confirmation of the theoretical development is found in

J. FRENZ and C. HORVATH, "High performance displacement development: calculation and experimental verification of zone development," *AIChE J.* **31**, 400–409 (1985).

The role of a desorbent characterized by adsorptivity intermediate between those of the components to be separated has been analyzed by

M. MORBIDELLI, G. STORTI, S. CARRA, G. NIEDERJAUFNER, and A. PONTOGLIO, "Role of the desorbent in bulk adsorption separations: application to a chlorotoluene isomer mixture," *Chem. Engrg. Sci.* **40**, 1155–1167 (1985).

Multicomponent Adsorption in Continuous Countercurrent Moving-Bed Adsorber

5

From an engineering point of view it is desirable to have continuous operation of the chromatographic process, and this may be effected by making the solid phase move countercurrently against the fluid phase. Indeed, such continuous countercurrent moving-bed processes are widely used on an industrial scale for some important separations.

In this chapter we would like to establish an equilibrium theory of multicomponent adsorption in a continuous countercurrent moving-bed adsorber by applying the mathematical theory developed in Chapters 3 and 4. The system to be considered is again subject to the equilibrium relations of Langmuir type and is treated only with constant initial and feed conditions, commonly referred to as the Riemann problem. Although the general theory is found to be essentially the same as for the fixed-bed problem, we shall observe here the new feature that waves can propagate either forward or backward, and the boundary can bear a discontinuity in concentration. Since the contacting region is finite in actual problems, the system obtains a steady state in finite time in contrast to the case of the semi-infinite fixed bed, and this we will examine with some emphasis. We shall also find that the flow rate ratio between the two phases plays an important role with respect to the structure of the solution as well as the performance of the system, and its effect is worth analyzing in some detail. Application will be illustrated with numerical examples.

The conservation equations are given in Sec. 5.1 and treated within the framework of the Riemann problem to show that the fundamental differential equations and the compatibility relations are invariant under the movement of the solid phase. We also examine the situation at boundaries and discuss the possibility of solutions being discontinuous at either boundary. The next section (Sec. 5.2) is concerned specifically with the Langmuir isotherm and develops the basic equations for the characteristic speed and the propagation speeed for a discontinuity. Here again the entropy condition gives rise to the shock conditions. After establishing various inequality conditions, we proceed to elucidate the overall structure of the solution in an infinite domain and the fundamental aspects of the wave interaction. The nature of wave propagation while in interaction is carefully examined for various cases that can arise.

In Sec. 5.3 we treat the problem in a semi-infinite domain with particular emphasis on the situation at the boundary. Critical values of the flow rate ratio are defined, and noting that this flow rate ratio has a significant effect on the structure of the solution, we classify the possibilities into four categories. For each of these categories, the overall structure of the solution is discussed along with the nature of the boundary. This is followed by Sec. 5.4, in which we are concerned with the problem in a finite domain. Although the solution scheme is basically the same, we observe here that, in general, waves will develop from both ends and be engaged in a sequence of interactions among themselves. After the interactions, all the waves eventually reach one of the boundaries (exceptional case apart) so that the system obtains a steady state. The steady states attainable in finite time are determined in terms of the feed conditions at both ends and the flow rate ratio, and this is illustrated by a numerical example. The possibility of a stationary shock is also exploited.

We then take up interaction analysis in Sec. 5.5. Various cases are treated separately by applying the same procedure as in Chapter 4. Since the simple waves in interaction are more likely to be based on a portion of a horizontal line, we find it more convenient to formulate the problem in terms of τ, the time variable. The states in and around the region of interaction are characterized by the three sets of characteristic parameters corresponding to the initial and the two feed states, respectively, and analytical expressions describing the wave interactions are developed in terms of these parameters. In Sec. 5.6 we present a numerical example to demonstrate the application. Starting with the steady-state analysis, we proceed to construct complete solutions for two different values of the flow rate ratio. Interactions are fully analyzed and the time when the system obtains a steady state is also calculated. We also examine a situation in which a shock becomes stationary inside the domain, and it turns out that the system performance reaches the maximum under these circumstances. This observation is found consistent with the result of the steady-state analysis.

5.1 Basic Formulation

Single-solute adsorption in a continuous countercurrent system was fully analyzed in Sec. 6.5 of Vol. 1 and the treatment was further elaborated in Chapter 8 of Vol. 1. Here we would like to extend the treatment to a multisolute system by applying the theoretical developments in Chapters 3 and 4.

Thus we shall consider a chromatographic column through which the adsorbent phase moves countercurrently at a speed V_s against the fluid phase flowing with interstitial velocity V_f. Let ϵ be the volume fraction of the fluid phase inside the column, and c_i and n_i be the concentrations of the solute A_i in the fluid and solid phases, respectively, both being expressed in moles per unit volume of their own phase. A schematic diagram of this chromatographic column is shown in Fig. 5.1.

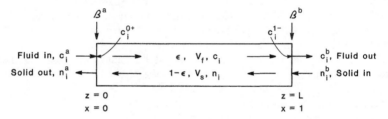

Figure 5.1

By considering the balance of the species A_i in a section between planes z and $z + dz$ from the fluid entrance to the column over a time interval $(t, t + dt)$, we can write a partial differential equation in terms of c_i and n_i. This was formulated in Sec. 1.5 of Vol. 1 to give

$$\epsilon V_f \frac{\partial c_i}{\partial z} - (1 - \epsilon)V_s \frac{\partial n_i}{\partial z} + \epsilon \frac{\partial c_i}{\partial t} + (1 - \epsilon)\frac{\partial n_i}{\partial t} = 0 \qquad (5.1.1)$$

As usual, we shall put

$$\nu = \frac{1 - \epsilon}{\epsilon} = \text{volume ratio} \qquad (5.1.2)$$

$$\mu = \frac{1 - \epsilon}{\epsilon}\frac{V_s}{V_f} = \text{volumetric flow rate ratio} \qquad (5.1.3)$$

and introduce the dimensionless independent variables defined as

$$x = \frac{z}{L}, \qquad \tau = \frac{V_f t}{L} \qquad (5.1.4)$$

where L denotes the total length of the column. Now Eq. (5.1.1) becomes

$$\frac{\partial}{\partial x}(c_i - \mu n_i) + \frac{\partial}{\partial \tau}(c_i + \nu n_i) = 0 \qquad (5.1.5)$$

and, clearly, this equation holds for any solute species present in the system.

Here we shall assume that local equilibrium is established everywhere at any time, and thus the solid-phase concentration n_i is related to the fluid-phase concentrations $\{c_i\}$ by a continuous function with as many derivatives as may be required; i.e.,

$$n_i = n_i(c_1, c_2, \ldots, c_m), \qquad i = 1, 2, \ldots, m \qquad (5.1.6)$$

Since these equations are essentially nonlinear, Eqs. (5.1.5) and (5.1.6) represent a quasilinear system of m partial differential equations of first order. In comparison to the equations for the fixed-bed problem, we have an additional parameter μ that represents the movement of the solid phase relative to the fluid phase, and obviously, the effect of this parameter μ on the structure of the solution is a major concern here.

This system of equations must be solved subject to the initial condition

$$\text{at } \tau = 0: \quad c_i = c_i^0(x), \qquad i = 1, 2, \ldots, m \qquad (5.1.7)$$

and the boundary conditions that describe the feed concentrations in the fluid stream at $x = 0$ and in the solid stream at $x = 1$; i.e., we have

$$\begin{aligned} \text{at } x = 0: \quad c_i &= c_i^a(\tau) \\ \text{at } x = 1: \quad n_i &= n_i^b(\tau) \end{aligned} \qquad (5.1.8)$$

for $i = 1, 2, \ldots, m$, in which the superscripts a and b represent the boundaries at $x = 0$ and $x = 1$, respectively.

From this point on we may follow the line laid down in Sec. 3.2 to obtain the set of characteristic differential equations and generate the solution by applying the iteration scheme. Just as in the case of fixed-bed adsorption, however, we shall confine our discussion to the Riemann problem, which includes the piecewise constant data problem. Hence all the initial and feed concentrations are given by constant values;

$$\begin{aligned} \text{at } \tau = 0: \quad c_i &= c_i^0 \, (\text{constant}) \\ \text{at } x = 0: \quad c_i &= c_i^a \, (\text{constant}) \\ \text{at } x = 1: \quad n_i &= n_i^b \, (\text{constant}) \end{aligned} \qquad (5.1.9)$$

for $i = 1, 2, \ldots, m$.

It was shown in Sec. 3.4 that the solution to a Riemann problem is given by a one-parameter family and thus the image in the hodograph space lies on a single curve Γ; i.e.,

$$c_i = c_i(\omega), \qquad i = 1, 2, \ldots, m \qquad (5.1.10)$$

where ω is the parameter running along the curve Γ. If we let $d/d\omega$ denote the differentiation in the direction of the curve Γ, Eq. (5.1.5) can be rewritten in the form

$$\left(1 - \mu\frac{\mathscr{D}n_i}{\mathscr{D}c_i}\right)\frac{dc_i}{d\omega}\frac{\partial\omega}{\partial x} + \left(1 + \nu\frac{\mathscr{D}n_i}{\mathscr{D}c_i}\right)\frac{dc_i}{d\omega}\frac{\partial\omega}{\partial\tau} = 0$$

or

$$\left\{\left(1 - \mu\frac{\mathscr{D}n_i}{\mathscr{D}c_i}\right)\frac{\partial\omega}{\partial x} + \left(1 + \nu\frac{\mathscr{D}n_i}{\mathscr{D}c_i}\right)\frac{\partial\omega}{\partial\tau}\right\}\frac{dc_i}{d\omega} = 0 \qquad (5.1.11)$$

in which we have

$$\frac{\mathscr{D}n_i}{\mathscr{D}c_i} = \sum_{j=1}^{m} n_{i,j}\frac{dc_j/d\omega}{dc_i/d\omega} \qquad (5.1.12)$$

with

$$n_{i,j} = \frac{\partial n_i}{\partial c_j} \qquad (5.1.13)$$

Clearly, c_i is not constant along the curve Γ, so $dc_i/d\omega \neq 0$. It then follows from Eq. (5.1.11) that

$$\left(\frac{d\tau}{dx}\right)_\omega = -\frac{\partial\omega/\partial x}{\partial\omega/\partial\tau} = \frac{1 + \nu\dfrac{\mathscr{D}n_i}{\mathscr{D}c_i}}{1 - \mu\dfrac{\mathscr{D}n_i}{\mathscr{D}c_i}} = \sigma \qquad (5.1.14)$$

This equation defines a specific direction in the (x, τ)-plane along which the parameter ω remains constant. This direction is determined in terms of the concentrations $c_1, c_2, \ldots, c_m$, but since ω is held constant, all the concentrations also remain fixed in that direction. This implies that the solution to a Riemann problem is given by a family of straight lines in the (x, τ)-plane, and each of these lines maps onto a single point on the curve Γ in the hodograph space.

We note that Eq. (5.1.14) holds for every species A_i, but $(d\tau/dx)_\omega = \sigma$ is independent of the subscript i. Therefore, we obtain the equation

$$\frac{\mathscr{D}n_1}{\mathscr{D}c_1} = \frac{\mathscr{D}n_2}{\mathscr{D}c_2} = \cdots = \frac{\mathscr{D}n_m}{\mathscr{D}c_m} \qquad (5.1.15)$$

which are the *fundamental differential equations* for the Riemann problem. Since this is the same as Eq. (3.4.19) which was established for the fixed-bed system, we observe that the image of the solution in the hodograph space is independent of the flow rate ratio μ and remains the same as that for the corresponding fixed-bed system.

Next, we would like to show that the direction σ defined by Eq. (5.1.14) is equivalent to the characteristic direction of the system (4.1.15). For this we shall rearrange Eq. (5.1.14) in the form

$$\frac{\mathscr{D}n_i}{\mathscr{D}c_i} = \frac{\sigma - 1}{\mu\sigma + \nu}, \qquad i = 1, 2, \ldots, m$$

and expand this by using Eq. (5.1.12) to obtain

$$n_{i,1}\frac{dc_1}{d\omega} + \cdots + n_{i,i-1}\frac{dc_{i-1}}{d\omega} + \left(n_{i,i} - \frac{\sigma - 1}{\mu\sigma + \nu}\right)\frac{dc_i}{d\omega} +$$

$$n_{i,i+1}\frac{dc_{i+1}}{d\omega} + \cdots + n_{i,m}\frac{dc_m}{d\omega} = 0, \qquad i = 1, 2, \ldots, m$$

These m equations may be put in vector form

$$\left(\nabla\mathbf{n} - \frac{\sigma - 1}{\mu\sigma + \nu}\mathbf{I}\right)\frac{d\mathbf{c}}{d\omega} = \mathbf{0} \tag{5.1.16}$$

where $\mathbf{c}$ and $\mathbf{n}$ denote the column vectors of m elements $\{c_i\}$ and $\{n_i\}$, respectively, and ∇ represents the gradient in the hodograph space. Equation (5.1.16) is a system of homogeneous equations for $d\mathbf{c}/d\omega$ and the existence of the curve Γ requires a nontrivial solution. Hence we have

$$\left|\nabla\mathbf{n} - \frac{\sigma - 1}{\mu\sigma + \nu}\mathbf{I}\right| = 0 \tag{5.1.17}$$

On the other hand, the system (5.1.5) can be put into vector form

$$(\mathbf{I} - \mu\nabla\mathbf{n})\frac{\partial\mathbf{c}}{\partial x} + (\mathbf{I} + \nu\nabla\mathbf{n})\frac{\partial\mathbf{c}}{\partial\tau} = \mathbf{0} \tag{5.1.18}$$

According to the discussion in Sec. 3.2, the characteristic direction $\sigma' = d\tau/dx$ is given by the eigenvalue of the matrix $(\mathbf{I} - \mu\nabla\mathbf{n})^{-1}(\mathbf{I} + \nu\nabla\mathbf{n})$ if $(\mathbf{I} - \mu\nabla\mathbf{n})$ is nonsingular. Hence σ' must satisfy the characteristic equation

$$|(\mathbf{I} - \mu\nabla\mathbf{n})^{-1}(\mathbf{I} + \nu\nabla\mathbf{n}) - \sigma'\mathbf{I}| = 0$$

or, by applying the rule for the product of matrices,

$$\left|\nabla\mathbf{n} - \frac{\sigma' - 1}{\mu\sigma' + \nu}\mathbf{I}\right| = 0 \tag{5.1.19}$$

Comparing this with Eq. (5.1.17), we confirm that σ defined by Eq. (5.1.14) represents the characteristic direction of the system (5.1.5).

Now we can picture the structure of the solution. In the hodograph space the image of the solution would consist of a sequence of Γ curves to be determined by the fundamental differential equations (5.1.15). To every point along the Γ curves there corresponds a straight characteristic C in the physical plane that bears constant values of the concentrations. If these characteristics fan clockwise as we proceed in the x-direction, they would span a simple wave region. If the characteristics fan in the opposite direction, we must introduce a discontinuity in the solution. Therefore, the overall picture of the solution is similar to that for the fixed-bed system. One feature to be noted is that the characteristic direction can be of negative sign for sufficiently large values of μ.

Suppose that the solution contains a discontinuity within the domain of interest. The principle of conservation of mass requires that the discontinuity must move with a speed such that there is no accumulation of material at the discontinuity. When applied to the species A_i, this principle gives rise to the equation

$$\frac{d\tau}{dx} = \frac{1 + \nu\dfrac{[n_i]}{[c_i]}}{1 - \mu\dfrac{[n_i]}{[c_i]}} = \bar{\sigma} \qquad (5.1.20)$$

which is the jump condition to be satisfied by a discontinuity. Here again the brackets represent the jump in the quantity enclosed across a discontinuity. Since adsorption equilibrium is established, it is obvious that if one of the concentrations $\{c_i\}$ is discontinuous, all other concentrations are also discontinuous. This implies that the discontinuity in the solution is an entity and must move as such. Now Eq. (5.1.20) gives the propagation direction or the reciprocal speed of the discontinuity, and since this must be the same for all the species, we obtain

$$\frac{[n_1]}{[c_1]} = \frac{[n_2]}{[c_2]} = \cdots = \frac{[n_m]}{[c_m]} \qquad (5.1.21)$$

which constitute the *compatibility relations* to be satisfied by the states on either side of the discontinuity. We note that Eq. (5.1.21) is identical to Eq. (3.9.5), so the compatibility relations are independent of the parameter μ. This implies that even if the solution contains one or more discontinuities, its image in the hodograph space is the same as that for the corresponding fixed-bed system and remains unchanged as the parameter μ varies.

If the solution contains discontinuities, we need as before an additional condition concerning the direction of the jump to ensure the uniqueness of the solution. Based on the argument above, we may presume that such a condition, known as the *entropy condition* of the problem, would be essentially identical to that for the corresponding fixed-bed system. Since, however, the propagation direction $\bar{\sigma}$ of a discontinuity can have negative values, the entropy condition must be expressed in terms of the propagation speeds λ and $\bar{\lambda}$ instead of the propagation direction σ and $\bar{\sigma}$, as we shall see in Sec. 5.2.

Before concluding the general discussion here, it seems necessary to pay attention to the determination of the states of the outgoing streams, i.e., $\{c_i^b\}$ and $\{n_i^a\}$. Since these streams are, in general, not in equilibrium with the corresponding incoming streams, the states $\{c_i^a\}$ and $\{n_i^b\}$ are not directly determined from the equations discussed above. Due to the nonequilibrium situation outside the boundaries, we expect to see a discontinuity at either boundary that remains stationary there. Indeed, by applying the conservation principle that there can be no accumulation of material at the boundary, we obtain the equations

$$\text{at } x = 0: \quad c_i^a - c_i^{0+} = \mu(n_i^a - n_i^{0+}) \tag{5.1.22}$$

$$\text{at } x = 1: \quad c_i^b - c_i^{1-} = \mu(n_i^b - n_i^{1-}) \tag{5.1.23}$$

for $i = 1, 2, \ldots, m$, where the superscripts $0+$ and $1-$ represent, respectively, the states at positions just inside the boundaries at $x = 0$ and $x = 1$ (see Fig. 5.1). Obviously, n_i^{0+} and n_i^{1-} are the equilibrium values corresponding to the states $\{c_i^{0+}\}$ and $\{c_i^{1-}\}$, respectively.

If there is no discontinuity at $x = 0$, then we have $c_i^{0+} = c_i^a$, which is specified, and $n_i^a = n_i^{0+}$ is in equilibrium with $\{c_i^{0+}\}$ and hence with $\{c_i^a\}$. A similar argument applies at $x = 1$ with respect to c_i^b. In this case the two equations (5.1.22) and (5.1.23) are satisfied identically. If not satisfied identically, they serve to determine the states, $\{n_i^a\}$ and $\{c_i^b\}$, of the outgoing streams by

$$n_i^a = n_i^{0+} + \frac{c_i^a - c_i^{0+}}{\mu} \tag{5.1.24}$$

$$c_i^b = c_i^{1-} + \mu(n_i^b - n_i^{1-}) \tag{5.1.25}$$

for $i = 1, 2, \ldots, m$. We note that although c_i^a and n_i^b remain constant, n_i^a and c_i^b can vary with time. In this sense, the conditions (5.1.22) and (5.1.23) are analogous to the free boundary conditions.

In contrast to the discontinuity appearing inside the domain, the discontinuity borne at either boundary remains fixed there. To distinguish the two, we will call the latter specifically a *boundary discontinuity* or B.D.

At steady state, Eqs. (5.1.5) and (5.1.6) indicate that the system will have a uniform state throughout the domain and thus $c_i^{0+} = c_i^{1-}$ and $n_i^{0+} = n_i^{1-}$. It then follows from Eqs. (5.1.22) and (5.1.23) that

$$c_i^a - c_i^b = \mu(n_i^a - n_i^b) \tag{5.1.26}$$

which is nothing more than the overall balance of the solute species A_i at the steady state. This implies that the existence of boundary discontinuities does not violate the principle of mass conservation. For convenience, we shall use the symbols $\mathcal{B}^a$ and $\mathcal{B}^b$ to denote the boundaries at $x = 0$ and $x = 1$, respectively, in the following sections (see Fig. 5.1).

Exercise

5.1.1. By comparing Eq. (5.1.5) with Eq. (3.2.1), identify the coefficients a_{ij} and b_{ij} in terms of the first-order derivatives of n_i. For a system with $m = 3$, find expressions for M_{kj}, $k = 1, 2$, and 3, appearing in Eq. (3.2.13).

5.2 Theoretical Development for the Langmuir Isotherm

We shall now assume that the equilibrium relationship (5.1.6) is given by the Langmuir adsorption isotherm

$$n_i = \frac{N_i K_i c_i}{1 + \sum_{j=1}^{m} K_j c_j} = \frac{\gamma_i c_i}{1 + \sum_{j=1}^{m} K_j c_j}, \qquad i = 1, 2, \ldots, m \qquad (5.2.1)$$

which was fully discussed in Sec. 3.5. With this isotherm a complete analysis of the fixed-bed system has been presented in Chapters 3 and 4. Since the fundamental differential equations (5.1.15), as well as the compatibility relations (5.1.21), are independent of the parameter μ, we know that the structure of the solution in the hodograph space will remain the same as the parameter μ varies. Therefore, those expressions established in Sec. 3.6 for the characteristic parameters and the Riemann invariants are equally valid here for the continuous countercurrent moving bed problem. However, the mapping of the solution from the hodograph space onto the physical plane of x and τ must be accomplished by means of Eqs. (5.1.14) and (5.1.20), both of which are dependent on the parameter μ, and hence proper modifications are necessary.

For the Langmuir isotherm the fundamental differential equations (5.1.15) have been solved to give m different kinds of solution. Of these, the kth kind, for example, is given by the one-parameter family

$$\phi_i - \phi_i^0 = J_i^k (\delta - \delta^0), \qquad i = 1, 2, \ldots, m \qquad (5.2.2)$$

where

$$\phi_i = K_i c_i \qquad (5.2.3)$$

$$\delta = 1 + \sum_{i=1}^{m} K_i c_i \qquad (5.2.4)$$

and the superscript 0 denotes a fixed state. The constants $\{J_i^k\}$ are the k-Riemann invariants defined by

$$J_i^k = \frac{K_i n_i}{\gamma_i - \omega_k}, \qquad i = 1, 2, \ldots, m \qquad (5.2.5)$$

for $k = 1, 2, \ldots, m$, in which the characteristic parameter ω_k is given by the kth root of the algebraic equation

$$\sum_{i=1}^{m} \frac{K_i n_i}{\gamma_i - \omega} = 1 \qquad (5.2.6)$$

It can be easily seen that this equation has m distinct, positive roots $\{\omega_k\}$ that can be arranged as

$$0 \le \omega_1 \le \gamma_1 \le \omega_2 \le \gamma_2 \le \cdots \le \gamma_{m-1} \le \omega_m \le \gamma_m \qquad (5.2.7)$$

(see Fig. 3.5). These roots $\{\omega_k\}$ constitute the set of characteristic parameters and there exists a one-to-one correspondence between the set $\{\omega_k\}$ and the set of concentrations $\{c_i\}$, the inverse transformation being given by Eq. (3.7.9).

Equation (5.2.2) represents a straight line passing through the fixed point $\{\phi_i^0\}$ in the hodograph space $\Phi(m)$, and this line is called the Γ^k. Out of the m characteristic parameters $\{\omega_i\}$, ω_k is the only parameter varying along the Γ^k, and thus the Γ^k becomes parallel to the ω_k-axis if it is mapped into the m-dimensional ω-space $\Omega(m)$.

According to the argument around Eq. (5.1.14), to every point along the Γ^k there corresponds a straight-line characteristic C in the physical plane. We shall call this the kth characteristic C^k and its slope will be denoted by σ_k. Since we have, from Eqs. (3.6.13) and (3.6.14),

$$\left(\frac{\mathcal{D}n_i}{\mathcal{D}c_i}\right)_{\Gamma^k} = \frac{\omega_k}{\delta}$$

along a Γ^k, Eq. (5.1.14) gives

$$\left(\frac{d\tau}{dx}\right)_{\omega_k} = \frac{1 + \nu\omega_k/\delta}{1 - \mu\omega_k/\delta} = \sigma_k \tag{5.2.8}$$

for $k = 1, 2, \ldots, m$. But δ can be expressed in terms of the characteristic parameters by Eq. (3.7.8); that is,

$$\delta = \prod_{i=1}^{m} \frac{\gamma_i}{\omega_i} \tag{5.2.9}$$

and thus we obtain

$$\left(\frac{d\tau}{dx}\right)_{\omega_k} = \sigma_k = \frac{1 + \nu p_k \omega_k^2}{1 - \mu p_k \omega_k^2} \tag{5.2.10}$$

for $k = 1, 2, \ldots, m$, where p_k is defined as

$$p_k = \frac{1}{\gamma_k} \prod_{i=1,k}^{m} \frac{\omega_i}{\gamma_i} \tag{5.2.11}$$

and remains constant along a Γ^k.

A kth characteristic C^k gets steeper as the parameter μ increases until it becomes vertical when

$$\mu = \mu_k = \frac{\delta}{\omega_k} = (p_k \omega_k^2)^{-1} \tag{5.2.12}$$

We shall call μ_k the critical ratio of flow rates for the kth characteristic field. For flow rate ratios above the critical value μ_k, the kth characteristic C^k is directed backward.

Since the characteristic direction σ_k can have both positive and negative values, quite often we shall find it more convenient to speak in terms of the characteristic speed λ_k, which is just the reciprocal of σ_k; that is,

$$\left(\frac{dx}{d\tau}\right)_{\omega_k} = \lambda_k = \frac{1 - \mu p_k \omega_k^2}{1 + \nu p_k \omega_k^2} \tag{5.2.13}$$

We note that λ_k is the propagation speed of a particular state of concentrations or a disturbance in the concentration field.

By applying the inequalities in Eq. (5.2.7) to Eq. (5.2.13), we find that

$$1 > \lambda_1 > \lambda_2 > \cdots > \lambda_{m-1} > \lambda_m > -\frac{\mu}{\nu} = -\frac{V_s}{V_f} \tag{5.2.14}$$

Therefore, any state of concentrations or any disturbance cannot propagate faster than the fluid phase when moving forward or than the solid phase when propagating backward.

Now we recall that the k-Riemann invariants $\{J_i^k\}$ remain constant along a Γ^k, and thus to every Γ^k in the hodograph space there corresponds a k-simple wave region in the physical plane. Such a region is covered by a family of straight-line characteristics C^k, each of which has a point image along the corresponding Γ^k in the hodograph space. For obvious reasons, however, the characteristics C^k must fan clockwise as we proceed in the x-direction, and this requires that

$$\frac{\partial \lambda_k}{\partial x} = \left(\frac{\partial \lambda_k}{\partial \omega_k}\right)_{P_k} \frac{\partial \omega_k}{\partial x} > 0 \tag{5.2.15}$$

Since it follows from Eq. (5.2.13) that

$$\left(\frac{\partial \lambda_k}{\partial \omega_k}\right)_{P_k} = \frac{-2(\nu + \mu) p_k \omega_k}{(1 + \nu p_k \omega_k^2)^2} < 0$$

for any k, we have the condition

$$\frac{\partial \omega_k}{\partial x} < 0 \tag{5.2.16}$$

to be satisfied in a k-simple wave region. Consequently, *the parameter ω_k is monotone decreasing in the x-direction.* We further notice that Eqs. (3.8.10) and (3.8.11) are independent of the parameter μ, so the Rule 3.1 is equally applicable to the countercurrent moving-bed problem; that is, *in a k-simple wave region, ϕ_i with $i < k$ decreases in the x-direction, whereas ϕ_i with $i \geq k$ increases in the x-direction.*

It is obvious that the region of a k-simple wave is bounded by a pair of characteristics C^k as depicted in Fig. 5.2. Therefore, it is appropriate to define two critical values of μ; one is the lower critical ratio

$$\mu_{k*} = [p_k(\omega_k^p)^2]^{-1} \tag{5.2.17}$$

and the other is the upper critical ratio

$$\mu_k^* = [p_k(\omega_k^q)^2]^{-1} \tag{5.2.18}$$

It then follows that for $\mu < \mu_{k*}$ the k-simple wave is forward-facing and for $\mu > \mu_k^*$ it is backward-facing, whereas for $\mu_{k*} \leq \mu \leq \mu_k^*$ there exists a vertical C^k so that the corresponding state of concentrations is stationary.

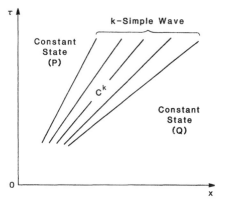

Figure 5.2

If the inequality sign in Eq. (5.2.16) is reversed, that is, if the parameter ω_k increases in the x-direction, the characteristics C^k would fan backward as we proceed in the x-direction and so a physically impossible situation would obtain. Under these circumstances we must introduce a discontinuity in the solution in such a way that the jump condition (5.1.20) and the compatibility relations (5.1.21) are satisfied.

For the Langmuir isotherm (5.2.1) it was shown in Sec. 3.9 that the compatibility relations (5.1.21) are satisfied by any pair of points located on a single piece of the curve Γ and so there are m different kinds of solution. It was also proven that there are no more than m kinds. If the states on both sides of a discontinuity have their images on a Γ^k, we shall call it a *k-discontinuity* after Definition 3.4. It then follows that for a k-discontinuity the states on both sides are related by a one-parameter family

$$[\phi_i] = J_i^k[\delta], \qquad i=1, 2, \ldots, m \tag{5.2.19}$$

By using this equation with the Langmuir isotherm, it can be shown that for a k-discontinuity, we have

$$\left(\frac{[n_i]}{[c_i]}\right)_{\Gamma^k} = \frac{\omega_k^l}{\delta^r} = \frac{\omega_k^r}{\delta^l} \tag{5.2.20}$$

in which, as usual, the superscripts l and r denote, respectively, the states on the left- and right-hand sides of a discontinuity [see Eq. (3.9.12)]. Substituting Eq. (5.2.20) into Eq. (5.1.20), we obtain the propagation direction $\bar{\sigma}_k$ of a k-discontinuity in the form

$$\frac{d\tau}{dx} = \bar{\sigma}_k = \frac{1 + \nu\omega_k^l/\delta^r}{1 - \mu\omega_k^r/\delta^l} = \frac{1 + \nu\omega_k^r/\delta^l}{1 - \mu\omega_k^r/\delta^l} \tag{5.2.21}$$

$$= \frac{1 + \nu p_k \omega_k^l \omega_k^r}{1 - \mu p_k \omega_k^l \omega_k^r} \tag{5.2.22}$$

for $k = 1, 2, \ldots, m$. Here p_k is defined as

$$p_k = \frac{1}{\gamma_k} \prod_{i=1,k}^{m} \frac{\omega_i^l}{\gamma_i} = \frac{1}{\gamma_k} \prod_{i=1,k}^{m} \frac{\omega_i^r}{\gamma_i} \qquad (5.2.23)$$

and it is invariant across a k-discontinuity, for the characteristic parameters except for ω_k remain constant along a Γ^k. The propagation speed $\tilde{\lambda}_k$ of a k-discontinuity is given by the reciprocal of $\tilde{\sigma}_k$; that is,

$$\frac{dx}{d\tau} = \tilde{\lambda}_k = \frac{1 - \mu \omega_k^l/\delta^r}{1 + \nu \omega_k^l/\delta^r} = \frac{1 - \mu \omega_k^r/\delta^l}{1 + \nu \omega_k^r/\delta^l} \qquad (5.2.24)$$

$$= \frac{1 - \mu p_k \omega_k^l \omega_k^r}{1 + \nu p_k \omega_k^l \omega_k^r} \qquad (5.2.25)$$

for $k = 1, 2, \ldots, m$. Just as for the characteristic speed λ_k, we would find it more convenient to think in terms of the propagation speed $\tilde{\lambda}_k$ because the propagation direction $\tilde{\sigma}_k$ can have both positive and negative values.

Suppose that we have two regions of constant state in the physical plane and that the two constant states have their images on two different points along a Γ^k. Within the scope of the Riemann problem the two regions can be connected either by a k-simple wave or by a k-discontinuity. If the condition (5.2.16) is satisfied, that is, if ω_k decreases in the x direction, it is obvious that the connection is made by a k-simple wave and thus the solution is continuous. Only when ω_k increases in the x-direction, we need to introduce a discontinuity, so the proper connection would be made by a k-discontinuity. This argument leads us to the *entropy condition: Across a k-discontinuity, the characteristic parameter ω_k increases from the left to the right;* that is, across a k-discontinuity, we must have

$$\omega_k^l < \omega_k^r \qquad (5.2.26)$$

for $k = 1, 2, \ldots, m$.

By making comparison between Eqs. (5.2.13) and (5.2.25) under the entropy condition, we immediately find that

$$\lambda_k^l > \tilde{\lambda}_k > \lambda_k^r \qquad (5.2.27)$$

and since $\omega_{k-1} < \omega_k < \omega_{k+1}$, we also note that

$$\lambda_{k+1}^l < \tilde{\lambda}_k < \lambda_{k-1}^r \qquad (5.2.28)$$

for $k = 1, 2, \ldots, m$. According to Lax (1957), the inequalitites (5.2.27) and (5.2.28) constitute the *shock conditions* that characterize a k-shock and hence we see that every discontinuity appearing inside the domain in moving-bed problems with the Langmuir isotherm is a shock.

In the physical plane the propagation path S^k of a k-shock gets steeper as the parameter μ increases and becomes vertical when

$$\mu = \bar{\mu}_k = \frac{\delta^l}{\omega_k^r} = \frac{\delta^r}{\omega_k^l} = [\, p_k \omega_k^l \omega_k^r \,]^{-1} \qquad (5.2.29)$$

For this critical ratio $\bar{\mu}_k$ the k-shock becomes stationary and for $\mu > \bar{\mu}_k$ the k-shock propagates backward.

If the state on one side, say $\{c_i^l\}$, is given and one concentration on the other side, say $c_m^r (< c_m^l)$, is further specified, we know that the compatibility relations (5.1.21) will give m different states for $\{c_i^r\}$ and each of these can be connected to the state $\{c_i^l\}$ by a shock of distinct index k. The propagation speeds of these shocks are given by Eq. (5.2.24), and by using Eq. (5.2.7) they are compared to give

$$1 > \tilde{\lambda}_1 > \tilde{\lambda}_2 > \cdots > \tilde{\lambda}_{m-1} > \tilde{\lambda}_m > -\frac{\mu}{\nu} = -\frac{V_s}{V_f} \qquad (5.2.30)$$

The same inequalities also hold for m different shocks having the state on the right-hand side in common. Equation (5.2.30) implies that no shock can propagate faster than the fluid phase when moving forward or than the solid phase when moving backward.

Finally, we shall examine the overall structure of the solution to a Riemann problem. For this purpose we would like to consider an infinite domain in which adsorption equilibrium is established, and discuss the solution scheme when there are two initial discontinuities, one at $x = 0$ and the other at $x = 1$. Thus the initial data are prescribed as

$$c_i(x, 0) = \begin{cases} c_i^a & \text{for } x < 0 \\ c_i^o & \text{for } 0 < x < 1 \\ c_i^b & \text{for } x > 1 \end{cases} \qquad (5.2.31)$$

Let A, O, and B represent the point images in the hodograph space of the constant states, $\{c_i^a\}$, $\{c_i^o\}$, and $\{c_i^b\}$, respectively. The two points A and O are to be connected by a path composed of a sequence of Γ's. If this path contains a segment of a Γ^k, the inverse mapping onto the physical plane will give a k-simple wave, if ω_k decreases in the x direction, or a k-shock, if ω_k increases in the x-direction. On either side of this k-wave there appears a region of constant state whose image falls on the corresponding end point of the Γ^k. The same argument holds for any k from 1 to m.

Now the inequalities in Eqs. (5.2.14) and (5.2.30) and the shock conditions (5.2.27) and (5.2.28) require that the path connecting A to O must contain no more than m segments of Γ's which are of distinct families and arranged in the order of descending index k as we move from A to O. It is readily seen that such a path always exists and is unique. The solution in the physical plane, therefore, consists of a succession of m waves that are of distinct families and arranged in the order of descending index k as we proceed in the x direction. Between any pair of adjacent waves a new constant state appears. The constant state appearing on the right-hand side of the k-wave will be called the k-constant state and denoted by a superscript in parentheses; for example, $c_i^{(k)}$ represents the concentration of A_i in the k-constant state.

Exactly the same argument applies to the discontinuity at $x = 1$, and thus it is not difficult to picture the general situation in the physical plane, at least for small time, as shown in Fig. 5.3. It is evident that the overall structure of the solution is the same as that for the fixed-bed problem except for the fact that some of the waves can be backward-facing. This effect is accentuated as the parameter μ increases.

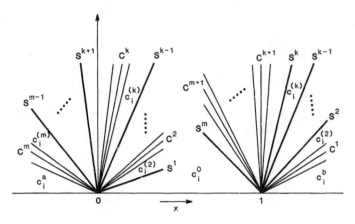

Figure 5.3

By simply looking at the layout of the characteristics and the shock paths, it is obvious that there will be a series of interactions of the waves as time goes on. The pairs of waves that are to interact can be easily discerned by using the inequality conditions imposed on the characteristic speeds λ_k and the shock speeds $\tilde{\lambda}_k$. But we note that Eqs. (5.2.14) and (5.2.30) are essentially the same as Eqs. (3.8.7) and (3.9.21) and further that the shock conditions (5.2.27) and (5.2.28) hold here and in the fixed-bed problem. Therefore, the argument leading to Rule 4.1 is equally applicable to the present system for the countercurrent moving bed.

Moreover, we recall that the image of the solution in the hodograph space is independent of μ and thus the nature of the wave interactions to be encountered in the present problem will be the same as that in the fixed-bed problem. Consequently, we can apply both Rules 4.1 and 4.2 here without modification.

When two waves interact, the state of concentrations varies along each characteristic or on one or both sides of the shock involved and thus the wave propagation is accelerated. Accordingly, in the physical plane the characteristics and the shock paths will become curved or refracted.

For a C^k in a k-simple wave region we have, from Eq. (5.2.13),

$$\lambda_k = \frac{1 - \mu p_j \omega_j \omega_k}{1 + \nu p_j \omega_j \omega_k} \tag{5.2.32}$$

where

$$p_j = \frac{1}{\gamma_j} \prod_{i=1,j}^{m} \frac{\omega_i}{\gamma_i} \qquad (5.2.33)$$

When a k-simple wave is transmitted across a j-shock with $j > k$, we see that along a C^k the parameter ω_j jumps up across the S^j while both p_j and ω_k remain fixed. This implies that

$$\lambda_k^l > \lambda_k^r \qquad (5.2.34)$$

and thus the C^k is refracted counterclockwise and the disturbance borne on the C^k decelerates.

If a k-simple wave is transmitted through a j-simple wave with $j > k$, the parameter ω_j now decreases along a C^k while p_j and ω_k remain unchanged, and hence λ_k increases. It then follows that every C^k becomes curved clockwise and any disturbance accelerates.

For a k-shock we have Eq. (5.2.25), and thus if two k-shocks are joined together upon interaction, the resultant shock takes an intermediate speed. When a k-shock interacts with a k-simple wave approaching from the right-hand side, the parameter ω_k^r decreases while p_k and ω_k^l remain fixed and hence the k-shock accelerates. But if the k-simple wave approaches from the left-hand side, we see that ω_k^l increases while p_k and ω_k^r remain unchanged. Therefore, the k-shock will decelerate.

On the other hand, Eq. (5.2.24) can be rewritten in the form

$$\tilde{\lambda}_k = \frac{1 - \mu p_j^l \omega_j^l \omega_k^r}{1 + \nu p_j^l \omega_j^l \omega_k^r} \qquad (5.2.35)$$

where

$$p_j^l = \frac{1}{\gamma_j} \prod_{i=1,j}^{m} \frac{\omega_i^l}{\gamma_i} \qquad (5.2.36)$$

If a k-shock is transmitted through a j-shock with $j < k$, ω_j^l jumps down while both p_j^l and ω_k^r remain constant along the S^k so that the k-shock accelerates. If we have $j > k$, ω_j^l jumps up and the k-shock decelerates. When a k-shock is transmitted through a j-simple wave with $j < k$, ω_j^l now increases along the S^k and thus the k-shock decelerates. In case $j > k$, ω_j^l decreases along the S^k so that the k-shock accelerates.

Exercise

5.2.1. Consider the width Δx of a k-simple wave region at a fixed time and by using Eq. (5.2.13), show that as the parameter μ increases, a simple wave becomes more expansive if it is forward-facing and becomes less expansive if it is backward-facing.

5.3 Analysis of Semi-Infinite Columns

In this section we shall assume that the contacting region extends indefinitely beyond the point $x = 0$ and that the initial and feed data are prescribed as

$$\text{at } \tau = 0: \quad c_i = c_i^0 \qquad \text{for } x > 0$$
$$\text{at } x = 0: \quad c_i = c_i^a \qquad \text{for } \tau > 0 \tag{5.3.1}$$

for $i = 1, 2, \ldots, m$, where $c_i^0 \neq c_i^a$ for some i. Thus we have a standard Riemann problem.

For $\mu = 0$ (fixed-bed system) the solution was completely established in Sec. 3.11 in terms of centered simple waves, shocks, and constant states. There are at most m waves of distinct families and these are ordered clockwise from the m-wave to the 1-wave, with the k-constant state appearing between the k- and $(k-1)$-waves. The regions of these m waves and the $(m-1)$ constant states in between represent the range of influence of the discontinuity at the origin.

From Eqs. (5.2.10) and (5.2.22), we observe that as the parameter μ increases, the range of influence rotates counterclockwise because every characteristic as well as the shock path becomes steeper. However, the image of the solution in the hodograph space is independent of μ and remains fixed even though the solid phase moves. This implies that the overall structure of the solution in the physical plane is maintained as μ varies. For sufficiently small values of μ, therefore, the solution can be determined by applying, in principle, the same procedure as for fixed-bed problems. It is clear that there is no discontinuity at $x = 0$ and Eq. (5.1.22) is satisfied identically.

Suppose that the parameter μ increases further. For certain values of μ, we expect that only one portion of the range of influence lies inside the domain while the remainder lies outside. Along the vertical line $x = 0+$ there corresponds a particular state of concentrations $\{c_i^{0+}\}$. Therefore, the initial discontinuity at the origin is split into two parts $\{c_i^0 \sim c_i^{0+}\}$ and $\{c_i^{0+} \sim c_i^a\}$. The first part influences the solution in the domain in such a way that Eq. (5.1.5) is satisfied, whereas the second part influences the state of the outgoing solid phase in such a way that Eq. (5.1.22) is satisfied. Application of Eq. (5.1.22) is straightforward since each of the m equations is independent, and this will determine the boundary discontinuity to appear at $x = 0$.

Now the question boils down to how to determine the state $\{c_i^{0+}\}$ for a given value of μ. Although the situation may vary over the whole spectrum of μ, it seems appropriate to classify the possibilities into four categories by making cuts at every critical value along the μ-spectrum. The critical values corresponding to the k-wave are defined by Eqs. (5.2.17) and (5.2.18) in case of a simple wave or by Eq. (5.2.29) in case of a shock.

For the present problem with the data given by Eq. (5.3.1), we first solve Eq. (5.2.6) for the states $\{c_i^a\}$ and $\{c_i^0\}$ to determine the sets of characteristic parameters $\{\omega_k^a\}$ and $\{\omega_k^0\}$, respectively. Recalling the image of the solution in the hodograph space $\Omega(m)$, we immediately find that the k-constant state, which is to appear on the right-hand side of the k-wave, is characterized by the set of characteristic parameters

$$[\omega_1^a, \omega_2^a, \ldots, \omega_{k-1}^a, \omega_k^o, \ldots, \omega_{m-1}^o, \omega_m^o] \qquad (5.3.2)$$

(see Table 3.3).

If $\omega_k^a > \omega_k^o$, the k-wave is a simple wave and it is represented by a family of straight characteristics C^k, for which we have

$$\frac{dx}{d\tau} = \lambda_k = \frac{1 - \mu p_k \omega_k^2}{1 + \nu p_k \omega_k^2}, \qquad \omega_k^o \le \omega_k \le \omega_k^a \qquad (5.3.3)$$

where

$$p_k = \frac{1}{\gamma_k} \prod_{i=1}^{k-1} \frac{\omega_i^a}{\gamma_i} \prod_{i=k+1}^{m} \frac{\omega_i^o}{\gamma_i} \qquad (5.3.4)$$

Hence the lower and upper critical values of μ are given by

$$\mu_{k*} = [p_k(\omega_k^a)^2]^{-1} \qquad (5.3.5)$$

and

$$\mu_k^* = [p_k(\omega_k^o)^2]^{-1} \qquad (5.3.6)$$

respectively, for $k = 1, 2, \ldots, m$.

In case $\omega_k^a < \omega_k^o$, the k-wave is a shock and its propagation speed is

$$\frac{dx}{d\tau} = \tilde{\lambda}_k = \frac{1 - \mu p_k \omega_k^a \omega_k^o}{1 + \nu p_k \omega_k^a \omega_k^o} \qquad (5.3.7)$$

where p_k is defined as in Eq. (5.3.4). There is now only one critical value of μ, given by

$$\tilde{\mu}_k = [p_k \omega_k^a \omega_k^o]^{-1} \qquad (5.3.8)$$

for $k = 1, 2, \ldots, m$.

In the following we shall discuss the structure of the solution for each range of μ with particular emphasis on the situation at the boundary $x = 0$.

Lower range of μ

This range spans the portion

$$0 \le \mu < \mu_{m*} \qquad (5.3.9)$$

of the μ-spectrum, so the fixed-bed problem is included here. All the waves are directed forward and the entire range of influence of the initial discontinuity at $x = 0$ lies inside the domain. Clearly, we have $c_i^{0+} = c_i^a$, so that Eq. (5.1.22) becomes trivial and simply yields $n_i^a = n_{i,eq}^a$, where $n_{i,eq}^a$ represents the equilibrium value of n_i corresponding to the state $\{c_i^a\}$. The solution is continuous at $x = 0$ and equilibrium is established at the boundary $\mathcal{B}^a$.

A schematic diagram of the solution is shown in Fig. 5.4(a). The boundary at $x = 0$ is inward space-like and the data specified there are fully transmitted

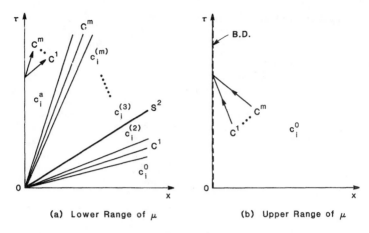

(a) Lower Range of μ (b) Upper Range of μ

Figure 5.4

into the domain. It is in this sense that we say the boundary $\mathscr{B}^a$ is now an *active* boundary.

Upper range of μ

Here the parameter has the range μ

$$\mu_1^* \leq \mu < \infty \tag{5.3.10}$$

Here all the waves become backward-facing and the range of influence of the initial discontinuity at $x = 0$ completely falls outside the domain. This implies that the feed data $\{c_i^a\}$ have no influence over the solution inside the domain. Thus we have $c_i^{0+} = c_i^o$ and $n_i^{0+} = n_{i,\text{eq}}^o$. The feed data, however, determine the state $\{n_i^a\}$ through the boundary discontinuity $\{c_i^a - c_i^o\}$ as given by Eq. (5.1.24).

In Fig. 5.4(b) it is shown that the boundary at $x = 0$ is now outward space-like and the data specified there cannot penetrate the boundary. In this regard, the boundary $\mathscr{B}^a$ may be referred to as a *passive boundary*. The dashed line along the τ axis in Fig. 5.4(b) represents the boundary discontinuity.

k-intermediate range of μ

Along the μ-spectrum this range covers the portion

$$\mu_k^* \leq \mu < \mu_{k-1,*} \tag{5.3.11}$$

for $k = 2, 3, \ldots, m$. In this case only the first $(k - 1)$ waves are directed forward and the state in the neighborhood of the τ axis is the k-constant state; that is, $c_i^{0+} = c_i^{(k)}$. The next $(m - k + 1)$ waves and the constant states in between all collapse at the boundary $\mathscr{B}^a$ to produce a boundary discontinuity there. Hence the solution is discontinuous at the boundary and the state $\{n_i^a\}$ is not given by the equilibrium relation but by Eq. (5.1.24) with $c_i^{0+} = c_i^{(k)}$ and $n_i^{0+} = n_{i,eq}^{(k)}$.

The physical plane portrait of the solution is illustrated in Fig. 5.5(a). The boundary at $x = 0$ is time-like with the first $(k - 1)$ characteristics $[C^1 \sim C^{k-1}]$ directed forward and the next $(m - k + 1)$ characteristics $[C^k \sim C^m]$ backward. A time-like boundary is only partially active in its data penetrating behavior.

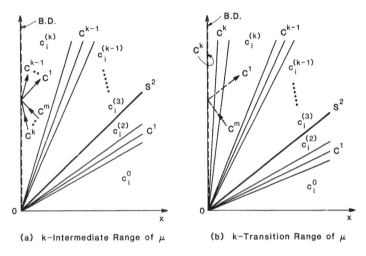

(a) k–Intermediate Range of μ (b) k–Transition Range of μ

Figure 5.5

Incidentally, we note from Eq. (5.3.2) that among the m characteristic parameters for the k-constant state the last $(m - k + 1)$ ones are identical to those for the initial state. This implies that if the solution is to be continuous at $x = 0$, we would be at liberty to specify only $(k - 1)$ conditions at $x = 0$, the number of the conditions being the same as the number of the characteristics directed inward there. Thus if a boundary is time-like with the first k characteristics directed inward, one can specify only k conditions at the boundary for continuity to exist there.[†] In the present problem, however, such a restriction is removed and the feed data $\{c_i^o\}$ may be specified arbitrarily because we admit a discontinuity at the boundary. Still, the state $\{c_i^{0+}\}$ just inside the boundary is subject to the restriction, so that it is not entirely independent of the initial state $\{c_i^o\}$. Apparently, the two states are related to each other by $(m - k + 1)$ conditions, so the two are partially dependent.

k-transition range of μ

This range corresponds to the portion between the lower and upper critical values for the k-wave:

[†] Compare this argument with the discussion in the last part of Sec. 1.2 along with Fig. 1.8.

$$\mu_{k*} \leq \mu < \mu_k^* \tag{5.3.12}$$

In case $\omega_k^a > \omega_k^o$, we have a k-simple wave and to the given value of μ in this range there corresponds a characteristic C_∞^k of infinite slope, which coincides with the τ-axis. This characteristic partitions the k-simple wave into two parts, of which only the portion to the right of C_∞^k comes into the domain along with the first $(k - 1)$ waves. The other portion collapses with the last $(m - k + 1)$ waves at the boundary $\mathcal{B}^a$ to produce a boundary discontinuity there. The situation is illustrated in Fig. 5.5(b).

The state borne on the characteristic C_∞^k remains stationary at $x = 0+$ and is characterized by the set of characteristic parameters

$$[\omega_1^a, \omega_2^a, \ldots, \omega_{k-1}^a, \omega_{k\infty}, \omega_{k+1}^o, \ldots, \omega_{m-1}^o, \omega_m^o] \tag{5.3.13}$$

in which we have

$$\omega_k = (\mu p_k)^{-1/2} \tag{5.3.14}$$

with p_k given by Eq. (5.3.4). By using these parameters in Eq. (3.7.9), we can determine the state $\{c_i^{0+}\}$. Since there is discontinuity at $x = 0$, the state $\{n_i^a\}$ is again determined by Eq. (5.1.24).

The boundary at $x = 0$ is now k-characteristic and has the nature of a partially active boundary. Looking at the set of characteristic parameters given by Eq. (5.3.13), we note that here the degree of freedom remains the same as for the case of the k-intermediate range because the value $\omega_{k\infty}$ borne on the characteristic C_∞^k is fixed.

If $\omega_k^a < \omega_k^o$, we have a k-shock and this transition range reduces to a single point $\mu = \tilde{\mu}_k$ for which the k-shock becomes stationary at $x = 0$ and coalesces with the boundary discontinuity. The state $\{c_i^{0+}\}$ is given by the k-constant state, and thus the situation will be the same as for the k-intermediate range. The case of $k = m$ is exceptional because the discontinuity at $x = 0$ is not a boundary discontinuity, but it is a stationary m-shock so that equilibrium is established across it.

In summary, the μ spectrum consists of one lower range, one upper range, $(m - 1)$ intermediate ranges, and m transition ranges. The state of the outgoing solid phase shows a piecewise continuous variation with respect to the parameter μ.

Analogous conclusions can be established without difficulty if a discontinuity is initially imposed at $x = 1$ and the contacting region extends from $x = 1$ to $x = -\infty$. In this case the initial and feed data are specified as

$$\begin{align} \text{at } \tau = 0: \quad c_i &= c_i^o \quad \text{for } -\infty < x < 1 \\ \text{at } x = 0: \quad n_i &= n_i^b \quad \text{for } \tau > 0 \end{align} \tag{5.3.15}$$

for $i = 1, 2, \ldots, m$, where $n_{i,\text{eq}}^o \neq n_i^b$ for some i. For both states $\{c_i^o\}$ and $\{n_i^b\}$, we first solve Eq. (5.2.6) to determine the sets of characteristic parameters, $\{\omega_k^a\}$ and $\{\omega_k^b\}$, and these two sets will take, respectively, the roles of $\{\omega_k^a\}$ and $\{\omega_k^o\}$ in the previous case.

In the lower range the boundary at $x = 1$ is outward space-like and remains passive. The boundary discontinuity is now characterized by Eq. (5.1.23), so the state $\{c_i^b\}$ of the outgoing fluid stream is given by Eq. (5.1.25). In the upper range the boundary becomes inward space-like and thus active. The intermediate and transition ranges can be treated similarly. The prime concern here is, of course, how to determine the state $\{c_i^{1-}\}$ just inside the boundary and thereby the state $\{c_i^b\}$ of the outgoing fluid stream.

Exercise

5.3.1. Consider the problem with the data given by Eq. (5.3.15) and determine the k-intermediate range of μ in terms of the characteristic parameters $\{\omega_k^a\}$ and $\{\omega_k^b\}$. For a given value of μ that belongs to the k-intermediate range, show that the state $\{c_i^{1-}\}$ just inside the boundary has only $(m - k + 1)$ degrees of freedom; that is, it is related to the initial state $\{c_i^o\}$ by $(k - 1)$ conditions.

5.4 Analysis of a Finite Column

In the practice of continuous countercurrent contact, the contacting region is of finite length and the initial and feed data would most likely be prescribed in the form of Eq. (5.1.9). Based on the discussions in previous sections, we expect that the basic features of the analysis remain the same except for the fact that the nature of the boundaries, $\mathcal{B}^a$ and $\mathcal{B}^b$, changes as the parameter μ increases. In this section, therefore, we will be mainly concerned with examining the nature of both boundaries over the whole spectrum of μ. Further treatment including the interaction analysis will be conducted and illustrated in the subsequent sections.

Just as shown in Fig. 5.3, let us suppose at the moment that the contacting region extends in both directions so that equilibrium is established everywhere. The initial and feed data will map on three different points in the hodograph space, which we denote as $O : \{\omega_k^o\}$, $A : \{\omega_k^a\}$, and $B : \{\omega_k^b\}$, respectively. In the beginning, the image of the solution is composed of m segments of distinct Γ^k's arranged in the order of decreasing k from A to O and the other m segments of distinct Γ^k's arranged in the order of decreasing k from O to B. In the physical plane, the solution can be determined by combining the two separate wave solutions, one developed from the origin and the other from the point $x = 1$ on the x-axis.

It is very clear from Fig. 5.3 that as time goes on, the waves will be involved in a series of interactions; that is, any two waves emanating from different points are bound to collide with each other and interact. After all the interactions are over, we can easily picture that the image of the solution in the hodograph space is given by m segments of distinct Γ^k's arranged again in the order of decreasing k from A directly to B. At this stage, the solution in the physical plane consists of m waves of distinct families, bounded by the two constant states, $\{c_i^a\}$ and $\{n_i^b\}$, and partitioned by $(m - 1)$ new constant states. The k-constant state $\{c_i^{(k)}\}$, which appears on the right-hand side of the k-wave, is characterized by the set of characteristic parameters

$$[\omega_1^a, \omega_2^a, \ldots, \omega_{k-1}^a, \omega_k^b, \ldots, \omega_{m-1}^b, \omega_m^b] \qquad (5.4.1)$$

and this set is independent of μ.

For a finite column, not all the waves propagate into the domain. Among the m waves to develop from the origin, only the forward-facing waves come into the domain and these are of lower families; that is, the wave index has lower numbers. Similarly, the waves emanating from the point $x = 1$ on the τ-axis come into the domain only when they are backward facing, so their wave indices are higher numbers. This implies that, in general, there are discontinuities at both boundaries, $\mathcal{B}^a$ and $\mathcal{B}^b$. Indeed, by using Rule 4.1, statement 2 which is also valid here, we can show that with the first $(k - 1)$ waves centered at the origin, only the next $(m - k + 1)$ waves can emanate from the other end. The only exception is when one or both of the $(k - 1)$-wave and the k-wave are shocks and in such a case the $(k - 1)$-wave and/or the k-wave can also emanate from the opposite end. It then follows that every forward (or backward)-facing wave will undergo $(m - k + 1)$ [or $(k - 1)$] transmissive interactions successively and finally reach the opposite boundary $\mathcal{B}^b$ [or $\mathcal{B}^a$] to encounter a boundary discontinuity there.

When the tail of the most slowly moving wave (either forward- or backward-facing) reaches the opposite end, the steady state is attained since no further change will be observed afterward. The steady state is, of course, given by the k-constant state $\{c_i^{(k)}\}$ in the domain and by Eqs. (5.1.24) and (5.1.25) for the outgoing streams.

The situation would be different for different values of μ, but just as for the case of a semi-infinite column, appropriate classification can be accomplished by making cuts at critical values of μ defined as follows: If $\omega_k^a > \omega_k^b$,

$$\mu_{k**} = [p_k(\omega_k^a)^2]^{-1} \qquad (5.4.2)$$

$$\mu_k^* = [p_k(\omega_k^b)^2]^{-2} \qquad (5.4.3)$$

and if $\omega_k^a < \omega_k^b$,

$$\bar{\mu}_k = [p_k \omega_k^a \omega_k^b]^{-1} \qquad (5.4.4)$$

for $k = 1, 2, \ldots, m$, where

$$p_k = \frac{1}{\gamma_k} \prod_{i=1}^{k-1} \frac{\omega_i^a}{\gamma_i} \prod_{i=k+1}^{m} \frac{\omega_i^b}{\gamma_i} \qquad (5.4.5)$$

[Compare these definitions to Eqs. (5.3.5), (5.3.6), and (5.3.8).] Now we shall discuss the general picture of the solution for μ values in different ranges with emphasis on the steady state.

Lower range: $0 \leq \mu < \mu_{m*}$

All the waves emanating from the origin are directed forward and the boundary $\mathscr{B}^a$ is inward space-like, whereas the boundary $\mathscr{B}^b$ is outward space-like as shown in Fig. 5.6(a). Since $\mathscr{B}^a$ is active, there is no discontinuity at $x = 0$, and we have $c_i^{0+} = c_i^a$ and $n_i^{0+} = n_{i,\text{eq}}^a$. The feed data at $x = 1$ has no influence over the solution inside the domain since the domain of dependence never contains any part of the boundary $\mathscr{B}^b$, but the data determine the state $\{c_i^b\}$ which is connected to the state inside by a boundary discontinuity.

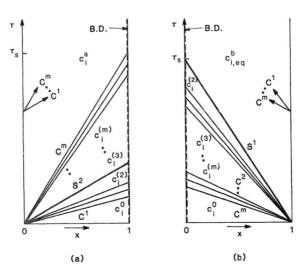

Figure 5.6

At $\tau = \tau_s$, when the tail of the m-simple wave reaches the boundary at $x = 1$, the steady state is attained, so we have

$$\tau_s = \sigma_m^a = \frac{1 + \nu p_m(\omega_m^a)^2}{1 - \mu p_m(\omega_m^a)^2} \tag{5.4.6}$$

where p_m is given by Eq. (5.4.5) with $k = m$. Hence τ_s is determined by the state $\{c_i^a\}$ alone. The concentration c_i^b changes with time from $c_i^a + \mu(n_i^b - n_{i,\text{eq}}^o)$ at $\tau = 0+$ to $c_i^a + \mu(n_i^b - n_{i,\text{eq}}^a)$ at $\tau = \tau_s$. The steady state may be described as

$$n_i^a = n_{i,\text{eq}}^a = \frac{\gamma c_i^a}{\delta^a}$$

$$c_i = c_i^a \qquad \text{for } 0 < x < 1 \tag{5.4.7}$$

$$c_i^b = c_i^a + \mu(n_i^b - n_{i,\text{eq}}^a)$$

If the m-wave is a shock, the lower range becomes $0 \le \mu < \bar{\mu}_m$ and the time to reach the steady state is

$$\tau_s = \tilde{\sigma}_m = \frac{1 + \nu p_m \omega_m^a \omega_m^o}{1 - \mu p_m \omega_m^a \omega_m^o} \tag{5.4.8}$$

Note that here τ_s depends on the initial state, but the steady state would be the same as before.

Upper range: $\mu_1^* < \mu < \infty$

With a μ-value in this range, all the waves develop from the point $x = 1$ on the τ-axis and are backward-facing. The boundary $\mathcal{B}^a$ is outward space-like while the boundary $\mathcal{B}^b$ is inward space-like, as shown in Fig. 5.6(b). The possibilities at the boundaries are reversed from those observed in the previous case. There is no discontinuity at $x = 1$, so that we have $n_i^{1-} = n_i^b$ and $c_i^b = c_{i,\text{eq}}^b$. The boundary at $x = 0$ bears a discontinuity from the beginning. As marked in Fig. 5.6(b), n_i^{0+} varies from n_i^o to n_i^b and accordingly n_i^a changes from $n_i^o + (c_i^a - c_i^o)/\mu$ to $n_i^b + (c_i^a - c_{i,\text{eq}}^b)/\mu$.

The steady state is given by

$$n_i^a = n_i^b + \frac{c_i^a - c_{i,\text{eq}}^b}{\mu}$$

$$c_i = c_{i,\text{eq}}^b \qquad \text{for } 0 < x < 1 \tag{5.4.9}$$

$$c_i^b = c_{i,\text{eq}}^b$$

and it is attained when the tail of the 1-simple wave reaches the boundary at $x = 0$, that is, at

$$\tau = \tau_s = -\sigma_1^b = -\frac{1 + \nu p_1 (\omega_1^b)^2}{1 - \mu p_1 (\omega_1^b)^2} \tag{5.4.10}$$

where p_1 is given by Eq. (5.4.5) with $k = 1$ and so purely dependent on the state $\{n_i^b\}$.

If the 1-wave happens to be a shock, the upper range becomes $\bar{\mu}_1 < \mu < \infty$ and we obtain

$$\tau_s = -\tilde{\sigma}_1 = -\frac{1 + \nu p_1 \omega_1^o \omega_1^b}{1 - \mu p_1 \omega_1^o \omega_1^b} \tag{5.4.11}$$

k-intermediate range: $\mu_{k*} < \mu < \mu_{k-1,*}$

Here the first $(k - 1)$ waves develop from the origin while the next $(m - k + 1)$ waves emanate from the opposite end. Both boundaries, $\mathcal{B}^a$ and $\mathcal{B}^b$, are time-like and bear boundary discontinuities from the beginning. A general situation is illustrated in Fig. 5.7(a).

After all the transmissive interactions are completed, there appears the region of the k-constant state characterized by Eq. (5.4.1), and this region gradu-

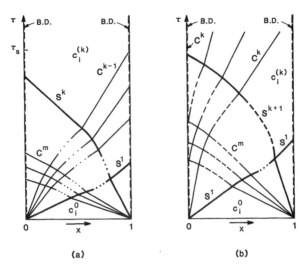

Figure 5.7

ally expands to span the entire domain. For every point in this region the domain of dependence consists of portions of $\mathscr{B}^a$ and $\mathscr{B}^b$ excluding the initial line $\tau = 0$, and this implies that the state there is determined by the feed data and independent of the initial data as expected.

At $\tau = \tau_s$, when the last trace of the waves reaches one of the boundaries, the steady state is attained. It is obvious that the steady state inside the domain is the k-constant state, and thus we have

$$n_i^a = n_{i,\text{eq}}^{(k)} + \frac{c_i^a - c_i^{(k)}}{\mu}$$

$$c_i = c_i^{(k)} \qquad \text{for } 0 < x < 1 \tag{5.4.12}$$

$$c_i^b = c_i^{(k)} + \mu(n_i^b - n_{i,\text{eq}}^{(k)})$$

at the steady state. The time τ_s can be determined only after analyzing all the interactions and it is apparent that τ_s depends on the initial state.

k-transition range: $\mu_{k*} \leq \mu \leq \mu_{k*}$

We shall consider two cases separately. First, if $\omega_k^a > \omega_k^b$, we expect to have a k-simple wave developing from the origin or from the opposite end, which is to be cut at the final stage either by the boundary $\mathscr{B}^a$ or by $\mathscr{B}^b$. Hence the corresponding boundary becomes k-characteristic and the particular state borne on the characteristic remains stationary there.

Figure 5.7(b) shows a situation in which the first $(k - 1)$ waves are directed forward and the next $(m - k + 1)$ waves backward, while the k-simple wave is cut by the boundary $\mathscr{B}^a$. In this case the state $\{c_i^{0+}\}$ just inside the boundary $\mathscr{B}^a$ changes with time but ultimately reaches the state $\{c_{i\infty}\}$ characterized by the set

$$[\omega_1^a, \omega_2^a, \ldots, \omega_{k-1}^a, \omega_{k\infty}, \omega_{k+1}^b, \ldots, \omega_{m-1}^b, \omega_m^b] \qquad (5.4.13)$$

where

$$\omega_{k\infty} = (\mu p_k)^{-1/2} \qquad (5.4.14)$$

with p_k given by Eq. (5.4.5). Both boundaries maintain discontinuities, and no steady state is attained in finite time. The state inside the domain approaches the state $\{c_{i\infty}\}$ only asymptotically.

Next, if $\omega_k^a < \omega_k^b$, this range degenerates to a single value $\mu = \bar{\mu}_k$. In the case when $\omega_k^a < \omega_k^o < \omega_k^b$, two k-shocks develop from both ends and are ultimately superposed to become stationary in the domain. Otherwise, the k-shock eventually becomes stationary at one of the boundaries and coalesces with the boundary discontinuity so that the situation will be the same as for the k- or $(k + 1)$-intermediate range. In either case both boundaries are time-like and bear boundary discontinuities.

With the stationary shock at $x = \tilde{x}$, $0 \leq \tilde{x} \leq 1$, we have

$$n_i^a = n_i^{(k+1)} + \frac{c_i^a - c_i^{(k+1)}}{\mu}$$

$$c_i = c_i^{(k+1)} \quad \text{for } 0 < x < \tilde{x}$$

$$c_i = c_i^{(k)} \quad \text{for } \tilde{x} < x < 1 \qquad (5.4.15)$$

$$c_i^b = c_i^{(k)} + \mu(n_i^b - n_{i,\text{eq}}^{(k)})$$

at the steady state. If $x = 0$ or $x = 1$, Eq. (5.4.15) becomes identical to Eq. (5.4.12) for k or $k + 1$, respectively.

In summary, the μ spectrum can be divided into one lower range, one upper range, $(m - 1)$ intermediate ranges, and m transition ranges. It turns out that a unique steady state corresponds to each of the lower, upper, and intermediate ranges. Consequently, there are $(m + 1)$ possible steady states for a given pair of feed data.[†] The steady state given by Eq. (5.4.15) is exceptional and not counted separately. These steady states are completely determined by the feed data alone, but the particular one that is attained depends also on the parameter μ. On the other hand, the time τ_s to reach the steady state depends on the initial and feed data, the parameters ν and μ, and the column length. If $\omega_k^a = \omega_k^b$ for any k, the k-intermediate range as well as the k-transition range does not exist, and accordingly, the number of possible steady states becomes one less.

To illustrate further the development in this section, we shall treat a numerical example for $m = 3$. Here we shall be concerned with the determination of various ranges of μ and the steady state for a given value of μ. Therefore, the initial data are arbitrary at the moment. The parameters are given as

---

[†] Here we consider only those steady states that are attained in finite time. Note that with a μ-value in one of the transition ranges, the steady state is attained only asymptotically.

$$\nu = 1.5$$

$$N_i = 1.0 \text{ mol/liter}, \quad i = 1, 2, 3$$

$$K_1 = 5.0 \text{ liters/mol}, \quad K_2 = 7.5, \quad K_3 = 10.0$$

When expressed in moles per liter of individual phase, the feed concentrations are specified as follows:

$$c_1^a = 0.150, \quad c_2^a = 0.060, \quad c_3^a = 0.020$$

$$n_1^b = 0.058, \quad n_2^b = 0.309, \quad n_3^b = 0.271$$

We note that the solid phase comes into the column with a surface coverage $\Theta^b = 0.638$.

The characteristic parameters are determined by solving Eq. (5.2.6) for each of the feed states and they are

$$\omega_1^a = 2.445, \quad \omega_2^a = 6.667, \quad \omega_3^a = 9.586$$

$$\omega_1^b = 2.797, \quad \omega_2^b = 5.409, \quad \omega_3^b = 8.974$$

It is immediately apparent that if the column were infinitely long, the ultimate form of the 1-wave would be a shock while the other two would be simple waves. Hence the 1-transition range degenerates to a single point.

We then calculate the critical values of μ by applying Eqs. (5.4.2) and (5.4.3) or Eq. (5.4.4) together with Eq. (5.4.5) and determine the ranges of μ:

Lower range	$0 \leq \mu < 0.250$
3-transition range	$0.250 \leq \mu \leq 0.286$
3-intermediate range	$0.286 < \mu < 0.385$
2-transition range	$0.385 \leq \mu \leq 0.584$
2-intermediate range	$0.584 < \mu < 1.130$
1-transition range	$\mu = 1.130$
Upper range	$1.130 < \mu < \infty$

There are four different steady states attainable in finite time. If, for instance, the μ value belongs to the 3-intermediate range, the steady state to be attained within the domain is the 3-constant state $\{c_i^{(3)}\}$, for which we have the set of characteristic parameters $[\omega_1^a, \omega_2^a, \omega_3^b]$. We first determine the state $\{c_i^{(3)}\}$ by using Eq. (3.7.9) or (3.11.6), with $k = 3$ and c_i^a for c_i^f, and then the states of the outgoing streams by applying Eq. (5.4.12). The steady state to be attained with $\mu = \frac{1}{3}$ is given in Table 5.1.

It is to be noted that as the solid phase passes through the column, the more strongly adsorbable solutes A_2 and A_3 are desorbed, whereas the least adsorbable solute A_1 is adsorbed more strongly. The solid phase experiences a slight decrease in the coverage. This example clearly demonstrates that one can achieve the displacement of a strongly adsorbed solute by a less strongly adsorbed solute. For this example such a displacement can be enhanced by using a lower value of μ within the 3-intermediate range.

TABLE 5-1. STEADY STATE TO BE ATTAINED WHEN $\mu = \dfrac{1}{3}$

	i			
	1	*2*	*3*	*Remark*
c_i^a	0.150	0.060	0.020	
$c_i(x)$	0.139	0.045	0.053	For $0 < x < 1$
c_i^b	0.068	0.104	0.074	
n_i^a	0.304	0.177	0.108	$\Theta^a = 0.589$
$n_i(x)$	0.271	0.132	0.207	$\Theta = 0.610$ for $0 < x < 1$
n_i^b	0.058	0.309	0.271	$\Theta^b = 0.638$

Exercise

5.4.1. In practical situations we may often have the condition $n_i^b = n_{i,\text{eq}}^o$ for all i. Under these circumstances, discuss the solution of the moving-bed problem for various ranges of the parameter μ. In particular, find an expression for the time τ_s when μ belongs to the k-intermediate range. Consider both cases, $\omega_k^a > \omega_k^b$ and $\omega_k^a < \omega_k^b$, and apply the result to the numerical example treated in this section to show that $\tau_s = 36.75$ for $\mu = \frac{1}{3}$.

5.5 Analysis of Wave Interactions

For a μ value in the lower range or in the upper range the waves all develop from a single point and are directed either forward or backward. The waves do not interact, so that it is a straightforward matter to construct the complete solution. If, however, the μ value belongs to one of the intermediate ranges or to the transition ranges, we have waves emanating from both ends and they certainly will interact. The patterns and principles of interactions are determined by Rules 4.1 and 4.2 as discussed in Sec. 5.2.

In this section we shall be concerned with the separate analysis of each interaction. Without loss of generality we shall always consider a case that can occur in the operation of a finite column when the value of μ belongs to the k-intermediate range. Obviously, it would be possible to employ the same approach here as for fixed-bed problems because the image in the hodograph space is independent of μ. For analysis, however, we may as well present the image in the space $\Omega(m)$ along with the physical plane portrait if we find it useful.

As exploited in Sec. 5.4, the solution for a finite column can be expressed in terms of the three sets of characteristic parameters, $\{\omega_k^a\}$, $\{\omega_k^b\}$, and $\{\omega_k^o\}$, corresponding to the feed and initial states, respectively. In every case, therefore, we shall identify the relevant characteristic parameters around the region of interaction and use these parameters as we develop the interaction analysis.

This implies that all the characteristic parameters appearing in the following development will bear one of the superscripts a, b, or o, if they are fixed.

Interaction between waves of the same family

Since the value of μ belongs to the k-intermediate range, we expect a k-wave to develop from the boundary $\mathcal{B}^b$. After a sequence of interactions this k-wave will reach the boundary $\mathcal{B}^a$ to give the k-constant state in the domain. If this k-wave is a shock, it may encounter another k-shock or a k-simple wave, developing from the boundary $\mathcal{B}^a$, at the last stage of a sequence of interactions, and this is the case that we will be concerned with here.

Here the waves in interaction are of the same family and their images in the hodograph space fall on a single Γ^k. Since ω_k is the only variable parameter along a Γ^k, we note that p_k defined by Eq. (5.2.11) or (5.2.23) remains fixed throughout the course of interaction and thus there exists a one-parameter representation of the interaction. In fact, at this stage the parameters ω_i, $i \neq k$, are given as

$$\omega_i = \omega_i^a \qquad \text{for } 1 \leq i \leq k - 1$$
$$\omega_i = \omega_i^b \qquad \text{for } k + 1 \leq i \leq m \tag{5.5.1}$$

so that we have

$$p_k = \frac{1}{\gamma_k} \prod_{i=1}^{k-1} \frac{\omega_i^a}{\gamma_i} \prod_{i=k+1}^{m} \frac{\omega_i^b}{\gamma_i} \tag{5.5.2}$$

When two k-shocks meet each other, the two are superposed instantaneously as shown in Fig. 5.8. This is what we call the confluence of two

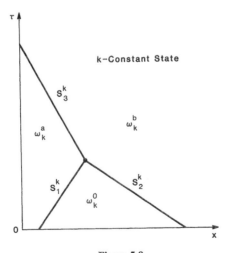

Figure 5.8

shocks. The characteristic parameter ω_k for each region of constant state can be easily identified and the relevant parameters are marked in Fig. 5.8. The two k-shocks before the interaction have the propagation speeds

$$\tilde{\lambda}_{k,1} = \frac{1 - \mu p_k \omega_k^o \omega_k^a}{1 + \nu p_k \omega_k^o \omega_k^a} \tag{5.5.3}$$

$$\tilde{\lambda}_{k,2} = \frac{1 - \mu p_k \omega_k^o \omega_k^b}{1 + \nu p_k \omega_k^o \omega_k^b} \tag{5.5.4}$$

respectively, and the new k-shock formed by interaction propagates with the speed

$$\tilde{\lambda}_{k,3} = \frac{1 - \mu p_k \omega_k^a \omega_k^b}{1 + \nu p_k \omega_k^a \omega_k^b} \tag{5.5.5}$$

Since $\omega_k^a < \omega_k^o < \omega_k^b$ in this case, we immediately find that

$$\tilde{\lambda}_{k,1} > \tilde{\lambda}_{k,3} > \tilde{\lambda}_{k,2} \tag{5.5.6}$$

and hence the new shock takes an intermediate speed. With μ in the k-intermediate range, the k-shock will be directed backward and eventually reach the boundary $\mathcal{B}^a$ as illustrated in Fig. 5.8.

Next, we shall consider a case when a k-shock, directed backward, encounters a forward-facing k-simple wave as shown in Fig. 5.9. While interacting, the two waves cancel each other so that the state on the right-hand side of the shock remains unchanged during the course of interaction. This implies that ω_k^l is the only variable of the system, so we should be able to represent the interaction in terms of ω_k^l alone. For the sake of convenience, we shall delete the superscript l here and let ω_k denote its value on the left-hand side of the k-shock. As the shock propagates, ω_k increases from ω_k^o to ω_k^a.

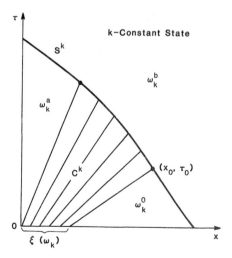

Figure 5.9

Since the k-simple wave might already have gone through a series of interactions, we do not expect it to be centered in general. It would instead be based on data distributed along a certain curve in the physical plane. Let us assume that the k-simple wave is based on data specified over a portion of the x-axis, as illustrated in Fig. 5.9, and the inverse of the data is determined as a function of ω_k only; that is, we have $\xi = \xi(\omega_k)$. At an arbitrary point (x, τ) on the shock path S^k, we can write two equations, one for the characteristic C^k and another for the shock path:

$$\lambda_k(\omega_k)\tau = x - \xi(\omega_k) \tag{5.5.7}$$

$$\frac{dx}{d\tau} = \frac{dx/d\omega_k}{d\tau/d\omega_k} = \tilde{\lambda}_k(\omega_k) \tag{5.5.8}$$

These equations give the parametric description of the shock path.

Obviously, here it is more convenient to eliminate x between Eqs. (5.5.7) and (5.5.8). This gives

$$(\tilde{\lambda}_k - \lambda_k)\frac{d\tau}{d\omega_k} - \frac{d\lambda_k}{d\omega_k}\tau = \frac{d\xi}{d\omega_k} \tag{5.5.9}$$

which is a linear differential equation for τ with ω_k as the independent variable. The initial condition is given by

$$\tau = \tau_0 \quad \text{at} \quad \omega_k = \omega_k^0 \tag{5.5.10}$$

where τ_0 is the ordinate of the intersection between the shock path S^k and the lowermost characteristic C^k carrying the value ω_k^0.

Now with p_k given by Eq. (5.5.2) we can write

$$\lambda_k(\omega_k) = \frac{1 - \mu p_k \omega_k^2}{1 + \nu p_k \omega_k^2} \tag{5.5.11}$$

$$\tilde{\lambda}_k(\omega_k) = \frac{1 - \mu p_k \omega_k^b \omega_k}{1 + \nu p_k \omega_k^b \omega_k} \tag{5.5.12}$$

and substituting these into Eq. (5.5.9) gives

$$(\omega_k - \omega_k^b)\frac{d\tau}{d\omega_k} + \frac{1 + \nu p_k \omega_k^b \omega_k}{1 + \nu p_k \omega_k^2}\tau$$
$$= \frac{(1 + \nu p_k \omega_k^2)(1 + \nu p_k \omega_k^b \omega_k)}{(\nu + \mu)p_k \omega_k}\frac{d\xi}{d\omega_k} \tag{5.5.13}$$

The left-hand side of this equation can be put into exact differential form. Then we obtain

$$\frac{d}{d\omega_k}\left\{\tau\frac{(\omega_k - \omega_k^b)^2}{1 + \nu p_k \omega_k^2}\right\} = \frac{(\omega_k - \omega_k^b)(1 + \nu p_k \omega_k^b \omega_k)}{(\nu + \mu)p_k \omega_k}\frac{d\xi}{d\omega_k} \tag{5.5.14}$$

Integrating this equation directly from the initial point (x_0, τ_0), where $\omega_k = \omega_k^0$, we get

$$\tau(\omega_k) = \tau_0 \frac{1 + \nu p_k \omega_k^2}{1 + \nu p_k (\omega_k^o)^2} \left(\frac{\omega_k^o - \omega_k^b}{\omega_k - \omega_k^b}\right)^2$$

$$+ \frac{1 + \nu p_k \omega_k^2}{(\nu + \mu) p_k (\omega_k - \omega_k^b)^2} \int_{\omega_k^o}^{\omega_k} \left(1 - \frac{\omega_k^b}{\omega_k}\right)(1 + \nu p_k \omega_k^b \omega_k) \frac{d\xi}{d\omega_k} \, d\omega_k \quad (5.5.15)$$

If the simple wave is centered, we can put $d\xi/d\omega_k = 0$, so

$$\tau(\omega_k) = \tau_0 \frac{1 + \nu p_k \omega_k^2}{1 + \nu p_k (\omega_k^o)^2} \left(\frac{\omega_k^o - \omega_k^b}{\omega_k - \omega_k^b}\right)^2 \quad (5.5.16)$$

Otherwise, we may take the point (x_0, τ_0) as the origin of the new coordinate system and adjust the function $\xi(\omega_k)$ accordingly. Let this be denoted by $\bar{\xi}(\omega_k)$, so that $\bar{\xi}(\omega_k^o) = 0$. Then, after applying integration by parts, Eq. (5.5.15) reduces to

$$\tau(\omega_k) = \frac{1 + \nu p_k \omega_k^2}{(\nu + \mu) p_k (\omega_k - \omega_k^b)} \left\{\left(\frac{1}{\omega_k} + \nu p_k \omega_k^b\right) \bar{\xi}(\omega_k)\right.$$

$$\left. - \frac{\omega_k^b}{\omega_k - \omega_k^b} \int_{\omega_k^o}^{\omega_k} \left(\frac{1}{\omega_k^2} + \nu p_k\right) \bar{\xi}(\omega_k) \, d\omega_k\right\} \quad (5.5.17)$$

and the integral can be readily evaluated if $\bar{\xi}$ is read along each of the characteristics C^k. Once τ is determined in terms of ω_k, we can use Eq. (5.5.7) to express x also as a function of ω_k.

Since ω_k increases as the shock propagates, it decelerates gradually until the interaction is completed. In any case, the interaction will terminate when $\omega_k = \omega_k^o$ and then the shock resumes a constant speed until it reaches the boundary $\mathcal{B}^a$.

In the $(k + 1)$-intermediate range, we may expect to have a case when a k-shock encounters a backward-facing k-simple wave. Interaction between these two waves can also be analyzed by applying a procedure similar to the previous one (see Exercise 5.5.2).

Interactions between waves of different families

Here the waves in interaction belong to different families and, according to Rule 4.2, the two waves are transmitted through each other. We shall consider the interaction between a j-wave and a k-wave, where $j < k$. This implies that the j-wave is forward-facing while the k-wave is directed backward and this arrangement is consistent with the situation for the k-intermediate range.

It is obvious that the image in the hodograph space $\Omega(m)$ falls completely on a plane parallel to the ω_j and ω_k axes, and thus the parameters ω_i, unless $i = j$ or $i = k$, remain constant during the course of interaction. Furthermore, only one of the two parameters ω_j and ω_k varies along a characteristic or a shock path while it passes through a wave of different family. All these features are the same as for the fixed-bed problems and provide the basis for a complete treatment of interaction analysis here. Just as for the fixed-bed problem, we shall treat three different cases separately.

Transmission between Two Shocks   In this case a j-shock, directed forward, collides with a backward-facing k-shock and these two waves are instantaneously transmitted across each other to produce two transmitted shocks of their respective families. In the hodograph space the images of these four shocks fall on a plane parallel to the ω_j- and ω_k-axes as shown in Fig. 5.10(a). The other $(m - 2)$ parameters are invariant on this plane.

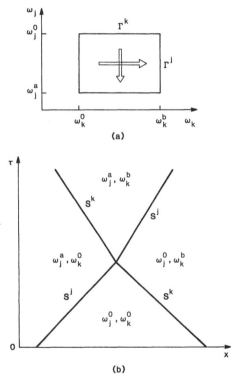

(a)

(b)

Figure 5.10

In the physical plane we have four regions of constant state around the point of interaction, all separated by the four shock paths, as illustrated in Fig. 5.10(b). These constant states are completely characterized by the appropriate pair of characteristic parameters ω_j and ω_k. The other $(m - 2)$ parameters are common to all four constant states and, indeed, their values can be identified as follows:

$$\omega_i = \omega_i^a \quad \text{for } 0 \le i \le j - 1$$

$$\omega_i = \omega_i^o \quad \text{for } j + 1 \le i \le k - 1 \qquad (5.5.18)$$

$$\omega_i = \omega_i^b \quad \text{for } k + 1 \le i \le m$$

Hence, we shall find it convenient to define the quantity

$$q_{jk} = \frac{1}{\gamma_j \gamma_k} \prod_{i=1}^{j-1} \frac{\omega_i^a}{\gamma_i} \prod_{i=j+1}^{k-1} \frac{\omega_i^o}{\gamma_i} \prod_{i=k+1}^{m} \frac{\omega_i^b}{\gamma_i} \qquad (5.5.19)$$

which remains fixed during the process of interaction.

The two shocks before interaction propagate with speeds

$$\tilde{\lambda}_j = \frac{1 - \mu q_{jk} \omega_j^o \omega_j^a \omega_k^o}{1 + \nu q_{jk} \omega_j^o \omega_j^a \omega_k^o} \qquad (5.5.20)$$

and

$$\tilde{\lambda}_k = \frac{1 - \mu q_{jk} \omega_j^o \omega_k^o \omega_k^b}{1 + \nu q_{jk} \omega_j^o \omega_k^o \omega_k^b} \qquad (5.5.21)$$

respectively. Upon interaction, the two shocks will pass through each other to produce two transmitted shocks of their respective families. This is indicated by the pair of arrows in Fig. 5.10(a). Their propagation speeds are given by

$$\tilde{\lambda}_j' = \frac{1 - \mu q_{jk} \omega_j^o \omega_j^a \omega_k^b}{1 + \nu q_{jk} \omega_j^o \omega_j^a \omega_k^b} \qquad (5.5.22)$$

$$\tilde{\lambda}_k' = \frac{1 - \mu q_{jk} \omega_j^a \omega_k^o \omega_k^b}{1 + \nu q_{jk} \omega_j^a \omega_k^o \omega_k^b} \qquad (5.5.23)$$

Since $\omega_k^o < \omega_k^b$ and $\omega_j^a < \omega_j^o$ [see the entropy condition (5.2.26)], we immediately find that

$$\tilde{\lambda}_j > \tilde{\lambda}_j', \qquad \tilde{\lambda}_k < \tilde{\lambda}_k' \qquad (5.5.24)$$

and thus the j-shock decelerates upon interaction while the k-shock accelerates.

Transmission between a Simple Wave and a Shock We shall consider the case when a j-shock meets a k-simple wave so that the image in the hodograph space and the physical plane portrait may be given as shown in Fig. 5.11(a) and (b), respectively. Here again q_{jk} defined by Eq. (5.5.19) remains constant. Just as in Fig. 5.11(a), the k-simple wave develops from a portion of the x-axis and it is assumed that the inverse of the data is defined as a function of ω_k alone; that is, $\xi = \xi(\omega_k)$.

As the interaction goes on, the k-simple wave is gradually transmitted across the j-shock, so the image in the hodograph space at an intermediate stage will be given by the dashed lines shown in Fig. 5.11(a), in which the arrow indicates the direction of the shock movement. It is then obvious that, as the shock passes through the k-simple wave, the parameter ω_k decreases from ω_k^o to ω_k^b while the parameter ω_j remains fixed on either side of the shock. Consequently, we expect to have a one-parameter representation for the shock path

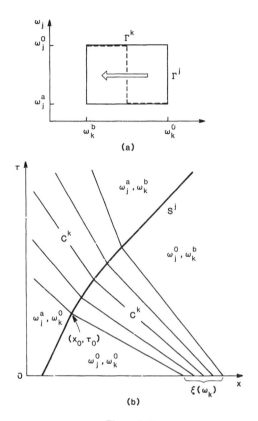

Figure 5.11

S^j. Indeed, at an arbitrary point (x, τ) on the shock path we can write two equations

$$\lambda_k(\omega_k)\tau = x - \xi(\omega_k) \tag{5.5.25}$$

$$\frac{dx}{d\tau} = \frac{dx/d\omega_k}{d\tau/d\omega_k} = \tilde{\lambda}_j(\omega_k) \tag{5.5.26}$$

Eliminating x between these equations gives the differential equation

$$(\tilde{\lambda}_j - \lambda_k)\frac{d\tau}{d\omega_k} - \frac{d\lambda_k}{d\omega_k}\tau = \frac{d\xi}{d\omega_k} \tag{5.5.27}$$

which is subject to the initial condition

$$\tau = \tau_0 \quad \text{at} \quad \omega_k = \omega_k^0 \tag{5.5.28}$$

where τ_0 is the ordinate of the intersection between the shock path S^j and the lowermost characteristic C^k as marked in Fig. 5.11(b).

Since ω_j is unchanged on either side of the shock path S^j, we shall define

$$p_k = q_{jk}\omega_j^o = \frac{1}{\gamma_k}\prod_{i=1}^{j-1}\frac{\omega_i^a}{\gamma_i}\prod_{i=j}^{k-1}\frac{\omega_i^o}{\gamma_i}\prod_{i=k+1}^{m}\frac{\omega_i^b}{\gamma_i} \tag{5.5.29}$$

which is invariant on the right-hand side of the shock path. We can then express λ_k and $\tilde{\lambda}_j$ in terms of ω_k as

$$\lambda_k(\omega_k) = \frac{1 - \mu p_k\omega_k^2}{1 + \nu p_k\omega_k^2} \tag{5.5.30}$$

$$\tilde{\lambda}_j(\omega_k) = \frac{1 - \mu p_k\omega_j^a\omega_k}{1 + \nu p_k\omega_j^a\omega_k} \tag{5.5.31}$$

and substituting these into Eq. (5.5.27) gives

$$(\omega_k - \omega_j^a)\frac{d\tau}{d\omega_k} + 2\frac{1 + \nu p_k\omega_j^a\omega_k}{1 + \nu p_k\omega_k^2}\tau$$
$$= \frac{(1 + \nu p_k\omega_k^2)(1 + \nu p_k\omega_j^a\omega_k)}{(\nu + \mu)p_k\omega_k}\frac{d\xi}{d\omega_k} \tag{5.5.32}$$

We immediately find that this equation is identical to Eq. (5.5.13) with ω_j^a for ω_k^b. Thus by following the same procedure we would obtain

$$\tau(\omega_k) = \tau_0\frac{1 + \nu p_k\omega_k^2}{1 + \nu p_k(\omega_k^o)^2}\left(\frac{\omega_k^o - \omega_j^a}{\omega_k - \omega_j^a}\right)^2$$
$$+ \frac{1 + \nu p_k\omega_k^2}{(\nu + \mu)p_k(\omega_k - \omega_j^a)^2}\int_{\omega_k^o}^{\omega_k}\left(1 - \frac{\omega_j^a}{\omega_k}\right)(1 + \nu p_k\omega_j^a\omega_k)\frac{d\xi}{d\omega_k}d\omega_k \tag{5.5.33}$$

In case the k-simple wave is centered, we have $d\xi/d\omega_k = 0$ and thus

$$\tau(\omega_k) = \tau_0\frac{1 + \nu p_k\omega_k^2}{1 + \nu p_k(\omega_k^o)^2}\left(\frac{\omega_k^o - \omega_j^a}{\omega_k - \omega_j^a}\right)^2 \tag{5.5.34}$$

In other cases, we take the point (x_0, τ_0) as the origin of the new coordinate system to have $\tau_0 = \bar{\xi}(\omega_k^o) = 0$, where $\bar{\xi}(\omega_k)$ is the adjusted form of the function $\xi(\omega_k)$. Hence we obtain

$$\tau(\omega_k) = \frac{1 + \nu p_k\omega_k^2}{(\nu + \mu)p_k(\omega_k - \omega_j^a)}\left\{\left(\frac{1}{\omega_k} + \nu p_k\omega_j^a\right)\bar{\xi}(\omega_k)\right.$$
$$\left. - \frac{\omega_j^a}{\omega_k - \omega_j^a}\int_{\omega_k^o}^{\omega_k}\left(\frac{1}{\omega_k^2} + \nu p_k\right)\bar{\xi}(\omega_k)d\omega_k\right\} \tag{5.5.35}$$

These equations can be substituted into Eq. (5.5.25) to determine x in terms of ω_k.

Since $\omega_k^o > \omega_k^b$, the parameter ω_k decreases as the j-shock passes through the k-simple wave and thus it accelerates while interacting. The interaction will be over when $\omega_k = \omega_k^b$ and then the shock obtains a constant speed. If we cross the j-shock along a C^k, the parameter ω_j jumps down from ω_j^o to ω_j^a, whereas ω_k remains constant. It then follows that, for a C^k on the left-hand side of the j-shock, we can write

$$\lambda_k' = \frac{1 - \mu p_k' \omega_k^2}{1 + \nu p_k' \omega_k^2} \qquad (5.5.36)$$

where

$$p_k' = q_{jk}\omega_j^a = \frac{1}{\gamma_k} \prod_{i=1}^{j} \frac{\omega_i^a}{\gamma_i} \prod_{i=j+1}^{k-1} \frac{\omega_i^o}{\gamma_i} \prod_{i=k+1}^{m} \frac{\omega_i^b}{\gamma_i}$$

and, since $\omega_j^o > \omega_j^a$, we have $\lambda_k < \lambda_k'$, so that every C^k is refracted upward upon crossing the shock path S^j. The state of concentrations borne on the refracted characteristic C^k is readily determined by using Eq. (3.7.9) because all the characteristic parameters are given. The totality of the refracted C^k's therefore constitutes the transmitted k-simple wave.

If we have a j-simple wave, facing forward, and a k-shock, directed backward, the two waves will be transmitted through each other by interaction. We note that the roles of the two waves are now interchanged in Fig. 5.11. Since here ω_j is the only variable parameter along the shock path S^k, we can apply the same procedure as before to determine the shock path in terms of ω_j (see Exercise 5.5.2). In particular, if the j-simple wave is centered at the origin, we would obtain

$$\tau(\omega_j) = \tau_0 \frac{1 + \nu p_j \omega_j^2}{1 + \nu p_j (\omega_j^o)^2} \left(\frac{\omega_k^b - \omega_j^o}{\omega_k^b - \omega_j} \right)^2 \qquad (5.5.37)$$

$$x(\omega_j) = \lambda_j \tau(\omega_j) = \tau_0 \frac{1 - \mu p_j \omega_j^2}{1 + \nu p_j (\omega_j^o)^2} \left(\frac{\omega_j^b - \omega_j^o}{\omega_k^b - \omega_j} \right)^2 \qquad (5.5.38)$$

where

$$p_j = q_{jk}\omega_k^o = \frac{1}{\gamma_j} \prod_{i=1}^{j-1} \frac{\omega_i^a}{\gamma_i} \prod_{i=j+1}^{k} \frac{\omega_i^o}{\gamma_i} \prod_{i=k+1}^{m} \frac{\omega_i^b}{\gamma_i} \qquad (5.5.39)$$

Transmission between Two Simple Waves Let us finally consider a case in which a j-simple wave facing forward collides with a k-shock directed backward, where $j < k$. Clearly, the two waves interact to form a penetration region in the physical plane, in which the solution, although continuous, is not given by a simple wave. Since, however, the image of the solution in the hodograph space $\Omega(m)$ falls on a plane parallel to the ω_j and ω_k axes as shown in Fig. 5.12(a), we expect both simple waves to emerge from the penetration region to give transmitted simple waves of their respective families. This is illustrated in Fig. 5.12(b), where the penetration region is marked as d–e–f–g.

The penetration region is covered by a network of the characteristics C^j and C^k which, if established, determines the solution. As we pass through the region along a specific C^j, ω_j remains fixed but ω_k decreases from ω_k^o to ω_k^b. Hence the C^j becomes less and less steep until it emerges across the boundary e–g and again becomes straight. Similarly, ω_j increases along a C^k inside the penetration region while ω_k remains unchanged and thus every C^k becomes

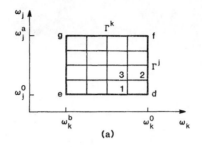

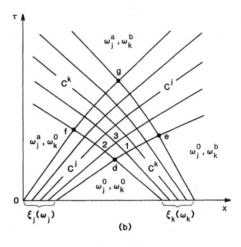

Figure 5.12

gradually curved downward until it emerges across the boundary f–g to become a straight line. The straight characteristics C^j and C^k, after emerging from the penetration region, compose the two transmitted simple waves, respectively.

First, we shall concentrate on the boundary d–e along which ω_k is the only variable. At an arbitrary point on d–e, therefore, we can write two equations in terms of ω_k; that is,

$$\lambda_k(\omega_k)\tau = x - \xi(\omega_k) \tag{5.5.40}$$

$$\frac{dx}{d\tau} = \lambda_j(\omega_k) \tag{5.5.41}$$

where

$$\lambda_k(\omega_k) = \frac{1 - \mu p_k \omega_k^2}{1 + \nu p_k \omega_k^2} \tag{5.5.42}$$

$$\lambda_j(\omega_k) = \frac{1 - \mu p_k \omega_j^o \omega_k}{1 + \nu p_k \omega_j^o \omega_k} \tag{5.5.43}$$

and

$$p_k = q_{jk}\omega_j^o \tag{5.5.44}$$

Comparing these equations with Eqs. (5.5.25)–(5.5.31), we observe that the two problems are identical except for the fact that we have ω_j^o here instead of ω_j^a. Hence the boundary d–e is described by Eq. (5.5.33) with ω_j^o in place of ω_j^a. If the k-simple wave is centered, we have, from Eq. (5.5.34),

$$\tau(\omega_j^o, \omega_k) = \tau_0 \frac{1 + \nu q_{jk}\omega_j^o\omega_k^2}{1 + \nu q_{jk}\omega_j^o(\omega_k^o)^2}\left(\frac{\omega_k^o - \omega_j^o}{\omega_k - \omega_j^o}\right)^2, \qquad \omega_k^o \geq \omega_k \geq \omega_k^b \tag{5.5.45}$$

in which Eq. (5.5.44) has been substituted for p_k and τ_0 denotes the ordinate of the point d. This equation may be substituted into Eq. (5.5.40) to give $x(\omega_j^o, \omega_k)$. The two equations then constitute the parametric description of the boundary of d–e in terms of ω_k.

For the boundary d–f we see that the situation would be the same as discussed in the last paragraph of the preceding subsection if the shock path S^k is now replaced by the characteristic C^k for ω_k^o (see also Exercise 5.5.3). This implies that here ω_k^o takes the role of ω_k^b there. Therefore, in case the k-simple wave is centered at the origin we obtain, from Eqs. (5.5.37) and (5.5.38),

$$\tau(\omega_j, \omega_k^o) = \tau_0 \frac{1 + \nu q_{jk}\omega_k^o\omega_j^2}{1 + \nu q_{jk}\omega_k^o(\omega_j^o)^2}\left(\frac{\omega_k^o - \omega_j^o}{\omega_k^o - \omega_j}\right)^2 \tag{5.5.46}$$

$$x(\omega_j, \omega_k^o) = \tau_0 \frac{1 - \mu q_{jk}\omega_k^o\omega_j^2}{1 + \nu q_{jk}\omega_k^o(\omega_j^o)^2}\left(\frac{\omega_k^o - \omega_j^o}{\omega_k^o - \omega_j}\right)^2 \tag{5.5.47}$$

for $\omega_j^o \leq \omega_j \leq \omega_j^a$, in which Eq. (5.5.39) has been substituted for p_j. This pair of equations describe the boundary d–f in terms of the parameter ω_j.

Now we come into the penetration region d–e–f–g and recall the fact that the parameter ω_j or ω_k is the only variable parameter along a C^k or a C^j, respectively. This implies that every C^k in the penetration region has its image on a Γ^j inside the rectangle d–e–g–f of Fig. 5.12(a), and similarly, every C^j has its image on a Γ^k inside the rectangle d–e–g–f. Consequently, the net of characteristics C^j and C^k inside the penetration region can always be constructed by mapping the rectangle d–e–g–f of Fig. 5.12(a) onto the physical plane. The correspondence between the hodograph space and the physical plane is illustrated with 25 points in Fig. 5.12. The characteristic network so constructed gives the solution in the penetration region because the characteristic parameters are known at every mesh point.

The mapping may be performed by employing the characteristic directions; that is, at every point in the penetration region we can calculate the two directions

$$C^j: \quad \frac{d\tau}{dx} = \sigma_j(\omega_j, \omega_k) = \frac{1 + \nu q_{jk}\omega_j^2\omega_k}{1 - \mu q_{jk}\omega_j^2\omega_k} \tag{5.5.48}$$

$$C^k: \quad \frac{d\tau}{dx} = \sigma_k(\omega_j, \omega_k) = \frac{1 + \nu q_{jk}\omega_j\omega_k^2}{1 - \mu q_{jk}\omega_j\omega_k^2} \tag{5.5.49}$$

where q_{jk} is a constant defined by Eq. (5.5.19). In the physical plane the portion $1 \rightarrow 3$ of a C^k may be approximated by a straight line of slope

$$\sigma_k = \frac{1}{2}(\sigma_{k,1} + \sigma_{k,3}) \qquad (5.5.50)$$

and the portion $2 \rightarrow 3$ can be drawn likewise to yield point 3 at the intersection. Iteration completes the mapping and generates the solution in the penetration region. The more mesh points we use, the more accurately the solution may be obtained.

The transmitted j-simple wave can be established by drawing a straight C^j from various points along the boundary $e-g$ with the slope σ_j given at the point. The same holds for the transmitted k-simple wave.

Aside from this mapping scheme, we also note that the net of the characteristics C^j and C^k can be used as a new coordinate system, so there exists a two-parameter representation of the solution inside the penetration region. Furthermore, since the solution is continuous, we may regard both x and τ as continuously differentiable functions of ω_j and ω_k. Consequently, at an arbitrary point (x, τ) we can write

$$\frac{\partial x}{\partial \omega_k} = \lambda_j(\omega_j, \omega_k)\frac{\partial \tau}{\partial \omega_k} = \frac{1 - \mu q_{jk}\omega_j^2\omega_k}{1 + \nu q_{jk}\omega_j^2\omega_k}\frac{\partial \tau}{\partial \omega_k} \qquad (5.5.51)$$

and

$$\frac{\partial x}{\partial \omega_j} = \lambda_k(\omega_j, \omega_k)\frac{\partial \tau}{\partial \omega_j} = \frac{1 - \mu q_{jk}\omega_j\omega_k^2}{1 + \nu q_{jk}\omega_j\omega_k^2}\frac{\partial \tau}{\partial \omega_j} \qquad (5.5.52)$$

where q_{jk} is a constant defined by Eq. (5.5.19).

Differentiating Eq. (5.5.51) with respect to ω_j and Eq. (5.5.52) with respect to ω_k and subtracting one from the other, we obtain the second-order partial differential equation for τ:

$$\frac{\partial^2 \tau}{\partial \omega_j \partial \omega_k} + \frac{2}{\omega_k - \omega_j}\left\{ \frac{1 + \nu q_{jk}\omega_j^2\omega_k}{1 + \nu q_{jk}\omega_j\omega_k^2}\frac{\partial \tau}{\partial \omega_j} \right.$$

$$\left. - \frac{1 + \nu q_{jk}\omega_j\omega_k^2}{1 + \nu q_{jk}\omega_j^2\omega_k}\frac{\partial \tau}{\partial \omega_k} \right\} = 0 \qquad (5.5.53)$$

This equation holds inside the rectangle $d-e-g-f$ of Fig. 5.12(a), in which we have $\omega_j < \omega_k$. We note that τ is given as a function of ω_k along the boundary $d-e$ and also as a function of ω_j along the boundary $d-f$. If both of the simple waves are centered, these are given by Eqs. (5.5.45) and (5.5.46), respectively. Since the boundaries $d-e$ and $d-f$ are both characteristics, here we have a characteristic initial value problem. We also observe that Eq. (5.5.53) is symmetric with respect to ω_j and ω_k and independent of the parameter μ. However, the solution $\tau(\omega_j, \omega_k)$ will certainly depend upon μ through the initial data.

Although Eq. (5.5.53) does not seem amenable to analytic solution we would like to treat the equation in two special cases. In the first case let us suppose that all the solutes are rather weakly adsorbed so that all the γ_i's are less than 1. Since ω_k is less than or equal to γ_k for $k = 1, 2, \ldots, m$, we find that the terms $\nu q_{jk}\omega_j\omega_k^2$ and $\nu q_{jk}\omega_j^2\omega_k$ would be much smaller than 1. However, the terms $\mu q_{jk}\omega_j\omega_k^2$ and $\mu q_{jk}\omega_j^2\omega_k$ are comparable to 1 because we are interested in the k-intermediate range. In this special case Eq. (5.5.53) reduces to the simple form

$$\frac{\partial^2 \tau}{\partial\omega_j\partial\omega_k} + \frac{2}{\omega_k - \omega_j}\left\{\frac{\partial\tau}{\partial\omega_j} - \frac{\partial\tau}{\partial\omega_k}\right\} = 0 \qquad (5.5.54)$$

which is identical to Eq. (4.3.48). If both of the simple waves are centered, the initial data given by Eqs. (5.5.45) and (5.5.46) are also simplified into the forms

$$\tau(\omega_j^o, \omega_k) = \tau_0\left(\frac{\omega_k^o - \omega_j^o}{\omega_k - \omega_j^o}\right)^2, \qquad \omega_k^o \geq \omega_k \geq \omega_k^b \qquad (5.5.55)$$

$$\tau(\omega_j, \omega_k^o) = \tau_0\left(\frac{\omega_k^o - \omega_j^o}{\omega_k^o - \omega_j}\right)^2, \qquad \omega_j^o \leq \omega_j \leq \omega_j^a \qquad (5.5.56)$$

The solution to this initial value problem was presented in Sec. 4.3.[†] From Eq. (4.3.49), therefore, we have

$$\tau(\omega_j, \omega_k) = \tau_0\left(\frac{\omega_k^o - \omega_j^o}{\omega_k - \omega_j}\right)^2\left\{1 - 2\frac{(\omega_k - \omega_k^o)(\omega_j - \omega_j^o)}{(\omega_k^o - \omega_j^o)(\omega_k - \omega_j)}\right\} \qquad (5.5.57)$$

This equation is then used in Eq. (5.5.51) or (5.5.52) to determine $x(\omega_j, \omega_k)$. Since ω_j remains fixed along a C^j, we can directly integrate Eq. (5.5.51) from $\omega_k = \omega_k^o$ to an arbitrary ω_k along a C^j to obtain

$$c(\omega_j, \omega_k) - x(\omega_j, \omega_k^o) = \int_{\omega_k^o}^{\omega_k}(1 - \mu q_{jk}\omega_j^2\omega_k)\frac{\partial\tau}{\partial\omega_k}\,d\omega_k$$

Applying integration by parts and substituting Eq. (5.5.47) for $x(\omega_j, \omega_k^o)$, we find the equation

$$x(\omega_j, \omega_k) = (1 - \mu q_{jk}\omega_j^2\omega_k)\tau(\omega_j, \omega_k)$$
$$+ 2\mu q_{jk}\omega_j^2\int_{\omega_k^o}^{\omega_k}\tau(\omega_j, \omega_k)d\omega_k \qquad (5.5.58)$$

In the second case we shall assume that all the solutes are strongly adsorbed so as to give γ_i much larger than 1 for all solutes. Then all the ω_k's would also be much larger than 1, and hence the term $\nu q_{jk}\omega_j^2\omega_k$ would become dominant in the factor $(1 + \nu q_{jk}\omega_j^2\omega_k)$. A similar argument would apply to the factor $(1 + \nu q_{jk}\omega_j\omega_k^2)$. Under these circumstances, Eq. (5.5.53) becomes

$$\frac{\partial^2 \tau}{\partial\omega_j\partial\omega_k} + \frac{2}{\omega_k - \omega_j}\left\{\frac{\omega_j}{\omega_k}\frac{\partial\tau}{\partial\omega_j} - \frac{\omega_k}{\omega_j}\frac{\partial\tau}{\partial\omega_k}\right\} = 0 \qquad (5.5.59)$$

[†] See Sec. 2.7 for the method of solution.

and Eqs. (5.5.45) and (5.5.46) are also reduced to

$$\tau(\omega_j^o, \omega_k) = \left\{ \frac{\omega_k(\omega_k^o - \omega_j^o)}{\omega_k^o(\omega_k - \omega_j^o)} \right\}^2, \qquad \omega_k^o \geq \omega_k \geq \omega_k^b \tag{5.5.60}$$

$$\tau(\omega_j, \omega_k^o) = \left\{ \frac{\omega_j(\omega_k^o - \omega_j^o)}{\omega_j^o(\omega_k^o - \omega_j)} \right\}^2, \qquad \omega_j^o \leq \omega_j \leq \omega_j^q \tag{5.5.61}$$

Based on the functional forms of these equations, we shall assume a solution of the form

$$\tau(\omega_j, \omega_k) = \left(\frac{\omega_j \omega_k}{\omega_k - \omega_j} \right)^2 u(\omega_j, \omega_k) \tag{5.5.62}$$

Then it can be shown that $u(\omega_j, \omega_k)$ must satisfy the equation[†]

$$\frac{\partial^2 u}{\partial \omega_j \partial \omega_k} + \frac{2}{\omega_k - \omega_j} u = 0 \tag{5.5.63}$$

This equation is identical to Eq. (2.7.48) for $\lambda = 2$, which was treated in detail in Sec. 2.7. Thus the solution can be expressed in the form

$$u(\omega_j, \omega_k) = \kappa \left\{ 1 - 2 \frac{(\omega_k - \omega_k^o)(\omega_j - \omega_j^o)}{(\omega_k^o - \omega_j^o)(\omega_k - \omega_j)} \right\}$$

where κ is an arbitrary constant. From Eq. (5.5.62), we have

$$\tau(\omega_j, \omega_k) = \kappa \left(\frac{\omega_j \omega_k}{\omega_k - \omega_j} \right)^2 \left\{ 1 - 2 \frac{(\omega_k - \omega_k^o)(\omega_j - \omega_j^o)}{(\omega_k^o - \omega_j^o)(\omega_k - \omega_j)} \right\}$$

and it is immediately apparent that both Eqs. (5.5.60) and (5.5.61) are satisfied if we choose

$$\kappa = \tau_0 \left(\frac{\omega_j^o \omega_k^o}{\omega_k^o - \omega_j^o} \right)^2$$

Consequently, the solution to the initial value problem of Eq. (5.5.59) is given by

$$\tau(\omega_j, \omega_k) = \tau_0 \left\{ \frac{\omega_j \omega_k (\omega_k^o - \omega_j^o)}{\omega_j^o \omega_k^o (\omega_k - \omega_j)} \right\}^2 \left\{ 1 - 2 \frac{(\omega_k - \omega_k^o)(\omega_j - \omega_j^o)}{(\omega_k^o - \omega_j^o)(\omega_k - \omega_j)} \right\} \tag{5.5.64}$$

By applying the same procedure as before to determine $x(\omega_j, \omega_k)$, we obtain

$$x(\omega_j, \omega_k) = \frac{\mu}{\nu} \left(\frac{1}{\mu q_{jk} \omega_j^2 \omega_k} - 1 \right) \tau(\omega_j, \omega_k)$$

$$- \frac{1}{\nu q_{jk} \omega_j^2} \int_{\omega_k^o}^{\omega_k} \omega_k^{-2} \tau(\omega_j, \omega_k) d\omega_k \tag{5.5.65}$$

[†] See also Exercise 2.7.5.

Exercises

5.5.1. For the situation shown in Fig. 5.9, assume that the k-simple wave is centered at the origin and the k-shock emanates from the point $x = 1$ on the x-axis, and determine the point (x_0, τ_0) to give

$$x_0 = \frac{(1 + \nu p_k \omega_k^b \omega_k^g)\{1 - \mu p_k (\omega_k^g)^2\}}{(\nu + \mu) p_k \omega_k^g (\omega_k^b - \omega_k^g)}$$

$$\tau_0 = \frac{(1 + \nu p_k \omega_k^b \omega_k^g)\{1 + \nu p_k (\omega_k^g)^2\}}{(\nu + \mu) p_k \omega_k^g (\omega_k^b - \omega_k^g)}$$

By applying this result to Eq. (5.5.16), show that the duration of interaction depends on μ only in such a way that it is inversely proportional to $(\nu + \mu)$ and thus the duration becomes shorter as μ increases.

5.5.2. Suppose that a k-shock, directed forward, collides with a backward-facing k-simple wave inside the domain. Noting that this situation can be realized only in the $(k + 1)$-intermediate range, discuss how the two waves cancel each other while interacting and determine the shock path S^k in terms of ω_k^g, ω_k^g, and ω_k^b.

5.5.3. When a j-simple wave facing forward interacts with a k-shock directed backward, sketch the physical plane portrait corresponding to Fig. 5.11, and carry out the interaction analysis to determine the parametric description of the shock path S^k. Show that your result reduces to Eqs. (5.5.37) and (5.5.38) if the j-simple wave is centered at the origin.

5.5.4. Develop the solution of the system given by Eqs. (5.5.40)–(5.5.44) and compare the result with Eq. (5.5.33).

5.5.5. Establish a parametric description of the curve d–f in Fig. 5.12(b) and compare the result with that of Exercise 5.5.3.

5.5.6. By integrating Eq. (5.5.52) instead of Eq. (5.5.51) determine (ω_j, ω_k) and compare the result with Eq. (5.5.65).

5.6 Illustration

In this section we would like to treat a numerical example involved with three solutes in a finite column so as to illustrate how to apply the theoretical development of this chapter. The parameters are given as

$$\epsilon = 0.4 \quad \text{or} \quad \nu = 1.5$$

$$N_i = 1.0 \text{ mol/liter for } i = 1, 2, 3$$

$$K_1 = 5 \text{ liters/mol}, \quad K_2 = 10, \quad K_3 = 15$$

The incoming fluid phase is a mixture containing all three solutes, A_1, A_2, and A_3, in the same concentration and the incoming solid phase is half saturated by the most strongly adsorbed solute A_3 alone. The initial state is taken arbitrarily. When concentrations are expressed in moles per liter of individual phase, the feed and initial data are specified as given in Table 5.2.

TABLE 5-2. FEED AND INITIAL STATES OF CONCENTRATIONS

	i			
	1	2	3	Θ
c_i^a	0.05	0.05	0.05	
c_i^b	0.025	0.036	0.075	0.617
n_i^b	0	0	0.5	0.5

Various ranges of μ

First, we determine the characteristic parameters for the feed and initial states by solving Eq. (5.2.6) with the equilibrium relation (5.2.1). These values are obtained as given in Table 5.3. Since $\omega_1^a < \omega_1^b$ and $\omega_2^a < \omega_2^b$, we expect that the 1- and 2-transition ranges are given by single points.

TABLE 5-3. CHARACTERISTIC PARAMETERS FOR THE FEED AND INITIAL STATES

	k		
	1	2	3
ω_k^a	3.387	7.050	12.563
ω_k^b	3.803	6.516	11.597
ω_k^b	5.0	7.5	10.0

Critical values of μ are calculated by applying Eqs. (5.4.2), (5.4.3), and (5.4.4). With these values we can divide the μ spectrum into various ranges as follows:

Lower range	$0 \le \mu < 0.199$
3-transition range	$0.199 \le \mu \le 0.314$
3-intermediate range	$0.314 < \mu < 0.419$
2-transition range	$\mu = 0.419$
2-intermediate range	$0.419 < \mu < 0.591$
1-transition range	$\mu = 0.591$
Upper range	$0.591 < \mu < \$*3I/$

Steady states attainable in finite time

It is clear that there are four different steady states attainable within the column in finite time and these are independent of the initial state. These state are given, respectively, by the feed states, $\{c_i^a\}$ and $\{c_{i,\text{eq}}^b\}$, the 3-constant state $\{c_i^{(3)}\}$, and the 2-constant state $\{c_i^{(2)}\}$. The last two are determined by using the sets of characteristic parameters, $[\omega_1^a, \omega_2^a, \omega_3^b]$ for the 2-constant state and $[\omega_1^a, \omega_2^b, \omega_3^b]$ for the 3-constant state, in Eq. (3.7.9). The four steady states are presented in Table 5.4.

TABLE 5-4. STEADY STATES ATTAINABLE INSIDE THE COLUMN

	\multicolumn{3}{c}{i}			
	1	*2*	*3*	*Range of μ*
c_i^a	0.05	0.05	0.05	Lower Range
$c_i^{(3)}$	0.0415	0	0.1289	3-intermediate range
$c_i^{(2)}$	0.0476	0	0.1143	2-intermediate range
$c_{i,\,eq}^b$	0	0	0.0667	Upper Range

The steady states of the outgoing streams are readily determined by using Eq. (5.4.7), (5.4.9), or (5.4.10) if the value of μ is given. In this way we can evaluate the steady-state coverage of the outgoing solid phase for various values of μ. The result is presented in Fig. 5.13, in which we observe the maximum coverage gain at $\mu = \mu_1 = 0.591$.

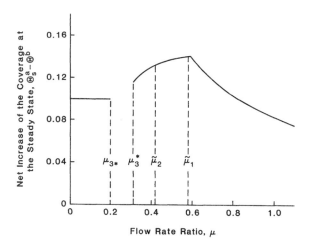

Figure 5.13

Solution for $\mu = 0.36$

We are in the 3-intermediate range of μ and so we expect to see 1- and 2-waves developing from the origin and a 3-wave from the other end. By looking at the table of characteristic parameters we can further tell that the waves developing from the other end is a 3-simple wave. These waves involve two transmissive interactions, which can be analyzed by applying the result of the previous section. Indeed, the shock path S^1 through the 3-simple wave region is determined by Eqs. (5.5.37) and (5.5.38) with $j = 1$ and $k = 3$. For the interaction between the 2- and 3-simple waves we have conveniently applied the step-by-step mapping procedure to construct the characteristic net inside the penetration region.

The physical plane portrait of the solution is presented in Fig. 5.14. The constant state I is the initial state and the state marked as (k) corresponds to the k-constant state. The other constant states can be easily identified by establishing the table of characteristic parameters as shown in Table 5.5. With the set of characteristic parameters we can use Eq. (3.7.9) to calculate the concentrations for each of the five new constant states, and these are given in Table 5.6.

In Fig. 5.14 a value of ω_k is marked along each of the characteristics C^k. *This value is kept constant along the characteristic C^k while the other two parameters remain the same as those for the adjacent constant state.* Therefore, we can determine the state along every characteristic by applying Eq. (3.7.9). Inside the penetration region of the 2- and 3-simple waves we have the value of ω_2 and ω_3 at each mesh point and ω_1 is given by ω_1^q everywhere so that the state is determined at every mesh point.

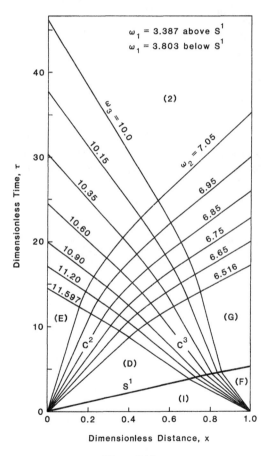

Figure 5.14

TABLE 5-5. CHARACTERISTIC PARAMETERS FOR THE CONSTANT STATES

Constant State	ω_1	ω_2	ω_3
I	ω_1^g	ω_2^g	ω_3^g
D	ω_1^q	ω_2^g	ω_3^g
E	ω_1^q	ω_2^g	ω_3^g
F	ω_1^q	ω_2^g	ω_3^b
G	ω_1^q	ω_2^g	ω_3^b
H	ω_1^q	ω_2^b	ω_3^b
B	ω_1^b	ω_2^b	ω_3^b
(3)	ω_1^q	ω_2^g	ω_3^b
(2)	ω_1^q	ω_2^b	ω_3^b

TABLE 5-6. CONCENTRATIONS FOR THE CONSTANT STATES

Constant State	c_1	c_2	c_3
D	0.0378	0.0431	0.0873
E	0.0472	0.0337	0.0756
F	0.0220	0	0.1278
G	0.0332	0	0.1488
H	0.0316	0	0.0980

The distribution of concentrations is presented in Fig. 5.15 in the form of concentrations profiles at five different times. The arrow on the profile indicates the propagation direction of the corresponding wave and here we can visualize the process of interaction from a different angle. The nonsimple wave solution in this particular example shows that the 3-simple wave is dominant over the 2-simple wave in the course of interaction. It is clearly shown that both boundaries bear discontinuities.

At $\tau = 24.9$, the interaction is over and the region of the 3-constant state stars to expand inside the domain. The tail of the 3-simple wave reaches the boundary $\mathcal{B}^2$ at $\tau = 48.8$ and then the operation obtains a steady state, which is given by Eq. (5.4.12) with $k = 3$. At this steady state 64.6% of A_1 and 100% of A_2 are removed from the incoming fluid phase by absorption while 20.7% of A_3 is desorbed from the incoming solid phase. The surface coverage increases from 0.5 at $x = 1$ to 0.6252 at $x = 0$.

Solution for $\mu = 0.54$

Since this is a case in the 2-intermediate range of μ, we naturally expect to see a 1-wave emanating from the origin and 2- and 3-waves from the other end. However, the 1-wave from the origin is a shock and thus there is a possibility that the 1-wave may also develop from the other end and come into the domain. This is exactly what we have in this case as shown in Fig. 5.16.

The 1-shock from the origin experiences two transmissive interactions with

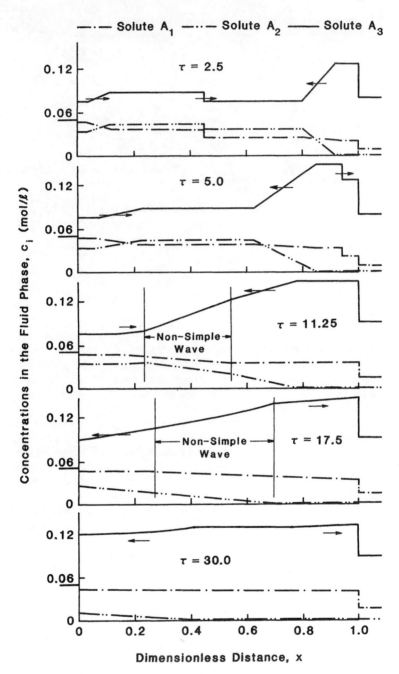

Figure 5.15

Continuous Countercurrent Adsorber   Chap. 5

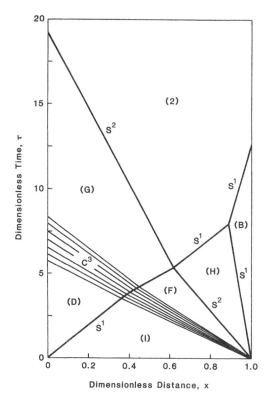

Figure 5.16

the 3-simple wave and the 2-shock originating from the opposite end and then collides with the 1-shock. The two 1-shock are superposed instantaneously and form a new 1-shock which continues to propagate forward and eventually reaches the boundary $\mathscr{B}^b$. All the interactions can be analyzed by directly applying the result of Sec. 5.5. In particular, while the 1-shock passes through the 3-simple wave region, its path is given by Eqs. (5.5.37) and (5.5.38) with $j = 1$ and $k = 3$.

For each of the constant states we can identify the set of characteristic parameters and thus determine the state of concentrations. These were given in Tables 5.5 and 5.6. We note that the constant state B of the incoming solid phase appears adjacent to the boundary $\mathscr{B}^b$ for $0 \leq \tau \leq 12.5$ so that the solution remains continuous across the boundary $\mathscr{B}^b$ in the meantime. Along each of the characteristics C^3 we must read the same value of ω_3 as that for the corresponding C^3 in Fig. 5.14. Consequently, the state of concentrations can be determined at every point inside the domain as well as at the boundaries.

The concentration profiles are constructed from Fig. 5.16 at five different times and presented in Fig. 5.17. With the propagation directions indicated by arrows, we can readily identify the pairs of interacting waves and observe the

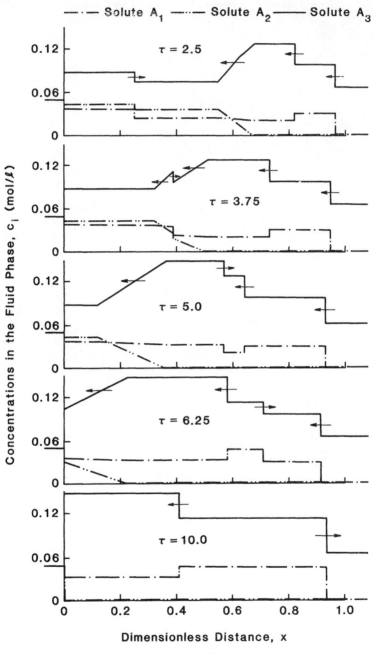

Figure 5.17

Continuous Countercurrent Adsorber   Chap. 5

process of interaction. It is clearly shown that the profiles are continuous at $x = 1$ in all the figures. At $\tau = 12.5$, however, the 1-shock reaches the boundary and then the profiles becomes discontinuous. We note that the operation attains a steady state in this part of the column, although the state in the earlier section is still unsteady due to the propagation of the 2-shock.

At $\tau = 19.18$, the 2-shock finally reaches the boundary $\mathcal{B}^a$ and the entire system obtains a steady state, which is given by Eq. (5.4.12) with $k = 2$ and contains boundary discontinuities at both ends. At this steady state, 7.62% desorption of A_3 is accompanied by 91.8% adsorption of A_1 and 100% adsorption of A_2 from the incoming fluid phase. The column performance is certainly improved from the for $\mu = 0.36$, if the objective is the removal of the solutes A_1 and A_2 from the fluid mixture.

As shown in Fig. 5.13, the surface coverage reaches its maximum when $\mu = \bar{\mu}_1 = 0.591$ and, correspondingly, the maximum adsorption yield is achieved. Under these circumstances the two 1-shocks meet together to be united and produce a stationary 1-shock. This takes place at $x = 0.83$ when $\tau = 8.3$. Otherwise, the physical plane portrait is similar to Fig. 5.16. The 2-shock reaches the boundary $\mathcal{B}^b$ at $\tau = 12.75$ and then the system obtains a steady state. One feature to be remarked is that the state to the right of the stationary 1-shock is identical to the state of the incoming solid phase so that the solution always remains continuous at $x = 1$. This implies that the outgoing fluid phase contains only the solute A_3, so solutes A_1 and A_2 are both completely removed from the incoming fluid phase. On the other hand, 0.56% of the solute A_3 is desorbed from the incoming solid phase. It is evident that this is the best performance that we could expect to obtain with respect to the given system.

Exercise

5.6.1. Consider the numerical example treated in this section and develop the solution for the case with $\mu = \bar{\mu}_2 = 0.419$. In particular, show that the 2-shock originating from the point at $x = 1$ on the τ-axis first comes into the domain, then turns forward after interaction with the 1-shock developing from the origin, and finally passes through the boundary $\mathcal{B}^b$. This implies that the 2-shock would become stationary only if the contacting region were extended beyond the line $x = 1$. Based on this argument, show that the overall picture of the solution remains the same as for the case with μ in the 3-intermediate range.

REFERENCES

Although elaborated here in more detail, most of the development in this chapter is taken from

H.-K. RHEE, R. ARIS, and N. R. AMUNDSON, "Multicomponent adsorption in continuous countercurrent exchangers," *Philos. Trans. Roy. Soc. London* **A267**, 187–215 (1971).

The shock conditions discussed in Sec. 5.2 were first defined in

P. D. LAX, "Hyperbolic systems of conservation laws:II," *Comm. Pure Appl. Math.* **10**, 537–566 (1957).

For the concept of a boundary discontinuity and the behavior of a finite column, reference may be made to the development in Sec. 6.5 of Vol. 1.

An expository discussion of the interaction between two simple waves of different families is to be found in

R. COURANT and K. O. FRIEDRICHS, *Supersonic Flow and Shock Waves,* Interscience, New York, 1948, Sec. 82.

The general subject of continuous countercurrent adsorption processes has been well discussed in

A. J. DEROSSET, R. W. NEUZIL, and D. B. BROUGHTON, "Industrial applications of preparative chromatography," in *Percolation Processes: Theory and Applications,* edited by A. E. Rodrigues and D. Tondeur, Sijthoff & Noordhoff, Alphen aan den Rijn, The Netherlands, 1981, pp. 249–281.

D. M. RUTHVEN, *Principles of Adsorption and Adsorption Processes,* Wiley-Interscience, New York, 1984, Chap. 12.

Although these are concerned with a nonequilibrium model for a single-solute system, the following articles treat the countercurrent moving-bed problem by using the method of characteristics:

M. A. JASWON and W. SMITH, "Countercurrent transfer processes in the non-steady state," *Proc. Roy. Soc. London* **A225**, 226–244 (1954).

K. S. TAN and I. H. SPINNER, "Numerical methods of solution for continuous countercurrent processes in the nonsteady state: Part I. Model equations and development of numerical methods and algorithms," *AIChE J.* **30**, 770–779 (1984).

K. S. TAN and I. H. SPINNER, "Numerical methods of solution for continuous countercurrent processes in the nonsteady state: Part II. Application of numerical methods," *AIChE J.* **30** 780–786 (1984).

The steady-state problem of multicomponent moving-bed adsorption for a system with mass transfer resistance and axial dispersion has been treated extensively by

B. R. LOCKE, "The effects of mass transfer resistance and axial dispersion on multicomponent moving-bed adsorption and ion exchange," M.S. thesis, University of Houston, 1982.

For a system involved with a chemical reaction on the solid phase (i.e., a countercurrent moving-bed chromatographic reactor), the steady-state problem has been thoroughly analyzed in

B. K. CHO, R. ARIS, and R. W. CARR, "The mathematical theory of a countercurrent catalytic reactor," *Proc. Roy. Soc. London* **A383**, 147–189 (1982).

T. PETROULAS, R. ARIS, and R. W. CARR, "Analysis of the counter-current moving-bed chromatographic reactor," *Comput. Math. Appl.* **11**, 5–34 (1985), Special issue on hyperbolic partial differential equations.

T. PETROULAS, R. ARIS, and R. W. CARR, "Analysis and performance of a countercurrent moving-bed chromatographic reactor," *Chem Engrg. Sci.* **40**, 2233–2240 (1985).

More on Hyperbolic Systems of Quasilinear Equations and Analysis of Adiabatic Adsorption Columns

6

In Chapters 3, 4, and 5 the nonlinear nature of the problem comes from the equilibrium relations which have been restricted to a special form, the Langmuir isotherm under isothermal conditions. Here we would like to extend the theoretical development to systems with arbitrary equilibrium relations and to treat, as a typical example, the adiabatic adsorption column extensively.

We recall that the first four sections of Chapter 3 were concerned with the problem with arbitrary equilibrium relations, but it was revealed that the existence of Riemann invariants is not guaranteed in general. It was then argued that the solution of the Riemann problem is given by a one-parameter family, which implies the existence of the Riemann invariants and thus makes the problem amenable to a systematic treatment. This is what we pursue further in this chapter.

In Sec. 6.1 we show that the energy conservation equation under adiabatic conditions can be put into the same form as the mass conservation equation, and thus both the fundamental differential equations as well as the compatibility relations are derived in the same form as those for isothermal systems. In Sec. 6.2 we deal with the formulation of the Riemann problem and the development of expressions for the characteristic directions and the propagation directions of discontinuities. We then proceed in Sec. 6.3 to discuss how to construct the solution, composed of constant states and simple waves, when it is continuous everywhere and naturally point out the possibility of the existence of discontinuities.

In Sec. 6.4 we introduce the extended entropy condition that must be satisfied across a discontinuity and elaborate on its implications for various types of nonlinearities of the characteristic field. More specifically, we classify various possibilities into three categories: shocks, contact discontinuities, and combined waves which consist of semishocks, simple waves, and/or contact discontinuities all of the same family. Brief discussions are also given for weak solutions and for the entropy condition. A survey of the literature is presented in Sec. 6.5 for the existence and uniqueness proofs for weak solutions and their structure as well as their asymptotic behavior. Also reviewed in this section are developments related to compressible fluid flow problems and the evaluation of numerical methods for the initial value problem. The subsequent section deals with the solution scheme for continuous countercurrent moving-bed problems under the restrictions that the initial bed is the same as the entering solid phase.

The last four sections are concerned with the analysis of adiabatic adsorption columns with the Langmuir isotherm. The temperature dependence of the isotherm parameters is so involved that we encounter various situations in which many features of the theoretical development of previous sections can be fully illustrated. We treat the adiabatic adsorption of a single solute in Sec. 6.7 and that of two solutes in Sec. 6.8. Basic equations are formulated and not only the mathematical features but also the physical aspects are examined by analyzing numerical examples. The next section (Sec. 6.9) deals with the situation in which adsorptivity reversal takes place. The emergence of a solute within the column is discussed from the mathematical point of view and the adsorptivity reversal is illustrated by using a numerical example. In Sec. 6.10 we introduce the shock layer theory developed in Sec. 2.10 to investigate the effect of axial dispersion of mass and energy.

6.1 Equations for the Adiabatic Adsorption Column

As a typical example of a hyperbolic system of quasilinear equations with arbitrary nonlinearity, we consider here an adiabatic adsorption column involving several solute species in which local equilibrium is established between the two phases everywhere at any time. Since it is evident that the fixed-bed problem may be regarded as a limiting case of the countercurrent moving-bed problem, we shall develop the treatment for a continuous countercurrent moving-bed system.

Let us consider an adiabatic adsorption column through which a fluid phase containing $(m - 1)$ different solutes flows with a volume fraction ϵ at an interstitial velocity V_f, whereas a solid adsorbent phase moves countercurrently against the fluid phase with a speed V_s. As usual, we shall use the symbols c_i and n_i to denote the concentrations of A_i in the fluid and solid phases, respectively, both being expressed in moles per unit volume of their own phase. The assumption of local equilibrium requires that at constant total pressure, there exist temperature-dependent functional relationships between the sets $\{c_i\}$ and $\{n_i\}$ and also that the temperature T be the same in both phases. A schematic diagram of the adsorption column is shown in Fig. 6.1.

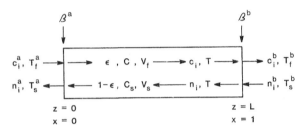

Figure 6.1

Since the velocity V_f is assumed constant, the mass balance equations previously derived are also valid in this case and thus we have, from Eq. (5.1.1),

$$\epsilon V_f \frac{\partial c_i}{\partial z} - (1 - \epsilon)V_s\frac{\partial n_i}{\partial z} + \epsilon\frac{\partial c_i}{\partial t} + (1 - \epsilon)\frac{\partial n_i}{\partial t} = 0 \qquad (6.1.1)$$

$$n_i = n_i(c_1, c_2, \ldots, c_{m-1}, T) \qquad (6.1.2)$$

for $i = 1, 2, \ldots, m - 1$, where Eq. (6.1.2) represents the equilibrium relationship. In addition to the usual requirements on the first-order partial derivatives; that is,

$$n_{i,j} = \frac{\partial n_i}{\partial c_j} \begin{cases} > 0 & \text{if } i = j \\ < 0 & \text{if } i \neq j \end{cases} \qquad (6.1.3)$$

we shall require, on physical grounds, the condition

$$\frac{\partial n_i}{\partial T} < 0 \qquad (6.1.4)$$

The energy balance equation was formulated with care in Sec. 1.4 of Vol. 1. Although it was done for a fixed-bed system, it can be readily extended to accommodate the present system. Thus from Eq. (1.4.8) of Vol. 1, we obtain[†]

[†] Note that the last term of Eq. (6.1.5) represents the total enthalpy change due to adsorption of solutes $A_1, A_2, \ldots,$ and A_{m-1}.

$$V_f C_f \frac{\partial T}{\partial z} - (1 - \epsilon)V_s C_s \frac{\partial T}{\partial z} + \epsilon C_f \frac{\partial T}{\partial t} + (1 - \epsilon)C_s \frac{\partial T}{\partial t}$$

$$-(1 - \epsilon) \sum_{i=1}^{m-1} (-\Delta H_i)\frac{\partial n_i}{\partial t} = 0 \tag{6.1.5}$$

in which C_f and C_s represent the heat capacities per unit volume of the fluid and solid phases, respectively, and ΔH_i denote the enthalpy change of adsorption per mole of A_i. All these parameters are assumed constant.

We shall now introduce new quantities c_m and n_m defined as

$$C_f c_m = C_f(T - T_0) = \begin{matrix} \text{energy content per unit volume of the} \\ \text{fluid phase} \end{matrix} \tag{6.1.6}$$

$$C_f n_m = C_s(T - T_0) - \sum_{i=1}^{m-1}(-\Delta H)_i n_i = \begin{matrix} \text{energy content per unit} \\ \text{volume of the solid phase} \end{matrix} \tag{6.1.7}$$

in which T_0 denotes the reference temperature. These definitions of c_m and n_m imply that the energy level of a solute is equal to zero when it is present in the fluid phase at temperature T_0. The choice of T_0, however, will turn out to be immaterial because we shall be always dealing with differences between two states or with derivatives of c_m and n_m. When Eqs. (6.1.6) and (6.1.7) are introduced, Eq. (6.1.5) can be rewritten in the form

$$\epsilon V_f \frac{\partial c_m}{\partial z} - (1 - \epsilon)V_s \frac{\partial n_m}{\partial z} + \epsilon \frac{\partial c_m}{\partial t} + (1 - \epsilon)\frac{\partial n_m}{\partial t} = 0 \tag{6.1.8}$$

$$n_m = n_m(c_1, c_2, \ldots, c_{m-1}, c_m) \tag{6.1.9}$$

where Eq. (6.1.9) immediately follows from Eqs. (6.1.2), (6.1.6), and (6.1.7). We also note that n_m defined by Eq. (6.1.7) satisfies the inequality conditions in Eq. (6.1.3) because $-\Delta H_i > 0$ for all solutes. It is now evident that the mathematical problem associated with the adiabatic adsorption column is essentially the same as the isothermal problem with one more solute.

As in Sec. 5.1, let us put

$$\nu = \frac{1 - \epsilon}{\epsilon}, \qquad \mu = \frac{1 - \epsilon}{\epsilon}\frac{V_s}{V_f} \tag{6.1.10}$$

and define the dimensionless independent variables as

$$x = \frac{z}{L}, \qquad \tau = \frac{V_f t}{L} \tag{6.1.11}$$

in which L denotes the total length of the column. Then Eqs. (6.1.1) and (6.1.8) are reduced to the form

$$\frac{\partial c_i}{\partial x} - \mu \frac{\partial n_i}{\partial x} + \frac{\partial c_i}{\partial \tau} + \nu \frac{\partial n_i}{\partial \tau} = 0, \qquad i = 1, 2, \ldots, m \tag{6.1.12}$$

and the equilibrium relations (6.1.2) and (6.1.9) can be written in the form

$$n_i = n_i(c_1, c_2, \ldots, c_m), \qquad i = 1, 2, \ldots, m \qquad (6.1.13)$$

which clearly satisfies the inequality conditions in Eq. (6.1.3) for $i, j = 1, 2, \ldots, m$.

The initial and boundary conditions will be specified as follows:

$$\text{at } \tau = 0: \quad c_i = c_i^0 \quad \text{and} \quad T = T^0 \qquad \text{for } 0 < x < 1$$
$$\text{at } x = 0: \quad c_i = c_i^a \quad \text{and} \quad T = T_f^a \qquad \text{for } \tau > 0 \qquad (6.1.14)$$
$$\text{at } x = 1: \quad n_i = n_i^b \quad \text{and} \quad T = T_s^b \qquad \text{for } \tau > 0$$

for $i = 1, 2, \ldots, m - 1$ as marked in Fig. 6.1. In accordance with Eqs. (6.1.6) and (6.1.7), we may put

$$c_m^0 = T^0 - T_0$$
$$C_f n_m^0 = C_s(T^0 - T_0) - \sum_{i=1}^{m-1} (-\Delta H_i) n_i^0$$
$$c_m^a = T_f^a - T_0 \qquad\qquad (6.1.15)$$
$$C_f n_m^b = C_s(T_s^b - T_0) - \sum_{i=1}^{m-1} (-\Delta H_i) n_i^b$$

where n_i^0 represents the initial state of the bed which is in equilibrium with the state $\{c_i^0\}$. We can then rewrite Eq. (6.1.14) in the form

$$\text{at } \tau = 0: \quad c_i = c_i^0 \ (n_i = n_i^0) \qquad \text{for } 0 < x < 1$$
$$\text{at } x = 0: \quad c_i = c_i^a \qquad\qquad \text{for } \tau > 0 \qquad (6.1.16)$$
$$\text{at } x = 1: \quad n_i = n_i^m \qquad\qquad \text{for } \tau > 0$$

for $i = 1, 2, \ldots, m$. Since we shall be mainly concerned with Riemann problems throughout this chapter, the three sets of data $\{c_i^0\}$, $\{c_i^a\}$, and $\{n_i^b\}$ will always be given by constant values.

The conditions in Eq. (6.1.16), however, are not sufficient to determine the states of the outgoing streams because equilibrium may not be established outside the contacting region. Thus we set up the mass and energy balances across each of the boundaries at $x = 0$ and $x = 1$ to obtain

$$c_i^a - c_i^{0+} = \mu\{n_i^a - n_i^{0+}\} \qquad (6.1.17)$$
$$c_i^b - c_i^{1-} = \mu\{n_i^b - n_i^{1-}\} \qquad (6.1.18)$$

for $i = 1, 2, \ldots, m$, in which n_i^{0+} and n_i^{1-} are the equilibrium values corresponding to the states $\{c_i^{0+}\}$ and $\{c_i^{1-}\}$, respectively. For $i = m$, these equations represent the energy balances and in terms of the original variables they must read

$$C_f(T_f^a - T^{0+}) = \mu\left\{ C_s(T_s^a - T^{0+}) - \sum_{i=1}^{m-1} (-\Delta H_i)(n_i^a - n_i^{0+}) \right\} \qquad (6.1.19)$$

$$C_f(T_f^b - T^{1-}) = \mu\left\{C_s(T_s^b - T^{1-}) - \sum_{i=1}^{m-1}(-\Delta H_i)(n_i^b - n_i^{1-})\right\} \quad (6.1.20)$$

respectively. If the solution is continuous across either boundary, Eqs. (6.1.17) and (6.1.18) are satisfied identically. If not satisfied identically, they will serve to determine the states $\{n_i^a\}$ and $\{c_i^b\}$ uniquely. It is evident that Eq. (6.1.17) or (6.1.18) is equivalent to a free boundary condition, since the states $\{n_i^a\}$ and $\{c_i^b\}$ may not be fixed, although the states $\{c_i^a\}$ and $\{n_i^b\}$ are maintained constant.

Under the conditions (6.1.17) and (6.1.18), either boundary may bear a discontinuity which implies an instantaneous exchange of mass and energy between the two phases. Such a discontinuity is non-propagating and will be called a *boundary discontinuity* (B.D.), as before, to distinguish it from the one appearing inside the domain. Here again we shall employ the symbol $\mathcal{B}^a$ and $\mathcal{B}^b$ to designate the boundaries at $x = 0$ and $x = 1$, respectively, as shown in Fig. 6.1.

6.2 Formulation for the Riemann Problem

We recall that Secs. 3.2, 3.3, and 3.4 were concerned with the general theory for hyperbolic systems of quasilinear equations, so the development there should be equally applicable here in this chapter. Indeed, the theoretical development that follows can be considered as a continuation from Sec. 3.4. Here we begin our discussion with the equations for the adiabatic adsorption column with countercurrent moving bed instead of the standard chromatographic equations used in Sec. 3.4.

Let us therefore consider the system of equations

$$\frac{\partial c_i}{\partial x} - \mu\frac{\partial n_i}{\partial x} + \frac{\partial c_i}{\partial \tau} + \nu\frac{\partial n_i}{\partial \tau} = 0 \quad (6.2.1)$$

$$n_i = n_i(c_i, c_2, \ldots, c_m) \quad (6.2.2)$$

for $i = 1, 2, \ldots, m$, as formulated in Sec. 6.1, and the initial and boundary conditions

$$
\begin{aligned}
&\text{at } \tau = 0: \quad c_i = c_i^0 \quad \text{and} \quad n_i = n_{i,eq}^0 \\
&\text{at } x = 0: \quad c_i = c_i^a
\end{aligned} \quad (6.2.3)
$$

for $i = 1, 2, \ldots, m$, where the subscript eq indicates that the quantity is an equilibrium value, i.e.,

$$n_{i,eq}^0 = n_i(c_1^0, c_2^0, \ldots, c_m^0) \quad (6.2.4)$$

Here it is assumed that the contacting region is semi-infinitely long, extending from $x = 0$. We also assume that c_i^0 and c_i^a are given by constant values and

$$c_i^0 \neq c_i^a \qquad (6.2.5)$$

for some i. Now we have a Riemann problem and it is clear that the mathematical problem remains identical to the one treated in Sec. 5.1.

By applying the argument of Sec. 3.4 (see also Sec. 5.1), therefore, we can show that the image of the solution in the hodograph space falls on a curve Γ, so the solution is given by a one-parameter family in terms of the parameter ω that runs along the curve Γ. Using this property of the solution, Eqs. (6.2.1) and (6.2.2) can be transformed into two expressions: one gives the *characteristic direction*

$$\sigma = \left(\frac{d\tau}{dx}\right)_\omega = \frac{1 + \nu \dfrac{\mathscr{D}n_i}{\mathscr{D}c_i}}{1 - \mu \dfrac{\mathscr{D}n_i}{\mathscr{D}c_i}} \qquad (6.2.6)$$

and the other the *fundamental differential equations*

$$\frac{\mathscr{D}n_i}{\mathscr{D}c_i} = \frac{\mathscr{D}n_2}{\mathscr{D}c_2} = \cdots = \frac{\mathscr{D}n_m}{\mathscr{D}c_m} \qquad (6.2.7)$$

in which $\mathscr{D}n_i/\mathscr{D}c_i$ represents the directional derivative of n_i with respect to c_i along the curve Γ^k. Thus we have

$$\frac{\mathscr{D}n_i}{\mathscr{D}c_i} = \frac{dn_i/d\omega}{dc_i/d\omega} = \sum_{j=1}^{m} \frac{\partial n_i}{\partial c_j} \frac{dc_j/d\omega}{dc_i/d\omega} = \sum_{j=1}^{m} n_{i,j} \frac{dc_j}{dc_i} \qquad (6.2.8)$$

where dc_j/dc_i is taken along the curve Γ.

On the other hand, we can put Eq. (6.2.1) into vector form,

$$\frac{\partial \mathbf{c}}{\partial x} + (\mathbf{I} - \mu \nabla \mathbf{n})^{-1}(\mathbf{I} + \nu \nabla \mathbf{n}) \frac{\partial \mathbf{c}}{\partial \tau} = \mathbf{0} \qquad (6.2.9)$$

in which $\mathbf{c}$ and $\mathbf{n}$ are vector functions of m elements, $\{c_i\}$ and $\{n_i\}$, respectively, and ∇ denotes the gradient in the m-dimensional hodograph space of dependent variables. Here we have assumed that the matrix $(\mathbf{I} - \mu \nabla \mathbf{n})$ is nonsingular. If we put

$$\mathbf{A} = (\mathbf{I} - \mu \nabla \mathbf{n})^{-1}(\mathbf{I} + \nu \nabla \mathbf{n}) \qquad (6.2.10)$$

it is clear from Sec. 5.1 that the characteristic direction σ defined by Eq. (6.2.6) is identical to the eigenvalue of the matrix $\mathbf{A}$; that is,

$$|\mathbf{A} - \sigma \mathbf{I}| = 0 \qquad (6.2.11)$$

Since we are concerned with a totally hyperbolic system, the matrix $\mathbf{A}$ has m real, distinct eigenvalues and this implies that there are m different characteristic directions for a given state $\{c_i\}$. Accordingly, there are m different curves Γ starting from every point in the hodograph space. We shall call each of them a Γ^k, $k = 1, 2, \ldots, m$, and the parameter running along a Γ^k will be denoted by ω_k. Then Eq. (6.2.6) may be rewritten in the form

$$\sigma_k = \left(\frac{d\tau}{dx}\right)_{\omega_k} = \frac{1 + \nu\left(\dfrac{\mathscr{D}n_i}{\mathscr{D}c_i}\right)_k}{1 - \mu\left(\dfrac{\mathscr{D}n_i}{\mathscr{D}c_i}\right)_k}, \qquad k = 1, 2, \ldots, m \qquad (6.2.12)$$

which defines the kth *characteristic direction* or, equivalently, the kth *characteristic field*. We can also write

$$\lambda_k = \left(\frac{dx}{d\tau}\right)_{\omega_k} = \frac{1 - \mu\left(\dfrac{\mathscr{D}n_i}{\mathscr{D}c_i}\right)_k}{1 + \nu\left(\dfrac{\mathscr{D}n_i}{\mathscr{D}c_i}\right)_k}, \qquad k = 1, 2, \ldots, m \qquad (6.2.13)$$

where λ_k, being the reciprocal of σ_k, represents the kth *characteristic speed*. The subscript k on $(\mathscr{D}n_i/\mathscr{D}c_i)$ is used to signify the fact that the derivative is taken in the direction of a Γ^k. As usual, we shall assign the index k in such a way that we have the following order:

$$\lambda_1 > \lambda_2 > \cdots > \lambda_{m-1} > \lambda_m \qquad (6.2.14)$$

Incidentally, we can rearrange Eq. (6.2.12) in the form

$$\left(\frac{\mathscr{D}n_i}{\mathscr{D}c_i}\right)_k = \frac{\sigma_k - 1}{\mu\sigma_k + \nu} = \beta_k \qquad (6.2.15)$$

and, by comparing this with Eq. (5.1.17), we observe that the directional derivative $\mathscr{D}n_i/\mathscr{D}c_i$ is given by the eigenvalue of the matrix $\mathbf{\nabla}\mathbf{n}$; that is, we have

$$|\mathbf{\nabla}\mathbf{n} - \beta\mathbf{I}| = 0 \qquad (6.2.16)$$

If the solution contains one or more discontinuities within the domain, we must apply the conservation principle requiring that the discontinuity propagates with such a speed that there is no accumulation of material and energy. Just as in the case of an isothermal system, this leads us to the *jump condition*

$$\tilde{\sigma} = \frac{d\tau}{dx} = \frac{1 + \nu\dfrac{[n_i]}{[c_i]}}{1 - \mu\dfrac{[n_i]}{[c_i]}} \qquad (6.2.17)$$

and the *compatibility relations*

$$\frac{[n_1]}{[c_1]} = \frac{[n_2]}{[c_2]} = \cdots = \frac{[n_m]}{[c_m]} \qquad (6.2.18)$$

in which the brackets represent the jump of the quantity enclosed across the discontinuity.

Equations (6.2.17) and (6.2.18) represent a system of m algebraic equa-

More on Hyperbolic Systems   Chap. 6

tions and thus it is clear that, given the state on one side of the discontinuity, the state on the other side can be completely determined if one of the values of $\{c_i\}$ there or the value of $\bar{\sigma}$ is further specified. If the state on the left-hand side $\{c_i^l\}$ is kept fixed, the state on the right-hand side $\{c_i^r\}$ forms a one-parameter family. Hence the trajectory of the feasible state is given by a single curve in the hodograph space and we will call this a Σ.

Let us consider a discontinuity with the state on the left-hand side $\{c_i^l\}$ fixed. The state on the right-hand side then falls on the curve Σ passing through the point $\{c_i^l\}$ in the hodograph space. On the other hand, we notice that in the limit as the discontinuity becomes infinitesimally weak, the compatibility relations (6.2.18) are reduced to the fundamental differential equations (6.2.7). This implies that the curve Σ tends to converge to one of the Γ's passing through the point $\{c_i^l\}$ as the discontinuity becomes weak, and thus it is clear that, in general, there are m different branches for the curve Σ. If the limit is a Γ^k, we shall call the curve a Σ^k and the corresponding discontinuity a *k-discontinuity*. A formal proof was established by Lax (1957), who showed that given a fixed point $\{c_i^0\}$, there are m smooth curves Σ^k issuing from the point and the curve Σ^k makes a third-order contact at the point with the curve Γ^k so that the two curves are tangent to each other at the point. (See also the discussion in the earlier part of Sec. 1.7.) The propagation direction of a k-discontinuity will be expressed as

$$\bar{\sigma}_k = \left(\frac{d\tau}{dx}\right)_k = \frac{1 + \nu\left(\frac{[n_i]}{[c_i]}\right)_k}{1 - \mu\left(\frac{[n_i]}{[c_i]}\right)_k} \tag{6.2.19}$$

and the propagation speed is given by the reciprocal; that is,

$$\bar{\lambda}_k = \left(\frac{dx}{d\tau}\right)_k = \frac{1 - \mu\left(\frac{[n_i]}{[c_i]}\right)_k}{1 + \nu\left(\frac{[n_i]}{[c_i]}\right)_k} \tag{6.2.20}$$

It is obvious that

$$\lim_{c_i^l \to c_i^r} \bar{\lambda}_k = \lambda_k^l, \qquad \lim_{c_i^l \to c_i^r} \bar{\lambda}_k = \lambda_k^r \tag{6.2.21}$$

We also understand that to ensure the uniqueness of the solution it is necessary to have an additional condition, which is known as the *entropy condition*. We will come back to this in Sec. 6.4 after studying the structure of the continuous solution in Sec. 6.3.

Unlike the case with a pair of equations, it is not obvious if the matrix $\mathbf{A}$ has m real, distinct eigenvalues, although the equilibrium relations (6.2.2) satisfy the inequality conditions in Eq. (6.1.3). More recently, however, Kvaalen et al. (1985) have shown by applying some thermodynamic considerations that the number of distinct characteristic directions is equal to the variance of the system, as given by Gibbs' phase rule. Since the fluid carrier and the solid adsorbent are not considered as independent components but the temperature is counted as a variable in their argument, the variance of the present system is equal to m. Consequently, it appears that with physically relevant isotherms the system given by Eqs. (6.2.1) and (6.2.2) is totally hyperbolic.

6.3 Construction of a Continuous Solution

In this section we assume that the solution is continuous everywhere in the physical plane and show that the solution for the Riemann problem is determined in terms of simple waves and constant states. Since the simple wave theory was established in a general context in Sec. 3.3, the theorems proved there are all valid here in this problem. We refer to those frequently in the following discussion.

Since the fundamental differential equations (6.2.7) represent the image Γ of the solution in the hodograph space, it is only natural to begin with these equations. We assume that there is at least one concentration, say c_m, that varies monotonically along the curve Γ and use c_m instead of the parameter ω. By using Eq. (6.2.8), we can expand the fundamental differential equations (6.2.7) and, after making rearrangements, obtain a system of $(m - 1)$ quadratic equations for the $(m - 1)$ derivatives dc_i/dc_m, $i = 1, 2, \ldots,$ $m - 1$. This system, however, is not subject to a direct solution for dc_i/dc_m unless $m = 2$ (see Exercise 6.3.1).

On the other hand, we may rearrange Eqs. (6.2.12) and (6.2.13) in the form

$$\left(\frac{\mathcal{D}n_i}{\mathcal{D}c_i}\right)_k = \frac{\sigma_k - 1}{\mu\sigma_k + \nu} = \frac{1 - \lambda_k}{\mu + \nu\lambda_k} \equiv \beta_k \tag{6.3.1}$$

and note that β_k so defined is given by the kth eigenvalue of the matrix $\nabla \mathbf{n}$ [see Eq. (6.2.16)]. Now, by substituting Eq. (6.2.8) into Eq. (6.3.1), we obtain the equation

$$n_{i,1}\frac{dc_1}{dc_m} + \cdots + n_{i,i-1}\frac{dc_{i-1}}{dc_m} + (n_{i,i} - \beta_k)\frac{dc_i}{dc_m}$$

$$+ n_{i,i+1}\frac{dc_{i+1}}{dc_m} + \cdots + n_{i,m-1}\frac{dc_{m-1}}{dc_m} = -n_{i,m} \tag{6.3.2}$$

in which it is implied that the ordinary derivatives are taken in the direction of a Γ^k. [Compare this equation with Eq. (5.1.16).] We can write Eq. (6.3.2) for $i = 1, 2, \ldots, m - 1$ to give a linear system of $(m - 1)$ equations for dc_i/dc_m, $i = 1, 2, \ldots, m - 1$. For a totally hyperbolic system, we know that the matrix $\nabla \mathbf{n}$ has only simple eigenvalues and thus the matrix of coefficients of the system (6.3.2) is nonsingular. Consequently, there exists a unique solution which is given by

$$\left(\frac{dc_i}{dc_m}\right)_k = \frac{\Delta_i^k}{\Delta_m^k}, \qquad i = 1, 2, \ldots, m - 1 \qquad (6.3.3)$$

where

$$\Delta_m^k = \begin{vmatrix} n_{1,1} - \beta_k & n_{1,2} & \cdots & n_{1,i} & \cdots & n_{1,m-1} \\ n_{2,1} & n_{2,2} - \beta_k & \cdots & n_{2,i} & \cdots & n_{2,m-1} \\ \vdots & \vdots & & \vdots & & \vdots \\ n_{i,1} & n_{i,2} & \cdots & n_{i,i} - \beta_k & \cdots & n_{i,m-1} \\ \vdots & \vdots & & \vdots & & \vdots \\ n_{m-1,1} & n_{m-1,2} & \cdots & n_{m-1,i} & \cdots & n_{m-1,m-1} - \beta_k \end{vmatrix} \qquad (6.3.4)$$

and Δ_i^k denotes the determinant, which is identical to Δ_m^k except that, in place of the elements in the ith column, $-n_{j,m}$ has been substituted, where j is the row number.

Given a fixed state $\{c_i^0\}$, we can now integrate Eq. (6.3.3) to generate the Γ emanating from the point image corresponding to the fixed state. Since there are m different values of β_k, it is clear that there exist m different kinds of Γ and the one that would be generated by using β_k in Eq. (6.3.3) will be called the Γ^k. The characteristic parameter running along a Γ^k will be denoted by ω_k.

Let us now consider a case in which a Γ^k is obtained as the image of a solution in the hodograph space. The solution is then determined by mapping the curve Γ^k onto the physical plane, and this can be done by using the characteristic directions. Thus we expect that the corresponding region in the physical plane is covered by a network of characteristics of various families $\{C^j\}$, but we are particularly interested in the family of the kth characteristics C^k, the directions of which are given by Eq. (6.2.12). Since it is implied that the parameter ω_k is held constant along it, every C^k has a point image on the Γ^k. It then follows that every C^k carries a constant state of concentrations and temperature and so it is a straight line in the physical plane.

It is immediately apparent from the argument above that a region in the physical plane, whose image falls completely on a single piece of Γ^k in the hodograph space, is covered by a family of straight characteristics C^k. This is illustrated for $m = 3$ in Fig. 6.2. According to Theorem 3.5, the solution in such a region is called a k-simple wave. It is evident that in a k-simple wave region the characteristics C^k must fan clockwise as we proceed in the x direction and this requires the condition

$$\frac{\partial \lambda_k}{\partial x} > 0 \qquad (6.3.5)$$

to be satisfied in a k-simple wave region. Here the condition is expressed in terms of the kth characteristic speed λ_k because the characteristic direction σ_k can have positive and negative values.

We also note that the solution in a region adjacent to a region of constant

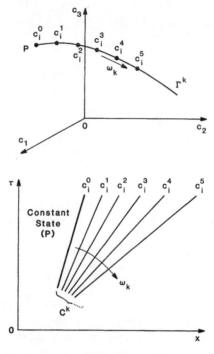

Figure 6.2

state is given by a simple wave, and if the boundary is a C^k, the solution is given by a k-simple wave. Therefore, any pair of constant states whose images fall on two different points along a single piece of Γ^k are connected by a k-simple wave, so the solution can be thought of as being given by a one-parameter family, such as

$$c_i = c_i(\omega_k), \qquad i = 1, 2, \ldots, m \tag{6.3.6}$$

or

$$c_i = c_i(c_m), \qquad i = 1, 2, \ldots, m - 1 \tag{6.3.7}$$

We shall now consider a problem for which the initial and boundary conditions are prescribed by Eq. (6.2.3) and assume that all the characteristics are directed forward. For this Riemann problem the images of the initial and boundary conditions fall on two different points O and A, respectively, in the hodograph space. It is then obvious that the image of the solution lies on a path that connects the point images O and A and consists of pieces of Γ's (see Fig. 3.4). The question is whether there exists such a path and further, whether it is unique if it exists.

The system of $(m - 1)$ equations (6.3.3) can be rearranged in the form

$$\frac{dc_1}{\Delta_1^k} = \frac{dc_2}{\Delta_2^k} = \cdots = \frac{dc_{m-1}}{\Delta_{m-1}^k} = \frac{dc_m}{\Delta_m^k} = d\omega_k \qquad (6.3.8)$$

which hold along a Γ^k. This implies that the curve Γ^k is tangent to the direction specified by the set

$$(\Delta_1^k, \Delta_2^k, \ldots, \Delta_i^k, \ldots, \Delta_m^k) \qquad (6.3.9)$$

On the other hand, it is not difficult to see that Δ_i^k is the cofactor of the ith element in the last row of the matrix $\nabla\mathbf{n} - \beta_k \mathbf{I}$, so the set $\{\Delta_i^k\}$ is given by the last column of the matrix adj $(\nabla\mathbf{n} - \beta_k\mathbf{I})$, which is essentially the same as the kth right eigenvector $\mathbf{r}_k$ of the matrix $\nabla\mathbf{n}$. Consequently, the curve Γ^k is everywhere tangent to the kth right eigenvector $\mathbf{r}_k$ of the matrix $\nabla\mathbf{n}$; that is, we can write

$$\frac{dc_1}{r_{1k}} = \frac{dc_2}{r_{2k}} = \cdots = \frac{dc_m}{r_{mk}} = d\omega_k \qquad (6.3.10)$$

along a Γ^k, where $r_{1k}, r_{2k}, \ldots, r_{mk}$ are the elements of the eigenvector $\mathbf{r}_k$. We note that this is the same as Eq. (3.3.20), which was derived in a general context.

There are m different eigenvectors $\{\mathbf{r}_k\}$ and these form a linearly independent set because the eigenvalues $\{\beta_k\}$ are all simple. An immediate consequence is that the set of eigenvectors can serve as a basis for the m-dimensional hodograph space. In this curvilinear coordinate system the curve Γ^k becomes parallel to the kth coordinate axis and thus the parameters $\{\omega_k\}$ are the curvilinear coordinates themselves. This argument leads us to the conclusion that there exists a topological transformation between the set $\{c_i\}$ and the set $\{\omega_k\}$, i.e.,

$$c_i = c_i(\omega_1, \omega_2, \ldots, \omega_m), \qquad i = 1, 2, \ldots, m \qquad (6.3.11)$$

and in the space $\Omega(m)$ of $\{\omega_k\}$ the image of a Γ^k becomes straight and parallel to the kth coordinate axis. This proves that there always exists a path that consists of pieces of Γ's and connects the points O and A in the hodograph space.

Such a path may consist of many pieces of Γ's of various kinds, to each of which there corresponds a simple wave in the physical plane. At first blush it appears that there are infinitely many possibilities, but by combining the inequality condition (6.2.14) and the structure of the simple wave, it can be shown that the path must contain at most m different pieces of Γ's of distinct kinds and these Γ^k's must be arranged in the descending order from Γ^m successively to Γ^1 as we pass from A to O. Obviously, such a path exists and it is unique. This establishes the existence of a unique solution.

The solution in the physical plane may be constructed by mapping the path obtained in the hodograph space. To each of the $(m - 1)$ vertices of the path there corresponds a constant state and hence there are $(m + 1)$ constant states, including the initial state and the feed state. These constant states are separated from one another by m simple waves, all centered at the origin. Since each of these simple waves corresponds to a single segment of Γ, they all belong to different families and arranged clockwise from the m-simple wave successively to the 1-simple wave. The feed state appears between the τ axis and the m-simple wave region, whereas the initial state is present between the 1-simple wave region and the x axis. Each of the $(m - 1)$ new constant states appears between two simple waves of consecutive indices and, as usual, the constant state appearing on the right-hand side of the k-simple wave will be called the k-constant state.

If we have an explicit form of the transformation (6.2.11), the $(m - 1)$ new constant states can be readily determined since the set $\{\omega_k\}$ can be specified for every constant state. It is then straightforward to construct the simple waves by applying the discussion in the earlier part of this section. Indeed, this is what we have done with isothermal chromatography with the Langmuir isotherm in previous chapters.

However, we cannot expect to have such an explicit form in general, and thus we may have to resort to an iteration scheme for the construction of the path in the hodograph space. For this purpose it appears appropriate to construct the projection of the path on the (c_i, c_m)-plane for $i = 1, 2, \ldots, m - 1$. Each of these planes contains two fixed points of the initial and feed states and the projection of the path connecting the two fixed points has $(m - 1)$ vertices on this plane. With $(m - 1)$ planes of this feature there are $(m - 1)$ sets of the corresponding vertices. When every set of the corresponding vertices is located at the same value of c_m, the iteration may be terminated and the path can be taken as the image of the solution.

We further observe that, just as in the case with isothermal systems, the path so constructed is independent of the parameters ν and μ, and depends solely on the matrix $\nabla \mathbf{n}$ and the initial and boundary conditions. On the other hand, the physical plane portrait of the solution will rotate counterclockwise as the parameter μ increases if the directional derivatives β_k are positive for all k [see Eqs. (6.2.12) and (6.3.1)]. For a certain value of μ, therefore, some of the simple waves may be located outside the domain. This implies that only one portion of the path remains meaningful, but it is important to note that the path as a whole is invariant under the variation of μ. This subject is treated further in Sec. 6.6.

In case $m = 2$, the path can be directly determined without any iteration because the path in the (c_1, c_2)-plane is the image of the solution itself. The situation here is exactly the same as the case treated in Sec. 1.5. This is illustrated in Sec. 6.7. For $m = 3$ or 4 we can easily imagine the iteration scheme and with experience a few iterations would be sufficient to determine the path. This is the subject of Sec. 6.8.

Exercise

6.3.1. Consider the fundamental differential equations (6.2.7) for $m = 3$. By using the last expression in Eq. (6.2.8), rearrange the fundamental differential equations into two equations for the derivatives dc_1/dc_3 and dc_2/dc_3. Can you solve these equations for dc_1/dc_3 and dc_2/dc_3?

6.4 Discontinuities, Weak Solutions, and the Entropy Condition

In Sec. 6.3 it was assumed that the characteristics C^k would fan forward as we proceed in the x direction in a k-simple wave region [see Eq. (6.3.5)]. Depending on the nonlinear nature of the system, however, we should also expect to have a case in which the characteristics C^k fan backward, or fan first forward and then backward, or vice versa, as we pass in the x direction through a k-simple wave region. In such a case, we encounter a physically impossible situation because three different states would correspond to every point in the region of interest, and thus we must introduce a discontinuity in the solution.

When the solution contains a discontinuity, the conservation principle gives the jump condition (6.2.17) and the compatibility relations (6.2.18), which together constitute the generalized Rankine–Hugoniot condition. We recall from Sec. 6.2 that there are m different kinds of discontinuities and the one with index k, called the k-discontinuity, propagates in the direction $\tilde{\sigma}_k$ with a speed $\tilde{\lambda}_k$ given by Eqs. (6.2.19) and (6.2.20), respectively. We also note that if the state on one side of the k-discontinuity is fixed, the state on the other side must fall on the curve Σ^k in the hodograph space which emanates from the point of the fixed state and is tangent to the curve Γ^k at the point.

The conditions (6.2.17) and (6.2.18), however, are not sufficient to determine the physically relevant solution because the ratio $[n_i]/[c_i]$ is symmetric with respect to the states on both sides. This is the same situation that we experienced in the cases of single equations and pairs of equations. To select the physically relevant solution, therefore, it is necessary to impose an additional condition, known as the *entropy condition*, that would specify the relevant direction of jumps.

Based on the argument around Eq. (6.3.5), we claim that this condition can be phrased as follows: Across a k-discontinuity we must have $\lambda_k^l > \lambda_k^r$, where the superscripts l and r denote the left- and right-hand sides of the discontinuity, respectively. As we shall see soon, this condition is valid if λ_k varies monotonically along every Γ^k. Otherwise, this fails to serve as the entropy condition.

In Sec. 1.7, we introduced the *extended entropy condition* due to Liu (1974), which is really an extension of Oleinik's celebrated *Condition E* to the case of pairs of equations. In fact, Liu (1975) has extended this condition further so that it can be applied to systems of quasilinear equations. We shall state his conclusions here without proof and interpret their implications for various types of nonlinearity. Since the development is similar to that for pairs of equations (see Sec. 1.7), we would like to make the discussion brief. Here it is convenient to speak in terms of the vector $\mathbf{c}$, whose elements are given by the concentrations $\{c_i\}$.

RULE 6.1 (Hypothesis): Let Σ_0^k be the curve Σ^k emanating from a fixed point $\mathbf{c}_0$. In the following we assume that for any $\mathbf{c}$ on Σ_0^k, $k = 1$, $2, \ldots, m$, $\mathbf{c} \neq \mathbf{c}_0$, the vectors $\mathbf{c} - \mathbf{c}_0$, $\mathbf{r}_1(\mathbf{c}), \ldots, \mathbf{r}_{k-1}(\mathbf{c})$, $\mathbf{r}_{k+1}(\mathbf{c}), \ldots,$

$\mathbf{r}_m(\mathbf{c})$ are linearly independent, where Σ_0^k, $k = 1, 2, \ldots, m$, are smooth curves tangent to Γ^k at $\mathbf{c}_0$.

RULE 6.2 (Extended Entropy Condition): Let Σ_0^k be the curve Σ^k emanating from the point $\mathbf{c}^l$ in the hodograph space and consider another point $\mathbf{c}^r$ on Σ_0^k. If the two states are to be connected by a k-discontinuity with $\mathbf{c}^l$ on the left-hand side and $\mathbf{c}^r$ on the right, we must have

$$\tilde{\lambda}_k(\mathbf{c}^l, \mathbf{c}^r) \leq \tilde{\lambda}_k(\mathbf{c}^l, \mathbf{c}) \qquad (6.4.1)$$

for every $\mathbf{c}$ on Σ_0^k between $\mathbf{c}^l$ and $\mathbf{c}^r$.

THEOREM 6.1: If two states $\mathbf{c}^l$ and $\mathbf{c}^r$, falling on the Σ_0^k satisfy the extended entropy condition, then λ_k satisfies the conditions

$$\lambda_k(\mathbf{c}^r) \leq \tilde{\lambda}_k(\mathbf{c}^l, \mathbf{c}^r) \leq \lambda_k(\mathbf{c}^l) \qquad (6.4.2)$$

Moreover, if the kth characteristic field is genuinely nonlinear for all $\mathbf{c}$, the extended entropy condition is equivalent to the shock conditions

$$\lambda_k(\mathbf{c}^r) < \tilde{\lambda}_k(\mathbf{c}^l, \mathbf{c}^r) < \lambda_k(\mathbf{c}^l) \qquad (6.4.3)$$

$$\lambda_{k+1}(\mathbf{c}^l) < \tilde{\lambda}_k(\mathbf{c}^l, \mathbf{c}^r) < \lambda_{k-1}(\mathbf{c}^r) \qquad (6.4.4)$$

In the following we shall discuss three different cases in view of the nonlinear nature of the kth characteristic field. Just as in the case of a single equation or a pair of equations, here again we shall see shocks, semishocks, and contact discontinuities.

Let us first consider the case in which the kth characteristic field is genuinely nonlinear. By Definition 3.3 we have

$$\mathbf{r}_k^T \nabla \lambda_k \neq 0 \qquad (6.4.5)$$

for all $\mathbf{c}$. We note that since $\partial \beta_k / \partial \lambda_k$ has a uniform sign, using the right-eigenvector $\mathbf{r}_k$ of $\nabla \mathbf{n}$ is equivalent to using that of the matrix $\mathbf{A}$ [see Eqs. (6.2.11) and (6.2.16)]. Clearly, Eq. (6.4.5) implies that λ_k varies monotonically along a Γ^k.

Suppose that we have the curve Σ_0^k passing through a fixed point $\mathbf{c}_0$ in the hodograph space and consider another point $\mathbf{c}_1$ on Σ_0^k. Since the sign of $\mathbf{r}_k^T \nabla \lambda_k$ is arbitrary, we shall assume that λ_k decreases along a Γ^k in the general direction from $\mathbf{c}_0$ toward $\mathbf{c}_1$. An immediate consequence is that λ_k decreases along Σ_0^k from $\mathbf{c}_0$ to $\mathbf{c}_1$ and thus the extended entropy condition is satisfied with $\mathbf{c}_0 = \mathbf{c}^l$ and $\mathbf{c}_1 = \mathbf{c}^r$. Hence the two states can be connected by a k-discontinuity. Since the kth characteristic field is genuinely nonlinear, we also find from Theorem 6.1 that the shock conditions (6.4.3) and (6.4.4) are satisfied. Consequently, if two states $\mathbf{c}_0$ and $\mathbf{c}_1$ lie on a curve Σ^k and satisfy the extended entropy condition (6.4.1), the two states are connected by a k-shock. The situation is depicted in Fig. 6.3, where c_i represents a typical concentration, and this figure is to be compared with Fig. 1.21.

Next, let us suppose that λ_k first decreases and then increases along every

More on Hyperbolic Systems   Chap. 6

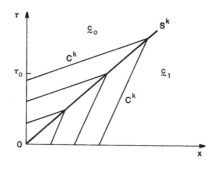

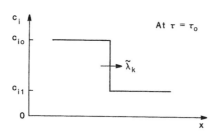

Figure 6.3

Γ^*. This implies that

$$\mathbf{r}_k^T \nabla \lambda_k = 0 \tag{6.4.6}$$

for a particular set $\mathbf{c}_0^*$. Here again we will consider the curve Σ_0^k passing through the point $\mathbf{c}_0$. In this case we can always find a point $\mathbf{c}_1$ on Σ_0^k at which the equality

$$\tilde{\lambda}_k(\mathbf{c}_0, \mathbf{c}_1) = \lambda_k(\mathbf{c}_1) \tag{6.4.7}$$

holds, where the point $\mathbf{c}_0^*$ falls between the points $\mathbf{c}_0$ and $\mathbf{c}_1$. Beyond this point we shall take the Γ_1^k that passes through the point $\mathbf{c}_1$ and along which λ_k increases. The two curves Σ_0^k and Γ_1^k connected together at $\mathbf{c}_1$ will be called the curve Λ_0^k (see Fig. 1.24).

If we think of a state $\mathbf{c}$ that falls on the portion of Σ_0^k between $\mathbf{c}_0$ and $\mathbf{c}_1$, it is obvious that the states $\mathbf{c}_0$ and $\mathbf{c}_1$ are connected by a k-shock with $\mathbf{c}_0 = \mathbf{c}^l$ and $\mathbf{c} = \mathbf{c}^r$. However, if the state $\mathbf{c}$ falls on Γ_1^k, we expect to have first a k-discontinuity connecting $\mathbf{c}_0$ to $\mathbf{c}_1$ and then a k-simple wave connecting $\mathbf{c}_1$ to $\mathbf{c}$ as illustrated in Fig. 6.4. Hence we find from Eq. (6.4.2) that the k-discontinuity satisfies the condition

$$\lambda_k^l > \tilde{\lambda}_k(\mathbf{c}^l, \mathbf{c}^r) = \lambda_k^r \tag{6.4.8}$$

where $\mathbf{c}_0 = \mathbf{c}^l$ and $\mathbf{c}_1 = \mathbf{c}^r$. The k-discontinuity is shock-like only on the left-hand side, while the state on the right-hand side propagates side by side with the k-simple wave without interval. Just as in the previous case, we shall call

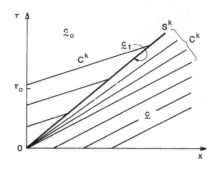

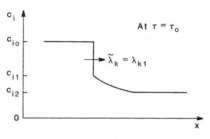

Figure 6.4

such a discontinuity a k-semishock, whereas Liu (1975) used the term *one-sided contact discontinuity*.

We can imagine a situation in which λ_k first increases and then decreases along every Γ^k. Then a similar argument can be made in the reversed direction to show that a k-semishock is joined by a k-simple wave on the left-hand side. The two waves joined together are obviously of the same family and constitute what we call a *combined wave*. Examples of this combined wave are discussed in Secs 6.7 and 6.8.

Finally, we would like to treat the general situation in which $\mathbf{r}_k^T \nabla \lambda_k$ changes its sign many times. Along every Γ^k, therefore, λ_k first decreases and then increases, and this pattern is repeated many times. Let us consider the fixed point $\mathbf{c}_0$ and move along Σ_0^k to meet the point $\mathbf{c}_1$, which was defined previously. Then we take Γ_1^k to continue but the nature of the kth characteristic field is such that there is a point $\mathbf{c}_1^*$ on Γ_1^k from which λ_k starts to decrease again. As long as we stay betwen $\mathbf{c}_0$ and $\mathbf{c}_1^*$ along the curve Λ_0^k, the situation remains the same as in the previous case.

Suppose now we are interested in a point $\mathbf{c}$ that is located on the opposite side but still in the neighborhood of the point $\mathbf{c}_1^*$. According to Liu (1975), in such a case there always exists a point $\mathbf{c}_2$ between the points $\mathbf{c}_1$ and $\mathbf{c}_1^*$ such that the curve Σ_2^k emanating from $\mathbf{c}_2$ passes through the point $\mathbf{c}$ and the condition

$$\lambda_k(\mathbf{c}_2) = \tilde{\lambda}_k(\mathbf{c}_2, \mathbf{c}) \tag{6.4.9}$$

is satisfied. The position of the point $\mathbf{c}_2$ is, of course, dependent on the location

More on Hyperbolic Systems   Chap. 6

of the point $\mathbf{c}$ and so is the location of the curve Σ_2^k. The curve Λ_0^k thus consists of Σ_0^k, Γ_1^k, and Σ_2^k and, starting from $\mathbf{c}_0$ on the left, we have a k-semishock to reach $\mathbf{c}_1$, then a k-simple wave to get to $\mathbf{c}_2$, and finally another k-semishock to make the connection to $\mathbf{c}$ on the right. Figure 6.5 illustrates this situation. With respect to the k-semishock on the left, we have the condition

$$\lambda_k(\mathbf{c}_0) > \tilde{\lambda}_k(\mathbf{c}_0, \mathbf{c}_1) = \lambda_k(\mathbf{c}_1) \tag{6.4.10}$$

and for the one on the right, the condition

$$\lambda_k(\mathbf{c}_2) = \tilde{\lambda}_k(\mathbf{c}_2, \mathbf{c}) > \lambda_k(\mathbf{c}) \tag{6.4.11}$$

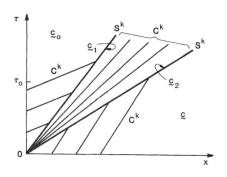

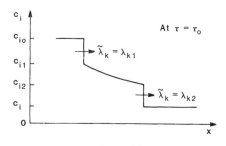

Figure 6.5

If we proceed further along the curve Σ_2^k, there may exist a point $\mathbf{c}_2^*$ from which λ_k begins to increase again. As before for the case of the k-semishock on the left, we can find a point $\mathbf{c}_3$ on Σ_2^k at which the equality condition

$$\tilde{\lambda}_k(\mathbf{c}_2, \mathbf{c}_3) = \lambda_k(\mathbf{c}_3) \tag{6.4.12}$$

is satisfied, where $\mathbf{c}_2^*$ lies between $\mathbf{c}_2$ and $\mathbf{c}_3$. From the point, we take the curve Γ_3^k that passes through the point $\mathbf{c}_3$ and along which λ_k increases. At this stage the curve Λ_0^k is composed of Σ_0^k, Γ_1^k, Σ_2^k, and Γ_3^k.

Let us now consider a point $\mathbf{c}$ on Γ_3^k and examine how the state $\mathbf{c}_0$ is connected to the state $\mathbf{c}$ on the right. Starting from $\mathbf{c}_0$, we first see the k-semishock connecting $\mathbf{c}_0$ to $\mathbf{c}_1$, then a k-simple wave making a bridge between $\mathbf{c}_1$ and $\mathbf{c}_2$, and a k-discontinuity that gives a jump to $\mathbf{c}_3$. This is then followed by another

k-simple wave making connection to **c**. This situation is depicted in Fig. 6.6, where c_i denotes a typical concentration.

The k-discontinuity on the right satisfies the conditions

$$\lambda_k(\mathbf{c}_2) = \tilde{\lambda}_k(\mathbf{c}_2, \mathbf{c}_3) = \tilde{\lambda}_k(\mathbf{c}_3) \qquad (6.4.13)$$

which are consistent with Eq. (6.4.2). Here the nature of a shock has disappeared on both sides and the discontinuity propagates side by side with k-simple waves on both sides. There is no tendency of overtaking or being overtaken from either side. As before, such a discontinuity will be referred to as a *contact discontinuity*, whereas Liu (1975) used the term *two-sided contact discontinuity*.

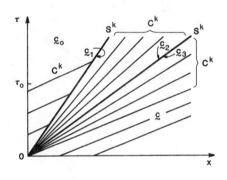

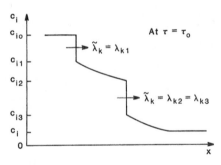

Figure 6.6

If we have

$$\mathbf{r}_k^T \nabla \lambda_k = 0 \qquad (6.4.14)$$

for all **c** of interest, λ_k remains invariant along each Γ^k. In such a case, we say that the kth characteristic field is *linearly degenerate*. It can be shown that any two states on a Γ^k satisfy both the fundamental differential equations and the compatibility relations and propagate with the same speed (see the last part of Sec. 1.7). An immediate consequence is that the two states are connected by a discontinuity which satisfies Eq. (6.4.13) and so it is a k-contact discontinuity.

We note that a completely linear system is a special case of linearly degenerate systems.

The argument connecting the state c_0 to a state c on Λ_0^k can be continued further in a similar manner if necessary. Based on this argument, Liu (1975) has established the following theorem:

> **THEOREM 6.2:** Suppose that Rule 6.1 holds and $\lambda_{k+1}(c) < \bar{\lambda}_k(c_0, c) < \lambda_{k-1}(c)$ for any c on Σ_0^k and that, along each Γ^k, $r_k^T \nabla \lambda_k$ vanishes at discrete points. Then there exists a curve Λ_0^k such that c_0 is connected to any c *on* Λ_0^k on the right by k-shocks, k-semishocks, k-simple waves, and k-contact discontinuities. Also, across any discontinuity (c^l, c^r), the extended entropy condition (6.4.1) is satisfied. Conversely, if c can be connected to c_0 by k-waves satisfying the extended entropy condition (6.4.1), then c lies on Λ_0^k and the solution has a unique form.

So far our discussion has been concentrated on the Riemann problem. Just as in the case of a single equation or a pair of equations, singularities can be developed from continuous initial data and this has been treated by various authors. We would like to introduce some of the references here. Jeffrey (1963) considered the quasilinear system of equations of the form

$$\frac{\partial u}{\partial t} + A \frac{\partial u}{\partial x} + b = 0 \qquad (6.4.15)$$

with Lipschitz continuous initial data

$$u(x, 0) = u_0(x) \qquad (6.4.16)$$

and determined equations for the time and distance at which a jump discontinuity first appears on the wave front. He also extended the analysis to the case of single nonlinear hyperbolic equations of order greater than unity (1966). For a genuinely nonlinear system (6.4.15) with $b = 0$, John (1974) treated sufficiently small initial data of compact support superimposed on a constant state to show that the first derivatives of u would become infinite in finite time. This result has been generalized and further refined by Liu (1979), who allowed some characteristic fields to be linearly degenerate and thus extended the application to important physical systems such as gas dynamic equations. For genuinely nonlinear systems, DiPerna (1975) has discussed the propagation of singularities in the solution and the limiting behavior at singularities. Although the analysis has been carried out for systems of two equations, the author has asserted that a straightforward generalization to systems of many equations is likely because his treatment did not involve Riemann invariants.

It is now evident that solutions to hyperbolic systems of quasilinear equations are not necessarily continuously differentiable, and furthermore, they can be discontinuous. Thus we need to introduce the concept of weak solutions due to Lax (1957). Since the definition of weak solutions is based on the conservation laws instead of the quasilinear equations in the form of Eq. (6.4.15), it is appropriate to begin with the system of conservation laws

$$\frac{\partial \mathbf{u}}{\partial t} + \frac{\partial}{\partial x}\mathbf{f(u)} = 0 \qquad (6.4.17)$$

which is subject to the initial data

$$\mathbf{u}(x, 0) = \mathbf{u}_0(x) \qquad \text{for } -\infty < x < \infty \qquad (6.4.18)$$

Here we note that, in general, Eq. (6.4.15) cannot be represented in the form of Eq. (6.4.17) (see, e.g., Rozdestvenskii, 1964). With respect to the system of conservation laws (6.2.1) and (6.2.2) of our immediate interest, we may put the equations in the form

$$\frac{\partial}{\partial x}\mathbf{f(c)} + \frac{\partial}{\partial \tau}\mathbf{g(c)} = 0 \qquad (6.4.19)$$

where

$$\mathbf{f(c)} = 1 - \mu\mathbf{n(c)} \qquad (6.4.20)$$

$$\mathbf{g(c)} = 1 + \nu\mathbf{n(c)} \qquad (6.4.21)$$

and consider this in parallel with Eq. (6.4.17).

Lax has defined a weak solution of the system (6.4.17) with the initial condition (6.4.18) as follows: If $\mathbf{u}$ and $\mathbf{f(u)}$ are integrable functions over every bounded set of the half-plane $t \geq 0$ and the integral relation

$$\int_0^\infty \int_{-\infty}^\infty \left\{ \frac{\partial \phi}{\partial t}\mathbf{u} + \frac{\partial \phi}{\partial x}\mathbf{f(u)} \right\} dx\, dt + \int_{-\infty}^\infty \phi(x, 0)\mathbf{u}_0(x, 0)\, dx = 0 \qquad (6.4.22)$$

is satisfied for all smooth functions ϕ that vanish for $|x| + t$ large enough, then $\mathbf{u}$ is called a weak solution of the system (6.4.17) subject to the initial condition (6.4.18).

Since this definition is in parallel with that for a single conservation law (see Sec. 7.4 of Vol. 1), here again we can show by applying integration by parts to Eq. (6.4.22) that if $\mathbf{u}$ is smooth in a region of the (x, t)-plane and satisfies Eq. (6.4.22), then $\mathbf{u}$ satisfies the system of conservation law (6.4.17). In case $\mathbf{u}$ is discontinuous, we take a contour containing a portion of the path of the discontinuity in the (x, t)-plane and apply Green's theorem as was done in Sec. 7.4 of Vol.1. This gives the generalized Rankine–Hugoniot condition

$$\tilde{\lambda}[\mathbf{u}] = [\mathbf{f}] \qquad (6.4.23)$$

which represents both the jump condition and the compatibility relations.

It was proved by Lax that if $\mathbf{u}^l$ is fixed, there are m different branches for the locus of the state $\mathbf{u}^r$ and each of these are tangent to one of the Γ^k's at the point $\mathbf{u}^l$. These branches are the curves Σ^k, $k = 1, 2, \ldots, m$, and in case $\mathbf{u}^r$ falls on the Σ^k, we put $\tilde{\lambda} = \tilde{\lambda}_k$ and then

$$\lim_{\mathbf{u}^r \to \mathbf{u}^l} \tilde{\lambda}_k = \lambda_k(\mathbf{u}^l) \qquad (6.4.24)$$

Clearly, it is necessary to impose an additional condition to ensure the uniqueness of the solution. This is the *entropy* condition and, for this, Lax has considered a genuinely nonlinear system and required the characteristics C of the same family to impinge on the path of the discontinuity from both sides in the positive t direction. This is motivated by the theory of free boundary value problems, according to which the number of relations to be imposed on the states on the two sides must be equal to the number of characteristics impinging on the path of the discontinuity from either side (see also the discussion in Sec. 5.3). This requirement immediately gives rise to the shock conditions

$$\lambda_k(\mathbf{u}^r) < \bar{\lambda}_k(\mathbf{u}^l, \mathbf{u}^r) < \lambda_k(\mathbf{u}^l) \tag{6.4.25}$$

$$\lambda_{k+1}(\mathbf{u}^l) < \bar{\lambda}_k(\mathbf{u}^l, \mathbf{u}^r) < \lambda_{k-1}(\mathbf{u}^r) \tag{6.4.26}$$

Here the index k on λ and $\bar{\lambda}$ is assigned starting from the largest value to the smallest value to make it consistent with our scheme, but Lax had it in the reversed order.

By imposing the shock conditions, which constitute the entropy condition for genuinely nonlinear systems, Lax was able to establish the uniqueness of the solution to the Riemann problem when the two initial constant states are close to each other. For general systems, Liu (1975) has extended Oleinik's Condition E (see Sec. 5.4 of Vol. 1) in the form of his extended entropy condition (Rule 6.2) and established Theorem 6.2. Although the formal proof was not presented, Liu asserted that the uniqueness of the solution for the Riemann problem can be proved with a pair of arbitrary constant states as the initial data.

The concept of entropy due to Friedrichs and Lax (1971) was discussed in some detail in Sec. 1.10 and it also applies to systems of conservation laws. Since the formulation is almost the same, we shall simply introduce important equations with brief comments. For the detail the reader may be referred to the discussion in the last part of Sec. 1.10, in which $\mathbf{w}$ and W were used, respectively, for $\mathbf{u}$ and U to be employed here (see also the references for this section).

If $\mathbf{u}$ is a solution of Eq. (6.4.17) which vanishes for $|x|$ sufficiently large, integrating Eq. (6.4.17) with respect to x gives

$$\int_{-\infty}^{\infty} \frac{\partial \mathbf{u}}{\partial t} \, dx = 0 \tag{6.4.27}$$

where we have assumed that $\mathbf{f}(0) = \mathbf{0}$. An immediate consequence is that

$$\int_{-\infty}^{\infty} \mathbf{u}(x, t) \, dx = \int_{-\infty}^{\infty} \mathbf{u}(x, 0) \, dx, \qquad t > 0 \tag{6.4.28}$$

so the integral of $\mathbf{u}$ is conserved. Let us then ask the question: When is a conservation law

$$\frac{\partial U}{\partial t} + \frac{\partial F}{\partial x} = 0 \tag{6.4.29}$$

where U and F are functions of $\mathbf{u}$, a consequence of the old ones, that is, the system (6.4.17)?

It was shown in Sec. 1.10 that the conservation law (6.4.29) holds if the *compatibility equation*

$$(\nabla U)^T \mathbf{A} = (\nabla F)^T \tag{6.4.30}$$

is satisfied, where

$$\mathbf{A} = \nabla \mathbf{f} \tag{6.4.31}$$

The compatibility equation (6.4.30) gives a system of m partial differential equations for U and F. For $m > 2$, therefore, the system is overdetermined and has no solution in general. There are, however, special cases in which a nontrivial solution exists. A general class of equations for which a solution exists are those with $\mathbf{f}$ given by a gradient; that is,

$$\mathbf{f} = \nabla \phi \tag{6.4.32}$$

In this case the matrix $\mathbf{A}$ is symmetric, so we have

$$\frac{\partial f_i}{\partial u_j} = \frac{\partial f_j}{\partial u_i} \tag{6.4.33}$$

Thus if we define U and F as

$$U = \frac{1}{2} \mathbf{u}^T \mathbf{u} \tag{6.4.34}$$

$$F = \mathbf{u}^T \mathbf{f} - \phi \tag{6.4.35}$$

it can be shown by direct substitution that Eq. (6.4.29) is satisfied. An important feature to note is that U is given by a convex function.

Just as in Sec. 1.10, we can add a small dissipative term to Eq. (6.4.17) to have

$$\frac{\partial \mathbf{u}}{\partial t} + \mathbf{A} \frac{\partial \mathbf{u}}{\partial x} = \mu \frac{\partial^2 \mathbf{u}}{\partial x^2}, \qquad \mu > 0 \tag{6.4.36}$$

and after some manipulations, take the limit as $\mu \to 0$ to obtain

$$\frac{\partial U}{\partial t} + \frac{\partial F}{\partial x} \leq 0 \tag{6.4.37}$$

The development above can be embodied by the following theorem.

> **THEOREM 6.3:** Let Eq. (6.4.17) be a system of conservation laws which implies an additional conservation law (6.4.29); suppose that U is strictly convex. Let $\mathbf{u}(x, t)$ be a weak solution of Eq. (6.4.17) which is the limit of solutions of the viscosity equation (6.4.36). Then $\mathbf{u}$ satisfies the inequality (6.4.37).

We can further argue that the integral

$$\int_{-\infty}^{\infty} U \, dx \qquad (6.4.38)$$

is a non-increasing function of t, and if **u** is piecewise continuous, the condition

$$\bar{\lambda}[U] - [F] \leq 0 \qquad (6.4.39)$$

is satisfied at every point of discontinuity. Since in compressible fluid flow the former corresponds to the increase of total negative entropy while the latter indicates that the entropy of particles increases upon crossing a shock, the function U is referred to as the *entropy* and the function F as the *entropy flux*. It is in this sense that Eq. (6.4.37) or (6.4.39) is called the *entropy condition*. It can be shown that, for genuinely nonlinear systems, Eq. (6.4.39) is equivalent to the shock conditions (6.4.25) and (6.4.26).

Exercises

6.4.1. Consider Eqs. (6.4.19), (6.4.20), and (6.4.21) with the initial data

$$\mathbf{c}(x, \tau) = \mathbf{c}^0(x), \qquad -\infty < x < \infty$$

and define weak solutions by writing an equation corresponding to Eq. (6.4.22). If **c** is a weak solution which is smooth in a region of the (x, τ)-plane, show that **c** satisfies Eq. (6.4.19) in the region.

6.4.2. Let us take a domain D large enough to occupy the whole of the half-plane $t \geq 0$, in which the path S of a discontinuity lies (see Fig. 1.35). Applying Green's theorem, and noting that the boundary must include both sides of S, carry out the integration of Eq. (6.4.22) to obtain Eq. (6.4.23) (see Sec. 7.4 of Vol. 1).

6.4.3. Apply the procedure of Exercise 6.4.2 to the resulting equation in integral form of Exercise 6.4.1 and show that Eqs. (6.2.17) and (6.2.18) are obtained.

6.4.4. Suppose that we have a weak solution of Eq. (6.4.17) containing a discontinuity which propagates along a smooth curve $x = y(t)$ and choose a and b so that the curve $x = y(t)$ intersects the interval $a \leq x \leq b$ at time t. Regarding U as an entropy for Eq. (6.4.17) and F as the entropy flux, apply the principle that the integral of U is a non-increasing function of t to obtain Eq. (6.4.39).

6.5 Existence, Uniqueness, Structure, and Asymptotic Behavior of Weak Solutions

For systems of more than two conservation laws, Lax (1957) had really laid the foundation of the mathematical theory and, since then, a number of papers dealing with various features of the solution have appeared in the mathematics literature. For pairs of equations, we have made an extensive survey in Sec.

1.11. Here again we would like to adopt a similar scheme to discuss some of the important findings that have been reported, although we must admit that our search may not have been complete.

We shall start with the system of conservation laws

$$\frac{\partial \mathbf{u}}{\partial t} + \frac{\partial}{\partial x}\mathbf{f(u)} = \mathbf{0} \qquad (6.5.1)$$

where $\mathbf{u}$ is a vector of m components and $\mathbf{f(u)}$ a vector-valued function of $\mathbf{u}$. When the differentiation in Eq. (6.5.1) is carried out, we obtain a system of quasilinear equations of first order

$$\frac{\partial \mathbf{u}}{\partial t} + \mathbf{A(u)}\frac{\partial \mathbf{u}}{\partial x} = \mathbf{0} \qquad (6.5.2)$$

where

$$\mathbf{A(u)} = \nabla\mathbf{f(u)} = d\mathbf{f} \qquad (6.5.3)$$

After having established the theory of simple waves and shocks under the assumption that the system (6.5.1) is totally hyperbolic, Lax (1957) considered the Riemann problem with the initial data

$$\mathbf{u}(x, 0) = \begin{cases} \mathbf{u}_0 & \text{for} \quad x < 0 \\ \mathbf{u}_1 & \text{for} \quad x > 0 \end{cases} \qquad (6.5.4)$$

where $\mathbf{u}_0$ and $\mathbf{u}_1$ are constant vectors. For systems for which each characteristic field is either genuinely nonlinear

$$\mathbf{r}_k^T \nabla \lambda_k \neq 0 \qquad \text{for all } \mathbf{u} \qquad (6.5.5)$$

or linearly degenerate

$$\mathbf{r}_k^T \nabla \lambda_k = 0 \qquad \text{for all } \mathbf{u} \qquad (6.5.6)$$

the shock conditions

$$\lambda_k(\mathbf{u}^r) < \bar{\lambda}_k(\mathbf{u}^l, \mathbf{u}^r) < \lambda_k(\mathbf{u}^l) \qquad (6.5.7)$$

$$\lambda_{k+1}(\mathbf{u}^l) < \bar{\lambda}_k(\mathbf{u}^l, \mathbf{u}^r) < \lambda_{k-1}(\mathbf{u}^r) \qquad (6.5.8)$$

are imposed as the entropy condition to establish the existence of a unique solution. Here λ_k, $\bar{\lambda}_k$, and $\mathbf{r}_k$ all have the usual meanings defined in previous sections. Lax's solution consists of $(m + 1)$ constant states connected by simple waves, shocks, and contact discontinuities all centered at the origin. According to Lax, such a solution exists if $\mathbf{u}_1$ lies in a neighborhood of $\mathbf{u}_0$, and it is unique, provided that all the intermediate states are restricted to lie in a neighborhood of $\mathbf{u}_0$.

This has been extended by Liu (1975) to the Riemann problem for general systems of conservation laws, in which the initial data $\mathbf{u}_0$ and $\mathbf{u}_1$ are given by arbitrary constant vectors. Here it is assumed that $\mathbf{r}_k \nabla \lambda_k$ vanishes at discrete points; that is, λ_k changes its direction of variation along a Γ^k at discrete points. By imposing the extended entropy condition (6.4.1) (see Rule 6.2),

Theorem 6.2 can be proved and, on the basis of the theorem, Liu asserts that the Riemann problem [Eqs. (6.5.1) and (6.5.4)] with two arbitrary constant initial vectors has a unique solution. In Liu's solution, therefore, the connection between constant states is made by simple waves, or shocks, or semishocks, or by contact discontinuities. Instead of presenting the general proof, he establishes the uniqueness theorem for the gas dynamics equations to be discussed later in this section.

The problem with distributed initial data

$$\mathbf{u}(x, 0) = \mathbf{u}_0(x), \qquad -\infty < x < \infty \qquad (6.5.9)$$

was treated first by Glimm (1965). For genuinely nonlinear systems, he shows that waves interact in almost the same manner as the wave interactions for single conservation laws except for an error term which is of the second order of the strengths of waves before interaction. This observation is then used to prove that the initial value problem [Eqs. (6.5.1) and (6.5.9)] has a global weak solution provided that the initial data $\mathbf{u}_0(x)$ have small total variation. Glimm's solution is given by the limit of difference approximations. His difference scheme has a probabilistic feature and it depends on a random sequence. Liu (1977) has shown that Glimm's scheme converges for any equidistributed sequence and thus established the deterministic version of the scheme.

A somewhat special system was taken up by Hurd (1969). He considers a case for which the matrix $\mathbf{A}$ is a symmetric and positive definite. By imposing the familiar condition that there exists a function $K(t)$ such that

$$\frac{u_i(x_1, t) - u_i(x_2, t)}{x_1 - x_2} \le K(t), \qquad i = 1, 2, \ldots, m \qquad (6.5.10)$$

holds, Hurd shows that the weak solution of the system (6.5.1) is uniquely determined by the initial condition $\mathbf{u}_0(x)$, which is given by an essentially bounded measurable function.

An important class of hyperbolic systems of conservations law is the one governing the flow of compressible, nonviscous, nonconductive fluids. In Lagrangian coordinates, the equations for one-dimensional, nonisentropic flow are as follows (see Courant and Friedrichs, 1948):

$$\text{(conservation of mass)} \quad \frac{\partial v}{\partial t} - \frac{\partial u}{\partial x} = 0$$

$$\text{(conservation of momentum)} \quad \frac{\partial u}{\partial t} + \frac{\partial p}{\partial x} = 0$$

$$\text{(conservation of energy)} \quad \frac{\partial E}{\partial t} + \frac{\partial (pu)}{\partial x} = 0 \qquad (6.5.11)$$

$$\text{(equation of state)} \quad p = p(v, e) = p(v, s)$$

where v, u, p, e, and s are, respectively, the specific volume, velocity, pressure, specific internal energy, and the entropy of the fluid and $E = e + \frac{1}{2}u^2$ is the total energy of the fluid. If a shock is contained, the Rankine–Hugoniot conditions

$$\lambda(v - v_0) = -(u - u_0)$$
$$\lambda(u - u_0) = p - p_0 \qquad (6.5.12)$$
$$\lambda(E - E_0) = pu - p_0 u_0$$

must be satisfied at a point of discontinuity from (v_0, u_0, E_0) to (v, u, E) (see also Exercise 1.7.4).

The special feature of the system (6.5.11) is that the 1- and 3-characteristic fields are nonlinear, but the 2-characteristic field is always linearly degenerate. These equations (6.5.11) and (6.5.12) gained attention some time ago along with the equations of isentropic flow (see, e.g., Godunov, 1956) and, in recent years, some important findings have been reported.

By imposing the thermodynamic requirements, Wendroff (1972) has shown how to construct the solution of the Riemann problem for the system (6.5.11) when there is no restriction on the sign of the derivative $(\partial^2 p / \partial v^2)_s$. This has been further elaborated by Liu (1976), who has established the uniqueness theorem for the Riemann problem with arbitrarily large initial jumps by introducing the extended entropy condition (6.4.1) (see also Liu, 1975). Liu's result does not involve the assumption that the equation of state is polytropic and applies to systems with several space variables. The occurrence of a vacuum region is also discussed.

The Cauchy problem for the system (6.5.11) with the initial data

$$v(x, 0) = v_0(x), \qquad u(x, 0) = u_0(x), \qquad E(x, 0) = E_0(x) \qquad (6.5.13)$$

has been examined by Liu (1977a). He considers a polytropic gas for which the equation of state is

$$p(v, s) = a^2 v^{-1} \exp \frac{(\gamma - 1)s}{R} \qquad (6.5.14)$$

where a and R are positive constants and $1 \leq \gamma \leq 5/3$ and shows that the Cauchy problem given by Eqs. (6.5.11) and (6.5.13) has a weak solution under the assumption that $v_0(x)$, $u_0(x)$, and $s_0(x) \equiv s(e_0(x), p_0(x))$ have bounded total variation and $v_0(x)$ is bounded away from zero. Liu's proof is based on the finite difference scheme of Glimm (1965).

Liu (1977b) has also treated the piston problem for the polytropic gas with the equation of state (6.5.14). The initial data are prescribed as

$$u(x, 0) = u_0(x), \qquad p(x, 0) = p_0(x), \qquad s(x, 0) = s_0(x) \qquad (6.5.15)$$

for some given $u_0(x)$, $p_0(x)$, and $s_0(x)$ such that $p_0(x) \geq p_0 > 0$. Liu first considers the problem in the quardrant $x > 0$, $t > 0$ with boundary data of the form

$$u(0, t) = u^l(t), \qquad t \geq 0 \qquad (6.5.16)$$

or

$$p(0, t) = p^l(t), \qquad t \geq 0 \tag{6.5.17}$$

where $p^l(t) \geq p_0 > 0$, and proves that the mixed problem of Eqs. (6.5.11), (6.5.15), and (6.5.16) or (6.5.17) has a global solution defined for all $t \geq 0$, provided that the total variation of $\{u_0(x), p_0(x), u^l(t), p^l(t); x > 0, t > 0\}$ does not exceed a constant depending only on γ, $s_0(x)$, and the function $p(v, s)$. For the problem in the strip $0 < x < 1$, $t > 0$, Liu specifies the boundary data (6.5.16) and (6.5.17) together with the condition

$$p(1, t) = p^r(t), \qquad t \geq 0 \tag{6.5.18}$$

where $p^r(t) \geq p_0 > 0$ and shows that, given $t_0 > 0$, the solution exists for $t_0 \geq t \geq 0$ if $(\gamma - 1)$ times the total variation of $\{u_0(x), p_0(x), u^l(t), p^l(t), p^r(t); x > 0, t_0 > 0, t > 0\}$ does not exceed a constant depending on $s_0(x)$, the function $p(v, s)$, and t_0. For the double-piston problem, for which velocities are given on the boundary, similar results are obtained, provided that the boundary data do not allow the pistons to come together or the vacuum to appear in finite time. Finally, the existence theorem for the Cauchy problem [Eqs. (6.5.11) and (6.5.15)] is also established under the assumption that $(\gamma - 1)$ times the total variation of $\{u_0(x), p_0(x)\}$ does not exceed a constant depending only on $s_0(x)$ and the function $p(v, s)$.

The question of how solutions behave for large values of t was first discussed by Lax (1963). Considering initial values $\mathbf{u}_0(x)$ which are constant outside a finite interval:

$$\mathbf{u}_0(x) = \mathbf{u}^0 \qquad \text{for } |x| > \alpha \tag{6.5.19}$$

Lax figured that the corresponding solution will tend to the constant state $\mathbf{u}^0$. Based on studies on the equations of gas dynamics, he further conjectured that solutions approach a constant state like $1/\sqrt{t}$ and that the asymptotic form of the solution consists of m distinct N-waves each propagating at one of the m distinct characteristic speeds of the state $\mathbf{u}^0$; that is,

$$\mathbf{u}(x, t) = \begin{cases} \mathbf{u}^0 + \left(\dfrac{x}{t} - \lambda_k\right)\mathbf{r}_k & \begin{array}{l}\text{for } a_k\sqrt{t} < x - \lambda_k t < b_k\sqrt{t}, \\ k = 1, 2, \ldots, m\end{array} \\ \mathbf{u}^0 & \text{otherwise} \end{cases} \tag{6.5.20}$$

Here λ_k and $\mathbf{r}_k$ are normalized so that

$$\mathbf{r}_k^T \nabla \lambda_k = 1 \tag{6.5.21}$$

The quantities a_k and b_k are functionals of the initial data $\mathbf{u}_0(x)$ and invariants of the system. Apparently, Lax's system is genuinely nonlinear.

For systems that admit linearly degenerate fields, DiPerna (1977) has studied the Cauchy problem with the initial data having compact support. The initial data need not have small variations. DiPerna derives three main estimates which govern the decay of solutions and thereby shows that the maximum strength of all shocks which exist in the solution at time t decays to zero as t approaches infinity, that a solution cannot support at arbitrarily large times both large oscillations and weak shocks, and that a solution cannot support both large total variation and small oscillation at arbitrarily large times.

The same system has also been investigated by Liu (1977c). He takes the initial data (6.5.9) which are assumed to have bounded total variation so that the limiting values of $\mathbf{u}_0(x)$ at $x = \pm \infty$ exist; that is, $\mathbf{u}^l = \mathbf{u}_0(-\infty)$ and $\mathbf{u}^r = \mathbf{u}_0(+\infty)$, and compares the solution with that of the corresponding Riemann problem with

$$\mathbf{u}(x, 0) = \begin{cases} \mathbf{u}^l & \text{for } x < 0 \\ \mathbf{u}^r & \text{for } x > 0 \end{cases} \qquad (6.5.22)$$

It is shown that the solution of the Cauchy problem converges to the linear superposition of traveling waves[†], shocks, and simple waves and that the strength and speed of these waves depend only on the values of the data at infinity. With the type of initial data that he considers, it is not expected that the solution will converge at algebraic rates. If the initial data have compact support, however, he predicts that the solution converges to that of the corresponding Riemann problem at algebraic rates.

Indeed, Liu (1977d) has taken up the case of the initial data having compact support; that is, the case of Eq. (6.5.19) with $\mathbf{u}^0 = \mathbf{0}$. He first proves that the solution of a genuinely nonlinear system decays at the rate of $1/\sqrt{t}$ in the total variation norm when the initial data have a small total variation, and then uses the decay law to prove the conjecture of Lax discussed in the above. The main reason for the decay of the solution is that the simple waves expand and thus cancel with the shock waves. According to Liu (1977d), although the error term for interactions of waves of different families is of second order, most of the wave interactions take place in a short time because waves of different families propagate at different speeds due to the strict hyperbolicity of the system. The interaction of waves of the same family may take place at arbitrarily large time, but those interactions only produce an error term of third order. Hence the solution will become uncoupled after a short time if we ignore a third order error. Liu applies this observation to establish the decay law for general genuinely nonlinear systems, and subsequently, derives equations for characteristic speeds to show that the solution converges to the superposition of m N-waves.

This approach has been further extended by Liu (1977e) to systems in which each characteristic field is either genuinely nonlinear or linearly degenerate with initial data of the form

$$\mathbf{u}(x, 0) = \begin{cases} \mathbf{u}^l & \text{for } x < -\alpha \\ \mathbf{u}_0(x) & \text{for } -\alpha \leq x < \alpha \\ \mathbf{u}^r & \text{for } x > \alpha \end{cases} \qquad (6.5.23)$$

where $\mathbf{u}_0(x)$ is a measurable function, $\mathbf{u}^l$ and $\mathbf{u}^r$ are two constant vectors, and $\alpha > 0$. The main idea is to locate precisely the distribution of interactions in the (x, t)-plane and thus obtain a rate of uncoupling of the solution fine enough

[†] The term *traveling wave* is used for the progressive wave defined in Sec. 1.11.

for the genuine nonlinearity effect to dominate. Liu shows that the potential amount of interaction after time t depends primarily on the amount of interaction of waves of the same family before time t. This is due to the strict hyperbolicity of the system, which implies that waves of different families travel with different speeds and thus the amount of interaction of waves of different families decays at high rate.

Consequently, waves of genuinely nonlinear fields cancel or unite and thus attain a very simple form as time goes to infinity. The asymptotic behavior of these waves is determined by the solution of the corresponding Riemann problem. Waves of linearly degenerate fields behave almost linearly, and converge to traveling waves as time tends to infinity. In all cases, Liu obtains algebraic rates of convergence in the norm of total variation. If the solution $\mathbf{u}^*(x, t)$ of the corresponding Riemann problem consists of only shocks and contact discontinuities, the solution $\mathbf{u}(x, t)$ converges to the superposition of shock waves in $\mathbf{u}^*(x, t)$ and traveling waves at the rate close to t^{-2}. In general, the rate of convergence of $u(x, t)$ to shock waves and traveling waves is $t^{-3/2}$ and to simple waves $t^{-1/2}$. When $\mathbf{u}^l = \mathbf{u}^r$, L_1 convergence is also obtained in addition to convergence in the norm of total variation. Waves of linearly degenerate fields converge to traveling waves at the rate $t^{-1/2}$, and waves of genuinely nonlinear fields converge to N-waves at the rate $t^{-1/6}$, both in L_1 norm.

More recently, Liu (1981) has also treated the large time behavior of the solution of the Cauchy problem for not necessarily genuinely nonlinear systems. With respect to the waves of the same family in interaction, he notes that the interaction would have an effect proportional to the angle between the waves before they come into interaction and further that the total amount of expansion waves may increase by the combination of two waves of the same family and the same direction, whereas it may decrease by the cancellation of two waves of the opposite direction. These phenomena are not observed in systems for which each characteristic field is either genuinely nonlinear or linearly degenerate. Using these results Liu treats the local interaction of elementary waves and shows that the interaction of waves of different families brings about a change in the wavelength and the wave pattern of the incoming waves. He then continues to show that the solution converges to the elementary waves contained in the solution of the Riemann problem with the data $\mathbf{u}^l = \mathbf{u}_0(-\infty)$ and $\mathbf{u}^r = \mathbf{u}_0(+\infty)$. The solution may include discontinuities across countable Lipschitz continuous curves, but in other regions it is continuous. The waves that come into interaction are compression waves and give rise to elementary waves which can be determined by the right and left limits of the solution.

One way to examine the structure of shocks is, of course, to study the viscous conservation laws and this approach was treated in some detail in Secs. 7.6 and 7.10 of Vol. 1 for single equations and in Sec. 1.11 for pairs of equations. With respect to systems of m equations, Kulikovskii (1962) considered the viscous conservation laws

$$\frac{\partial \mathbf{u}}{\partial x} + \frac{\partial}{\partial t}\mathbf{f}(\mathbf{u}) = \mu\frac{\partial}{\partial x}\left(\mathbf{B}(\mathbf{u})\frac{\partial \mathbf{u}}{\partial x}\right), \qquad -\infty < x < \infty, \quad t > 0 \qquad (6.5.24)$$

where the corresponding nonviscous system (6.5.1) is genuinely nonlinear, and by assuming that the viscosity matrix **B** is positive definite, he established the following theorem:

> **THEOREM 6.4:** Suppose that the matrix **B** is positive definite. If $\mathbf{u}^l$ and $\mathbf{u}^r$ are two constant states, close to each other, and satisfy the jump condition (6.4.23) with $\bar{\lambda} = \lambda_k > \lambda_k(\mathbf{u}^r)$, Eq. (6.5.24) has a solution of the type
>
> $$\mathbf{u}_k(x, t) = \mathbf{u}_k(x - \lambda_k t) \qquad (6.5.25)$$
>
> taking on values of $\mathbf{u}^l$ at $x = -\infty$ and $\mathbf{u}^r$ at $x = +\infty$. If $\mathbf{B} \to \mathbf{0}$, the continuous solution given by Eq. (6.5.25) tends to the corresponding k-shock.

For the same system Foy (1964) applied a bifurcation technique based on an expansion procedure to obtain essentially the same conclusion as Kulikovskii. Although the formal proof was given for the identity viscosity matrix, Foy suggested that a constant matrix or a diagonal matrix would also be suitable for the viscosity matrix **B**.

A class of matrices suitable for $\mathbf{B}(\mathbf{u})$ has been discussed by Smoller and Conley (1972), who have also shown that Theorem 6.4 could be proven in a rather simple way if $\mathbf{f}$ is a gradient. Their conclusion is as follows: Let $\mathbf{u}^l$ and $\mathbf{u}^r$ be two states connected by a shock of system (6.5.1) and suppose that **B** is any positive definite matrix which leaves the three subspaces spanned by $(\mathbf{r}_1, \ldots, \mathbf{r}_{k-1})$, $(\mathbf{r}_k)$, $(\mathbf{r}_{k+1}, \ldots, \mathbf{r}_m)$ invariant; then **B** is suitable if the shock is sufficiently weak.

Conley and Smoller (1971) have considered modifications of Eq. (6.5.1) in the form

$$\frac{\partial \mathbf{u}}{\partial t} + \frac{\partial}{\partial x}\mathbf{f}(\mathbf{u}) = \alpha\mu\frac{\partial^2 \mathbf{u}}{\partial x^2} + \beta\mu\frac{\partial^3 \mathbf{u}}{\partial x^3} \qquad (6.5.26)$$

in which the second- and third-order derivative terms represent the dissipation and dispersion, respectively. For genuinely nonlinear systems it is shown that weak shocks can be realized as limits of *progressive wave solutions*[†] of the associated dissipative system; that is, Eq. (6.5.26) with $\beta = 0$. In the case $\mathbf{f}(\mathbf{u}) = \nabla\phi(\mathbf{u})$, however, weak shocks are never realized as limits of progressive wave solutions of the associated dispersive system; that is, Eq. (5.1.26) with $\alpha = 0$.

The nonlinear stability of progressive waves (or equivalently traveling waves) has most recently been taken up by Liu (1985). He has coined the term *viscous shock wave* to represent the traveling wave solution of the viscous conservation laws (6.5.24). Liu considers a case in which each characteristic field of the corresponding nonviscous system is either genuinely nonlinear or linearly degenerate and shows that when the initial data are a perturbation of

---

[†] For the definition of a progressive wave solution, see Eq. (1.11.24) and the subsequent paragraph in Sec. 1.11.

weak viscous shock waves, the solution tends to appropriately translated viscous shock waves in the limit as time goes to infinity. The perturbation is not necessarily weak in comparison to viscous shock waves. The motivation here comes from the fact that the system is conservative, and a general solution is ultimately decomposed into normal modes, that is, viscous shock waves, linear and nonlinear diffusion waves, and an error term. Liu then combines the characteristic method with the energy method, which is a standard technique for parabolic systems, to establish the nonlinear stability of viscous shock waves.

Numerical methods for the system (6.5.1), employing finite difference schemes, have been treated and applied by various authors and has received more attention in recent years. In Sec. 3.2 we briefly discussed how to construct the solution of a system of conservation laws of the form (3.2.1) by a finite difference scheme using strips and characteristic directions. This was originally due to Courant and Friedrichs (1948) and further elaborated by Courant et al. (1952), who also described another finite difference scheme based on a rectangular net. These authors determine criteria for selecting the size of mesh widths, the type of difference quotients, and the proper number of decimal places to ensure convergence of the numerical procedure and prove that either scheme provides an approximate numerical solution accurate within an error of the same order of magnitude as the mesh width of the net.

In their pioneering work, Lax and Wendroff (1960) developed the scheme that has become one of the standard tools for the numerical calculations of compressible gas dynamics. With respect to the system (6.5.1) they define a vector-valued function of two vector arguments as

$$\mathbf{g(a, b)} = \frac{1}{2}\{\mathbf{f(a)} + \mathbf{f(b)}\} + \frac{1}{2}\gamma A^2(\mathbf{b} - \mathbf{a}) \qquad (6.5.27)$$

in which γ abbreviates the quotient of the time and space increments, i.e.,

$$\gamma = \frac{\Delta t}{\Delta x} \qquad (6.5.28)$$

and take the difference analog of Eq. (6.5.1) in the form

$$\frac{\Delta \mathbf{u}}{\Delta t} = \frac{\Delta \mathbf{g}}{\Delta x} \qquad (6.5.29)$$

where $\Delta \mathbf{u}$ denotes the forward difference

$$\Delta \mathbf{u} = \mathbf{u}(x, t + \Delta t) - \mathbf{u}(x, t) \qquad (6.5.30)$$

and $\Delta \mathbf{g}$ the symmetric space difference

$$\Delta \mathbf{g} = \mathbf{g}\left(x + \frac{1}{2}\Delta x\right) - \mathbf{g}\left(x - \frac{1}{2}\Delta x\right) \qquad (6.5.31)$$

in which

$$\mathbf{g}\left(x + \frac{1}{2}\Delta x\right) = \mathbf{g}(\mathbf{u}_0, \mathbf{u}_1) \qquad (6.5.32)$$

$$g\left(x - \frac{1}{2}\Delta x\right) = g(\mathbf{u}_{-1}, \mathbf{u}_0) \qquad (6.5.33)$$

where $\mathbf{u}_k$, $k = -\infty, \ldots, +\infty$, represents the value of $\mathbf{u}(x, t)$ at the lattice points $x + k\,\Delta x$ and at time t.

From Eq. (6.5.29) Lax and Wendroff determine

$$\mathbf{u}(x, t + \Delta t) = \mathbf{u}(x, t) + \gamma\,\Delta g \qquad (6.5.34)$$

and show that the truncation error in the difference scheme (6.5.34) is $O(\Delta^3)$ plus terms which are $O(|\mathbf{a} - \mathbf{b}|^2)$ for $(\mathbf{a} - \mathbf{b})$ small. Given the initial values of $\mathbf{u}$, therefore, one can determine successively the values of $\mathbf{u}$ at all times which are integer multiples of Δt.

In their development the quantity $\mathbf{A}^2$ in Eq. (6.5.27) is taken as

$$\frac{1}{2}\{\mathbf{A}^2(\mathbf{a}) + \mathbf{A}^2(\mathbf{b})\}$$

for the sake of symmetry; any other choice would make a difference that is quadratic in $(\mathbf{a} - \mathbf{b})$. Denoting the function (6.5.27) by g_0, they redefine the function $\mathbf{g}$ in the form

$$\mathbf{g} = g_0 + \frac{1}{2}\mathbf{Q}(\mathbf{a}, \mathbf{b})(\mathbf{b} - \mathbf{a}) \qquad (6.5.35)$$

where $\mathbf{Q}(\mathbf{a}, \mathbf{b})$ is a matrix that vanishes when its two vector arguments are identical. Substituting this for $\mathbf{g}$ in Eq. (6.5.34) gives

$$\mathbf{u}(x, t + \Delta t) = \mathbf{u}(x, t) + \gamma\Delta'\,\mathbf{f} + \frac{1}{2}\gamma^2\,\Delta\mathbf{A}^2\,\Delta\mathbf{u} + \frac{1}{2}\gamma\,\Delta\mathbf{Q}\,\Delta\mathbf{u} \qquad (6.5.36)$$

in which Δ' denotes the operator $\frac{1}{2}[T(\Delta x) - T(-\Delta x)]$ and Δ the operator $T(\frac{1}{2}\Delta x) - T(-\frac{1}{2}\Delta x)$, $T(s)$ representing translation of the independent variable by the amount s. The quantity $\mathbf{Q}$ is called the artificial viscosity, since it appears in the difference equations in a way that is similar to the artificial viscosity term. Lax and Wendroff demonstrate that the finite difference scheme (6.5.36) is the best one which has the smallest truncation error, in which the discontinuities are confined to a narrow band of two to three meshpoints, and which is stable under a mild strengthening of the Courant–Friedrichs–Lewy criterion:

$$\gamma^{-1} \geq |\lambda|_{\text{max}} \qquad (6.5.37)$$

where $|\lambda|_{\text{max}}$ is the largest eigenvalue of $\mathbf{A}$ at any point within the relevant range of values.

The initial value problem for hyperbolic systems of quasilinear equations involving many space variables has been treated by Shampine and Thompson (1970). These authors discuss the forward time difference scheme which is conditionally stable and the backward time difference scheme which is unconditionally stable, and illustrates the application by treating single equations with two independent variables.

More recently, Majda and Ralston (1979) have suggested an independent difference equation whose solution connecting two constant states is called a *discrete shock profile*. A discrete shock profile is an entropy satisfying shock if the entropy condition is satisfied. Otherwise, it is called an entropy violating shock. These authors establish the existence of discrete shock profiles for a difference scheme approximating a genuinely nonlinear system of conservation laws and necessary and sufficient conditions which give rise to entropy satisfying or entropy violating shocks. They also show some theoretical examples of an entropy violating shock profile.

Experimental and theoretical examples have been reported in which solutions by the Lax–Wendroff scheme converged to a weak solution that violates the entropy condition. To resolve this, Majda and Osher (1979) have developed the modified Lax–Wendroff scheme which has the important feature that if solutions of the difference scheme converge, the limiting solution satisfies the entropy condition. In addition, this modified scheme retains the desirable computational advantages; that is, it is of conservation form, a three-point scheme and has second-order accuracy on smooth solutions.

These authors define the standard Lax–Wendroff difference scheme as

$$\mathbf{u}_j^{n+1} = \mathbf{u}_j^n - \frac{1}{2} \gamma_n \Delta_0 \mathbf{f}(\mathbf{u}_j^n) + \frac{1}{2} \gamma_n^2 \Delta_- (\mathbf{A}_{j+1/2}^n \Delta_+ \mathbf{f}(\mathbf{u}_j^n)) \qquad (6.5.38)$$

where

$$\mathbf{u}_j^n = \mathbf{u}\left(\sum_{l=0}^{n-1} \Delta t_{l,j} \, \Delta x \right) \qquad (6.5.39)$$

$$\mathbf{A}_{j+1/2}^n = \mathbf{A}\left(\frac{\mathbf{u}_{j+1}^n + \mathbf{u}_j^n}{2} \right) \qquad (6.5.40)$$

$$\gamma_n = \frac{\Delta t_n}{\Delta x} \qquad (6.5.41)$$

and

$$\Delta_+ \mathbf{u}_j = \mathbf{u}_{j+1} - \mathbf{u}_j, \qquad \Delta_0 \mathbf{u}_j = \mathbf{u}_{j+1} - \mathbf{u}_{j-1}, \qquad \Delta_- \mathbf{u}_j = \mathbf{u}_j - \mathbf{u}_{j-1} \qquad (6.5.42)$$

The modified scheme has the form

$$\mathbf{u}_j^{n+1} = \mathbf{u}_j^n - \frac{1}{2} \gamma_n \Delta_0 \mathbf{f}(\mathbf{u}_j^n) + \frac{1}{2} \gamma_n^2 \Delta_- \left[\frac{\Delta_+ \mathbf{f}(\mathbf{u}_j^n)}{\Delta_+ \mathbf{u}_j^n} \Delta_+ \mathbf{f}(\mathbf{u}_j^n) \right]$$

$$+ \gamma_n \Delta_- \left[C\theta\left(\frac{|\Delta_+ \mathbf{u}_j^n|}{h^\alpha} \right) \Delta_+ \mathbf{A}(\mathbf{u}_j^n) \Delta_+ \mathbf{u}_j^n \right] \qquad (6.5.43)$$

where $h = \Delta x$, C is an appropriate constant, $\alpha > 0$, and θ is a function defined by

$$\theta(s) = \begin{cases} 0 & \text{for } |s| < 1 \\ 1 & \text{for } |s| \geq 1 \end{cases} \qquad (6.5.44)$$

When compared to Eq. (6.5.38), this variant contains two modifications. One is the replacement of $A_{j+1/2}^n$ by $\Delta_+ f(u_j^n)/\Delta_+ u_j^n$ and the other is the addition of a nonlinear artificial viscosity term which is the last term of Eq. (6.5.44). The role of the function θ is to turn on the artificial viscosity only in regions of rapid discrete transition.

In their study, the time steps Δt_n are chosen to satisfy the Courant–Friedrichs–Lewy criterion in the form

$$\gamma_n \max |A(u_j^n)| \le \epsilon_0 \le 1 \qquad (6.5.45)$$

The result is embodied in the following theorem:

> **THEOREM 6.5:** Suppose that u_j^n is computed by the modified Lax–Wendroff scheme (6.5.43) with ϵ_0 sufficiently small, $C(\epsilon_0)$ chosen appropriately, and $1 \ge \alpha \ge \frac{1}{3}$. Assume that u_j^n converges boundedly almost everywhere to u. Then u is a weak solution of Eq. (6.5.1) and u satisfies the entropy condition (6.4.38) with
>
> $$U(u) = \frac{1}{2} u^T u$$
>
> $$F(u) = \int_0^u s A(s)\, ds$$
>
> In particular, wherever u is piecewise smooth, the shock conditions (6.5.7) and (6.5.8) are satisfied.

The stability in the total variation norm and convergence for general finite difference scheme has been recently treated by DiPerna (1982) for hyperbolic systems in which each characteristic field is either genuinely nonlinear or linearly degenerate.

6.6 Solution Scheme for a Moving-Bed Problem

Returning to the problem of adiabatic adsorption in a continuous countercurrent moving-bed system, we have the initial and boundary conditions specified as in Eq. (6.1.16). Let us suppose at the moment that the contacting region extends indefinitely beyond both boundaries. Clearly, the solution will be given by a pair of wave sets, one developing from the origin and the other from the point $x = 1$ on the τ axis. Each wave set is composed of m waves of different families which are partitioned by constant states. The individual wave may be a centered simple wave, a shock, or a combined wave discussed in Secs. 6.3 and 6.4. The totality of those m waves and $(m - 1)$ constant states in between constitute the range of influence of the initial discontinuity located at $x = 0$ or at $x = 1$.

Clearly, the situation will be similar to that shown in Fig. 5.3. After a finite period of time, the two ranges of influence are bound to collide with each other, and afterward the states are influenced by the boundary conditions and the initial condition at the same time. Here the waves emanating from different points of discontinuity will be engaged in a sequence of interactions, and thus the solution will determined by analyzing all the wave interactions involved. In Chapters 4 and 5 we have been concerned with the isothermal system, and there it was possible to make a complete analysis of wave interactions because the characteristic parameters could be determined explicitly. Although we have discussed some of the features of wave interactions in Sec. 6.5, the wave interactions in general are not amenable to analytical treatment. Here we do not attempt to treat the general problem but rather confine our discussion to a simplified problem.

We now claim that the case with $c_i^0 = c_{i,eq}^b$ is representative of many practical situations, where $\{c_{i,eq}^b\}$ denotes the state that would be in equilibrium with the state $\{n_i^b\}$. Since there is no discontinuity at $x = 1$ on the x axis, the solution inside the domain is completely determined by the wave set developing from the origin.

We recall from Sec. 6.3 that the wave set developing from the origin rotates counterclockwise as the parameter μ increases, maintaining the overall features since the image of the solution in the hodograph space is independent of μ. For a certain value of μ, therefore, one can readily think of a situation when the region of the wave set is divided into two parts by the τ axis, one part forward-facing and the other facing backward. It is then obvious that only the forward-facing part is meaningful to the solution inside the domain. The backward-facing part will collapse on the boundary $\mathscr{B}^a$ at $x = 0$ and form a boundary discontinuity there.

Since the image of the solution in the hodograph space is given by a path connecting the initial state $\{c_i^0\}$ to the feed state $\{c_i^a\}$, the argument above implies that the path is divided into two portions by the state $\{c_i^{0+}\}$, which is present just inside the boundary $\mathscr{B}^a$. The portion between $\{c_i^{0+}\}$ and $\{c_i^0\}$ generates the forward-facing part of the wave set and thus determines the solution inside the domain in such a way that Eq. (6.1.12) is satisfied. On the other hand, the portion between $\{c_i^a\}$ and $\{c_i^{0+}\}$ gives the boundary discontinuity at $x = 0$ and determines the state $\{n_i^a\}$ to satisfy Eq. (6.1.17).

When each of the forward-facing waves reaches the boundary $\mathscr{B}^b$ at $x = 1$, the mass balance is no longer described by Eq. (6.1.12) but by Eq. (6.1.18). This implies that the boundary $\mathscr{B}^b$ also bears a discontintuity and hence the state $\{c_i^b\}$ is determined by applying Eq. (6.1.18). When the tail of the uppermost wave reaches the boundary $\mathscr{B}^b$, no further change will be observed and the system obtains a steady state. At this stage the state inside the domain is given by $\{c_i^{0+}\}$.

From the argument above it is evident that the prime issue is how to determine the state $\{c_i^{0+}\}$ for a given set of initial and boundary conditions and for a given value of μ. Since the state $\{c_i^{0+}\}$ corresponds to a single point on the image of the solution in the hodograph space and the image is independent of μ, one can regard the image as the locus of the state $\{c_i^{0+}\}$ under the variation of μ. As μ increases, therefore, the point $\{c_i^{0+}\}$ moves along the image from the point $\{c_i^a\}$ to the point $\{c_i^0\}$. When this picture is related to the physical plane portrait of the solution, one can easily imagine various situations to be encountered and, just as discussed in Sec. 5.3, classify the possibilities into four categories by defining cuts along the μ spectrum whenever one side of a simple wave or of a combined wave or a shock path itself becomes vertical.

Without loss of generality, we shall assume that both the directional derivatives

$$\beta_k = \left(\frac{\mathscr{D}n_i}{\mathscr{D}c_i}\right)_k \tag{6.6.1}$$

and the ratios

$$\tilde{\beta}_k = \left(\frac{[n_i]}{[c_i]}\right)_k \tag{6.6.2}$$

for $k = 1, 2, \ldots, m$, are all nonnegative for the relevant values of $\{c_i\}$. Here again the constant state between the k-wave and the $(k-1)$-wave will be called the k-constant state and designated by the superscript (k), where $c_i^{(m+1)} = c_i^a$ and $c_i^{(1)} = c_i^0 = c_{i,\text{eq}}^b$. With this scheme of notations the above-mentioned cuts for the k-wave, $k = 1, 2, \ldots, m$, can be expressed as follows.

If the k-wave is a simple wave, the corresponding region is covered by a family of straight characteristics C^k whose slope is

$$\sigma_k = \frac{1 + \nu\beta_k}{1 - \mu\beta_k} \tag{6.6.3}$$

Thus the two cuts on the μ spectrum are given by

$$\mu_{k*} = \frac{1}{\beta_k^{(k+1)}} \tag{6.6.4}$$

and

$$\mu_k^* = \frac{1}{\beta_k^{(k)}} \tag{6.6.5}$$

In case of a k-shock, we have

$$\tilde{\sigma}_k = \frac{1 + \nu\tilde{\beta}_k}{1 - \mu\tilde{\beta}_k} \tag{6.6.6}$$

for the direction of the shock path, and hence there is only one cut defined by

$$\bar{\mu}_k = \frac{1}{\tilde{\beta}_k} = \frac{c_i^{(k+1)} - c_i^{(k)}}{n_i^{(k+1)} - n_i^{(k)}} \tag{6.6.7}$$

This also holds for a k-contact discontinuity connecting the $(k+1)$-constant state directly to the k-constant state.

For a combined wave in which a k-simple wave is joined by a k-semishock on the right-hand side, we have

$$\mu_k^* = \frac{1}{\bar{\beta}_k} = \frac{c_i^J - c_i^{(k)}}{n_i^J - n_i^{(k)}} \tag{6.6.8}$$

in which the superscript J denotes the state at the junction of the two waves. The other cut μ_{k*} is given by Eq. (6.6.4). If the order is reversed between the simple wave and the semishock, we must have

$$\mu_{k*} = \frac{1}{\bar{\beta}_k} = \frac{c_i^{(k+1)} - c_i^J}{n_i^{(k+1)} - n_i^J} \tag{6.6.9}$$

and μ_k^* is given by Eq. (6.6.5).

In general, there are $2m$ cuts so defined along the μ spectrum and these give rise to $(2m + 1)$ disjoint ranges of μ. Clearly, the structure of the solution for each range of μ will be similar to that discussed in Sec. 5.3 for isothermal adsorption in a semi-infinite column. In the following, therefore, we shall discuss briefly the solution scheme for each range of μ.

Lower range of μ

Here the parameter μ spans the interval

$$0 \leq \mu < \mu_{m*} \tag{6.6.10}$$

All the waves are forward-facing and the whole range of influence of the initial discontinuity at the origin falls inside the domain. Since there is no discontinuity at the boundary $\mathcal{B}^a$, we have $c_i^{0+} = c_i^a$ and Eq. (6.1.17) simply gives $n_i^a = n_i^{0+} = n_{i,\text{eq}}^a$. The boundary $\mathcal{B}^a$ is inward space-like, so active in its data-transmitting character. The situation is shown schematically in Fig. 6.7(a).

On the other hand, the boundary $\mathcal{B}^b$ is outward space-like and remains passive. Hence it starts to bear a discontinuity when the head of the 1-wave reaches there. The boundary discontinuity is governed by Eq. (6.1.18). While each of the m waves arrives at the boundary $\mathcal{B}^b$ successively, c_i^{1-} varies from c_i^0 to c_i^a and, accordingly, c_i^b changes from $c_i^0 + \mu(n_i^b - n_i^0)$ to $c_i^a + \mu(n_i^b - n_i^a)$.

After the tail of the m-wave reaches the boundary $\mathcal{B}^b$ at

$$\tau = \tau_s = \frac{1 + \nu/\mu_{m*}}{1 - \mu/\mu_{m*}} \tag{6.6.11}$$

no further change is expected and the system obtains a steady state. At this stage we have

$$n_i^a = n_{i,\text{eq}}^a$$
$$c_i = c_i^a \qquad \text{for} \quad 0 < x < 1 \tag{6.6.12}$$
$$c_i^b = c_i^a + \mu(n_i^b - n_{i,\text{eq}}^a)$$

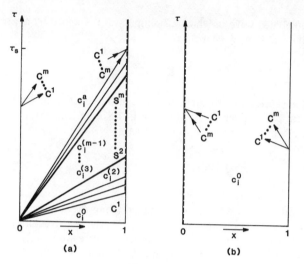

Figure 6.7

Upper range of μ

This range corresponds to the portion

$$\mu_1^* \le \mu < \infty \qquad (6.6.13)$$

Every wave that is expected to develop from the origin would be directed backward and hence the initial discontinuity at the origin has no influence over the solution inside the domain. This implies that the boundary $\mathscr{B}^a$ remains passive and bears a discontinuity for all time. Clearly, we have $c_i^{0+} = c_i^0$, and from Eq. (6.1.17), $n_i^a = n_i^0 + (c_i^a - c_i^0)/\mu$.

The boundary $\mathscr{B}^b$ is inward space-like as shown in Fig. 6.7(b) and thus the solution remains continuous across the boundary. It is evident that the system immediately attains a steady state which may be described as

$$n_i^a = n_i^0 + \frac{c_i^a - c_i^0}{\mu}$$

$$c_i = c_i^0 \qquad \text{for} \quad 0 < x < 1 \qquad (6.6.14)$$

$$c_i^b = c_i^0$$

k-intermediate range of μ

In this range the parameter μ covers the interval

$$\mu_k^* \le \mu < \mu_{k-1,*} \qquad (6.6.15)$$

for $k = 2, 3, \ldots, m$. Only the first $(k - 1)$ waves propagate into the domain

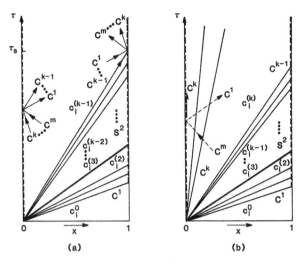

Figure 6.8

as depicted in Fig. 6.8(a), whereas the other $(m - k + 1)$ waves collapse at the boundary $\mathcal{B}^a$. Therefore, the solution in the region adjacent to the τ axis is given by the k-constant state; that is, $c_i^0 = c_i^{(k)}$, and from Eq. (6.1.17) we obtain $n_i^a = n_i^{(k)} + (c_i^a - c_i^{(k)})/\mu$. The boundary $\mathcal{B}^a$ bears a discontinuity from the beginning. It is now time-like with the first $(k - 1)$ characteristics, $C^1 \sim C^{k-1}$, directed forward and the remaining $(m - k + 1)$ characteristics, $C^k \sim C^m$ backward.

When the head of the 1-wave reaches the boundary $\mathcal{B}^b$, the solution begins to become discontinuous there. As the subsequent waves reach the boundary $\mathcal{B}^b$, c_i^{1-} changes from c_i^0 to $c_i^{(k)}$ and, correspondingly, c_i^b varies from $c_i^0 + \mu(n_i^b - n_i^0)$ to $c_i^{(k)} + \mu(n_i^b - n_i^{(k)})$. The boundary $\mathcal{B}^b$ is also time-like as shown in Fig. 6.8(a) and we note that the two boundaries are complementary to each other in their data-transmitting character.

Clearly, the system attains a steady state when

$$\tau = \tau_s = \frac{1 + \nu/\mu_{k-1,*}}{1 - \mu/\mu_{k-1,*}} \qquad (6.6.16)$$

at which time the tail of the $(k - 1)$-wave reaches the boundary $\mathcal{B}^b$. The steady state is described by

$$n_i^a = n_i^{(k)} + \frac{c_i^a - c_i^{(k)}}{\mu}$$

$$c_i = c_i^{(k)} \qquad \text{for} \quad 0 < x < 1 \qquad (6.6.17)$$

$$c_i^b = c_i^{(k)} + \mu(n_i^b - n_i^{(k)})$$

k-transition range of μ

This range corresponds to the portion

$$\mu_{k*} \leq \mu < \mu_k^* \qquad (6.6.18)$$

for $k = 1, 2, \ldots, m$. Here the k-wave is partitioned into two parts by the τ axis and only the right-hand part propagates into the domain along with the first $(k - 1)$ waves. Hence there exists a k-characteristic C^k that coincides with the vertical line $x = 0+$ and the state $\{c_i^{0+}\}$ borne on this particular characteristic C^k becomes stationary there. This implies that the system would not reach a steady state in finite time if μ belongs to this range.

The boundary $\mathcal{B}^a$ bears a discontinuity from the beginning, whereas the boundary $\mathcal{B}^b$ becomes discontinuous when the head of the 1-wave reaches there as illustrated in Fig. 6.8(b). The former is k-characteristic while the latter is time-like with the first k characteristics directed forward and the next $(m - k)$ characteristics backward.

If the k-wave is a discontinuous type, either a shock or a contact discontinuity, the k-transition range of μ degenerates to a single point and the situation becomes the same as for the k-intermediate range of μ. In case $k = m$, we have an m-shock or an m-contact discontinuity and thus equilibrium is established across the boundary $\mathcal{B}^a$.

The discussion of this section may be summarized as follows. For a given set of data, the μ spectrum is composed of one upper range, one lower range, $(k - 1)$ intermediate ranges, and k transition ranges. The system attains a steady state in finite time unless μ belongs to one of the transition ranges. The steady state inside the domain is completely determined by the feed states and there are $(m + 1)$ different ones. Each of these steady states corresponds to one of the upper range, the lower range, and the intermediate ranges of μ. The states of the outgoing streams are determined by the feed states and the parameter μ. On the other hand, the time τ_s, at which the steady state is attained, is dependent on the given set of data and parameters ν and μ.

6.7 Adiabatic Adsorption of a Single Solute

Here and in the subsequent sections we would like to consider the adsorption of one or two solutes in an adiabatic column, which uses either a fixed bed or a countercurrently moving bed, and illustrate how to apply the theoretical development of this chapter. For this purpose we shall treat several numerical examples.

In this section we shall be concerned exclusively with the problem of single-solute adsorption, for which Eq. (6.1.12) gives rise to a pair of quasilinear equations, one mass balance equation, and one energy balance equation. Since the mathematical problem of a pair of equations was fully explored in a general context in Chapter 1, it is certain that the line of approach developed there can be directly applied to treat the present problem, just as we treated the example related to the polymer flood in Sec. 1.8.

For the Riemann problem, however, we believe that the line of approach established in this chapter would be more convenient to apply even in the case with a pair of equations. Therefore, treating the problem of single-solute adsorption along this line will not only be instructive to demonstrate the application but also have the merit of making a bridge to the discussion in Sec. 6.8, which is involved with a system of three equations.

Since we are concerned with the Riemann problem for a pair of equations, it is appropriate to begin with the fundamental differential equation (6.2.7) for $m = 2$. For this we shall introduce Eq. (6.2.8) and write

$$n_{1,1} + n_{1,2}\left(\frac{dc_2}{dc_1}\right)_k = n_{2,1}\left(\frac{dc_1}{dc_2}\right)_k + n_{2,2}$$

or

$$n_{2,1}\left(\frac{dc_1}{dc_2}\right)_k^2 - (n_{1,1} - n_{2,2})\left(\frac{dc_1}{dc_2}\right)_k - n_{1,2} = 0 \qquad (6.7.1)$$

for $k = 1, 2$, in which we have

$$c_1 = c, \qquad c_2 = T \qquad (6.7.2)$$

$$n_1 = n = n(c, T) \qquad (6.7.3)$$

and

$$n_2 = \kappa - \frac{(-\Delta H)}{C_f}n(c, T) \qquad (6.7.4)$$

where

$$\kappa = \frac{C_s}{C_f} = \text{heat capacity ratio} \qquad (6.7.5)$$

For a physically relevant isotherm $n = n(c, T)$, both $n_{1,2}$ and $n_{2,1}$ are negative [see Eqs. (6.1.3) and (6.1.4)]. Hence Eq. (6.7.1) always has two distinct real roots for $(dc_1/dc_2)_k$: one is positive and the other negative.

We shall now introduce the familiar Langmuir adsorption isotherm

$$n = \frac{NKc}{1 + Kc} \qquad (6.7.6)$$

in which the saturation value N may be regarded as being fixed. The parameter K is no longer a constant but is given by a strong function of the temperature T, which we will assume has the form

$$K = K(T) = K_0 T^{1/2}e^{-\Delta H/RT} \qquad (6.7.7)$$

where K_0 is a constant and $\Delta H\,(< 0)$ denotes the enthalpy change of adsorption per mole of solute. We note that k decreases with increasing temperature and thus

$$\frac{\partial n}{\partial T} = \frac{Nc}{(1 + Kc)^2}\frac{dK}{dT} < 0 \tag{6.7.8}$$

Substituting Eqs. (6.7.6) and (6.7.7) into Eq. (6.7.1) gives

$$\zeta_1 = \left(\frac{dc}{dT}\right)_1 = a + \sqrt{a^2 + b} \geq 0 \tag{6.7.9}$$

for the direction of Γ^1 and

$$\zeta_2 = \left(\frac{dc}{dT}\right)_2 = a - \sqrt{a^2 + b} \leq 0 \tag{6.7.10}$$

for the direction of Γ^2, where

$$a = a(c, T) = \frac{1}{2}\left\{\frac{C_s(1 + Kc)^2}{(-\Delta H)NK} + \left(\frac{-\Delta H}{RT} - \frac{1}{2}\right)\frac{c}{T} - \frac{C_f}{-\Delta H}\right\} \tag{6.7.11}$$

and

$$b = b(c, T) = \frac{C_f}{-\Delta H}\left\{\frac{-\Delta H}{RT} - \frac{1}{2}\right\}\frac{c}{T} \tag{6.7.12}$$

With the assignment above the Γ^1 and Γ^2 here correspond, respectively, to the Γ_+ and Γ_- of Chapters 1 and 2. The network of the Γ^1 and Γ^2 in the hodograph plane can be constructed by integrating Eqs. (6.7.9) and (6.7.10).

The T axis on which $c = 0$ deserves special attention since we find

$$a = a(0, T) = \frac{C_f}{2(-\Delta H)NK}(\kappa - NK) \tag{6.7.13}$$

and

$$b = b(0, T) = 0 \tag{6.7.14}$$

This implies that we have $\zeta_1 = 0$ and $\zeta_2 < 0$ if $NK > \kappa$, but $\zeta_1 > 0$ and $\zeta_2 = 0$ if $NK < \kappa$. Hence the T axis consists of two portions: one for $NK > \kappa$ is a Γ^1 and the other for $NK < \kappa$ is a Γ^2. The point B on the T axis at which we obtain $NK = \kappa$ is equivalent to the watershed point of Sec. 2.2.

Now the characteristic directions in the physical plane can be determined by using Eq. (6.2.12) as

$$\sigma_k = \left(\frac{d\tau}{dx}\right)_{\omega_k} = \frac{1 + \nu\beta_k}{1 - \mu\beta_k} \tag{6.7.15}$$

for $k = 1, 2$, where

$$\beta_k = \left(\frac{\mathcal{D}n_2}{\mathcal{D}c_2}\right)_k = \kappa + \frac{(-\Delta H)NK}{C_f(1 + Kc)^2}\left\{\left(\frac{-\Delta H}{RT} - \frac{1}{2}\right)\frac{c}{T} - \zeta_k\right\} \tag{6.7.16}$$

for $k = 1, 2$. Since $\beta_1 < \beta_2$, we have the inequality

$$\frac{1}{\sigma_1} > \frac{1}{\sigma_2} \qquad (6.7.17)$$

which is consistent with Eq. (6.2.14).

We note that along the T axis of the hodograph plane, Eq. (6.7.16) gives

$$\beta_k = \kappa = \text{constant} \qquad (6.7.18)$$

and hence

$$\sigma_k = \frac{1 + \nu\kappa}{1 - \mu\kappa} \qquad (6.7.19)$$

This implies that if $c = 0$, the k-characteristic field becomes linearly degenerate. Hence we expect to see a contact discontinuity across which the temperature jumps while the concentration remains zero. This will be referred to as a *pure thermal wave* in the sequel.

If the solution contains one or more discontinuities, we must consider the compatibility relation (6.2.18), which reads

$$\frac{n - n^0}{c - c^0} = \kappa - \frac{-\Delta H}{C_f} \frac{n - n^0}{T - T^0} \qquad (6.7.20)$$

where the superscript 0 denotes a fixed state, say, the state on one side of a discontinuity. Introducing the Langmuir isotherm (6.7.6) for n and making rearrangement, we obtain

$$p(c - c^0)^2 - 2q(c - c^0) - r = 0 \qquad (6.7.21)$$

where

$$p = \frac{-\Delta H}{C_f(T - T^0)} - \frac{\kappa}{N}(1 + K^0 c^0) \qquad (6.7.22)$$

$$2q = p\left(\frac{1}{K} + c^0\right) + \frac{(-\Delta H)(1 + K^0 c^0)}{C_f K(T - T^0)} - 1 \qquad (6.7.23)$$

and

$$r = c^0\left(\frac{K^0}{K} - 1\right) \qquad (6.7.24)$$

Here K^0 represents the value of K evaluated for $T = T^0$. Given the state (c^0, T^0), we first specify various values of $(T - T^0)$ and solve Eq. (6.7.21) for $(c - c^0)$ to establish the curves Σ^1 and Σ^2 emanating from the point (c^0, T^0). The two branches can be distinguished by examining the relationship with the Γ's passing through the point.

Once the curve Σ^k is constructed, the state on the other side of the discontinuity must fall on the Σ^k. If the state is given by (n, T), the propagation direction of this k-discontinuity is given by Eq. (6.2.19) in the form

$$\bar{\sigma}_k = \left(\frac{d\tau}{dx}\right)_k = \frac{1 + \nu\tilde{\beta}_k}{1 - \mu\tilde{\beta}_k} \qquad (6.7.25)$$

for $k = 1, 2$ where

$$\tilde{\beta}_k = \left(\frac{n - n^0}{c - c^0}\right)_k \qquad (6.7.26)$$

Along the T axis of the hodograph plane, we see that Eq. (6.7.20) gives $(n - n^0)/(c - c^0) = \kappa = \tilde{\beta}_k$ for any change in T and thus Eq. (6.7.25) becomes identical to Eq. (6.7.19). This confirms that a pure thermal wave is a contact discontinuity.

To illustrate the application we shall treat a numerical example for which the data are taken to apply to the adsorption of benzene vapor carried by nitrogen gas on charcoal at a total pressure of 10 atm. The parameters are

$$\nu = 1.0, \quad C_s = 405.0 \text{ cal/liter} \cdot {}^\circ K, \quad C_f = 2.7 \text{ cal/liter} \cdot {}^\circ K$$

$$N = 5.50 \text{ mol/liter} \quad -\Delta H = 10.4 \text{ kcal/mol}$$

$$K_0 = 3.88 \times 10^{-5} \text{ liter/mol} \cdot {}^\circ K^{1/2}$$

With these values Eqs. (6.7.9) and (6.7.10) are integrated from various points in the hodograph plane to construct the network of the Γ^1 and Γ^2 as shown in Fig. 6.9.

All the curves Γ^1 approach the T axis asymptotically as T (and c also) decreases. Along each Γ^1, the parameter β_1 decreases monotonically in the direction of increasing T (and c also). This implies that the 1-characteristic field is genuinely nonlinear, a 1-wave is a simple wave if c or T increases from the left

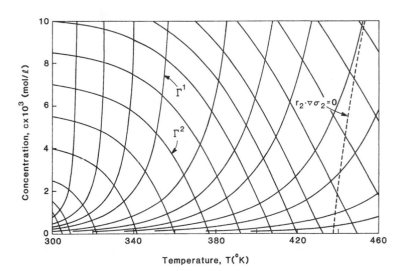

Figure 6.9

More on Hyperbolic Systems   Chap. 6

to the right across the wave region, and it is a shock if c or T changes in the reversed direction.

We have $NK = \kappa = 150$ when $T = 505.5°K$. Hence the T axis shown in Fig. 6.9 is a Γ^1, along which $\zeta_1 = 0$, $\beta_1 = 150$, and

$$\sigma_1 = \frac{1 + \nu\kappa}{1 - \mu\kappa} = \frac{151}{1 - 150\mu}$$

It then follows that $r_1 \cdot \nabla\sigma_1 = 0$ along this portion of the T axis, so a pure thermal wave is a 1-contact discontinuity.

The curves Γ^2 intersect the T axis. If we move along a Γ^2 in the direction of increasing T (or decreasing c), we see the parameter β_2 increasing, but this is so only up to the point of intersection with the dashed line shown in Fig. 6.9. Beyond the point, β_2 decreases along a Γ^2 as T increases. In fact, the dashed line is the locus of the point at which $r_2 \cdot \nabla\sigma_2 = 0$. It is clear that the 2-characteristic field remains genuinely nonlinear only in the region of low to moderate temperature. In the higher-temperature range it is not genuinely nonlinear. As long as the two states to be connected by a 2-wave fall on one side of the dashed line, the 2-wave is either a simple wave or a shock. If the two states lie on different sides, the 2-wave can be given by a combined wave.

As we discuss practical examples in the following, each state will be represented in the form $A(c \times 10^3$ in mol/liter, n in mol/liter, T in °K), where n is the equilibrium value corresponding to the state (c, T).

Example 6.1: Elution of a fixed bed by pure solvent

Here the initial bed assumes the state $A(10, 5.267, 350)$ and this is to be eluted by using a stream of pure nitrogen at the state $B(0, 0, 350)$. The image of the solution in the hodograph plane is shown in Fig. 6.10. As we pass through the column in the x direction, the state changes along $B \rightarrow D (a \ \Gamma^2)$ and then along $D \rightarrow A (a \ \Gamma^1)$. Along $B \rightarrow D$, β_2 decreases, whereas β_1 decreases along $D \rightarrow A$. Consequently, we expect to have a 2-simple wave and a 1-simple wave separated by the new constant state D $(2.467, 4.964, 338.1)$.

Since $\mu = 0$, Eq. (6.7.15) simply gives

$$\sigma_k = 1 + \nu\beta_k$$

for $k = 1, 2$, and thus the solution in the physical plane of x and τ can be readily constructed as shown in the upper diagram of Fig. 6.11. The solute distribution and the temperature profile at $\tau = 16.0$ are also presented in Fig. 6.11. We observe that the 1-wave carries the major portion of the desorbed solute (75.33%) and cools the bed by 11.9°K while the 2-wave desorbs the major portion of the solute (94.06%) and transfers the energy for desorption from the feedstream.

Example 6.2: Saturation of a fresh bed

Initially, the bed consists of fresh charcoal at the state $B(0, 0, 350)$, and this is to be saturated by introducing a stream of benzene carried by nitrogen at the state $A(10.0, 5.267, 350)$. In the hodograph plane we first try with the path composed

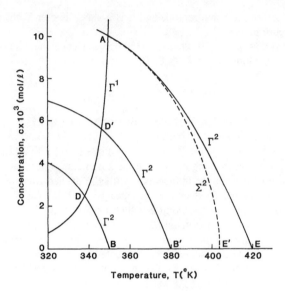

Figure 6.10

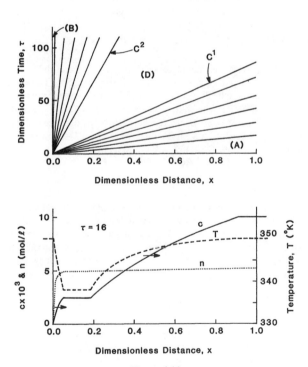

Figure 6.11

More on Hyperbolic Systems   Chap. 6

of $A \to E(a\ \Gamma^2)$ and $E \to B(a\ \Gamma^1)$, as marked in Fig. 6.10. Along $A \to E$, however, β_k increases and hence the characteristic C^2 will fan backward in the physical plane as we proceed in the x direction (see the upper diagram of Fig. 6.12). This implies that the 2-wave is a shock. We then apply Eq. (6.7.21) to construct the curve $A \to E'(a\ \Sigma^2)$ and locate the point E' at the intersection with the T axis. Along $E' \to B(a\ \Gamma^1)$ we have $\sigma_1 = 151$ at every point, so the 1-wave is a contact discontinuity.

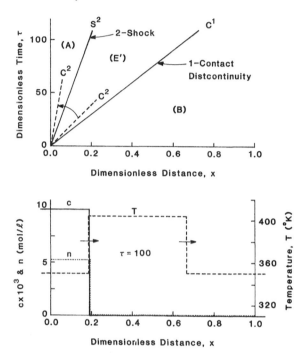

Figure 6.12

Figure 6.12 shows the physical plane portrait of the solution and the profiles of c, n, and T at $\tau = 100$. The slope of the 2-shock path is determined by Eq. (6.7.25) to give $\bar{\sigma}_2 = 1 + (5.267 - 0)/(10 \times 10^{-3} - 1) = 527.7$. Between the two paths of the 2-shock and the 1-contact discontinuity we see the region of the new constant state $E'(0, 0, 403.9)$ developing. While the 2-wave corresponds to the adsorption front, the 1-wave which is a pure thermal wave takes the role of the carrier of the energy evolved by the adsorption process. The bed will be fully saturated at $\tau = \tau_s = 527.7$.

Example 6.3: Elution by countercurrent contact

Suppose that the initial bed is at the state $A(10.0, 5.267, 350)$ and the incoming solid phase has the same state A. The incoming fluid stream is pure nitrogen at the state $B'(0, 0, 380)$ and flows countercurrently against the solid stream. As shown in Fig. 6.10, the image of the solution in the hodograph plane consists of

$B' \rightarrow D'(a\ \Gamma^2)$ and $D' \rightarrow A(a\ \Gamma^1)$. To each of these there corresponds a simple wave. Between the two simple wave regions we expect to have a region of the new constant state $D'(5.58, 5.151, 346.3)$.

Using Eqs. (6.6.4) and (6.6.5), we can determine various ranges of μ as follows:

Lower range	$0 \leq \mu \times 10^3 < 0.251$
2-transition range	$0.251 \leq \mu \times 10^3 < 4.50$
2-intermediate range	$4.50 \leq \mu \times 10^3 < 25.253$
1-transition range	$25.253 \leq \mu \times 10^3 < 56.883$
Upper range	$56.883 \leq \mu < \infty$

For $\mu = 0.01$, which belongs to the 2-intermediate range, only the 1-simple wave propagates into the domain as shown in the left diagram of Fig. 6.13. Hence we have the state D' at $x = 0+$ and the boundary $\mathscr{B}^a$ bears a discontinuity from the beginning. From Eq. (6.1.17) we obtain $n^a = 4.593$ mol/liter and Eq. (6.1.19) gives $T_s^a = 354.5°K$. The head of the 1-simple wave reaches the boundary $\mathscr{B}^b$ and, thenceforth, Eqs. (6.1.18) and (6.1.20) determine the state (c^b, T_f^b) of the outgoing fluid phase, which varies from $c^b = 0.01$ mol/liter and $T_f^b = 350°K$ to $c^b = 6.74 \times 10^3$ mol/liter and $T_f^b = 347.4°K$. The latter state is obtained at $\tau = \tau_s = 67.2$, at which the system attains the steady state. The right diagram of Fig. 6.13 shows the profiles of c, n and T at $\tau = 20$ and for $\tau \geq \tau_s = 67.2$.

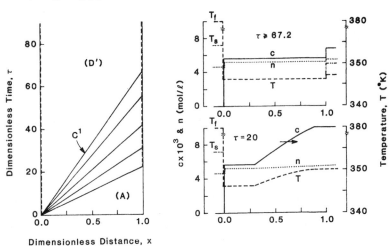

Figure 6.13

Repeating the procedure above for various values of μ, we can determine the steady-state values, n_s^a and c_s^b, and the time τ_s at which the system behavior reaches a steady state. The results are presented in Fig. 6.14. For μ in transition ranges, c^b cannot have a steady-state value because $\tau_s = \infty$. Since the operation is for the elution of the solid phase, we would prefer having n_s^a as low and c_s^b as high as possible, whereas τ_s may be considered as a secondary factor. From Fig. 6.14 we readily observe that the optimum operation obtains when $\mu =$

More on Hyperbolic Systems   Chap. 6

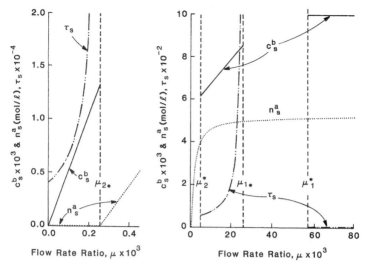

Figure 6.14

0.251×10^{-3}. With this value of μ, the solid phase will be completely eluted; that is, $n_s^a = 0$, but c^b approaches asymptotically to a limiting value of 1.32×10^{-3} mol/liter.

Example 6.4: Saturation by countercurrent contact

The initial bed contains a small amount of the solute at the state B'' (0.1, 1.015, 350) and the incoming solid phase takes the same state B''. The solid phase is to be further saturated by passing it countercurrently against the fluid stream which is fed at the state A' (10.0, 5.438, 320). As shown in Fig. 6.15, the hodograph plane portrait of the solution consists of $A' \rightarrow G'$ (a Σ^2) and $G' \rightarrow B''$ (a Σ^1) instead of $A' \rightarrow G$ (a Γ^2) and $G \rightarrow B''$ (a Γ^1). We note that the curve $B'' \rightarrow G$ (a Γ^1) is almost linear and thus the curve $B'' \rightarrow G'$ (a Σ^1) is nearly coincident with the curve $B'' \rightarrow G$. To each of the portions $A' \rightarrow G'$ and $G' \rightarrow B''$ there corresponds a shock and we expect to have the new constant state G' (0.280, 1.041, 376) between the two shocks. The fact that 1-wave is also given by a shock presents a contrast to Example 6.1, in which the bed was initially fresh and the 1-wave was a contact discontinuity (a pure thermal wave). This implies that in saturation problems the initial state of the bed has a significant influence on the structure of the solution.

Since the two transition ranges of μ collapse to single values, $\bar{\mu}_1$ and $\bar{\mu}_2$, respectively, the μ spectrum is now partitioned by applying Eq. (6.6.7) as follows:

Lower range	$0 \le \mu \times 10^3 < 2.21$
2-intermediate range	$2.21 \le \mu \times 10^3 < 6.845$
Upper range	$6.845 \le \mu \times 10^3 < \infty$

The solution for $\mu = 5 \times 10^{-3}$, which belongs to the 2-intermediate range,

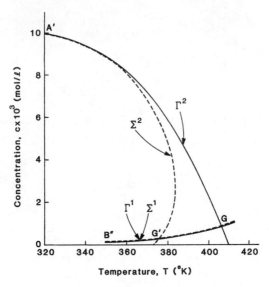

Figure 6.15

is depicted in Fig. 6.16. From the left diagram we see that only the 1-shock propagates into the domain to reach the boundary $\mathscr{B}^b$ at $\tau = 546.0$, and then the system obtains a steady state. At $x = 0+$ we have the state G' (0.28, 1.041, 376), so that the boundary $\mathscr{B}^a$ bears a discontinuity for all time and the state of the outgoing solid phase is determined by Eqs. (6.1.17) and (6.1.19) to give $n^a = 2.985$ mol/liter and $T_s^a = 351.3°$K. The boundary $\mathscr{B}^b$ also bears a discontinuity from $\tau = 546.0$, at which the state of the outgoing fluid phase jumps discontinuously from $c^b = 10^{-4}$ mol/liter and $T_f^b = 350°$K to $c^b = 1.5 \times 10^{-4}$

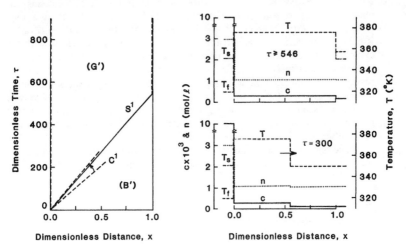

Figure 6.16

More on Hyperbolic Systems   Chap. 6

mol/liter and $T_f^b = 357°K$. The right diagram of Fig. 6.16 shows the profiles of c, n, and T at $\tau = 300$ and for $\tau \geq \tau_s = 546$.

The analysis above can be carried out for any value of μ to obtain the steady-state values of n^a and c^b as well as τ_s as functions of μ. The results are summarized in Fig. 6.17. Since this is a saturation problem, we would naturally require that n_s^a be as high as possible and c_s^b be as low as possible. It is obvious from Fig. 6.17 that the requirements are best met when $\mu = 2.21 \times 10^{-3}$. For this optimum value of μ, the solid phase leaves the column at a highly saturated state ($n_s^a = 5.438$ mol/liter, surface coverage = 98.9%) while the outgoing fluid phase contains a negligible amount of solute ($c_s^b = 2.25 \times 10^{-4}$ mol/liter or 2.25% of c^a). Moreover, the steady state is attained rather quickly ($\tau_s = 217.3$). Incidentally, under this condition the 2-shock remains stationary at $x = 0$, and hence the outgoing solid phase is in equilibrium with the incoming fluid phase.

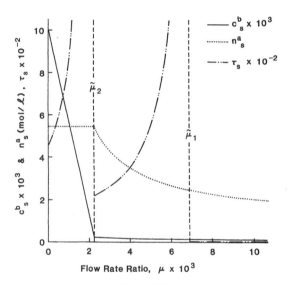

Figure 6.17

From these four examples, we realize that the structure of the solution can take various forms depending on the prescribed states of the initial bed and the incoming fluid stream. Now it must be expected that the possibilities become even more diverse if the process is run in the higher-temperature range because the T axis of the hodograph plane switches its role from a Γ^1 to a Γ^2 at the temperature $T_w = 505.5°K$ for which $NK = \kappa = 150$. We would like to examine the nature of the Γ's in the neighborhood of this temperature to see if there might be any advantage of practical interest by treating a couple of examples involved with high temperatures.

By integrating Eqs. (6.7.9) and (6.7.10) the network of the Γ^1 and Γ^2 may be constructed in the high-temperature range as shown in Fig. 6.18. Here again the dashed line represents the locus of the point at which β_2 reaches a maximum value along each Γ^2. In other words, we have the condition $\mathbf{r}_2 \cdot \nabla \sigma_2 = 0$ satisfied on the dashed line. As expected, the Γ^1 along the T axis takes off at $T = T_w$ to

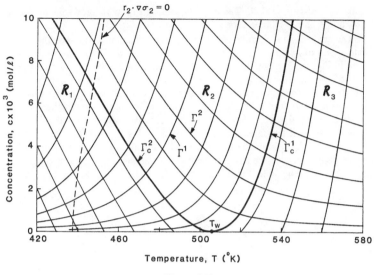

$$r_2 \cdot \nabla \sigma_2 = 0$$

Figure 6.18

give the heavier curve designated by Γ_c^1. On the other hand, the heavier curve Γ_c^2 touches the T axis at $T = T_w$ and beyond this temperature the T axis becomes a Γ^2. The two curves Γ_c^1 and Γ_c^2 divide the hodograph plane into three regions, $\mathfrak{R}_1$, $\mathfrak{R}_2$, and $\mathfrak{R}_3$, as marked in Fig. 6.18. The layout of the Γ's is rather symmetrical across the region $\mathfrak{R}_2$ and this feature may give rise to some unusual phenomena of practical interest. Here we would like to treat two such examples.

Example 6.5: Consecutive elution and saturation

Suppose that the initial bed is at the state Q (5, 0.320, 550). This represents a bed which was presaturated and then heated to a high temperature. Now this bed is to be treated by feeding a fluid stream at the state P (5, 2.054, 440). The image of the solution in the hodograph plane is constructed as shown in Fig. 6.19. It consists of $P \to J$ (a Σ^2), $J \to U$ (a Γ^2), $U \to T_w$ (a Γ^1), $T_w \to V$ (a Γ^2), and $V \to Q$ (a Γ^1).

The portion $P \to U$ is the image of the 2-combined wave, which is composed of a 2-semishock $(P \to J)$ joined by a 2-simple wave $(J \to U)$ on the right. The portions $U \to T_w$ and $T_w \to V$ represent the 1- and 2-contact discontinuities, respectively, but since their propagation speeds are the same, the two form a single discontinuity and move as such. To the portion $V \to Q$ there corresponds a 1-simple wave.

Obviously, the point T_w separates the image of the solution into two parts. The first part from P to T_w represents the saturation process while the second part from T_w to Q stands for the elution process. In the physical plane, therefore, we see two disjoint regions separated by the contact discontinuity line as shown in the upper diagram of Fig. 6.20. The region in the left is for saturation and the one in the right is for elution. This feature is better visualized in the lower diagram of Fig. 6.20, in which the profiles of c, n, and T at $\tau = 50$ are presented. The elution wave propagates far ahead of the saturation wave and between the

More on Hyperbolic Systems   Chap. 6

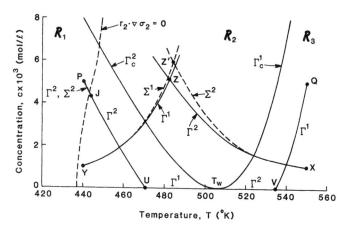

Figure 6.19

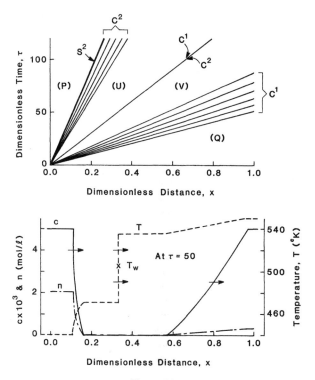

Figure 6.20

two intervenes a pure thermal wave which takes the role of the carrier of excess energy.

We note that the present problem is essentially nothing more than a simple combination of an elution problem and a saturation problem. The two processes take place without having any influence upon each other. Such a feature, which is peculiar to hyperbolic systems of equations, is allowed by virtue of the existence of the point T_w on the T axis.

If we have the initial state on the Γ_c^1 and the feed state on the Γ_c^2, the overall feature of the solution would remain the same except for the fact that the head of the saturation wave would just touch the tail of the elution wave. The pure thermal wave would degenerate to a single point at which we have $T = T_w$ and the two waves are joined. In such a case the energy gain of the initial bed over the feedstream is just balanced by the sum of the heat evolved by adsorption and the heat required for desorption. This represents the minimum amount of energy required to realize elution and saturation in sequence in a single bed. The minimum amount of energy increases with the solute content since it is represented by the horizontal distance between the Γ_c^2 and the Γ_c^1 in Fig. 6.18.

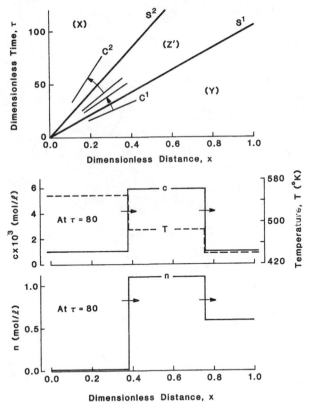

Figure 6.21

Example 6.6: Formation of a high concentration band

Here the initial bed at the state Y (1, 0.586, 440) is being treated by a hot feed-stream at the state X (1, 0.062, 550). The image of the solution in the hodograph plane is given by the path $X \to Z'$ (a Σ^2) and the path $Z' \to Y$ (a Σ^1) as shown in Fig. 6.19. Therefore, we have two shocks whose paths of propagation are depicted in the upper diagram of Fig. 6.21. Between the two paths lies the region of the new constant state Z' (5.91, 1.102, 484). The lower diagram of Fig. 6.21 clearly shows how the higher concentration band of the state Z' is formed between the initial and feed states. It is observed that the high-energy state of the feedstream tends to push the solute ahead across the 2-shock, whereas the 1-shock is an adsorption front that holds the solute supplied from the back side. This is an interesting example showing that the energy state can affect predominantly the adsorption process. It is evident that the higher the energy gain of the feed state over the initial state, the higher the concentration of the intermediate band becomes.

6.8 Adiabatic Adsorption of Two Solutes

We are now in a position to treat the adiabatic adsorption of two solutes, for which Eq. (6.1.12) gives three quasilinear equations. Here again we shall be concerned with the Riemann problem and hence we may begin with the fundamental differential equations (6.2.7) for $m = 3$, which can be rearranged in the form of Eqs. (6.3.3) and (6.3.4). Thus we have

$$
\begin{aligned}
\left(\frac{dc_1}{dT}\right)_k &= \frac{n_{1,3}(n_{2,2} - \beta_k) - n_{1,2}n_{2,3}}{(n_{1,1} - \beta_k)(n_{2,2} - \beta_k) - n_{1,2}n_{2,1}} \\
\left(\frac{dc_2}{dT}\right)_k &= \frac{n_{2,3}(n_{1,1} - \beta_k) - n_{2,1}n_{1,3}}{(n_{1,1} - \beta_k)(n_{2,2} - \beta_k) - n_{1,2}n_{2,1}}
\end{aligned}
\tag{6.8.1}
$$

for $k = 1, 2, 3$, where β_k, representing the directional derivative $(\mathcal{D}n_i / \mathcal{D}c_i)_k$, is determined by the kth eigenvalue of the matrix $\nabla \mathbf{n}$ [see Eq. (6.2.16)].

For a given set of the concentrations and the temperature we first determine the roots β_k, $k = 1, 2, 3$, of the characteristic equation

$$
|\nabla \mathbf{n} - \beta \mathbf{I}| = 0
$$

or

$$
\beta^3 - P_2\beta^2 + P_1\beta - P_0 = 0
\tag{6.8.2}
$$

where

$$
P_2 = \nabla \cdot \mathbf{n} = n_{1,1} + n_{2,2} + \kappa - \frac{-\Delta H_1}{C_f}n_{1,3} - \frac{-\Delta H_2}{C_f}n_{2,3}
\tag{6.8.3}
$$

$$P_1 = n_{1,1}n_{2,2} - n_{1,2}n_{2,1} + \kappa(n_{1,1} + n_{2,2})$$

$$+ \frac{-\Delta H_1}{C_f}(n_{1,2}n_{2,3} - n_{2,2}n_{1,3}) + \frac{-\Delta H_2}{C_f}(n_{2,1}\,n_{1,3} - n_{1,1}n_{2,3}) \tag{6.8.4}$$

$$P_0 = |\nabla \mathbf{n}| \tag{6.8.5}$$

and

$$\kappa = \frac{C_s}{C_f} \tag{6.8.6}$$

Let us now introduce the Langmuir isotherm

$$n_i = \frac{N_i K_i c_i}{\delta} \tag{6.8.7}$$

for $i = 1, 2$, in which we have

$$\delta = 1 + K_1 c_1 + K_2 c_2 \tag{6.8.8}$$

Here N_i, denoting the saturation value of n_i, is considered to be independent of the temperature but may vary from one solute to the other depending on the surface area occupied by a molecule. Hence the total surface coverage of the adsorbent surface is determined as

$$\Theta = \frac{n_1}{N_1} + \frac{n_2}{N_2} = 1 - \frac{1}{\delta} \tag{6.8.9}$$

The equilibrium constant K_i for the adsorption of the solute A_i is given by a strong function of the temperature:

$$K_i = K_{0i} T^{1/2} e^{-\Delta H_i/RT} \tag{6.8.10}$$

for $i = 1, 2$, in which K_{0i} is a constant and ΔH_i represents the enthalpy change of adsorption per mole of the solute A_i. The relative adsorptivity may be defined as

$$\gamma = \frac{n_2/c_2}{n_1/c_1} = \frac{N_2 K_2}{N_1 K_1} = \frac{N_2 K_{02}}{N_1 K_{01}} e^{-(\Delta H_2 - \Delta H_1)/RT} \tag{6.8.11}$$

The ratio is normally controlled by the exponential factor so that the values of $-\Delta H_i$ provide a direct measure to the relative adsorptivity.

For the Langmuir isotherm (6.8.7) we have

$$n_{i,i} = \frac{N_i K_i(1 + K_i c_i)}{\delta^2} \tag{6.8.12}$$

$$n_{i,j} = \frac{-N_i K_i c_i K_j}{\delta^2} \tag{6.8.13}$$

$$n_{i,3} = \frac{N_i K_i c_i}{\delta^2}\left\{1 + K_j c_j\left(1 - \frac{d\ln K_j}{d\ln K_i}\right)\right\}\frac{d\ln K_i}{dT} \tag{6.8.14}$$

$$\frac{d\ln K_i}{dT} = -\frac{1}{T}\left(\frac{-\Delta H_i}{RT} - \frac{1}{2}\right) \tag{6.8.15}$$

for $i, j = 1, 2$ $(i \neq j)$ and

$$|\nabla \mathbf{n}| = \frac{\kappa N_1 K_1 N_2 K_2}{\delta^3} \qquad (6.8.16)$$

The curve Γ^k in the hodograph space can be constructed by integrating Eq. (6.8.1) in connection with Eq. (6.8.2). As discussed in the last part of Sec. 6.3, it is more convenient to construct the projections of the curve Γ^k on the (c_1, T)- and (c_2, T)-planes, respectively. If $c_1 = 0$ or $c_2 = 0$, it is convenient to follow the procedure discussed in Sec. 6.7.

The characteristic directions in the physical plane are determined by applying Eq. (6.2.12) as

$$\sigma_k = \left(\frac{d\tau}{dx}\right)_k = \frac{1 + \nu\beta_k}{1 - \mu\beta_k} \qquad (6.8.17)$$

for $i = 1, 2, 3$ and, if we put

$$\beta_1 < \beta_2 < \beta_3 \qquad (6.8.18)$$

we have

$$\frac{1}{\sigma_1} > \frac{1}{\sigma_2} > \frac{1}{\sigma_3} \qquad (6.8.19)$$

which is consistent with Eq. (6.2.14).

If the solution involves one or more discontinuities, we must consider the compatibility relations (6.2.18), which may be rewritten in the form

$$\frac{n_1 - n_1^0}{c_1 - c_1^0} = \frac{n_2 - n_2^0}{c_2 - c_2^0} = \kappa - \frac{-\Delta H_1}{C_f} \frac{n_1 - n_1^0}{T - T^0} - \frac{-\Delta H_2}{C_f} \frac{n_2 - n_2^0}{T - T^0} \qquad (6.8.20)$$

where the superscript 0 represents a fixed state; for example, the state on one side of a discontinuity. For the Langmuir isotherm (6.8.7), the compatibility relations (6.8.20) give rise to a pair of quadratic equations

$$\begin{aligned} B_1 c_1^2 + B_2 c_1 c_2 + A_{11} c_1 + A_{12} c_2 + A_{10} = 0 \\ B_2 c_2^2 + B_1 c_1 c_2 + A_{21} c_1 + A_{22} c_2 + A_{20} = 0 \end{aligned} \qquad (6.8.21)$$

in which

$$B_i = K_i\{(-\Delta H_i) N_i - \phi\} \qquad (6.8.22)$$

$$A_{i0} = \phi c_{i0} - (T - T^0) n_i^0 \qquad (6.8.23)$$

$$A_{ii} = K_i A_{i0} - \frac{N_i K_i c_i^0 (-\Delta H_i)}{C_f} + N_i K_i (T - T^0) - \phi \qquad (6.8.24)$$

$$A_{ij} = K_j A_{i0} - \frac{N_j K_j c_i^0 (-\Delta H_i)}{C_f} \qquad (6.8.25)$$

for $i, j = 1, 2$ $(i \neq j)$ and

$$\phi = \kappa(T - T^0) + \frac{(-\Delta H_1) n_1^0 + (-\Delta H_2) n_2^0}{C_f} \qquad (6.8.26)$$

Suppose that the state (c_1^0, c_2^0, T^0) is given. For various values of $(T - T^0)$, Eq. (6.8.21) may be solved to determine the pair (c_1, c_2) corresponding to each value of T. There can be at most three different solutions. These solutions will give the curves Σ^1, Σ^2, and Σ^3, where the indices 1, 2, and 3 must be assigned in the relationship with the Γ^k constructed a priori. The curve Σ^k so determined is the locus of the state that can be connected to the state (c_1^0, c_2^0, T^0) by a k-shock or a k-semishock. Once the state on the other side is determined, the propagation direction of this k-discontinuity is given by Eq. (6.2.19):

$$\tilde{\sigma}_k = \left(\frac{d\tau}{dx}\right)_k = \frac{1 + \nu\tilde{\beta}_k}{1 - \mu\tilde{\beta}_k} \qquad (6.8.27)$$

for $k = 1, 2, 3$, in which

$$\tilde{\beta}_k = \left(\frac{n_i - n_i^0}{c_i - c_i^0}\right)_k \qquad (6.8.28)$$

represents the common value of the ratios in Eq. (6.8.20).

Let us now consider the adsorption on charcoal of benzene and cyclohexane, the mixture of which is carried by a stream of nitrogen through an adiabatic column under the total pressure of 10 atm. This may be regarded as an extension of the system treated in Sec. 6.7, so the physical properties are given as

$$\nu = 1.0, \quad C_s = 405.0 \text{ cal/liter} \cdot {}^\circ\text{K}, \quad C_f = 2.7 \text{ cal/liter} \cdot {}^\circ\text{K}$$

The isotherm parameters and the heats of adsorption are taken from the literature to give

$N_1 = 3.77$ mol/liter,

$$K_{01} = 1.037 \times 10^{-3} \text{ liter/mol} \cdot {}^\circ\text{K}^{1/2}, \quad -\Delta H_1 = 7.8 \text{ kcal/mol}$$

$N_2 = 5.50$ mol/liter,

$$K_{02} = 3.880 \times 10^{-5} \text{ liter/mol} \cdot {}^\circ\text{K}^{1/2}, \quad -\Delta H_2 = -10.4 \text{ kcal/mol}$$

where the subscripts 1 and 2 represent cyclohexane and benzene, respectively. The relative adsorptivity of benzene to cyclohexane is given by Eq. (6.8.11) and it is observed that

$$\gamma = \frac{N_2 K_2}{N_1 K_1} \begin{cases} \geq 1 & \text{for } T \leq 449.9^\circ\text{K} \\ < 1 & \text{for } T > 449.9^\circ\text{K} \end{cases} \qquad (6.8.29)$$

Therefore, the adsorptivity reversal occurs between the two solutes at $T = 449.9^\circ\text{K}$. In this section we limit our discussion so that this phenomenon will not be involved. Since, however, the adsorptivity reversal will certainly have an important bearing with respect to the structure of the solution, we will come back to this subject in Sec. 6.9.

For all values of the concentrations and the temperature Eq. (6.8.2) gives three distinct, real, and positive values for β. This implies that the system of equations is totally hyperbolic and, if $\mu = 0$ (i.e., fixed bed), the characteristics C^k in the physical plane are directed forward. Given the initial state $I(c_1^0, c_2^0, T^0)$ of the column and the state $F(c_1^q, c_2^q, T_f^q)$ of the incoming fluid phase, we shall employ, in principle, the following scheme to construct the solution.

1. Locate the points I and F on the (c_1, T)- and (c_2, T)-planes.
2. Integrate Eq. (6.8.1) for $k = 1$ from the point I to obtain the projections of the Γ^1 on the two planes.
3. Repeat step 2 with $k = 3$ to establish the projections of the Γ^3 emanating from the point F.
4. From an arbitrary point on the Γ^1, integrate Eq. (6.8.1) for $k = 2$ to determine the projections of the Γ^2 that makes a bridge between the Γ^1 and the Γ^3. Repeat this procedure until the projections of the Γ^2 intersect the projections of the Γ^3 at the same value of the temperature on both planes.
5. If along the Γ^k, $k = 1, 2, 3$, so determined the value of β_k increases as we pass from I to F, calculate the values of σ_k by Eq. (6.8.17) at various points along the Γ^k and construct the k-simple wave in the physical plane.
6. If along the Γ^k the value of β_k decreases monotonically in the direction from I to F, apply Eq. (6.8.21) in the corresponding step among steps 2, 3, and 4 to construct the projections of the curve Σ^k. Find the value of σ_k by Eq. (6.8.27) and draw the propagation path of the k-shock in the physical plane.
7. If the parameter β_k takes extremum values at one or more points along the Γ^k, employ the procedure described in Sec. 6.4 to establish the curve Λ_0^k and construct the k-combined wave in the physical plane.
8. From the physical plane portrait of the three waves and four constant states, obtain the profiles of c_1, c_2, and T at any fixed time.

If $c_1 = 0$ and $c_2 = 0$ at the starting point in step 2, 3, or 4, use Eq. (6.7.9) or (6.7.10) to initiate the numerical integration. For the case in which only one solute is involved in the k-wave, we naturally apply the procedure discussed in Sec. 6.7.

If both solutes are absent from a certain region, we have

$$P_2 = N_1 K_1 + N_2 K_2 + \kappa$$
$$P_1 = N_1 K_1 N_2 K_2 + \kappa (N_1 K_1 + N_2 K_2) \qquad (6.8.30)$$
$$P_0 = \kappa N_1 K_1 N_2 K_2$$

and thus Eq. (6.8.2) generates κ, $N_1 K_1$, and $N_2 K_2$ for β_k, $k = 1, 2, 3$. For the present example, $\kappa = 150$ and it can be shown that

$$
\begin{array}{lll}
\kappa < N_1 K_1 \le N_2 K_2 & \text{for} & 0 < T \le 449.9°\text{K} \\
\kappa \le N_2 K_2 < N_1 K_1 & \text{for} & 449.9°\text{K} < T \le 505.5°\text{K} \\
N_2 K_2 < \kappa \le N_1 K_1 & \text{for} & 505.5°\text{K} < T \le 529°\text{K} \\
N_2 K_2 < N_1 K_1 < \kappa & \text{for} & T < 529.0°\text{K}
\end{array}
\qquad (6.8.31)
$$

Along the T axis in the hodograph space, therefore, we find from Eqs. (6.8.1) and (6.8.18) that

$$\left(\frac{dc_i}{dT}\right)_1 = 0, \qquad \beta_1 = \kappa = 150 \quad \text{for} \quad 0 < T < 505.5°\text{K}$$

$$\left(\frac{dc_i}{dT}\right)_2 = 0, \qquad \beta_2 = \kappa = 150 \quad \text{for} \quad 505.5°\text{K} < T < 529.0°\text{K}$$

(6.8.32)

and

$$\left(\frac{dc_i}{dT}\right)_3 = 0, \qquad \beta_3 = \kappa = 150 \quad \text{for} \quad T > 529.0°\text{K}$$

for $i = 1, 2$. This implies that in the hodograph space the T axis is composed of three different portions. The portion for $0 < T < 505.5°\text{K}$ is a Γ^1 to which there corresponds a 1-contact discontinuity in the physical plane because β_1 remains constant across the wave. Similarly, the portion for $505.5°\text{K} < T < 529.0°\text{K}$ or $T > 529.0°\text{K}$ takes the role of a Γ^2 or a Γ^3, respectively, and there corresponds a 2- or 3-contact discontinuities in the physical plane. Hence a pure thermal wave is necessarily given by a contact discontinuity. This feature of the T axis must have a significant influence on the structure of the solution especially for the conventional problems of saturation and elution. In particular, if the adsorption process involves a temperature higher than $505.5°\text{K}$, it is expected that some unusual situation may arise. We shall illustrate this by treating an example later in this section.

In the following we would like to discuss several examples of practical interest and, in doing so, we shall denote a state of the concentrations and the temperature in the form $A(c_1 \times 10^3$ in mol/liter, $c_2 \times 10^3$ in mol/liter, T in °K; n_1 in mol/liter, n_2 in mol/liter), where n_i represents the equilibrium value corresponding to the state (c_1, c_2, T).

Example 6.7: Elution of a fixed bed by pure solvent

Here the initial bed is a finite chromatogram at the state A (5, 10, 380; 1.029, 3.514), and this is to be eluted by passing a stream of nitrogen at state B (0, 0, 350; 0, 0). By employing the procedure discussed in the above, we first construct the image of the solution in the hodograph space as shown in Fig. 6.22. In the direction of the fluid flow, therefore, the state changes successively along $B \rightarrow D$ (a Γ^3), $D \rightarrow E$ (a Γ^2) and then $E \rightarrow A$ (a Γ^1). Along each of these portions, the corresponding β_k decreases and so they will give three simple waves of distinct families. The states D (0, 0.699, 377.2; 0, 1.973) and E (1.305, 2.355, 363.2; 0, 885, 3.197) represent the 3- and 2-constant states, respectively.

The solution in the physical plane is constructed by using Eq. (6.8.17) to give the three simple waves and the four constant states as depicted in the upper diagram of Fig. 6.23. From this the distribution of solutes and the temperature profile at $\tau = 20$ are determined and presented in the middle and lower diagrams of Fig. 6.23. We observe that the 1-simple wave propagating in the front carries away the major portion of the desorbed solutes (73.2% of A_1 and 76.45%

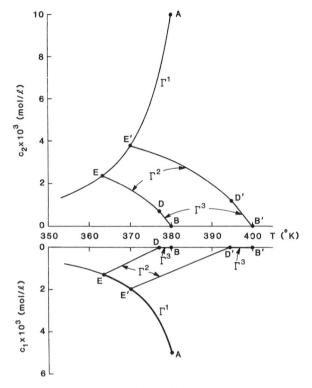

Figure 6.22

of A_2) and cools the column by 16.8°K. The 2-simple wave attains a complete desorption of A_2 (85.2%) and 34.8% desorption of A_1, whereas the 3-simple wave realizes a complete desorption of A_1 (56.2% of the total amount). It is clearly seen that the energy required for desorption is supplied from the feedstream through the 3- and 2-simple waves. The initial chromatogram will be completely eluted when $\tau = \tau_s = 3985.2$, at which the tail of the 3-simple wave reaches the opposite end of the column.

If the same initial bed is treated by a stream of nitrogen at 400°K, which corresponds to the state B' (0, 0, 400; 0, 0), the hodograph space portrait consists of the three segments $B' \to D'$ (a Γ^3), $D' \to E'$ (a Γ^2), and $E' \to A$ (a Γ^1). These are the images of three simple waves of distinct families and the points D' (0, 1.218, 394.8; 0, 1.921) and E' (1.984, 3.777, 369.9; 0.926, 3.310) represent the 3- and 2-constant states, respectively. The physical plane portrait of the solution is similar to Fig. 6.23 but the 2- and 3-simple waves are considerably accelerated and their role as a desorption wave becomes more prominent. Thus the complete elution is accomplished much earlier at $\tau = \tau_s = 2054.1$.

For elution problems, therefore, we may call the 2- and 3-waves active desorption waves, whereas the 1-wave remains passive, and it is suggested that the feed temperature be kept higher than the initial temperature of the bed.

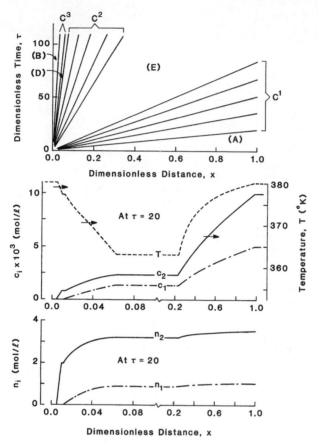

Figure 6.23

Example 6.8 Saturation of a fresh fixed bed

Suppose that the column contains initially fresh charcoal at the state B (0, 0, 380; 0, 0), and this is to be saturated by using a stream of feed mixture under the total pressure of 10 atm, which is at the state A (5, 10, 380; 1.029, 3.514). For the image of the solution in the hodograph space we first obtain $A \rightarrow G$ (a Γ^3), $G \rightarrow H$ (a Γ^2), and $H \rightarrow B$ (a Γ^1) as given by the solid line in Fig. 6.24. Along the arc $A \rightarrow G$, however, the parameter β_3 increases from A to I and then decreases from I to G. Hence we encounter a situation discussed in Sec. 6.4 [see the discussion around Eq. (6.4.7)]. This implies that the 3-wave is a combined type that consists of a 3-semishock joined by a 3-simple wave on the right. The proper junction is made at the point J (8.685, 1.761, 416.6; 2.333, 0.597). The curve $A \rightarrow J$ (a Σ^3) is nearly coincident with the curve $A \rightarrow G$ (a Γ^3).

Along the portion $G \rightarrow H$ (a Γ^2) the parameter β_2 increases from G to I' and then decreases from I' to H. Although this is similar to the situation along the arc $A \rightarrow G$, it turns out that the 2-wave is given by a single shock which corre-

478   More on Hyperbolic Systems   Chap. 6

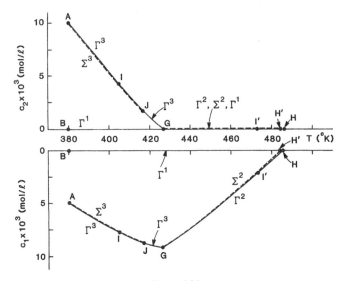

Figure 6.24

sponds to the dashed curve $G \rightarrow H'$ (a Σ^2) of Fig. 6.24. The portion $H' \rightarrow B$ (a Γ^1) represents a pure thermal wave which, according to Eq. (6.8.32), is a 1-contact discontinuity. The states G (9.15, 0, 426.4; 2.492, 0) and H' (0, 0, 485.2; 0, 0) denote the 3- and 2-constant states, respectively.

As shown in the upper diagram of Fig. 6.25, the solution in the physical plane consists of the 3-combined wave, the 2-shock, the 1-contact discontinuity, and the four constant states, A, G, H', and B. The slope of the characteristic C^3 or C^1 is given by Eq. (6.8.17) while that of the shock path S^3 or S^2 is determined by Eq. (6.8.27).

The distribution of solutes and the temperature profiles are examined at $\tau = 140$ and presented in the middle and lower diagrams of Fig. 6.25. It is observed that the 3-combined wave accomplishes the complete adsorption of A_2 while passing the solute A_1 to the front and pushing the heat evolved by the adsorption process to the front. The solute A_1 is completely adsorbed across the 2-shock and the heat evolved by adsorption is carried away by the 1-contact discontinuity, which is a pure thermal wave.

We note that the chromatogram of the solute A_1 is contaminated by the solute A_2 in the rear due to the diffuse portion of the 3-combined wave. Further operation pushes the A_1-chromatogram out of the column to establish at $\tau = \tau_s = 355$ the steady state, which is the equilibrium state corresponding to the feed state.

Example 6.9 Saturation by countercurrent contact

Here we treat a continuous countercurrent moving-bed problem. The initial bed is at the state P (0.05, 0.1, 380; 0.106, 0.361) and the incoming solid phase assumes the same state while the entering fluid stream is at the state F (5, 10, 350; 0.880, 4.038). This implies that the solid stream, containing slight amounts of both solutes A_1 and A_2, is to be further saturated by passing it through the column against the fluid stream containing both A_1 and A_2.

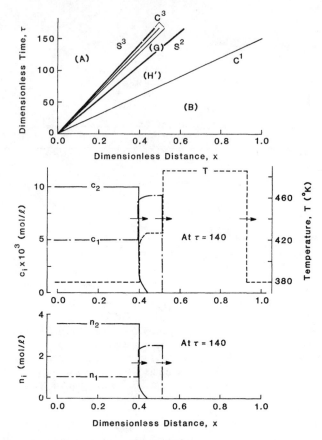

Figure 6.25

For the hodograph space portrait of the solution we first establish the path $F \to Q$ (a Γ^3), $Q \to R$ (a Γ^2), and $R \to P$ (a Γ^1) as shown in Fig. 6.26. Along each arc Γ^k, however, the corresponding parameter β_k increases monotonically as we pass in the direction from F to P. Hence we introduce Eq. (6.8.21) to construct the path $F \to Q'$ (a Σ^3), $Q' \to R'$ (a Σ^2), and $R' \to P$ (a Σ^1), and it is clear that there would correspond three shocks. The portion $R' \to P$ (a Σ^1) is almost completely overlapping over the arc $R \to P$ (a Γ^1). The points Q' (10.213, 1.074, 381.5; 2.943, 0.521) and R' (0.276, 0.920, 444; 0.138, 0.477) represent the 3- and 2-constant states.

Since all three waves are shocks, the three transition ranges of μ degenerate to single values and thus the μ spectrum is partitioned by applying Eq. (6.6.7) as follows:

Lower range	$0 \leq \mu \times 10^3 < 2.527$
3-intermediate range	$2.527 \leq \mu \times 10^3 < 3.555$
2-intermediate range	$3.555 \leq \mu \times 10^3 < 7.064$
Upper range	$7.064 \leq \mu \times 10^3 < \infty$

More on Hyperbolic Systems   Chap. 6

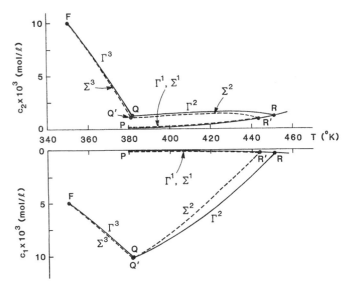

Figure 6.26

The solution for $\mu = 3 \times 10^{-3}$, which belongs to the 3-intermediate range, is presented in Fig. 6.27. The left diagram shows the physical plane portrait of the solution. The 1- and 2-shocks propagate into the domain, so we have at $x = 0+$ the state Q' (10.213, 1.074, 381.5; 2.934, 0.521). The 3-shock collapses on the boundary $\mathscr{B}^a$ which bears a discontinuity from the beginning. The state of the outgoing solid phase is determined by applying Eqs. (6.1.17) and (6.1.19) to give $n_1^a = 1.196$ mol/liter, $n_2^a = 3.496$ mol/liter, and $T_s^a = 354.5°$K. This state remains unchanged for all time.

At $\tau = 248$ when the 1-shock arrives at the opposite end at $x = 1$, the boundary $\mathscr{B}^b$ begins to bear a discontinuity and the state of the outgoing fluid phase, being governed by Eqs. (6.1.18) and (6.1.20), changes discontinuously to $c_1^b = 1.8 \times 10^{-4}$ mol/liter, $c_2^b = 5.7 \times 10^{-4}$ mol/liter, and $T_f^b = 416.8°$K. The 2-shock reaches the boundary $\mathscr{B}^b$ at $\tau = \tau_s = 1865$ and then the system obtains the steady state. The state of the outgoing fluid phase again changes discontinuously to $c_1^b = 3.73 \times 10^{-3}$ mol/liter, $c_2^b = 5.94 \times 10^{-4}$ mol/liter, and $T_f^b = 407.2°$K. The right diagram of Fig. 6.27 shows the profiles of c_1, c_2, and T at $\tau = 200$ and $\tau = 1000$, and for $\tau \geq \tau_s = 1865$.

Repeating the analysis above for various values of μ, we can determine the steady-state values of n_1^a, n_2^a, T_f^a, c_1^b, c_2^b, and T_s^b as well as the time τ_s as functions of the parameter μ. Among these, n_{1s}^a, n_{2s}^a, and τ_s are plotted against μ in Fig. 6.28. If the objective is the recovery of the more strongly adsorbed solute A_2, the optimum value of μ is given by $\mu = \bar{\mu}_3 = 2.527 \times 10^{-3}$. With this value μ the outgoing solid phase shows the A_2 coverage of 73.42% and the A_1 coverage of 23.34% and the steady state is attained at $\tau = 976$. On the other hand, if the recovery of the solute A_1 is considered more important, it is readily observed from Fig. 6.28 that the operation is best performed with $\mu = \bar{\mu}_2 = 3.555 \times 10^{-3}$. For this value of μ the outgoing solid phase bears the A_2 coverage of 55.11% and the A_1 coverage of 38.91%.

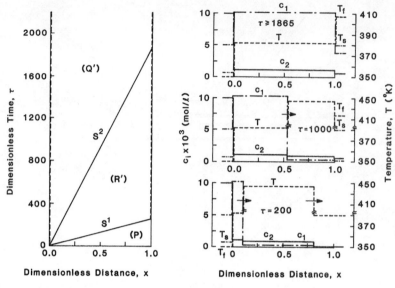

Figure 6.27

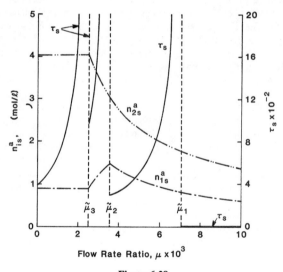

Figure 6.28

Example 6.10 Elution of a high-temperature chromatogram

Because of the peculiar character of the temperature axis [see Eq. (6.8.32)], it is to be expected that a high-temperature chromatogram may present a contrasting elution performance just as observed in the case with a single solute (see Exam-

　　　　　　　　　　　　　　More on Hyperbolic Systems　　Chap. 6

ple 6.5) and it seems worthwhile to put some effort in this matter. Suppose that the chromatogram at the state V (5, 10, 575; 0.361, 0.383) is being eluted by a stream of pure nitrogen at 480°K. Thus the feedstream is at the state U (0, 0, 480; 0, 0) and the hodograph space portrait of the solution is composed of $V \rightarrow Y$ (a Γ^1), $Y \rightarrow X$ (a Γ^2), and $X \rightarrow U$. In fact, the portion $X \rightarrow U$ consists of three segments, $X \rightarrow T_{w1}$ (a Γ^3), $T_{w1} \rightarrow T_{w2}$ (a Γ^2), and $T_{w2} \rightarrow U$ (a Γ^1) as depicted in Fig. 6.29. Although these three segments all give rise to three contact discontinuities of distinct families, they all propagate with the same speed $\lambda = (1 + \nu\kappa)^{-1} = 1/151$, and thus they are simply superimposed to form a single pure thermal wave and propagate as such. To the portions $V \rightarrow Y$ and $Y \rightarrow X$ there corresponds the 1- and 2-simple waves.

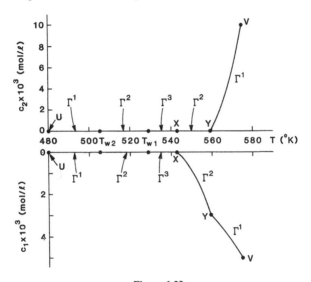

Figure 6.29

Mathematically, we have three independent problems partitioned by the points T_{w1} and T_{w2}, and since the three problems have no influence upon one another, they can be solved separately to establish the physical plane portrait and the profiles at $\tau = 25$ as shown in Fig. 6.30. In contrast to the case of Example 6.7, the 3- and 2-simple waves are the desorption waves of the solutes A_2 and A_1, respectively, whereas the pure thermal wave, propagating in the rear, simply cools the eluted bed with no influence on the desorption process. Obviously, the high-energy state of the initial bed not only accelerates the simple waves considerably but also keeps them active in their desorption performance. We note that in the temperature range of this example the solute A_1 is more strongly adsorbed and thus tends to be retained in the bed. This is readily observed from Fig. 6.30.

Since here the pure thermal wave forms a complete partition between the elution process in the front and the situation in the rear, we may attempt to introduce a second process in the rear. Indeed, by using a feed mixture at the state A (5, 10, 380; 1.029, 3.514), one can establish the consecutive elution and saturation as shown in Fig. 6.31. In essence, this is nothing more than a combination of Example 6.9 with Example 6.8. The two problems can be solved separately

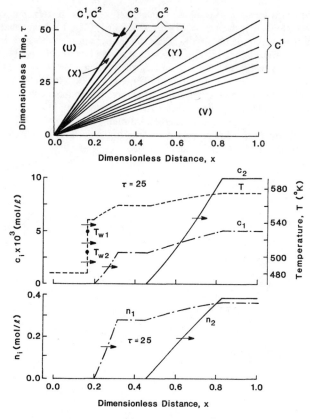

Figure 6.30

and then put together simply by making the contact discontinuity lines overlap. Since the adsorptivity reversal takes place somewhere inside the column, the solute A_1 is more strongly adsorbed in the front, while the solute A_2 adsorbs more strongly in the rear. Consequently, the solute A_2 tends to accumulate on both ends of the column.

6.9 Adiabatic Adsorption with Adsorptivity Reversal

Conventional elution or saturation processes employ a pure solvent as the feed or a fresh bed in the initial column, respectively, and thus its analysis is necessarily involved with the emergence or the exhaustion of each solute within the column. In case of isothermal adsorption with the Langmuir isotherm, which is representative of cases with constant relative adsorptivities, we have already established in Sec. 3.11 that each wave allows only one emergence or one ex-

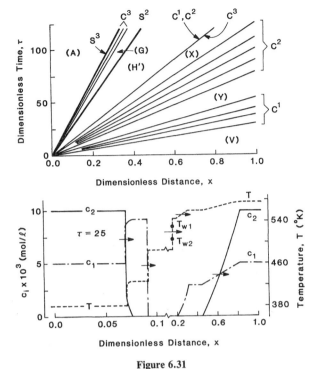

Figure 6.31

haustion of a single solute and these are arranged in the order of decreasing adsorptivity in the direction of the fluid flow.

On the other hand, we have observed in Sec. 6.8 that if the relative adsorptivity is dependent on the temperature, which is one of the state variables, the adsorptivity may reverse between some pairs of solutes and the exchange performance could become quite abnormal.[†] In addition, the situation appears to be even more involved because the energy as a pseudosolute competes in a unique way with real solute species, as illustrated by examples in Sec. 6.7 and 6.8.

With general equilibrium relations, therefore, we should pay more attention in numbering the solute species, noting that the solute species emerging through a particular wave does not necessarily bear the same index as the wave (see Figs. 6.20 and 6.30). This implies that, in general, the emergence of solute species must be examined from the mathematical point of view, which immediately suggests a problem with singularities. Here we would like to treat such problems first in a general context and then illustrate the features by analyzing a numerical example involved with the adsorptivity reversal.

The discussion that follows may be regarded as a direct supplement to Sec.

[†] A similar feature has been briefly discussed by Helfferich and Klein (1970) with respect to the heterovalent ion-exchange system.

6.3, in which the solution to the Riemann problem was constructed. Let us now consider the problem of eluting or saturating an adsorption column under adiabatic conditions, which gives rise to the question of emergence or exhaustion of solute species inside the column. In this regard, the adsorptivity is of great importance and its temperature dependence demands careful investigation. We shall use the symbols $(\cdot)_{i0}$ and $(\cdot)_0$ to represent the state without the solute A_i and the state of no solute at all, respectively. The totality of the state $( )_{i0}$ is an $(m - 1)$-dimensional subspace $\mathcal{H}(m - 1 : c_i = 0)$ of the m-dimensional hodograph space $\mathcal{H}(m)$ and the totality of the state $(\cdot)_0$, which corresponds to the T-axis, is a subspace $\mathcal{H}(1)$ of the $\mathcal{H}(m - 1 : c_i = 0)$. On physical grounds, we claim that

$$(n_i)_{i0} = 0 \qquad (6.9.1)$$

$$(n_{i,i})_{i0} > 0 \qquad (6.9.2)$$

and

$$(n_{i,j})_{i0} = 0 \qquad \text{if } i \neq j \qquad (6.9.3)$$

for $i, j = 1, 2, \ldots, m - 1$. Recalling the definitions of c_m and n_m [see Eqs. (6.1.6) and (6.1.7)], we also find that

$$(n_{i,m})_{i0} = 0 \qquad \text{if } i \neq m \qquad (6.9.4)$$

$$(n_{m,m})_0 = \kappa = \frac{C_s}{C_f} \qquad (6.9.5)$$

and

$$(n_{m,j})_0 = 0 \qquad \text{if } j \neq m \qquad (6.9.6)$$

which are consistent with Eqs. (6.9.2) and (6.9.3).

If the solution is everywhere continuous, its image in the hodograph space $\mathcal{H}(m)$ consists of m different Γ's with distinct indices which are arranged successively from Γ^m to Γ^1 as we pass in the direction of the fluid flow. The direction of each Γ is determined by Eq. (6.3.2), which may be rewritten in the form

$$\sum_{j=1}^{m} (n_{i,j} - \beta \delta_{ij}) \, dc_j = 0, \qquad i = 1, 2, \ldots, m \qquad (6.9.7)$$

where δ_{ij} denotes the Kronecker delta and β is one of the eigenvalues of the matrix $\nabla \mathbf{n} = [n_{i,j}]$; that is,

$$\det[n_{i,j} - \beta \delta_{ij}] = 0 \qquad (6.9.8)$$

When the k-th eigenvalue β_k is used, Eq. (6.9.6) determines the curve Γ^k.

We shall first consider the case when the feed mixture does not contain the solute A_p, so that the solution is involved with the emergence of A_p. In the hodograph space $\mathcal{H}(m)$, the emergence of A_p corresponds to departing from the subspace $\mathcal{H}(m - 1 : c_p = 0)$ to the outside along a Γ and thus we must ask which Γ would be the relevant one. Within the subspace

$\mathcal{H}(m - 1 : c_p = 0)$ we have $(n_{p,j})_{p0} = 0, j = 1, 2, \ldots, m$, from Eqs. (6.9.3) and (6.9.4), and hence Eq. (6.9.7) for $i = p$ takes the simple form

$$\{(n_{p,p})_{p0} - \beta\}\, dc_p = 0 \qquad (6.9.9)$$

while Eq. (6.9.8) may be rewritten as

$$\{(n_{p,p})_{p0} - \beta\} \det_{i,\, j \neq p}\, [n_{i,j} - \beta\delta_{ij}] = 0 \qquad (6.9.10)$$

An immediate consequence is that $(n_{p,p})_{p0}$ is one of the roots of Eq. (6.9.10), say the kth one, so that $\beta_k = (n_{p,p})_{p0}$. It then follows from Eq. (6.9.9) that the concentration c_p remains unchanged unless $\beta = \beta_k$. Therefore, the departure from the subspace $\mathcal{H}(m - 1 : c_p = 0)$ can take place only along a Γ^k. This establishes the conclusion that if $\beta_k = (n_{p,p})_{p0}$, the solute A_p emerges through the k-simple wave, and the line of emergence is given by the characteristic C^k of slope σ_k evaluated with $\beta_k = (n_{p,p})_{p0}$.

Let us next suppose that two solutes A_p and A_q are absent from the feed mixture. We are now interested in the departure from the subspace $\mathcal{H}(m - 2 : c_p = c_q = 0)$, in which Eq. (6.9.7) for $i = p$ and $i = q$ reduces to

$$\{(n_{p,p})_{p0} - \beta\}\, dc_p = 0 \qquad (6.9.11)$$

and

$$\{(n_{q,q})_{q0} - \beta\}\, dc_q = 0 \qquad (6.9.12)$$

respectively, and Eq. (6.9.8) may be expressed in the form

$$\{(n_{p,p})_{p0} - \beta\}\{(n_{q,q})_{q0} - \beta\} \det_{i,\, j \neq p,q}\, [n_{i,j} - \beta\delta_{ij}] = 0 \qquad (6.9.13)$$

If Eq. (6.9.13) gives $\beta_k = (n_{p,p})_{p0}$ and $\beta_h = (n_{q,q})_{q0}$, where $h < k$, the departure from the subspace $\mathcal{H}(m - 2 : c_p = c_q = 0)$ is made by a Γ^k, along which only c_p can change because of Eqs. (6.9.11) and (6.9.12). This implies that as we pass in the direction of the fluid flow, we would first meet the k-simple wave through which the solute A_p emerges while the solute A_q remains at zero concentration. Therefore, if two solutes A_i and A_j are to emerge from zero concentrations and $(n_{i,i})_{i0} < (n_{j,j})_{j0}$, the solute A_j emerges first through a simple wave in the direction of the fluid flow. In case when every solute is different from other solutes adsorptivity, the quantity $(n_{i,i})_{i0}$ can serve as a measure of the adsorptivity of the solute A_i, and thus the statement above implies that the more strongly adsorbed solute would emerge first as we proceed in the direction of the fluid flow.

Coming into the space $\mathcal{H}(m - 1 : c_q = 0)$ along a Γ^k, we find that Eq. (6.9.12) remains unchanged through the corresponding k-simple wave. We also have the equation

$$\{(n_{q,q})_{q0} - \beta\} \det_{i,\, j \neq q}\, [n_{i,j} - \beta\delta_{ij}] = 0 \qquad (6.9.14)$$

Therefore, unless $(n_{q,q})_{q0}$ becomes the kth root of Eq. (6.9.14), c_q remains zero throughout the k-simple wave. We expect that this will be the case under normal circumstances but, if the adsorptivities reverse between the solutes A_p and A_q so that $(n_{q,q})_{q0}$ becomes the kth root of Eq. (6.9.14) at some point along the

Γ^k, the solute A_q emerges within the k-simple wave. This leads us to the conclusion that without a reversal between the adsorptivities, a simple wave can accommodate only one emergence of a single solute at its left margin.

In the elution problem, the feedstream consists of the pure solvent and thus its image in the hodograph space falls on the subspace $\mathcal{H}(1)$, which represents the T-axis. In this case Eqs. (6.9.7) and (6.9.8) reduce to

$$\{(n_{i,i})_0 - \beta\}\, dc_i = 0, \qquad i = 1, 2, \ldots, m - 1 \qquad (6.9.15)$$

$$(\kappa - \beta)\, dT = 0 \qquad (6.9.16)$$

and

$$(\kappa - \beta) \prod_{i=1}^{m-1} \{(n_{i,i})_0 - \beta\} = 0 \qquad (6.9.17)$$

respectively. If the largest root of Eq. (6.9.17) is given by one of the $(n_{i,i})_0$, say $(n_{p,p})_0$, so that $\beta_m = (n_{p,p})_0$, Eq. (6.9.15) asserts that the solute A_p, which is the most strongly adsorbed solute at the temperature, emerges through the m-simple wave. Without adsorptivity reversals, it should be clear from the discussion above that each simple wave except for one gives rise to the emergence of a single solute. Hence there are $(m - 1)$ waves involved with the emergence of a solute, respectively, and these $(m - 1)$ emergences are arranged in the order of decreasing adsorptivity as we proceed in the direction of the fluid flow.

If Eq. (6.9.17) gives $\beta_m = \kappa$, we see that Eq. (6.9.15) prohibits any change in concentrations through the m-wave. Since only the temperature is allowed to change, the m-wave becomes a pure thermal wave across which the characteristic direction remains fixed. Hence the m-characteristic field is linearly degenerate and the m-wave is given by a contact discontinuity. In practice, this situation may occur at a very high temperature where all the adsorptivities would become very small (see Examples 6.5 and 6.10).

So far we have been dealing with the elution problem. Clearly, the same argument should be valid for the saturation problem if we proceed in the reverse direction. Therefore, the least strongly adsorbed solute will be encountered first, and so on. In particular, the 1-wave often becomes a pure thermal wave, for under normal conditions we would obtain $\beta_1 = \kappa$ from Eq. (6.9.17).

Next, we shall examine the emergence of a solute across a discontinuity, for which we have the compatibility relations (6.2.18):

$$\frac{[n_1]}{[c_1]} = \frac{[n_2]}{[c_2]} = \cdots = \frac{[n_m]}{[c_m]} = \tilde{\beta} \qquad (6.9.18)$$

We have shown in Sec. 6.2 that given the state on one side of a discontinuity, Eq. (6.9.18) determines m different curves Σ for the locus of the feasible state across the discontinuity. Choosing the one that is tangent to the Γ^k passing through the point of the given state, we call it the Σ^k, and then every point on the Σ^k can be connected to the given state by a k-discontinuity. Also, to every point on the Σ^k there corresponds a unique value $\tilde{\beta}_k$ determined by Eq. (6.9.18).

Suppose that the solute A_p is absent from the feedstream so that the concen-

More on Hyperbolic Systems   Chap. 6

tration field involves the emergence of the solute A_p at some position inside the column. In the hodograph space, this is represented by the departure from the subspace $\mathcal{H}(m - 1 : c_p = 0)$. Let us now consider a case when Eq. (6.9.10) requires this departure to be made by a Γ^k along which, however, σ_k increases as we move away from the subspace $\mathcal{H}(m - 1 : c_p = 0)$. In such a case the k-wave, through which the solute A_p is to emerge, is given by a discontinuity and hence we must construct the curve Σ^k. Since the curve Σ^k is tangent to the Γ^k at the point of departure, it is directed away from the subspace $\mathcal{H}(m - 1 : c_p = 0)$. At every point along the curve Σ^k, the compatibility relations (6.9.18) may be rewritten in the form

$$\frac{[n_1]}{[c_1]} = \frac{[n_2]}{[c_2]} = \cdots = \frac{n_p^r}{c_p^r} = \cdots = \frac{[n_m]}{[c_m]} = \tilde{\beta}_k \qquad (6.9.19)$$

where the superscript r denotes the state on the right side of the discontinuity and the last equality to $\tilde{\beta}_k$ signifies the fact that the curve Σ^k must be chosen so as to be tangent to the Γ^k at the point of departure. The argument above implies that, depending on the nature of the k-characteristic field, it is possible to have a sudden emergence of a solute across a k-discontinuity.

Next we shall consider a case when Eq. (6.9.13) predicts that the solutes A_p and A_q emerge through the k-wave and h-wave, respectively, where $h < k$. This implies that A_p is more strongly adsorbed than A_q. Suppose that the entropy condition requires the k-wave to be a discontinuity. If both solutes are to emerge simultaneously across the k-discontinuity, we must have

$$\frac{n_p^r}{c_p^r} = \frac{n_q^r}{c_q^r} = \tilde{\beta}_k \qquad (6.9.20)$$

which cannot be satisfied unless the two solutes happen to have identical adsorptivities for the state on the right of the k-discontinuity [see Eq. (6.8.11)]. Since A_p is more strongly adsorbed for the state on the left, we can write Eq. (6.9.19) with the ratio $[n_q]/[c_q]$ deleted and thus A_p emerges across the k-discontinuity while A_q remains at zero concentration.

If the feedstream consists of pure solvent, the compatibility relations applied to the m-discontinuity give rise to the equation

$$\frac{n_p^r}{c_p^r} = \kappa - \frac{-\Delta H_p}{C_f} \frac{n_p^r}{T^r - T^l} = \tilde{\beta}_m \qquad (6.9.21)$$

in which A_p is assumed to be the most strongly adsorbed and T^l corresponds to the feed temperature. We note that $T^l > T^r$ unless $n_p^r = 0$, and thus the expression in the middle is greater than κ. An immediate consequence is that the m-discontinuity allows the solute A_p to emerge only if the ratio $n_p^r/c_p^r > \kappa$, and otherwise it is a pure thermal wave because $n_p^r = 0$ and $\tilde{\beta}_m = \kappa$.

For a saturation problem, left and right with respect to the discontinuity must be interchanged and so we must interchange the roles of the superscripts r and l to establish a similar argument.

From the discussion above, it is evident that for a system involved with adsorptivity reversals the wave, which would give the emergence of a particular solute cannot be fixed, and furthermore, it could happen that a simple wave contains several emergence lines of different solutes. It is also expected that if a particular solute switches its order of adsorptivity within a simple wave region, its concentration profile may assume a local extremum.

If the heat capacity ratio κ is less than any of the $(m - 1)$ quantities $(n_{i,i})_0$, $i = 1, 2, \ldots, m - 1$, evaluated at the initial bed temperature, the energy may be considered as the least strongly adsorbed species, and thus the saturation process will develop a pure thermal wave which propagates the fastest (see Figs. 6.12 and 6.25). On the other hand, if κ is larger than any of the $(m - 1)$ quantities $(n_{i,i})_0$, $i = 1, 2, \ldots, m - 1$, evaluated at the feed temperature, we may regard the energy as the most strongly adsorbed species, and thus the elution process gives rise to a pure thermal wave propagating with the least speed (see Figs. 6.20 and 6.30).

To be more specific we shall introduce the Langmuir isotherm

$$n_i = \frac{N_i K_i c_i}{1 + \sum\limits_{j=1}^{m-1} K_j c_j} \tag{6.9.22}$$

in which

$$K_i = K_{0i} T^{1/2} e^{-\Delta H_i / RT} \tag{6.9.23}$$

for $i = 1, 2, \ldots, m - 1$. Clearly, we see that the conditions specified by Eqs. (6.9.1)–(6.9.6) are all satisfied.

The conventional relative adsorptivity is defined as

$$\gamma_{ij} = \frac{n_i/c_i}{n_j/c_j} = \frac{N_i K_{0i}}{N_j K_{0j}} e^{-(\Delta H_i - \Delta H_j)/RT} \tag{6.9.24}$$

and it turns out that this is the same as the ratio of $n_{i,i}$ to $n_{j,j}$ when both solutes A_i and A_j are absent. If $N_i K_{0i} \geq N_j K_{0j}$ and $-\Delta H_i > -\Delta H_j$, then $\gamma_{ij} > 1$ for all temperature values. On the other hand, if $N_i K_{0i} < N_j K_{0j}$ and $-\Delta H_i > -\Delta H_j$, γ_{ij} decreases monotonically as the temperature increases and becomes less than one above a specific temperature T_{ij}^R, at which an adsorptivity reversal takes place between the solutes A_i and A_j.

Along the temperature axis in the hodograph space, we have $(n_m/c_m)_0 = (n_{m,m})_0 = \kappa$ and the ratio $(\gamma_{im})_0 = N_i K_i / \kappa$ may be regarded as the relative adsorptivity between the solute A_i and the energy as the pseudosolute. The ratio $(\gamma_{im})_0$ decreases as the temperature increases and the temperature at which $(\gamma_{im})_0 = 1$ will be called T_{wi} as in previous sections. There are $(m - 1)$ different temperatures of this kind and these values divide the temperature axis into m disjoint portions. Applying Eqs. (6.9.15), (6.9.16), and (6.9.17), we see that each of the m portions corresponds to a Γ^k of distinct index and they are ordered successively from Γ^1 to Γ^m as the temperature increases. This implies that the effect of the energy state becomes more significant as it reaches a higher level. It becomes predominant if the temperature gets higher than the maximum of T_{wi}, $i = 1, 2, \ldots, m - 1$.

For the purpose of illustration we shall now consider the numerical example which was treated in Sec. 6.8; that is, the adiabatic adsorption of benzene and

More on Hyperbolic Systems   Chap. 6

cyclohexane on charcoal, which will be assumed fixed in the column. With the data provided there, it was shown that $T^R = T^R_{12} = 449.9°K$, $T_{w2} = 505.5°K$, and $T_{w1} = 529°K$, where cyclohexane and benzene are identified as the solutes A_1 and A_2, respectively. This implies that for $T < T^R$, benzene is more strongly adsorbed, whereas for $T > T^R$, cyclohexane is more strongly adsorbed.

For Example 6.8 (see Figs. 6.24 and 6.25) we observe that when an initially fresh bed is saturated, the 3- and 2-waves are the adsorption fronts, respectively, for the more and less strongly adsorbed solutes; that is, A_2 and A_1, respectively, while the 1-wave simply carries away the heat evolved. The feed temperature was 380°K and such a high temperature caused the 3-wave to be a combined wave, which was partially diffuse and thus unfavorable. Taking into account the adsorptivity reversal, we expect that the saturation characteristics would become more involved as the feed temperature increases. We shall make a further investigation in this regard. Here a specific state of concentrations and temperature will be denoted by the same scheme as in Sec. 6.8.

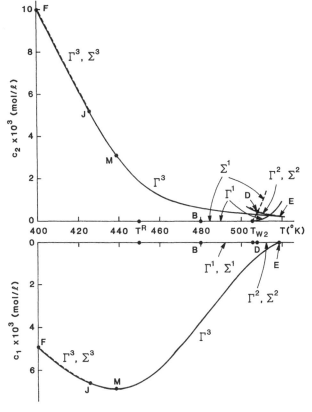

Figure 6.32

Suppose that a feedstream at the state F (5, 10, 400; 1.078, 3.098) is introduced to saturate a fresh bed at 480°K; that is, at the state B. As constructed in Fig. 6.32, the image of the solution in the hodograph space consists of $F \rightarrow J \rightarrow M \rightarrow E$ (a Σ^3 and a Γ^3 joined together at J), $E \rightarrow D$ (a Σ^2), and $D \rightarrow T_{w2} \rightarrow B$ (a Σ^1). The physical plane portrait of the solution is shown in the upper diagram of Fig. 6.33 and the profiles of concentrations and temperature at $\tau = 120$ are presented in the middle and lower diagrams.

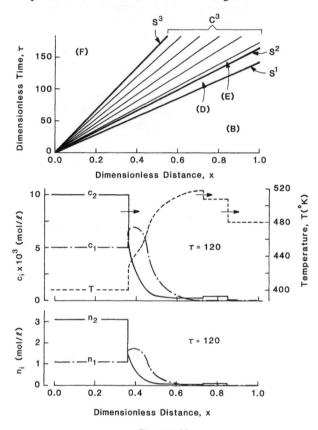

Figure 6.33

If we start from the feed state F and proceed in the direction of fluid flow, the 3-semishock ($F \rightarrow J$) adsorbs the solute A_2 partially and pushes the solute A_1 ahead. The semishock is joined at the state J by the 3-simple wave ($J \rightarrow M \rightarrow E$), which adsorbs A_2 further. Through this simple wave c_1 reaches a maximum when $T = 438.5°K$, beyond which the solute A_1 appears to be more strongly adsorbed than the solute A_2. At the right edge of the simple wave, therefore, the solute A_1 is completely adsorbed, whereas a small amount of A_2 is passed forward.

The 3-constant state E (0, 0.202, 518.8; 0, 0.24) assumes a very high en-

More on Hyperbolic Systems   Chap. 6

ergy state, and thus the 2-shock ($E \rightarrow D$) pushes the solute A_2 further instead of adsorbing it. The solute A_2 is completely adsorbed across the 1-shock ($D \rightarrow B$). Here the 1-wave cannot be a pure thermal wave because the temperature jumps across $T_{w2} = 505.5°K$, at which the curve Σ^1 departs from the T axis.

With a feed mixture at a higher temperature, that is, at the state F' (5, 10, 420; 1.075, 2.644), the image of the solution in the hodograph space is given by the segments, $F' \rightarrow J' \rightarrow M' \rightarrow G$ (a Σ^3 and a Γ^3 joined at J), $G \rightarrow H$ (a Σ^2), and $H \rightarrow T_{w2} \rightarrow B$ (a Σ^1) as depicted in Fig. 6.34. The upper diagram of Fig. 6.35 shows the physical plane portrait of the solution, while in the middle and lower diagrams the profiles of concentrations and temperature at $\tau = 120$ are presented.

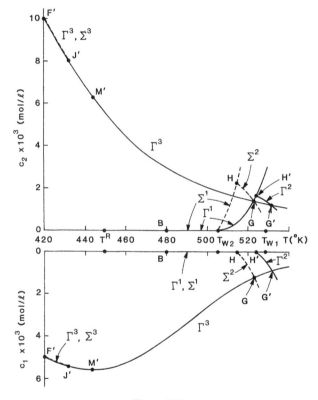

Figure 6.34

The overall features remain unchanged except for the fact that the Γ^3 ($J \rightarrow G$) does not reach the (c_2, T)-plane as in the previous case (see point E of Fig. 6.32). This is because the T axis for $T > T_{w1} = 529°K$ takes the role of a Γ^3. Therefore, the 3-combined wave cannot achieve complete adsorption of either solute. Instead, the solutes A_1 and A_2 are completely adsorbed across the 2-shock ($G \rightarrow H$) and the 1-shock ($H \rightarrow B$), respectively, as we can see more clearly in Fig. 6.35. Comparing Fig. 6.35 with Fig. 6.33, we note that the effect of the high energy introduced by the feedstream becomes more pronounced as the feed temperature increases.

In both Figs. 6.32 and 6.33, it is observed that the projections of the Γ^3 onto the (c_1, T)-plane show peak points M (6.95, 3.11, 438.5; 1.722, 0.831) and M' (5.56, 6.291, 444; 1.263, 1.476), respectively. This feature then gives rise to the unusual variation in c_1 as depicted in Figs. 6.33 and 6.35. We would expect that when the c_1 profile passes through the maximum point, the adsorptivities reverse between the two solutes and thus to the right of the maximum point the solute A_1 is more strongly adsorbed. The totality of those peak points (e.g., the points M and M') will be given by a surface in the hodograph space which intersects the T axis at $T_{12}^R = 449.9°K$. This implies that the conventional relative adsorptivity defined by Eq. (6.9.24) may not predict the adsorptivity reversal adequately in general.

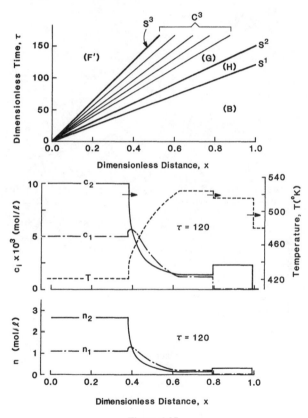

Figure 6.35

6.10 Shock Layer Analysis of Adiabatic Adsorption

So far we have been concerned with adiabatic adsorption under equilibrium conditions. In actual situations, however, we expect to have not only dispersive effects in the fluid phase but also interphase transfer resistances and these will definitely smear out any discontinuity in the profiles of concentrations and temperature, as we have already observed in Secs. 7.10 of Vol. 1 and 2.10 of the present volume. Here again we employ the shock layer approach to investigate the effect of axial dispersions of mass and energy.

Let us consider the adiabatic adsorption of a single solute in a fixed bed, in which local equilibrium is established between the two phases. If we let E_1 denote the axial dispersion coefficient for the solute and take the axial position z in the direction of the fluid flow, the conservation equation for the solute at time t can be written in the form

$$E_1 \frac{\partial^2 c}{\partial z^2} = V \frac{\partial c}{\partial z} + \frac{\partial c}{\partial t} + \frac{1 - \epsilon}{\epsilon} \frac{\partial n}{\partial t} \qquad (6.10.1)$$

$$n = f(c, T) \qquad (6.10.2)$$

where f represents the equilibrium relation. Since thermal equilibrium is established between the two phases, the energy conservation equation is given in the form

$$E_2 \frac{\partial^2 T}{\partial z^2} = V \frac{\partial T}{\partial z} + \frac{\partial T}{\partial t} + \frac{1 - \epsilon}{\epsilon} \left\{ \frac{C_s}{C_f} T - \frac{-\Delta H}{C_f} n \right\} \qquad (6.10.3)$$

in which E_2 denotes the effective thermal diffusivity.

Just as in Secs. 6.1 and 6.7, we shall put

$$c_1 = c, \qquad c_2 = T$$

$$n_1 = n, \qquad n_2 = \frac{C_s}{C_f} T - \frac{-\Delta H}{C_f} n \qquad (6.10.4)$$

$$f_1 = f(c_1, c_2), \qquad f_2 = \frac{C_s}{C_f} c_2 - \frac{-\Delta H}{C_f} f(c_1, c_2)$$

We shall introduce the dimensionless parameters defined as

$$\nu = \frac{1 - \epsilon}{\epsilon} \qquad (6.10.5)$$

$$Pe_i = \frac{VZ}{E_i} \qquad (6.10.6)$$

for $i = 1, 2$, where Z is a characteristic length of the system, and the usual dimensionless variables

$$x = \frac{z}{Z}, \qquad \tau = \frac{Vt}{Z} \qquad (6.10.7)$$

Then Eqs. (6.10.1), (6.10.2), and (6.10.3) can be rearranged into the form

$$\frac{1}{Pe_i} \frac{\partial^2 c_i}{\partial x^2} = \frac{\partial c_i}{\partial x} + \frac{\partial c_i}{\partial \tau} + \nu \frac{\partial n_i}{\partial \tau} \qquad (6.10.8)$$

$$n_i = f_i(c_1, c_2) \qquad (6.10.9)$$

for $i = 1, 2$, where the functions f_i satisfy the conditions

$$\frac{\partial f_i}{\partial c_j} \begin{cases} > 0 & \text{if } i = j \\ < 0 & \text{if } i \neq j \end{cases} \qquad (6.10.10)$$

We note that Eq. (6.10.8) takes the same form as Eq. (2.10.7).

Suppose now the bed is extended indefinitely in both directions and the boundary conditions are specified as

$$\begin{aligned} c_i &= c_i^l \quad \text{at} \quad x = -\infty \\ c_i &= c_i^r \quad \text{at} \quad x = +\infty \end{aligned} \qquad (6.10.11)$$

We then ask if there exists a moving coordinate ξ defined as

$$\xi = x - \tilde{\lambda}\tau \qquad (6.10.12)$$

with constant $\tilde{\lambda}$, in which the solution to Eqs. (6.10.8) and (6.10.9) can be expressed in the form

$$c_i(x, \tau) = c_i(\xi) \qquad (6.10.13)$$

for $i = 1, 2$ and satisfies the conditions

$$\begin{aligned} c_i = c_i^l \quad \text{and} \quad \frac{dc_i}{d\xi} = 0 \quad \text{at} \quad \xi = -\infty \\ c_i = c_i^r \quad \text{and} \quad \frac{dc_i}{d\xi} = 0 \quad \text{at} \quad \xi = +\infty \end{aligned} \qquad (6.10.14)$$

If such a real value $\tilde{\lambda}$ exists, Eq. (6.10.13) represents the shock layer of the system (6.10.8) and (6.10.9) and $\tilde{\lambda}$ is the propagation speed of the shock layer.

If there exists a shock layer, Eqs. (6.10.8) and (6.10.9) can be transformed into a pair of ordinary differential equations

$$\frac{1}{Pe_i} \frac{d^2 c_i}{d\xi^2} = (1 - \tilde{\lambda}) \frac{dc_i}{d\xi} - \nu \tilde{\lambda} \frac{df_i}{d\xi} \qquad (6.10.15)$$

Integrating from $\xi = -\infty$, we obtain

$$\frac{1}{Pe_i} \frac{dc_i}{d\xi} = (1 - \tilde{\lambda})(c_i - c_i^l) - \nu \tilde{\lambda}(f_i - f_i^l)$$

or

$$\frac{1}{\nu\tilde{\lambda}\,Pe_i}\frac{dc_i}{d\xi} = \tilde{\beta}(c_i - c_i^l) - (f_i - f_i^l) \tag{6.10.16}$$

for $i = 1, 2$, in which $f_i^l = f_i(c_1^l, c_2^l)$ and

$$\tilde{\beta} = \frac{1 - \tilde{\lambda}}{\nu\tilde{\lambda}} \tag{6.10.17}$$

Applying the second condition of Eq. (6.10.14) to Eq. (6.10.16) gives

$$\frac{1}{\tilde{\lambda}} = 1 + \nu\frac{f_i^l - f_i^r}{c_i^l - c_i^r} \tag{6.10.18}$$

where $f_i^r = f_i(c_1^r, c_2^r)$. Since $\tilde{\lambda}$ is independent of the subscript i, we must have

$$\frac{f_1^l - f_1^r}{c_1^l - c_1^r} = \frac{f_2^l - f_2^r}{c_2^l - c_2^r} = \tilde{\beta} \tag{6.10.19}$$

or, in terms of the original variables,

$$\frac{n^l - n^r}{c^l - c^r} = \frac{C_s}{C_f} - \frac{-\Delta H}{C_f}\frac{n^l - n^r}{T^l - T^r} \tag{6.10.20}$$

which is identical to the compatibility relation (6.7.20). This implies that if there exists a shock layer, its end states $L(c^l, T^l)$ and $R(c^r, T^r)$ must fall on the same curve Σ in the hodograph plane. If the end states fall on a Σ^1 (or Σ^2), the shock layer propagates with the same speed $\tilde{\lambda}_1$ (or $\tilde{\lambda}_2$) as the corresponding shock, which is independent of Pe_i.

A shock layer, if it exists, must satisfy Eq. (6.10.16) and the second condition of Eq. (6.10.14). We note that Eq. (6.10.16) is exactly the same as Eq. (2.10.21), for which the existence as well as the uniqueness of the solution has been established in Sec. 2.10 for arbitrary equilibrium relations f_1 and f_2. The only adjustments required are as follows:

$$\Gamma_- \longleftrightarrow \Sigma^2, \quad \Gamma_+ \longleftrightarrow \Sigma^1, \quad \zeta_- \longleftrightarrow \zeta_2, \quad \zeta_+ \longleftrightarrow \zeta_1, \quad \sigma_+ \longleftrightarrow \sigma_2$$

$$\sigma_- \longleftrightarrow \sigma_1, \quad \tilde{\sigma}_+ \longleftrightarrow \tilde{\sigma}_2, \quad \tilde{\sigma}_- \longleftrightarrow \tilde{\sigma}_1, \quad \tilde{\lambda}_+ \longleftrightarrow \tilde{\lambda}_2, \quad \tilde{\lambda}_- = \tilde{\lambda}_1$$

Therefore, we shall not repeat the proof here but simply rephrase the conclusions in the present terms:

1. There exists a shock layer if and only if the end states satisfy the shock conditions

$$\sigma_2^l < \tilde{\sigma}_2 < \sigma_2^r, \qquad \tilde{\sigma}_2 > \sigma_1^l, \sigma_1^r \tag{6.10.21}$$

or

$$\sigma_1^l < \tilde{\sigma}_1 < \sigma_1^r, \qquad \tilde{\sigma}_1 < \sigma_2^l, \sigma_2^r \tag{6.10.22}$$

2. For a given state $L(c_1^l, c_2^l)$ there can be two different shock layers, one having the end states on the curve Σ^2 and the other on the curve Σ^1. In the for-

mer case L is an unstable node and R a saddle point so that the shock layer can be constructed by backward integration of Eq. (6.10.16). In the latter case, however, Eq. (6.10.16) must be integrated forward because L is a saddle point and R a stable node.

3. If $\bar{\lambda}$ or one of c^r and T^r is further specified, the shock layer of either kind is unique.

4. A shock layer propagates with the same speed as the corresponding shock, which is independent of the dispersion coefficients.

To illustrate the application we shall consider the same example with the Langmuir isotherm as in Sec. 6.7, which is concerned with the adiabatic adsorption of benzene vapor on a charcoal bed with nitrogen as carrier at 10 atm. Thus we have

$$\nu = 1.0, \qquad C_s = 405.0 \text{ cal/liter} \cdot {}^\circ K, \qquad C_f = 2.7 \text{ cal/liter} \cdot {}^\circ K$$

$$N = 5.50 \text{ mol/liter}, \quad K_0 = 3.88 \times 10^{-5} \text{ liter/mol} \cdot {}^\circ K^{1/2},$$

$$-\Delta H = 10.4 \text{ kcal/mol}$$

First, we shall solve numerically the transient equations [Eqs. (6.10.8) and (6.10.9)] with the Langmuir isotherm

$$f(c, T) = \frac{NK(T)c}{1 + K(T)c}$$

$$K(T) = K_0 T^{1/2} e^{-\Delta H/RT}$$

(6.10.23)

and compare the shock layer with the transient solution to demonstrate the validity and usefulness of the shock layer analysis. For convenience, we shall put

$$\text{Pe}_1 = \text{Pe}_2 = 40$$

Since we are interested in the saturation process, the initial and boundary conditions are prescribed as

at $\tau = 0$: $c = 0$ and $T = 350{}^\circ K$

at $x = 0$: $c = 0.01$ mol/liter   and $T = 350{}^\circ K$

at $x = X$: $\dfrac{dc}{dx} = 0$ and $\dfrac{dT}{dx} = 0$

The fixed boundary condition at $x = 0$ is considered reasonable in the analysis of fixed-bed adsorption processes. The numerical scheme briefly discussed in Sec. 2.10 is introduced here to generate the profiles of concentrations and temperature at successive times.

The profiles at $\tau = 100$ are shown in Fig. 6.36. The pure thermal wave propagating in the front has a spreading tendency and this is just what we have expected because the pure thermal wave corresponds to a contact discontinuity in a nondispersive system, and thus no shock layer exists in this case. However, the position of the middle point, $T \cong 376{}^\circ K$ at $x = 0.662$, could have

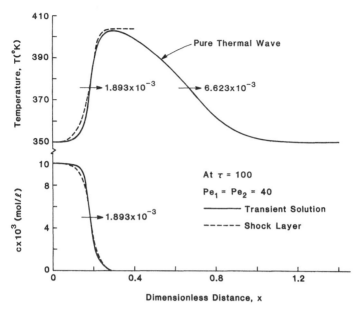

Figure 6.36

been predicted by the propagation speed $\lambda = (1 + \nu C_s / C_f)^{-1} = 6.623 \times 10^{-3}$ multiplied by the time $\tau = 100$.

Although the plateau is not fully established, probably due to the spreading tendency of the pure thermal wave, the peak temperature (403°K) comes quite close to the plateau value (403.85°K) that can be determined a priori by Eq. (6.10.20), with $c^r = 0$, $c^l = 0.01$ mol/liter, and $T = 350$°K. Here again the position of the middle point (0.005 mol/l) in the adsorption front is nearly coincident with the location predicted by the propagation speed $\bar{\lambda} = 1.893 \times 10^{-3}$, which is determined from Eq. (6.10.18).

It is obvious that in the corresponding nondispersive system, there would be a shock corresponding to the adsorption front which has the end states on a Σ^2. This implies that there exists a shock layer corresponding to the adsorption front and it can be constructed by integrating Eq. (6.10.16) backward from the state, $c = 0$ and $T = 403.85$°K, with $\bar{\lambda} = 1.893 \times 10^{-3}$. The shock layer so determined is compared with the transient solution by matching the middle point of the concentration profile as depicted in Fig. 6.36. Clearly, the transient profile can be well approximated by the shock layer located by using the propagation speed $\bar{\lambda}$. As the time increases, however, the peak temperature of the transient profile decreases due to the spreading tendency of the pure thermal wave and thus the adsorption front cannot maintain a constant pattern profile.

We have observed from Example 6.4 that if the initial bed is slightly saturated and if it is to be further saturated, the equilibrium model predicts two shocks in the solution. To examine a case with two shock layers, therefore, we shall treat another example in which the boundary conditions remain the same as before but the initial condition is specified as

$$\text{at } \tau = 0: \quad c = 5 \times 10^{-4} \text{mol/liter} \quad \text{and} \quad T = 350°\text{K}$$

The profiles of concentrations and temperature at $\tau = 145$ are presented in Fig. 6.37. We note that between the two adsorption fronts a plateau develops on which we have $c = 3.66 \times 10^{-3}$ mol/liter and $T = 397.2°\text{K}$.

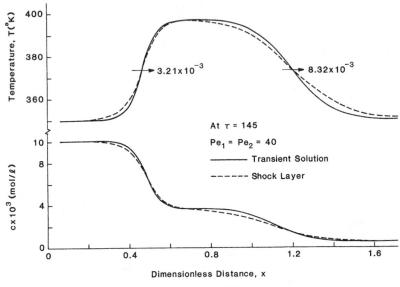

Figure 6.37

If the corresponding nondispersive system is analyzed, we expect to see the same plateau developing between two shocks which propagate with speeds $\tilde{\lambda}_2 = 3.21 \times 10^{-3}$ and $\tilde{\lambda}_1 = 8.32 \times 10^{-3}$, respectively. If the two shocks are located in Fig. 6.37, we would find that the position of each shock coincides with that of the middle point of the temperature profile in each front.

To obtain the shock layer corresponding to the fast-moving front, we start from the state of the plateau and integrate Eq. (6.10.16) forward with $\tilde{\lambda} = 8.32 \times 10^{-3}$. On the other hand, the other shock layer can be constructed by integrating the same equation backward with $\tilde{\lambda} = 3.21 \times 10^{-3}$ starting from the state of the plateau. These shock layers are then located in Fig. 6.37 by matching the middle points of the temperature profile in both fronts. The shock layers are in fair agreement with the transient profile, and indeed it can be seen that the transient profiles converge to the shock layer as time goes on. We may therefore conclude that whenever two shocks are expected from a nondispersive system, the transient behavior for large time of the corresponding dispersive system can be estimated very well by applying the shock layer analysis.

More on Hyperbolic Systems   Chap. 6

Next we consider each of the shock layers from the same system as above separately and make one of the Peclet numbers vary while the other remains unchanged to investigate the individual effect of dispersion of mass or energy.

Let us first take the case with the end states ($c^l = 0.01$ mol/liter, $T^l = 350°K$) and ($c^r = 3.66 \times 10^{-3}$ mol/liter, $T^l = 397.2°K$) that fall on a Σ^2 with $\bar{\lambda} = 3.21 \times 10^{-3}$ and satisfy the shock condition. Integrating Eq. (6.10.16) backward, we obtain the shock layers for various sets of Pe_1 and Pe_2, and these are plotted in Figs. 6.38 and 6.39 by using the middle point of the temperature profile as a common fixed point.

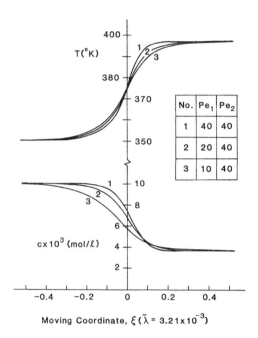

No.	Pe_1	Pe_2
1	40	40
2	20	40
3	10	40

Moving Coordinate, ξ ($\bar{\lambda} = 3.21 \times 10^{-3}$)

Figure 6.38

Shock layers for three different values of the mass dispersion coefficient with the effective thermal diffusivity fixed are compared in Fig. 6.38. We clearly observe that the profiles tend to become steeper as the Peclet number for mass increases and that the dispersion of mass has a much stronger influence on the concentration profile than on the temperature profile. It is also noticed that as the Peclet number for mass decreases, its effect on the concentration profile becomes more significant.

In Fig. 6.39 we present the opposite situation, in which the Peclet number for energy changes while that for mass remains fixed. Here the effect turns out to be much more pronounced on both the temperature profile and the concentration profile, although the overall features remain the same. It appears that the thermal dispersion tends to bring the adsorption front forward.

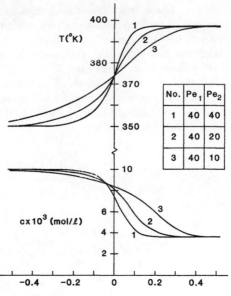

Figure 6.39

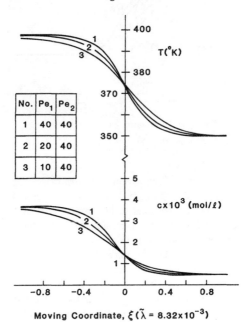

Figure 6.40

More on Hyperbolic Systems   Chap. 6

In the second case we have the end states ($c^l = 3.66 \times 10^{-3}$ mol/liter, $T^l = 397.2°$K) and ($c^r = 5 \times 10^{-4}$ mol/liter, $T^r = 350°$K) that fall on a Σ^1 with $\bar{\lambda} = 8.32 \times 10^{-3}$ and satisfy the shock condition. By integrating Eq. (6.10.16) forward we generate the shock layers for various sets of values of Pe_1 and Pe_2. The profiles are plotted in Figs. 6.40 and 6.41 for comparison, from which we can make similar observations with the thermal dispersion having more pronounced influence than the mass dispersion. However, it appears that both the thermal and mass dispersions tend to make the adsorption front more retained.

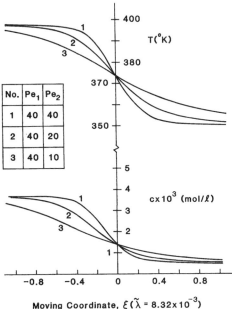

No.	Pe_1	Pe_2
1	40	40
2	40	20
3	40	10

Figure 6.41

REFERENCES

6.1–6.4 and **6.6** The theoretical development of these sections follows somewhat on the lines laid down in

H.-K. RHEE and N. R. AMUNDSON, "An analysis of an adiabatic adsorption column: Part I. Theoretical development," *Chem. Engrg. J.* **1**, 241–254 (1970).

6.2. The discussion on the nature of the curve Σ^k is to be found in

P. D. LAX, "Hyperbolic systems of conservation laws: II," *Comm. Pure Appl. Math.* **10**, 537–566 (1957).

For the thermodynamic argument mentioned at the end of this section, see

E. KVAALEN, L. NEEL, and D. TONDEUR, "Directions of quasi-static mass and energy transfer between phases in multicomponent open systems," *Chem. Engrg. Sci.* **40**, 1191–1204 (1985).

6.4. The extended entropy condition has been discussed fully in

T.-P. LIU, "The Riemann problem for general 2 × 2 conservation laws," *Trans. Amer. Math. Soc.* **199,** 89–112 (1974).

T.-P. LIU, "The Riemann problem for general systems of conservation laws," *J. Differential Equations* **18,** 218–234 (1975).

Concerning the development of singularities and their limiting behavior, see

A. JEFFREY, "The development of jump discontinuities in nonlinear hyperbolic systems of equations in two independent variables," *Arch. Rational Mech. Anal.* **14,** 27–37 (1963).

A. JEFFREY, "The developments of singularities of solutions of nonlinear hyperbolic equations of order greater than unity," *J. Math. Mech.* **15,** 585–598 (1966).

F. JOHN, "Formation of singularities in one-dimensional nonlinear wave propagation," *Comm. Pure Appl. Math.* **27,** 377–405 (1974).

T.-P. LIU, "Development of singularities in the nonlinear waves for quasilinear hyperbolic partial differential equations," *J. Differential Equations* **33,** 92–111 (1979).

R. DIPERNA, "Singularities of solutions of nonlinear hyperbolic systems of conservation laws," *Arch. Rational Mech. Anal.* **60,** 75–100 (1975).

This definition of weak solutions is given in the 1957 paper by Lax cited previously for Sec. 6.2.

An example of a system of equations that cannot be represented in the form of conservation laws is to be found in

B. C. ROZDESTVENSKII, "Conservativeness of systems of quasi-linear equations," *Uspehi Mat. Nauk (N.S).* **14,** 217–218 (1959); English translation in *Amer. Math. Soc. Transl.,* Ser. 2, **42,** 37–39 (1964).

This concept of entropy is discussed in some detail in

K. O. FRIEDRICHS and P. D. LAX, "Systems of conservation equations with a convex extension," *Proc. Nat. Acad. Sci. USA* **68,** 1686–1688 (1971).

See also

P. D. LAX, "Shock waves and entropy," in *Contributions to Nonlinear Functional Analysis,* edited by E. H. Zarantonello, Academic Press, New York, 1971, pp. 603–634.

P. D. LAX, "Hyperbolic systems of conservation laws and the mathematical theory of shock waves," *Regional Conference Series in Applied Mathematics,* No. 11, SIAM, Philadelphia, 1973.

P. D. LAX, "On the notion of hyperbolicity," *Comm. Pure Appl. Math.* **33,** 395–397 (1980).

J. SMOLLER, *Shock Waves and Reaction-Diffusion Equations,* Springer-Verlag, New York, 1983, pp. 397–403.

6.5. For the Riemann problem, the existence of a unique solution is established in the 1957 paper by Lax cited previously for Sec. 6.2 and in the 1975 paper by Liu cited previously for Sec. 6.4.

The first existence proof of global solutions to quasilinear systems with distributed initial data is found in

J. GLIMM, "Solutions in the large for nonlinear hyperbolic systems of equations," *Comm. Pure Appl. Math.* **18,** 697–715 (1965).

See also

T.-P. Liu, "The deterministic version of the Glimm scheme," *Comm. Math. Phys.* **57**, 135–148 (1977).

A. E. Hurd, "A uniqueness theorem for weak solutions of symmetric quasilinear hyperbolic systems," *Pacific J. Math.* **28**, 555–559 (1969).

For the equations of nonisentropic flow, see, for example,

R. Courant and K. O. Friedrichs, *Supersonic Flow and Shock Waves,* Wiley-Interscience, New York, 1948.

The equations of isentropic flow are treated in

S. K. Godunov, "On the uniqueness of the solution of the equations of hydrodynamics," *Mat. Sb.* **82**, (N.S. 40), 467–478 (1956), in Russian.

The Riemann problem for the equations of nonisentropic flow [Eq. (6.5.11)] is discussed in

B. Wendroff, "The Riemann problem for materials with nonconvex equations of state: II. General case," *J. Math. Anal. Appl.* **38**, 640–658 (1972).

T.-P. Liu, "Shock waves in the non-isentropic gas flow," *J. Differential Equations* **22**, 442–452 (1976).

See also the 1975 paper by Liu cited for Sec. 6.4.

The problem with distributed initial data and the piston problem for the equations of nonisentropic flow is treated in

T.-P. Liu, "Solutions in the large for the equations of non-isentropic gas dynamics," *Indiana Univ. Math. J.* **26**, 147–177 (1977a).

T.-P. Liu, "Initial-boundary value problems for gas dynamics," *Arch. Rational Mech. Anal.* **64**, 137–168 (1977b).

For the asymptotic solution and the decay of solutions, see

P. D. Lax, "Nonlinear hyperbolic systems of conservation laws," in *Nonlinear Problems,* edited by R. E. Langer, University of Wisconsin Press, Madison, Wis., 1963, pp. 3–12.

R. DiPerna, "Decay of solutions of hyperbolic systems of conservation laws with a convex extension," *Arch. Rational Mech. Anal.* **64**, 1–46 (1977).

T.-P. Liu, "Large-time behavior of solutions of initial and initial-boundary value problems of a general system of hyperbolic conservation laws," *Comm. Math. Phys.* **55**, 163–177 (1977c).

T.-P. Liu, "Decay to N-waves of solutions of general systems of nonlinear hyperbolic conservation laws," *Comm. Pure Appl. Math.* **30**, 585–610 (1977d).

T.-P. Liu, "Linear and nonlinear large-time behavior of solutions of general systems of hyperbolic conservation laws," *Comm. Pure Appl. Math.* **30**, 767–796 (1977e).

T.-P. Liu, "Admissible solutions of hyperbolic conservation laws," *Mem. Amer. Math. Soc.* **30**, No. 240, 1–78 (1981).

The treatment of the viscous conservation laws [Eq. (6.5.24)] may be found in

A. G. Kulikovskii, "The structure of shock waves," *PMM J. Appl. Math. Mech.* **26**, 950–964 (1962).

L. R. Foy, "Steady state solutions of hyperbolic systems of conservation laws with viscosity terms," *Comm. Pure Appl. Math.* **17**, 177–188 (1964).

C. C. Conley and J. A. Smoller, "Shock waves as limits of progressive wave solutions of higher order equations," *Comm. Pure Appl. Math.* **24,** 459–472 (1971).

J. A. SMOLLER and C. C. CONLEY, "Shock waves as limits of progressive wave solutions of higher order equations: II," *Comm. Pure Appl. Math.* **25,** 133–146 (1972).

T.-P. LIU, "Nonlinear stability of shock waves for viscous conservation laws," *Mem. Amer. Math. Soc.* **56,** No. 328, 1–108 (1985).

Concerning the numerical methods employing finite difference schemes, see the book by Courant and Friedrichs (1948) cited previously for this section and

R. COURANT, E. ISAACSON, and M. REES, "On the solution of nonlinear hyperbolic differential equations by finite differences," *Comm. Pure Appl. Math.* **5,** 243–255 (1952).

P. D. LAX and B. WENDROFF, "Systems of conservation laws," *Comm. Pure Appl. Math.* **13,** 217–237 (1960).

L. F. SHAMPINE and R. J. THOMPSON, "Difference methods for nonlinear first-order hyperbolic systems of equations," *Math. Comp.* **24,** 45–56 (1970).

A. MAJDA and J. RALSTON, "Discrete shock profiles for systems of conservation laws," *Comm. Pure Appl. Math.* **32,** 445–482 (1979).

A. MAJDA and S. OSHER, "Numerical viscosity and the entropy condition," *Comm. Pure Appl. Math.* **32,** 797–838 (1979).

R. J. DiPERNA, "Finite difference scheme for conservation laws," *Comm. Pure Appl. Math.* **35,** 379–450 (1982).

6.7. This treatment of the adiabatic adsorption of a single solute is taken from

H.-K. RHEE, E. D. HEERDT, and N. R. AMUNDSON, "An analysis of an adiabatic adsorption column: Part II. Adsorption of a single solute," *Chem. Engrg. J.* **1,** 279–290 (1970).

H.-K. RHEE and N. R. AMUNDSON, "An analysis of an adiabatic adsorption column: Part IV. Adsorption in the high temperature range," *Chem. Engrg. J.* **3,** 121–135 (1972).

See also

N. R. AMUNDSON, R. ARIS, and R. SWANSON, "On simple exchange waves in fixed beds," *Proc. Royal Soc. London.* **A286,** 129–139 (1965).

C. Y. PAN and D. BASMADJIAN, "An analysis of adiabatic sorption of single solutes in fixed beds: pure thermal wave formation and its practical implications," *Chem. Engrg. Sci.* **25,** 1653–1664 (1970).

C. Y. PAN and D. BASMADJIAN, "An analysis of adiabatic sorption of single solutes in fixed beds: equilibrium theory," *Chem. Engrg. Sci.* **26,** 45–57 (1971).

D. BASMADJIAN, K. D. HA, and C. Y. PAN, "Nonisothermal desorption by gas purge of single solutes in fixed-bed adsorbers: I. Equilibrium theory," *Indust. Engrg. Chem. Process Des. Dev.* **14,** 328–340 (1975).

P. JACOB and D. TONDEUR, "Nonisothermal adsorption: separation of gas mixtures by modulation of feed temperature," *Sep. Sci. Tech.* **15,** 1563–1577 (1980).

D. BASMADJIAN, "Rapid procedures for the prediction of fixed-bed adsorber behavior: 2. Adiabatic sorption of single gases with arbitrary isotherms and transport modes," *Indust. Engrg. Chem. Process Des. Dev.* **19,** 137–144 (1980).

S. SIRCAR and R. KUMAR, "Equilibrium theory for adiabatic desorption of bulk binary gas mixtures by purge," *Indust. Engrg. Chem. Process Des. Dev.* **24,** 358–364 (1985).

Some experimental work on adiabatic adsorption in fixed beds may be found in

H. G. GRAYSON, "Dynamic adiabatic air drying with bead-type desiccant," *Indust. Engrg. Chem.* **47**, 41–45 (1955).

F. W. LEAVITT, "Non-isothermal adsorption in large fixed beds," *Chem. Engrg. Progr.* **58**, 54–59 (1962).

R. J. GETTY and W. P. ARMSTRONG, "Drying air with activated alumina under adiabatic conditions," *Indust. Engrg. Chem. Process Des. Dev.* **3**, 60–65 (1964).

D. BASMADJIAN, K. D. HA, and D. P. PROULX, "Nonisothermal desorption by gas purge of single solutes from fixed-bed adsorbers: II. Experimental verification of equilibrium theory," *Indust. Engrg. Chem. Process Des. Dev.* **14**, 340–347 (1975).

L. MARCUSSEN and C. VINDING, "Experimental breakthrough curves determined under carefully controlled conditions for the adsorption of water vapor on alumina in adiabatic fixed beds," *Chem. Engrg. Sci.* **37**, 311–317 (1982).

For the analysis of adiabatic countercurrent, staged adsorption columns, see

E. T. KVAALEN and P. C. WANKAT, "Analysis of multicomponent and adiabatic countercurrent columns," *Indust. Engrg. Chem. Fund.* **23**, 14–19 (1984).

6.8. The formulation for the adiabatic adsorption of two solutes as well as the examples treated here is taken from

H.-K. RHEE, E. D. HEERDT, and N. R. AMUNDSON, "An analysis of an adiabatic adsorption column: Part III. Adiabatic adsorption of two solutes," *Chem. Engrg. J.* **3**, 22–34 (1972).

and the 1972 paper by Rhee and Amundson (Part IV) cited for Sec. 6.7.

For the experimental work on adiabatic adsorption of two solutes, see

D. BASMADJIAN and D. W. WRIGHT, "Nonisothermal sorption of ethane-carbon dioxide mixtures in beds of 5Å molecular sieves," *Chem. Engrg. Sci.* **36**, 937–940 (1981).

6.9. The theoretical development and the illustration in this section are taken from the 1972 paper by Rhee and Amundson (Part IV) cited for Sec. 6.7.

The adsorptivity reversal in heterovalent ion-exchange systems is discussed briefly in

F. HELFFERICH and G. KLEIN, *Multicomponent Chromatography: Theory of Interference,* Marcel Dekker, New York, 1970.

The existence of the temperature T_{wi} (the reversal temperature) and its practical implication is further exploited in

P. JACOB and D. TONDEUR, "Non-isothermal gas adsorption in fixed beds: I. A simple linearized equilibrium model," *Chem. Engrg. J.* **22**, 187–202 (1981).

P. JACOB and D. TONDEUR, "Non-isothermal gas adsorption in fixed beds: II. Nonlinear equilibrium theory and 'Guillotine' effect," *Chem. Engrg. J.* **26**, 41–58 (1983a).

P. JACOB and D. TONDEUR, "Adsorption non-isotherme de gas en lit fixe: III. Étude expérimental des effects de guillotine et de focalisation séparation *n*-pentane/isopentane sur tamis 5Å," *Chem. Engrg. J.* **26**, 143–156 (1983b), in French.

6.10. The formulation here is in parallel with that of Sec. 2.10 and the examples discussed here are taken from

H.-K. RHEE, N. R. AMUNDSON, and R. P. LEWIS, "Shock layer analysis of adiabatic adsorption in fixed beds with axial dispersion," *Chem. Engrg. J.* **23**, 167–176 (1983).

For the existence and uniqueness proof for the shock layer, see Sec. 2.10 and the references for Sec. 2.10 listed at the end of Chapter 2.

The shock layer in adiabatic fixed-bed adsorption systems has been recognized as a constant pattern profile and treated as such in the 1962 paper by Leavitt cited for Sec. 6.7 and also in

C. Y. PAN and D. BASMADJIAN, "Constant-pattern adiabatic fixed-bed adsorption," *Chem. Engrg. Sci.* **22**, 285–297 (1967).

Although studies for the dispersion model are scarce, the nonequilibrium model with heat and mass transfer resistances has been investigated extensively by various authors. See, for example,

O. A. MEYER and T. W. WEBER, "Nonisothermal adsorption in fixed beds," *AIChE J.* **13**, 457–465 (1967).

J. W. CARTER, "Isothermal and adiabatic adsorption in fixed beds," *Trans. Inst. Chem. Engrs. London.* **46**, T213–T222 (1968).

C. W. CHI and D. T. WASAN, "Fixed bed adsorption drying," *AIChE J.* **16**, 23–31 (1970).

D. O. COONEY, "Numerical investigation of adiabatic fixed-bed adsorption," *Indust. Engrg. Chem. Process Des. Dev.* **13**, 368–373 (1974).

D. M. RUTHVEN, D. R. GARG, and R. M. CRAWFORD, "The performance of molecular sieve adsorption columns: non-isothermal systems," *Chem. Engrg. Sci.* **30**, 803–810 (1975).

K. IKEDA, "Performance of the non-isothermal fixed bed adsorption column with nonlinear isotherms," *Chem. Engrg. Sci.* **34**, 941–949 (1979).

M. PADEREWSKI, A. MAJKENT, and A. JEDRZEJAK, "Simplified model of adiabatic fixed-bed adsorption," *Inz. Chem.* **9**, 169–178 (1979); English translation in *Internt. Chem. Engrg.* **21**, 129–134 (1981).

L. MARCUSSEN, "Comparison of experimental and predicted breakthrough curves for adiabatic adsorption in fixed bed," *Chem. Engrg. Sci.* **37**, 299–309 (1982).

S. SIRCAR, R. KUMAR, and K. J. ANSELMO, "Effects of column nonisothermality or nonadiabaticity on the adsorption breakthrough curves," *Indust. Engrg. Chem. Process Des. Dev.* **22**, 10–15 (1983).

H. YOSHIDA and D. M. RUTHVEN, "Dynamic behavior of an adiabatic adsorption column: I," *Chem. Engrg. Sci.* **38**, 877–884 (1983).

Numerical analysis of the nonequilibrium, axial dispersion model is found in

P. S. K. CHOI, L. T. FAN, and H. H. HSU, "Modeling and simulation of an adiabatic adsorber," *Sep. Sci.* **10**, 701–721 (1975).

R. KUMAR and G. R. DISSINGER, "Nonequilibrium, nonisothermal desorption of single adsorbate by purge," *Indust. Engrg. Chem. Process Des. Dev.* **25**, 456–464 (1986).

The nonequilibrium model for the adiabatic adsorption of two solutes is treated numerically in

J. H. HARWELL, A. I. LIAPIS, R. LITCHFIELD, and D. T. HANSON, "A non-equilibrium model for fixed-bed multi-component adiabatic adsorption," *Chem. Engrg. Sci.* **35**, 2287–2296 (1980).

A. I. LIAPIS and O. K. CROSSER, "Comparison of model predictions with non-isothermal sorption data for ethane-carbon dioxide mixtures in beds of 5Å molecular sieves," *Chem. Engrg. Sci.* **37**, 958–961 (1982).

See also

S. KAGUEI, Q. YU, and N. WAKAO, "Thermal waves in an adsorption column: parameter estimation," *Chem. Engrg. Sci.* **40**, 1069–1076 (1985).

S. KAGUEI and N. WAKAO, "Validity of infinite bed assumption in the estimation of parameters from thermal waves measured in a non-isothermal adsorption column," *Chem. Engrg. Sci.* **40**, 1851–1853 (1985).

Adiabatic adsorption columns involved with solute condensation are treated in

D. K. FRIDAY and M. D. LEVAN, "Solute condensation in adsorption beds during thermal regeneration," *AIChE J.* **28**, 86–91 (1982).

G. R. SCHOOFS, "Observations and consequences of adsorbate condensation during thermal regeneration of industrial adsorbers," *Indust. Engrg. Chem. Process Des. Dev.* **25**, 800–804 (1986).

The shock layer analysis applied to adiabatic, catalytic fixed-bed reactors may be found in

H.-K. RHEE, D. FOLEY, and N. R. AMUNDSON, "Creeping reaction zone in a catalytic, fixed-bed reactor: a cell model approach," *Chem. Engrg. Sci.* **28**, 607–615 (1973).

H.-K. RHEE, R. P. LEWIS, and N. R. AMUNDSON, "Creeping profiles in catalytic fixed bed reactors: continuous models," *Indust. Engrg. Chem. Fund.* **13**, 317–323 (1974).

Chemical Reaction in a Countercurrent Reactor   **7**

The countercurrent adsorber can also be used with good effect as a chemical reactor by designing it so that its separating power enhances the reaction that is taking place. We have already seen in the first volume (see Sec. 8.5) that the reaction of a single component leads to an interesting variant of the basic chromatographic equations, and in this chapter we consider similar problems for two chemical species. We shall find, for instance, that a reversible reaction A $\rightleftharpoons$ B taking place on the solid may be driven past the equilibrium limitation and from a fluid phase feed of pure A a product of pure B may be obtained. There is a price of course—in this case, a loss of A by removal with the solid phase—but it may well be advantageous, as would be the case if A were very toxic.

7.1 General Formulation

We use, as far as possible, the same notation as is used in Chapter 5 but make the assumptions that the two phases are not necessarily at equilibrium and that chemical reaction can take place in either or both phases. In particular, the rate of adsorption of the species A_i is r_{ai} and of desorption r_{di}, while its rates of formation by the reactions taking place in the two phases are R_{fi} and R_{si}, respectively. Then, with c_i and n_i as the concentrations in the two phases, V_f and V_s the two velocities, we have

$$\epsilon V_f \frac{\partial c_i}{\partial z} + \epsilon \frac{\partial c_i}{\partial t} + s(r_{ai} - r_{di}) - \epsilon R_{fi} = 0 \qquad (7.1.1)$$

and

$$-(1 - \epsilon)V_s \frac{\partial n_i}{\partial z} + (1 - \epsilon)\frac{\partial n_i}{\partial t} - s(r_{ai} - r_{di}) - (1 - \epsilon)R_{si} = 0 \qquad (7.1.2)$$

where s is the adsorption area per unit volume of the bed and ϵ fractional volume of the fluid phase.

These equations are quite general but to make any progress we need to be more specific. If we adopt the rates of adsorption and desorption that will, at equilibrium, lead to the Langmuir isotherm, we can write (see Sec. 3.5)

$$sr_{ai} = k_{ai}c_i\left(1 - \sum_{j=1}^{m} \frac{n_j}{N_j}\right) \qquad (7.1.3)$$

and

$$sr_{di} = \frac{k_{di} n_i}{N_i} \qquad (7.1.4)$$

The rates of formation of A_i in the two phases depend on the reactions that are taking place there. The most general formation would be to say that reactions

$$\sum_{i=1}^{m} \alpha_{fji} A_i = 0, \quad j = 1, \ldots, M_f, \quad \text{and} \quad \sum_{i=1}^{m} \alpha_{ski} A_i = 0, \quad k = 1, \ldots, M_s$$

take place in the two phases and that their rates per unit volume of their respective phases are $\bar{R}_{fj}(c_1, \ldots, c_m)$ and $\bar{R}_{sk}(n_1, \ldots, n_m)$. Then

$$R_{fi} = \sum_{j=1}^{M_f} \alpha_{fji} \bar{R}_{fj} \quad \text{and} \quad R_{si} = \sum_{k=1}^{M_s} \alpha_{ski} \bar{R}_{sk} \qquad (7.1.5)$$

and we thus have a system of $2m$ equations for $c_1, \ldots, n_m$.

At the boundaries of the countercurrent reactor we can specify the concentration of the species in the phase that is introduced there (see Fig. 5.1). Thus

$$c_i(0, t) = c_i^a$$

and

$$n_i(L, t) = n_i^b$$

where the superscripts a and b denote the boundaries at $z = 0$ and at $z = L$, respectively. The initial conditions will specify $c_i(z, 0)$ and $n_i(z, 0)$.

If the rates of adsorption and desorption are very fast, we may presume that the equilibrium relationship $r_{ai} = r_{di}$ obtains and that this allows the n_i to be determined as functions of the c_i. Then adding Eqs. (7.1.1) and (7.1.2) and substituting for the n_i gives m equations in the m unknown c_i.

If a reaction is very fast, it imposes an equilibrium relationship between some of the concentrations and so leads to a reduction of the number of equations. The relationship is of the form

$$\prod_{i=1}^{m} c_i^{\alpha_{fji}} = K_{fj} \quad \text{or} \quad \prod_{i=1}^{m} n_i^{\alpha_{ski}} = K_{sk} \tag{7.1.6}$$

7.2 Case of Two Reactants

We turn now to the case of two reactants and one reaction which we will assume takes place on the solid surface by catalysis. The mechanism is $A_1 + S \rightleftharpoons A_1 S \rightleftharpoons A_2 S \rightleftharpoons A_2 + S$, where S denotes an adsorption site and $A_1 S$ the adsorbed species. Thus

$$R_{f1} = R_{f2} = 0, \qquad -R_{s1} = R_{s2} = \hat{k}_1 n_1 - \hat{k}_2 n_2 \tag{7.2.1}$$

and

$$\epsilon V_f \frac{\partial c_1}{\partial z} + \epsilon \frac{\partial c_1}{\partial t} + k_{a1} c_1 \left(1 - \frac{n_1}{N_1} - \frac{n_2}{N_2}\right) - \frac{k_{d1} n_1}{N_1} = 0 \tag{7.2.2}$$

$$\epsilon V_f \frac{\partial c_2}{\partial z} + \epsilon \frac{\partial c_2}{\partial t} + k_{a2} c_2 \left(1 - \frac{n_1}{N_1} - \frac{n_2}{N_2}\right) - \frac{k_{d2} n_2}{N_2} = 0 \tag{7.2.3}$$

$$-(1 - \epsilon)V_s \frac{\partial n_1}{\partial z} + (1 - \epsilon) \frac{\partial n_1}{\partial t} - k_{a1} c_1 \left(1 - \frac{n_1}{N_1} - \frac{n_2}{N_2}\right)$$
$$+ \frac{k_{d1} n_1}{N_1} + (1 - \epsilon)(\hat{k}_1 n_1 - \hat{k}_2 n_2) = 0 \tag{7.2.4}$$

$$-(1 - \epsilon)V_s \frac{\partial n_2}{\partial z} + (1 - \epsilon) \frac{\partial n_2}{\partial t} - k_{a2} c_2 \left(1 - \frac{n_1}{N_1} - \frac{n_2}{N_2}\right)$$
$$+ \frac{k_{d2} n_2}{N_2} - (1 - \epsilon)(\hat{k}_1 n_1 - \hat{k}_2 n_2) = 0 \tag{7.2.5}$$

Since the presence of a reaction is of the essence in this chapter, we take a slightly different mode of nondimensionalizing the equations by using $\hat{k}_1$ as a characteristic time and write

$$x = \frac{z\hat{k}_1}{V_f \epsilon}, \qquad y = \frac{t\hat{k}_1}{\epsilon}, \qquad X = \frac{L\hat{k}_1}{V_f \epsilon} \tag{7.2.6}$$

In keeping with Chapter 5, we set

$$\nu = \frac{1 - \epsilon}{\epsilon} \quad \text{and} \quad \mu = \frac{(1 - \epsilon)V_s}{\epsilon V_f} \tag{7.2.7}$$

and take the dimensionless dependent variables to be

$$u_1 = \frac{k_{a1}c_1}{k_{d1}} = K_1 c_1, \qquad u_2 = \frac{k_{a2}c_2}{k_{d2}} = K_2 c_2, \qquad v_1 = \frac{n_1}{N_1}, \qquad v_2 = \frac{n_2}{N_2}$$

$$(7.2.8)$$

When these are substituted in the equations, the following dimensionless parameters emerge:

$$\text{capacity ratios:} \quad \alpha_1 = \gamma_1 \nu, \qquad\qquad \alpha_2 = \gamma_2 \nu \qquad (7.2.9)$$

$$\text{flux ratios:} \quad \beta_1 = \gamma_1 \mu, \qquad\qquad \beta_2 = \gamma_2 \mu \qquad (7.2.10)$$

$$\text{adsorptivities:} \quad \gamma_1 = N_1 K_1, \qquad\qquad \gamma_2 = N_2 K_2 \qquad (7.2.11)$$

$$\text{adsorption rate parameters:} \quad \phi_1 = \frac{k_{a1}}{\hat{k}_1(1-\epsilon)}, \qquad \phi_2 = \frac{k_{a2}}{\hat{k}_1(1-\epsilon)}$$

$$(7.2.12)$$

Then the equations become

$$\frac{\partial u_1}{\partial x} + \frac{\partial u_1}{\partial y} + \phi_1\{u_1(1 - v_1 - v_2) - v_1\} = 0 \qquad (7.2.13)$$

$$\frac{\partial u_2}{\partial x} + \frac{\partial u_2}{\partial y} + \phi_2\{u_2(1 - v_1 - v_2) - v_2\} = 0 \qquad (7.2.14)$$

$$-\beta_1 \frac{\partial v_1}{\partial x} + \alpha_1 \frac{\partial v_1}{\partial y} - \phi_1\{u_1(1 - v_1 - v_2) - v_1\} + \gamma_1\left(v_1 - \frac{v_2}{K_e}\right) = 0 \qquad (7.2.15)$$

$$-\beta_2 \frac{\partial v_2}{\partial x} + \alpha_2 \frac{\partial v_2}{\partial y} - \phi_2\{u_2(1 - v_1 - v_2) - v_2\} - \gamma_2\left(v_1 - \frac{v_2}{K_e}\right) = 0 \qquad (7.2.16)$$

where $K_e = \hat{k}_1 N_1 / \hat{k}_2 N_2$ is the equilibrium constant for the reaction. The adsorptivities, γ_1 and γ_2, are the slopes of the equilibrium isotherms at low concentrations for at adsorption equilibrium

$$n_i = \frac{\gamma_i c_i}{1 + K_1 c_1 + K_2 c_2} \qquad (7.2.17)$$

Without loss of generality we can take

$$\kappa = \frac{\gamma_1}{\gamma_2} < 1 \qquad (7.2.18)$$

The physical meaning of the parameters becomes apparent, for ν is the ratio of the volumes of the two phases, so that $\alpha_i = \nu\gamma_i$ is the ratio of the capacities of the two phases to hold A_i. It is made definite by being the maximum of that ratio for $0 < \partial n_i/\partial c_i < \gamma_i$. Similarly, μ is the ratio of the flow rates, so $\beta_i = \mu\gamma_i$ is the maximum flux ratio between the two phases. It is worth noting that $\kappa = \gamma_1/\gamma_2 = \alpha_1/\alpha_2 = \beta_1/\beta_2$.

In dimensionless form the adsorption isotherm is

$$v_i = \frac{u_i}{1 + u_1 + u_2} = \frac{u_i}{\delta} \qquad (7.2.19)$$

If this relation obtains, it implies that ϕ_1 and ϕ_2 tend to infinity and the expressions they multiply vanish. Thus to use the equations we must add Eqs. (7.2.13) and (7.2.15) and Eqs. (7.2.14) and (7.2.16) and substitute from Eq. (7.2.19), to give the two equations

$$\left\{1 - \beta_1 \frac{1 + u_2}{\delta^2}\right\} \frac{\partial u_1}{\partial x} + \left\{1 + \alpha_1 \frac{1 + u_2}{\delta^2}\right\} \frac{\partial u_1}{\partial y} + \beta_1 \frac{u_1}{\delta^2} \frac{\partial u_2}{\partial x} \tag{7.2.20}$$

$$- \alpha_1 \frac{u_1}{\delta^2} \frac{\partial u_2}{\partial y} + \gamma_1 \frac{u_1 - u_2/K_e}{\delta} = 0$$

$$\beta_2 \frac{u_2}{\delta^2} \frac{\partial u_1}{\partial x} - \alpha_2 \frac{u_2}{\delta^2} \frac{\partial u_1}{\partial y} + \left\{1 - \beta_2 \frac{1 + u_1}{\delta^2}\right\} \frac{\partial u_2}{\partial x} \tag{7.2.21}$$

$$+ \left\{1 + \alpha_2 \frac{1 + u_1}{\delta^2}\right\} \frac{\partial u_2}{\partial y} - \gamma_2 \frac{u_1 - u_2/K_e}{\delta} = 0$$

For future manipulations it is convenient to write these as

$$\frac{\partial g_1}{\partial x} + \frac{\partial h_1}{\partial y} + \gamma_1 \frac{u_1 - u_2/K_e}{\delta} = 0 \tag{7.2.22}$$

$$\frac{\partial g_2}{\partial x} + \frac{\partial h_2}{\partial y} - \gamma_2 \frac{u_1 - u_2/K_e}{\delta} = 0 \tag{7.2.23}$$

where

$$g_i(u_1, u_2) = u_i\left(1 - \frac{\beta_i}{\delta}\right) \quad \text{and} \quad h_i(u_1, u_2) = u_i\left(1 + \frac{\alpha_i}{\delta}\right) \tag{7.2.24}$$

are the countercurrent analogs of the column isotherms in chromatography (see Sec. 1.2 of Vol. 1). h_1 and h_2 are convex functions with a positive Jacobian, but the Jacobian

$$\frac{\partial(g_1, g_2)}{\partial(u_1, u_2)} = [\delta^3 - \{\beta_1(1 + u_2) + \beta_2(1 + u_1)\}\delta + \beta_1 \beta_2]\delta^{-3} \tag{7.2.25}$$

can change sign for a range of values of β_1 and β_2. It is in fact this change of sign that gives rise to the interesting behavior of the countercurrent reactor.

We should also consider the equations for the steady state, obtained by putting the derivatives with respect to y equal to zero. This gives four ordinary differential equations in the case of finite adsorption rate:

$$\frac{du_1}{dx} = -\phi_1\{u_1(1 - v_1 - v_2) - v_1\} \tag{7.2.26}$$

$$\frac{du_2}{dx} = -\phi_2\{u_2(1 - v_1 - v_2) - v_2\} \tag{7.2.27}$$

$$-\beta_1 \frac{du_1}{dx} = \phi_1\{u_1(1 - v_1 - v_2) - v_1\} - \gamma_1\left(v_1 - \frac{v_2}{K_e}\right) \tag{7.2.28}$$

$$-\beta_2 \frac{du_2}{dx} = \phi_2\{u_2(1 - v_1 - v_2) - v_2\} + \gamma_2\left(v_1 - \frac{v_2}{K_e}\right) \tag{7.2.29}$$

With assumption of adsorption equilibrium this gives two equations

$$\frac{1}{\gamma_1} \frac{dg_1}{dx} = -\frac{1}{\gamma_2} \frac{dg_2}{dx} = -\frac{u_1 - u_2/K_e}{\delta} \quad (7.2.30)$$

The partial differential equations need initial conditions of the form

$$u_i(x, 0) = u_i^0, \qquad v_i(x, 0) = v_i^0 \quad (7.2.31)$$

which, in the case of adsorption equilibrium, are also constrained by

$$v_i = \frac{u_i}{\delta^0}, \qquad \delta^0 = 1 + u_1^0 + u_2^0 \quad (7.2.32)$$

The boundary conditions must recognize that in the case of adsorption equilibrium, there may be a discontinuity at either the boundary at $x = 0$ or the boundary at $x = X$ and that boundary conditions specifying

$$u_i(0, y) = u_i^a$$
$$v_i(X, y) = v_i^b \quad (7.2.33)$$

may not be appropriate. The boundary conditions then derive from the fact that there must be no accumulation in the planes $x = 0$ or $x = X$ and that since they have no volume the reaction has no effect there. Thus at $x = 0$

$$u_i^a - \beta_i v_i^a = g_i(u_1(0), u_2(0)) \quad (7.2.34)$$

while at $x = X$,

$$u_i^b - \beta_i v_i^b = g_i(u_1(X), u_2(X)) \quad (7.2.35)$$

where $u_i(0)$ and $u_i(X)$ denote values that must be present within the reactor at $x = 0$ and $x = X$, respectively.

The four models that we have developed are summarized in the following table.

Model	Assumptions		Equations	Initial Condition	Boundary Condition
I	Transient, finite adsorption rate	4 partial differential equations	(7.2.13)–(7.2.16)	(7.2.31)	(7.2.33)
II	Transient, adsorption equilibrium	2 partial differential equations	(7.2.20) and (7.2.21) (7.2.22)–(7.2.24)	(7.2.31) and (7.2.32)	(7.2.34) and (7.2.35)
III	Steady state, finite adsorption rate	4 ordinary differential equations	(7.2.26)–(7.2.29)	—	(7.2.33)
IV	Steady state, adsorption equilibrium	2 ordinary differential equations	(7.2.30)	—	(7.2.34) and (7.2.35)

Their relationship is entirely similar to that between the models for a single reactant in Chapter 8 (see p. 456, Vol. 1).

7.3 Characteristics and Discontinuities for the Case of Two Reactants with Adsorption Equilibrium

We continue with the analysis of model II and use the theory of pairs of partial differential equations developed in Chapter 1. We know that the characteristics C have slopes λ^{-1}, where the characteristic speeds λ are given by the roots of the quadratic

$$a\lambda^2 - 2b\lambda + c = 0 \tag{7.3.1}$$

Here

$$\lambda = \frac{dx}{dy}$$

$$a = \frac{\partial(h_1, h_2)}{\partial(u_1, u_2)}$$

$$= [\delta^3 + \{\alpha_1(1 + u_2) + \alpha_2(1 + u_1)\}\delta + \alpha_1\alpha_2]\delta^{-3}$$

$$2b = \frac{\partial(h_1 g_2)}{\partial(u_1, u_2)} + \frac{\partial(g_1, h_2)}{\partial(u_1, u_2)} \tag{7.3.2}$$

$$= [2\delta^3 + \{(\alpha_1 - \beta_1)(1 + u_2) + (\alpha_2 - \beta_2)(1 + u_1)\}\delta - (\alpha_2\beta_1 + \alpha_1\beta_2)]\delta^{-3}$$

$$c = \frac{\partial(g_1, g_2)}{\partial(u_1, u_2)}$$

$$= [\delta^3 - \{\beta_1(1 + u_2) + \beta_2(1 + u_1)\}\delta + \beta_1\beta_2]\delta^{-3}$$

It can be shown that

$$b^2 - ac = \left\{\frac{(\alpha_2 + \beta_2)(1 + u_2)}{2\delta}\right\}^2$$
$$\left[\left\{\kappa + \frac{(1 + u_1)(1 + u_2) - 2}{(1 + u_2)^2}\right\}^2 + \frac{4\delta}{(1 + u_2)^4}\right] \tag{7.3.3}$$

so that the equations are always hyperbolic.

The hodograph transformation does not linearize the equations because they are not homogeneous and the interchange of independent and dependent variables throws up the Jacobian $\partial(x, y)/\partial(u_1, u_2)$, which multiplies the reaction rate term. It is, however, profitable to work out the characteristics in the (u_1, u_2)-plane in the case of no reaction for the following reason. Although these characteristics do not usually have a special relationship to the discontinuities, we know that in the case of Langmuir isotherms the image of a discontinuity does lie on a characteristic because the latter are straight lines. Now, a discontinuity has no volume, so the reaction makes no difference to the conditions that it must satisfy, so we may take over the theory of chromatography without reaction as it applies to discontinuities and hence can use the characteristics in the hodograph plane. These directions are given by the roots of

$$a'\zeta^2 - 2b'\zeta + c' = 0 \qquad (7.3.4)$$

where

$$\zeta = \frac{du_1}{du_2}$$

$$a' = \frac{-u_2(\alpha_2 + \beta_2)}{\delta^2}$$

$$2b' = \frac{(\alpha_1 + \beta_1)(1 + u_2) - (\alpha_2 + \beta_2)(1 + u_1)}{\delta^2} \qquad (7.3.5)$$

$$c' = \frac{u_1(\alpha_1 + \beta_1)}{\delta^2}$$

As in Sec. 2.1, the quadratic, which may be written as

$$u_1 = \zeta u_2 - \frac{(1 - \kappa)\zeta}{\kappa + \zeta} \qquad (7.3.6)$$

is an ordinary differential equation of Clairaut's form whose solution is the family of straight lines obtained by giving ζ a constant value. These were seen, in Sec. 2.1, to be the tangents of the parabola

$$(u_1 + \kappa u_2 + 1 - \kappa)^2 = 4\kappa(1 - \kappa)u_2 \qquad (7.3.7)$$

The usual balance over a shock or discontinuity shows that if nothing is to accumulate in the plane, the speed of a discontinuity in A_i must be

$$\frac{[g_i]}{[h_i]}$$

where the brackets denote the difference in value over the discontinuity. The compatibility of the shocks in the two concentrations implies that their common velocity is

$$\dot{\lambda} = \frac{[g_1]}{[h_1]} = \frac{[g_2]}{[h_2]} \qquad (7.3.8)$$

For a stationary shock $[g_1] = [g_2] = 0$ and this may be written as

$$[u_1] - \beta_1[v_1] = [u_2] - \beta_2[v_2] = 0 \qquad (7.3.9)$$

Since $[u_1] = u_1^+ - u_1^-$, we have $[v_1] = [u_1] + d/\delta^+\delta^-$ with

$$d = u_1^+ u_2^- - u_1^- u_2^+ = u_2^-[u_1] - u_1^-[u_2] = u_2^+[u_1] - u_1^+[u_2]$$

where the $^+$ and $^-$ denote values on the right and left sides of the discontinuity, respectively. This allows the two equations to be written

$$u_2^\pm[u_1]^2 + \{\kappa(1 + u_2^\pm) - (1 + u_1^\pm)\}[u_1][u_2] - \kappa u_1^\pm[u_2]^2 = 0 \qquad (7.3.10)$$

where $u_1^\pm$ denotes that either u_1^+ or u_1^- may be used. But Eq. (7.3.6) can be written as

$$u_2\zeta^2 + [\kappa(1 + u_2) - (1 + u_1)]\zeta - \kappa u_1 = 0 \qquad (7.3.11)$$

$$\frac{[u_1]}{[u_2]} = \zeta(u_1^\pm, u_2^\pm) \qquad (7.3.12)$$

Since the characteristics are straight, the two points (u_1^+, u_2^+) and (u_1^-, u_2^-) that represent conditions on either side of the shock must lie on a characteristic in the hodograph plane.

The entropy condition requires that the characteristic speeds should bracket the speed of the shock, so that it is needed to prevent characteristics from crossing. Thus either

$$\lambda_-(u_1^+, u_2^+) < \bar{\lambda} < \lambda_-(u_1^-, u_2^-), \qquad \bar{\lambda} < \lambda_+(u_1^+, u_2^+)$$

or

$$\lambda_-(u_1^-, u_2^-) < \bar{\lambda}, \qquad \lambda_+(u_1^+, u_2^+) < \bar{\lambda} < \lambda_+(u_1^-, u_2^-) \qquad (7.3.13)$$

7.4 The Steady State

With the assumption of adsorption equilibrium the equations for the two concentrations are

$$-\frac{1}{\gamma_1}\frac{d}{dx}\left\{u_1 - \frac{\beta_1 u_1}{\delta}\right\} = \frac{1}{\gamma_2}\frac{d}{dx}\left\{u_2 - \frac{\beta_2 u_2}{\delta}\right\} = \left(u_1 - \frac{u_2}{K_e}\right)\frac{1}{\delta} \qquad (7.4.1)$$

subject to the boundary conditions

$$u_i^a - \beta_i v_i^a = \left(u_i - \frac{\beta_i u_i}{\delta}\right)_0$$

and

$$u_i^b - \beta_i v_i^b = \left(u_i - \frac{\beta_i u_i}{\delta}\right)_x \qquad (7.4.2)$$

We will return to the boundary conditions in a moment after first looking at continuous solutions within $0 < x < X$. From the first two terms in Eq. (7.4.1),

$$\frac{d}{dx}\left[u_1 + \kappa u_2 - \frac{\beta_1(u_1 + u_2)}{\delta}\right] = 0$$

or

$$(u_1 + \kappa u_2 - w)(1 + u_1 + u_2) + \beta_1 = 0 \qquad (7.4.3)$$

where w is a constant. Thus the solution passing through (u_{10}, u_{20}) is the hyperbola

$$(u_1 + \kappa u_2 - w_0)(1 + u_1 + u_2) + \beta_1 = 0$$

with

$$w_0 = u_{10} + \kappa u_{20} + \frac{\beta_1}{1 + u_{10} + u_{20}} \qquad (7.4.4)$$

The possible forms this can take depend on the values of κ and β_1, and are shown in Fig. 7.1. If (κ, β_1) lies in the region A, i.e., $\beta_1 < \kappa < 1$, the hyperbolas of Eq. (7.4.4) all intersect the positive quadrant of the (u_2, u_1)-plane in monotonic arcs, as shown in Fig. 7.1(a). If $\kappa < \beta_1 < 1$ [region B, Fig. 7.1(b)], the hyperbolas have a horizontal tangent on $u_1 + u_2 = \beta_2^{1/2} - 1$, and the hyperbola through the origin also passes through $u_1 = 0$, $u_2 = \beta_2 - 1$. If $\beta_1 > 1$ [region C, Fig. 7.1(c)], the preceding statements for B are true, although the arc through the origin goes first to the second quadrant and appears

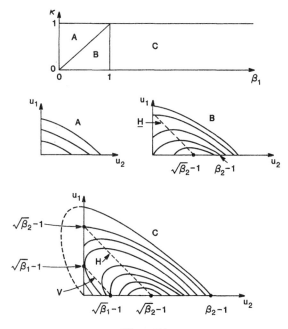

Figure 7.1

in the first quadrant between $u_1 = \beta_1 - 1$, $u_2 = 0$ and $u_1 = 0$, $u_2 = \beta_2 - 1$. In addition there is a line of vertical tangencies on $u_1 + u_2 = \beta_1^{1/2} - 1$. These lines of horizontal and vertical tangency, denoted by H and V, respectively, allow the trajectory of a continuous part of the solution to be sketched in very readily.

If, however, we do the full differentiations in Eq. (7.4.1),

$$\left\{\frac{1}{\gamma_1} - \mu\frac{1 + u_2}{\delta^2}\right\}\frac{du_1}{dx} + \mu\frac{u_1}{\delta^2}\frac{du_2}{dx} = -\frac{u_1 - u_2/K_e}{\delta}$$

$$\mu\frac{u_2}{\delta^2}\frac{du_1}{dx} + \left\{\frac{1}{\gamma_2} - \mu\frac{1 + u_1}{\delta^2}\right\}\frac{du_2}{dx} = \frac{u_1 - u_2/K_e}{\delta}$$

and these equations can be solved to give

$$\frac{du_1}{dx} = -\gamma_1(\delta^2 - \beta_2)\frac{e(u_1, u_2)}{f(u_1, u_2)} \qquad (7.4.5)$$

$$\frac{du_2}{dx} = \gamma_2(\delta^2 - \beta_1)\frac{e(u_1, u_2)}{f(u_1, u_2)} \qquad (7.4.6)$$

where

$$e(u_1, u_2) = u_1 - \frac{u_2}{K_e} \qquad (7.4.7)$$

and

$$f(u_1, u_2) = \delta^3 - \delta\{\beta_1(1 + u_1) + \beta_2(1 + u_2)\} + \beta_1\beta_2 \qquad (7.4.8)$$

If neither $e(u_1, u_2)$ nor $f(u_1, u_2)$ vanish,

$$\frac{du_1}{du_2} = -\kappa\frac{\delta^2 - \beta_2}{\delta^2 - \beta_1} \qquad (7.4.9)$$

of which, of course, the hyperbolas of Eq. (7.4.4) are the solution. But the straight line, E,

$$E: \quad e(u_1, u_2) = 0 \qquad (7.4.10)$$

is a line of critical points, while

$$F: \quad f(u_1, u_2) = 0 \qquad (7.4.11)$$

is the locus of infinite derivatives, i.e., where the curves $u_1(x)$ and $u_2(x)$ turn back on themselves. To see the disposition of these curves, we must further divide the region C into C, $1 < \beta_1 < 1/\kappa$ and D, $1/\kappa < \beta_1$. Then note in Fig. 7.2 the relative disposition of the points $P = (0, \beta_1^{1/2} - 1)$, $Q = (\beta_1 - 1, 0)$, $R = (\beta_2^{1/2} - 1, 0)$, and $S = (0, \beta_2 - 1)$ in the (u_2, u_1)-plane. The diagonals through these points PP', $Q'Q$, $R'R$, and SS' are, respectively, the lines along which $\delta = \beta_1^{1/2}$, β_1, $\beta_2^{1/2}$, and β_2. The cubic curve F goes through the points P, Q, R, and S, its two branches F_+ and F_- lying in the regions carved

Chemical Reaction in a Countercurrent Reactor   Chap. 7

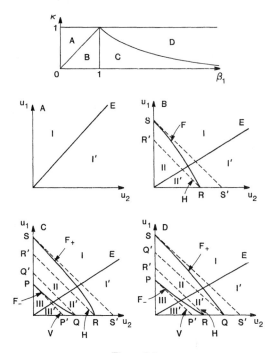

Figure 7.2

out by these diagonals. For (β_1, κ) in region A, f does not appear in the first quadrant; if (β_1, κ) is in B, only one branch appears. With (β_1, κ) in C or D both branches cross the first quadrant and with the equilibrium locus E divide the quadrant into six regions. The signs of e and f and consequently of λ_+ and λ_- for the several regions are shown in the following table.

Region	Sign of			
	e	f	λ_+	λ_-
I	+	+	+	+
I'	+	+	+	+
II	+	−	+	−
II'	−	−	+	−
III	+	+	−	−
III'	−	+	−	−

We can now see how the phase plane of u_1 and u_2 allows us to build up a picture of the solutions. E being a continuous line of critical points implies that one eigenvalue at any equilibrium point is zero and it may be shown that the other is

$$-\frac{1 + K_e \kappa}{K_e f(u_1, u_2)}(\delta^2 - \tilde{\beta}) \qquad \text{where} \qquad \frac{1 + K_e}{\tilde{\beta}} = \frac{1}{\beta_1} + \frac{K_e}{\beta_2} \qquad (7.4.12)$$

This is negative if f and $\delta^2 - \tilde{\beta}$ are both of the same sign and, since $\beta_1 < \tilde{\beta} < \beta_2$, the eigenvalue changes sign at

$$\tilde{u}_2 = \tilde{u}_1 K_e = \frac{\tilde{\beta}^{1/2} - 1}{1 + K_e} K_e \qquad (7.4.13)$$

a point that lies on E between H and V.

If (β_1, κ) is in region A, β_2 is also less than 1 and the bed behaves very much as a fixed bed. All points on the equilibrium line are stable, so we may put arrows on the monotonic curves in Fig. 7.1(a) pointing inward from the axes to E. A long reactor with an inlet (u_1^a, u_2^a) such as is represented by point B in Fig. 7.3 would allow the concentration $u_1(x)$ to fall asymptotically to equilibrium at E and $u_2(x)$ would rise correspondingly. Similarly, if the inlet conditions at $x = 0$ were represented by B', the approach to equilibrium would be from the other side. E is the point satisfying Eqs. (7.4.4) and (7.4.10), namely,

$$(1 - \kappa K_e)u_{1e} - w_0(1 + K_e)u_{1e} + 1 + \beta_1 = 0$$
$$w_0 = u_1^a + \kappa u_2^a + \frac{\beta_1}{1 + u_1^a + u_2^a} \qquad (7.4.14)$$

The length of bed necessary to reach any point $u_1(X)$, $u_2(X)$ on the path would be

$$\gamma_1 X = \int_{u_1(X)}^{u_1^a} f(u_1, u_2)(\delta^2 - \beta_2)\frac{du_1}{u_1 - u_2/K_e} \qquad (7.4.15)$$

Finally, the emerging concentrations of A_1 and A_2 at $x = X$ would depend on what was put in with the solid feed, for a boundary discontinuity would provide the adjustment necessary and give

$$u_i^b = \beta_i v_i^b + u_i(X)\left\{1 - \frac{\beta_i}{1 + u_1(X) + u_2(X)}\right\} \qquad (7.4.16)$$

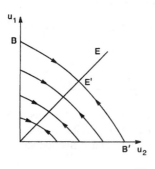
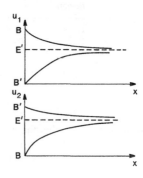

Figure 7.3

The whole behavior is that of a fixed bed with the minor variation of a boundary discontinuity at $x = X$. The same situation obtains for paths in regions I or I' in the other cases.

When (β_1, κ) is in region B of Fig. 7.2 and a branch of F is in the first quadrant more interesting variants obtain. A point such as T in Fig. 7.4 would be of no use as a feed condition at $x = 0$, for the trajectory leads away from it to negative values of u_2. It could, however, be used for a condition at $x = X$ by integrating backward in x from X. But if this were done along a path TFE leading to the equilibrium point E, the sign of du_1/dx and du_2/dx would change at F, giving the physically unacceptable solution shown at the right. This indicates that we shall only be able to get solutions by taking proper account of discontinuities.

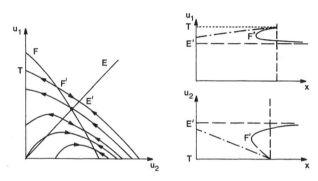

Figure 7.4

We have already seen that the image of a shock must lie on a Γ characteristic of the hodograph plane and that

$$\frac{[u_1] - d}{\delta^+ \delta^-} = \frac{[u_1]}{\beta_1}, \qquad \frac{[u_2] + d}{\delta^+ \delta^-} = \frac{[u_2]}{\beta_2} \qquad (7.4.17)$$

where $d = u_1^+ u_2^- - u_2^+ u_1^-$. But the hyperbola on which a segment of continuous solution lies [see Eq. (7.4.3)] can be written

$$\frac{u_1}{\beta_1} + \frac{u_2}{\beta_2} - \frac{u_1 + u_2}{\delta} = \text{constant}$$

If (u_1^+, u_2^+) and (u_1^-, u_2^-) were on the same hyperbola, we would have

$$\frac{u_1^+}{\beta_1} + \frac{u_2^+}{\beta_2} - \frac{u_1^+ + u_2^+}{\delta^+} = \frac{u_1^-}{\beta_1} + \frac{u_2^-}{\beta_2} - \frac{u_1^- + u_2^-}{\delta^-}$$

which can be rearranged to give

$$\frac{u_1^+}{\delta^+} - \frac{u_1^-}{\delta^-} - \frac{[u_1]}{\beta_1} = \frac{u_2^-}{\delta^-} - \frac{u_2^+}{\delta^+} + \frac{[u_2]}{\beta_2} \qquad (7.4.18)$$

and both sides are zero if the stationary shock condition (7.4.17) obtains. We thus have the important information that *a stationary shock is represented by a chord of the hyperbola representing the continuous solutions it connects.*

The types of stationary shock are best seen from the geometry of Fig. 7.5. Section (a) of that figure shows the disposition of the Γ_+ and Γ_- characteristics and the subsequent sections show that for (β_1, κ) in region B of Fig. 7.1 (i.e., $\kappa < \beta_1 < 1$) members of the family Γ_- can be chords, while for $\beta_1 > 1$ (regions C of Fig. 7.1 and C and D of Fig. 7.2) both families of characteristics can serve. From Fig. 7.5(d) it is clear that ζ_- must be between the values corresponding to tangency and intersection with the hyperbola and the u_2 axis.

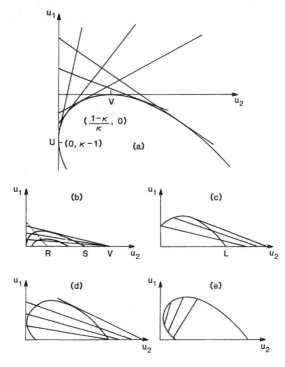

Figure 7.5

It is clear that the shocks will tend to cross F since it is the infinite slope on F that causes the breakdown of the continuous solution. If a shock goes from right to left across F_+ (the RS or QS branch of F in Fig. 7.2) it satisfies the first entropy condition (7.3.13), for $\tilde{\lambda} = 0$ and from the table $\lambda_-^{\pm} < 0 < \lambda_-^-$, $\lambda_+^{\pm} > 0$. Similarly, the second condition is satisfied by crossing F_- from right to left.

To understand more clearly how shocks are formed, it is useful to look at the nonequilibrium case where they are thin layers rather than actual discontinuities. The steady-state equations (7.2.26)–(7.2.29) may be divided one by another in pairs to give

Chemical Reaction in a Countercurrent Reactor   Chap. 7

$$\frac{a v_1}{d u_1} = \frac{1}{\beta_1} - \frac{1}{\phi_1 \mu} \frac{v_1 - v_2/K_e}{u_1(1 - v_1 - v_2) - v_1}$$

$$\frac{d v_2}{d u_2} = \frac{1}{\beta_2} + \frac{1}{\phi_2 \mu} \frac{v_1 - v_2/K_e}{u_2(1 - v_1 - v_2) - v_2} \qquad (7.4.19)$$

If the adsorption rates ϕ_i are large,

$$\frac{d v_1}{d u_1} \sim \frac{1}{\beta_1} \quad \text{and} \quad \frac{d v_2}{d u_2} \sim \frac{1}{\beta_2} \qquad (7.4.20)$$

which are just the differential forms of the stationary shock conditions

$$\frac{[v_1]}{[u_1]} = \frac{1}{\beta_1} \quad \text{and} \quad \frac{[v_2]}{[u_2]} = \frac{1}{\beta_2} \qquad (7.4.21)$$

for when the changes take place over a vanishingly short interval the derivatives become difference quotients. The exception to this limit is when the denominators in the second terms vanish, i.e., at adsorption equilibrium. We can thus assert that for very large ϕ_i the solution either lies in the equilibrium surface or has projections on the (u_1, v_1)- and (u_2, v_2)-planes that are straight lines of slope $1/\beta_1$ and $1/\beta_2$. Below the adsorption equilibrium surface these trajectories are so directed that u_1 and v_1 (or u_2 and v_2) both decrease, while above the equilibrium surface, both increase.

The trajectory does, in fact, lie in a three-dimensional subspace of the four-dimensional (u_1, u_2, v_1, v_2)-space since $(u_1/\beta_1 + u_2/\beta_2 - v_1 - v_2)$ is constant along a solution. This does not help very much in visualizing the solution, as the constant value of this linear combination varies from trajectory to trajectory. A calculation example will help at this point, and we choose the parameters and conditions to be

$$u_1^a = v_1^b = v_2^b = 0 \qquad u_2^a = 7.5, \qquad X = 8.4$$

$$\kappa = 0.2, \qquad \beta_1 = 1.5, \qquad \frac{\phi_1}{\phi_2} = \gamma_1 = 0.2, \qquad K_e = 2 \qquad (7.4.22)$$

$$\phi_1 = 0.5, 1, 2, \infty$$

We thus look for solutions of the equations (7.2.26)–(7.2.29) that link $(0, 7.5, v_1^a, v_2^a)$ with $(u_1^b, u_2^b, 0, 0)$. The unknown conditions at the two ends must be determined by iteration, using a method that will be described later.

In Fig. 7.6 the projections of the solution onto the (u_1, u_2)-plane are shown for $\phi_1 = 0.1, 0.5, 1, 2$, and it is clear that as ϕ_1 increases, the trajectories get closer to the equilibrium trajectory

$$(u_1 + 0.2u_2 - 1.676)(1 + u_1 + u_2) + 1.5 = 0 \qquad (7.4.23)$$

given by Eq. (7.4.4). If the value of u_2 is plotted against x (as in Fig. 7.7), the curves grow steeper as ϕ_1 increases, although it is evident that a rather large value of ϕ_1 is necessary before the shock is at all well approximated. In Fig. 7.8, curves for $\phi_1 = 0.01, 0.05, 0.1, 0.2, 0.5, 0.75, 1, 2$ in the space of

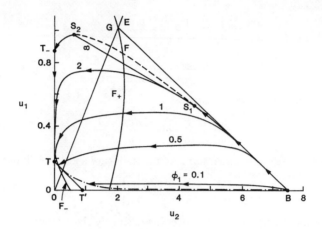

Figure 7.6

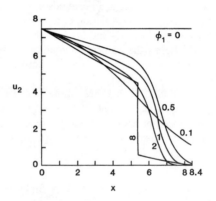

Figure 7.7

u_1, u_2 and v_2 are shown as solid lines. They all start from the vertical line BB' ($u_1 = 0$, $u_2 = 7.5$) since this represents the conditions at $x = 0$, but at different points corresponding to the value of v_2 found by iteration. The dashed lines are lines of constant x corresponding to the values 1, 2, 3, 4, 5, 5.5, 6, 6.5, 7, 7.5, and 8.4. These all emanate from the point B since for $\phi_1 = 0$ the whole trajectory degenerates to the point B. The curve for $\phi_1 = \infty$, PS_1S_2Q, is part of the hyperbola BFT of Fig. 7.6 cut off with the discontinuity (shown with short and long dashes) S_1S_2. This curve meets the u_1 axis at $u_1 = 0.882$, $u_2 = 0$ (the point Q), and since $v_1^b = 0$, Eq. (7.4.16) shows that the emerging concentration of u_1 is

$$u_1^b = u_1(X)\left\{1 - \frac{\beta_1}{1 + u_1(X)}\right\} = 0.1790 \qquad (7.4.24)$$

We can now return to the equilibrium case and see how the discontinuities develop in a reactor as the length of the reactor is increased. To do this we

Chemical Reaction in a Countercurrent Reactor   Chap. 7

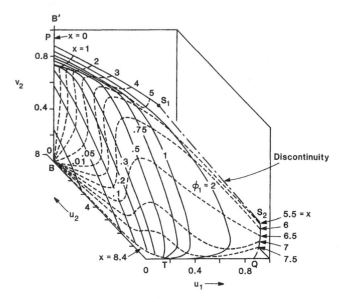

Figure 7.8

shall take constant feed conditions corresponding to the point V on the u_2 axis in the top part of Fig. 7.9; thus

$$v_1^b = v_2^b = u_1^a = 0, \qquad u_2^a = U_2 \qquad (7.4.25)$$

The value of W that gives the continuous trajectory through V is

$$W = \kappa U_2 + \frac{\beta_1}{1 + U_2} \qquad (7.4.26)$$

and the equation of the trajectory is

$$\left\{ u_1 + \kappa(u_2 - U_2) - \frac{\beta_1}{1 + U_2} \right\}(1 + u_1 + u_2) + \beta_1 = 0 \qquad (7.4.27)$$

The point U at which this meets the u_1 axis is

$$u_1 = U_1 = \frac{1}{2}(W - 1) + \left[\frac{1}{2}(W + 1)^2 - \beta_1 \right]^{1/2} \qquad (7.4.28)$$

These are shown in Fig. 7.9 and it will be useful to refer to the value of the variables at the various points by using the name of the point as argument, e.g., $u_1(D)$, $u_1(V) = 0$, $u_2(V) = U_2$, etc.

The geometry of the figure is important and is the key to understanding the system. The line RT is the locus of u_1^b, u_2^b for a boundary discontinuity at $x = X$ with $v_1^b = v_2^b = 0$. For

$$u_i^b = u_i(X)\left\{ 1 - \frac{\beta_1}{\delta(X)} \right\}$$

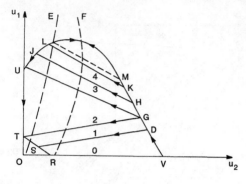

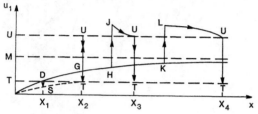

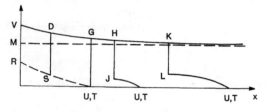

Figure 7.9

where $u_i(X)$ is the value just within the reactor, and

$$\frac{u_1^b}{\beta_1} + \frac{u_2^b}{\beta_2} = \frac{u_1(X)}{\beta_1} + \frac{u_2(X)}{\beta_2} - \frac{u_1(X) + u_2(X)}{\delta(X)} = \frac{W}{\beta_1} - 1$$

so

$$\frac{u_1^b}{\beta_1} + \frac{u_2^b}{\beta_2} = \frac{W - \beta_1}{\beta_1} \qquad (7.4.29)$$

which is the equation of RT. TG is the Γ_+ characteristic through T, so

$$u_1(T) = W - \beta_1 = U_2\left\{\kappa - \frac{\beta_1}{1 + U_2}\right\} = (1 - \kappa)u_1(G)$$

and

Chemical Reaction in a Countercurrent Reactor   Chap. 7

$$u_2(G) = \left\{ \frac{\beta_2 - W}{1 - \kappa} \right\} - 1, \qquad \delta(G) = \beta_2 \qquad (7.4.30)$$

It can further be shown that GU is the Γ_- characteristic through G. The characteristics of the Γ_+ or Γ_- families shown in Fig. 7.9 are VR, DS, GT of the former and GU, HJ, KL, ME of the latter. The arrows indicate the direction of the jump.

Consider now a sequence of reactors of increasing height. If $X = 0$, no reaction can take place, so no A_1 can be formed. However, the exit state must lie on RT and so must be R since $u_1 = 0$. It can get there from V by a discontinuity VR, since the equilibrium theory implies equilibration of adsorption even in a column of zero length. Thus, although no reaction can take place, the feed of A_2 is split, some of it emerging in the fluid stream at concentration

$$u_2^b = U_2 - \frac{\beta_2 U_2}{1 + U_2} \qquad (7.4.31)$$

and the remainder goes out on the solid with

$$v_2^q = \frac{U_2}{1 + U_2}$$

If the reactor is short, it can sustain a continuous solution corresponding to the arc VD, but again exit conditions must lie on RT, so the state gets there by a Γ_+ shock DS which is stationary at $x = X$. The corresponding profiles of $u_1(X)$ and $u_2(X)$ are shown in the lower parts of Fig. 7.9. This type of solution obtains until the reactor reaches a length of X_2 for which the continuous solution just reaches G. Beyond this point a Γ_+ shock will not finish up on RT. However, we notice a second set of permissible discontinuities links G to T: the Γ_- shock followed by UT. Then for slightly longer reactor (say X_3) the Γ_- shock becomes internal as at HJ and the discontinuity at $x = X$ is always UT. The lower parts of the figure show that u_1 increases to $u_1(H)$, suffers a discontinuous increase to $u_1(J)$, declines continuously to $U_1 = u_1(U)$, and emerges as $u_1^b = u_1(T) = W - \beta_1$. We observe that Γ_- shocks such as HJ jump the reaction equilibrium locus E and so permit "complete" conversion in the fluid stream which emerges with $u_2^b = 0$. Of course, the price that is paid is that some A_2 is carried away with the solid.

As the reactor length is increased, the internal shock moves upward in the (u_2, u_1)-plane and approaches EM, the Γ_- shock through the intersection of the continuous solution and the equilibrium locus E. Near E the reaction rate is so low that a very long reactor can be accommodated. However, once the length exceeds

$$\gamma_1 X_* = \int_{U_2}^{u_2(G)} \frac{f(u_1(u_2), u_2)\, du_2}{e(u_1(u_2), u_2)(\delta^2(u_2) - \beta_1)} \qquad (7.4.32)$$

the exit conditions become independent of length. X_* in the example just given is 2.1. The solution for $X = 8.4$ is shown in Fig. 7.10.

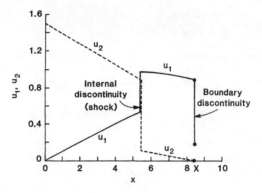

Figure 7.10

To see what this really means in the evaluation of countercurrent reactors, we need a measure of performance. With a fixed bed and only one reaction there is simply the conversion, but for the chromatographic reactor there are two feed and two product streams. We can therefore talk about the conversion of A_i as

$$\eta_i = 1 - \frac{u_i^b + \beta_i v_i^a}{u_i^a + \beta_i v_i^b} \qquad (7.4.33)$$

provided that A_i can be regarded as a reactant; i.e., it is indeed fed to the reactor or u_i^a or $\beta_i v_i^b = 0$. The purity of the streams from the boundaries at $x = 0$ and at $x = X$ can be defined as

$$P_i^a = \frac{v_i^a}{v_1^a + v_2^a}$$

$$P_i^b = \frac{u_i^b}{u_1^b + u_2^b} \qquad (7.4.34)$$

In the example given above, A_2 was regarded as the reactant and

$$\eta_2 = 1 - \frac{u_2^b + \beta_2 v_2^a}{U_2} = 1 - \frac{\beta_2}{1 + U_2} - \frac{u_2^b}{U_2} \qquad (7.4.35)$$

Figure 7.11(a) shows that the conversion rises to a maximum of 12% as the length of the reactor goes from zero to $X_* = 2.1$ and thereafter remains constant. The purity of the stream from $x = X$ rises to 100% in the same interval [Fig. 7.11(b)]. If, therefore, a 12% conversion were adequate, the countercurrent reactor need not be much longer than X_*. Although the fixed-bed reactor could achieve a greater conversion if it were long enough, it could never attain better than 33% A_1 purity in the stream leaving the boundary at $x = X$.

Chemical Reaction in a Countercurrent Reactor   Chap. 7

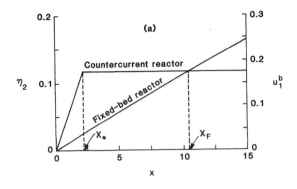

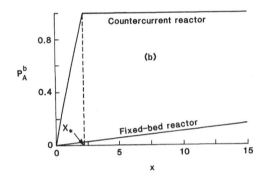

Figure 7.11

7.5 General Procedure for Mapping Out the Steady-State Solution

We start with data on the physical and chemical properties (giving κ, γ_i, ϕ_i, K_e) and feed and operating conditions (u_1^q, v_1^b, β_i). Within the reactor

$$\frac{u_1}{\beta_1} + \frac{u_2}{\beta_2} + \frac{1}{\delta} = \frac{w}{\beta_1} \tag{7.5.1}$$

is a constant and this constancy is maintained across an internal discontinuity, which is represented by a chord of this hyperbola belonging to the family of straight lines

$$u_1 = \zeta\left[u_2 - \frac{1 - \kappa}{\kappa + \zeta}\right] \tag{7.5.2}$$

In fact,

$$\frac{u_1}{\beta_1} + \frac{u_2}{\beta_2} - v_1 - v_2 = \frac{w}{\beta_1} - 1 \tag{7.5.3}$$

is continuous also over either end of the reactor, but it is only within the reactor that equilibrium prevails and

$$v_i = \frac{u_i}{\delta} \tag{7.5.4}$$

reducing Eq. (7.5.3) to Eq. (7.5.1). Note that Eq. (7.5.3) allows us to put bounds on w, for the lines $(u_1^b/\beta_1) + (u_2^b/\beta_2) = (w/\beta_1) - 1 + v_1^b + v_2^b$ and $v_1^a + v_2^a = (u_1^a/\beta_1) + (u_2^a/\beta_2) - (w/\beta_1) + 1$ must intersect in the positive quadrant, so that

$$w_* = \beta_1[1 - v_1^b + v_2^b] < w < \beta_1 + u_1^a + u_2^a = w^* \tag{7.5.5}$$

The conditions at either boundary

$$\frac{[v_i]}{[u_i]} = \frac{1}{\beta_i} \tag{7.5.6}$$

then reduce to

$$u_i^b = u_i(X)\left[1 - \frac{\beta_i}{\delta(X)}\right] + \beta_i v_i^b \tag{7.5.7}$$

and

$$v_i^a = \frac{u_i^a}{\beta_i} - u_i(0)\left[\frac{1}{\beta_i} - \frac{1}{\delta(0)}\right] \tag{7.5.8}$$

A second example may usefully be introduced at this point. Its parameters are

$$\beta_1 = 1.5, \qquad \kappa = 0.2, \qquad \frac{\phi_1}{\phi_2} = 0.2, \qquad K_e = 0.2$$

The inequality (7.5.5) gives $1.2 = w_* < w < w^* = 2.7$. Clearly, we are in the situation C of Fig. 7.2 with this second example. The (u_2, u_1)-plane should be drawn, and a few representative trajectories for w between w_* and w^* with arrows in the direction of increasing x. (This is left as an exercise for the reader, since we shall be more concerned with enlargements of small parts of the plane.)

Consider first the limit $X = 0$ and ignore the unlikely event of v_i^b being in equilibrium with u_i^a when nothing would happen. Since $X = 0$, $u_i(0) = u_i(X)$ and $v_i(0) = v_i(X) = u_i(0)/\delta(0)$. If the concentrations are continuous at $x = 0$, $u_i^a = u_i(0) = u_i(X)$ and the isotherms in the (u_1, v_1)- and (u_2, v_2)-planes are

$$v_1 = \frac{u_1}{1 + u_1 + u_2^a}, \qquad v_2 = \frac{u_2}{1 + u_1^a + u_2}$$

These curves are shown in Fig. 7.12(a) and (b), where the trajectories for this limiting solution are single points $(u_i^a, u_i^a/\delta^a)$ (corresponding to the vanishing small continuous solution) and connected by line segments of slope $1/\beta_i$ to the horizontals $v_i = v_i^b$ (corresponding to the discontinuity). Let u_{i1} be the point

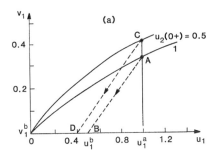

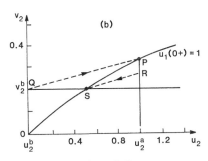

Figure 7.12

where the isotherm has slope $1/\beta_i$ and u_{i2} the upper end of the chord of slope $1/\beta_i$ passing through the origin:

$$u_{i1} = \{\beta_i(1 + u_j^a)\}^{1/2} - (1 + u_j^a), \qquad j \neq i \qquad (7.5.9)$$

$$u_{i2} = \beta_i - 1 - u_j^a, \qquad j \neq i \qquad (7.5.10)$$

If $u_i > u_{i2}(u_j^a)$, the horizontal line $v_i = v_i^b$ can be reached by a discontinuity for all v_i^b. If $u_{i1}(u_j^a) < u_i^a < u_{i2}(u_j^a)$, then v_i^b must be greater than $1 - \delta^a/\beta_i$, the ordinate of the other end of the chord of slope $1/\beta_i$. Finally, if $u_i^a < u_{i2}(u_j^a)$, no solution can be found and the assumption of continuity at $x = 0$ is false. But conditions that invalidate a discontinuity at $x = 0$ permit a discontinuity at $x = X$ so that we have a way of knowing where the discontinuity will be in this limiting case.

Figure 7.13 shows how this may be represented graphically. For each i the curve of Eq. (7.5.9) and the line of Eq. (7.5.10) can be drawn in the (u_1^a, u_2^a)-plane, $v_i^b = 0$. The cylinder erected on the former between $v_i^b = 0$ and $v_i^b = 1$ intersects the plane $u_1^a + u_2^a + \beta_i v_i^b = \beta_i - 1$ in the curve BC. If the feed conditions are such that (u_1^a, u_2^a, v_i^b) lies behind the surface $ABCDE$, a discontinuity at $x = 0$ is appropriate; otherwise, the discontinuity is at $x = X$. Of course, the boundaries at $x = 0$ and at $x = X$ are identical in this limiting case, but the analysis is necessary if we are to get off on the right foot.

Notice that a discontinuity at $x = 0$ may be appropriate for one species and one at $x = X$ for the other, as is the case in the second example. If we test the

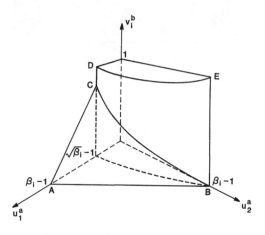

Figure 7.13

consistency of assuming continuity at $x = 0$ in Fig. 7.12, the internal state will correspond to A in Fig. 7.12(a) and P in Fig. 7.12(b), since $u_1(0) = u_1^a = 1$, $u_2^a = 1$ at these points. To accommodate the conditions at $x = X$; $v_1^b = 0$, $v_2^b = 0.2$, these points must be followed by jumps to the horizontals $v_1 = 0$ and $v_2 = 0.2$, respectively. But when the arrows are put on the lines of slope $1/\beta_i$ to indicate permissible directions, A is consistent with a discontinuity at $x = X$ but P is not. In fact, P can only be reached from the horizontal $v_2^b = 0.2$ by a discontinuity at $x = 0$. But then the internal state would be the point S, which is capable of being joined to a feed condition $u_2 = u_2^a$ by the discontinuity RS between $u_2 = 1$ and $u_2 = 0.5$. In Fig 7.12(a) we have to abandon the assumption that the relevant isotherm is OA, which was drawn for $u_2(0) = 1$, and draw instead the isotherm OC for $u_2(0) = 0.5$. The point C on the vertical through $u_1^a = 1$ can now represent the internal state and be followed by a discontinuity at $x = X$ to D on $v_1^b = 0$. The solution for zero length of reactor is

$$u_1^a = 1 = u_1(0), \qquad u_1^b = \frac{2}{5}$$

$$u_2^a = 1, \qquad u_2(0) = u_2^b = \frac{1}{2}$$

$$v_1^a = \frac{2}{5} = v_1(0), \qquad v_1^b = 0$$

$$v_2^a = \frac{4}{15}, \qquad v_2(0) = v_2^b = \frac{1}{5}$$

We are now in a position to consider the family of reactors of increasing length since we understand the limiting member $X = 0$. In Fig. 7.14 we return to the (u_2, u_1)-plane, shown for convenience with broken scales, in which the

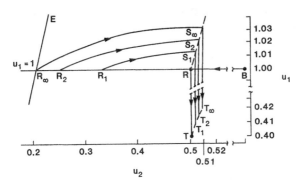

Figure 7.14

solution for $X = 0$ is the set of points B (the feed conditions at $x = 0$; $u_1^a = u_2^a = 1$), R (the internal state: $u_1(0) = 1$, $u_2(0) = 0.5$), and T (the state at $x = X$: $u_1^b = 0.4$, $u_2^b = 0.5$). As X increases there will be the continuous solutions between the discontinuities at $x = 0$ and at $x = X$ that will grow away from R as a sequence of arcs $R_1 S_1$, $R_2 S_2$, . . . of the family of hyperbolas of Eq. (7.5.1). Since the arrows on these all point to the right and since the equilibrium point is to the left at R, they should form a complete family because the contribution to

$$X = \frac{1}{\gamma_1} \int_{u_2(R)}^{u_2(S)} \frac{f(u_1(u_2), u_2)}{e(u_1(u_2), u_2)[\delta^2(u_2) - \beta_1]} du_2 \qquad (7.5.11)$$

from an arc $u_1 = u_1(u_2)$ can be arbitrarily large by starting it close enough to R_∞.

The discontinuity in u_2 at $x = 0$ is represented by a segment of the horizontal line BR_∞ since u_1 is continuous. If there is no discontinuity in u_2 at $x = X$, the discontinuity at $x = X$ will be a vertical line between a point S_i on a curve through R to a point T_i on a curve through T. To determine these curves we use the continuity of u_2, i.e.,

$$v_2^b = v_2(X) = \frac{u_2(X)}{1 + u_1(X) + u_2(X)}$$

to see that S_i must lie on the line

$$u_1(X) - \left[\frac{1}{v_2^b} - 1\right] u_2(X) + 1 = 0 \qquad (7.5.12)$$

The locus of T_i is obtained from the condition $[u_1] = \beta_1[v_1]$ or

$$u_i(T_i) = u_1^b = \beta_1 v_1^b + \frac{(\nu u_2(X) - 1)(u_2(X) - \beta_1 u_2^b)}{u_2(X)} \qquad (7.5.13)$$

$$\nu = \frac{1 - v_2^b}{v_2^b} \qquad (7.5.14)$$

These are the dashed lines RS_∞ and TT_∞ in Fig. 7.14. An enlargement of the upper half of the figure is shown in Fig. 7.15, where the dashed lines are contours of constant X, which converge on $R_\infty S_\infty$ as $X \to \infty$, and the arcs $R_1 S_1, \ldots, R_6 S_6$ correspond to $X = 0.13, 0.3, 0.5, 0.8, 1.4,$ and 3, respectively. By contrast to the first example, there is no possibility of crossing the equilibrium line.

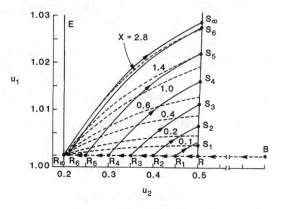

Figure 7.15

The effect of a finite adsorption rate can be seen by taking a particular member of this family of solutions and integrating the four equations (7.2.26)–(7.2.29) for various values of ϕ_1 and fixed ϕ_2/ϕ_1. Because the boundary conditions are split, this requires an iterative technique, but since the solutions are continuous, the projection of any trajectory on the (u_2, u_1)-plane begins at B, $u_2 = u_2^a$, $u_1 = u_1^a$ and finishes on a locus (shown as a dashed line in Fig. 7.16) between $B(\phi_1 = 0)$ and $T(\phi_1 \doteq \infty)$), the condition at $x = X$ for the equilibrium case. Figure 7.16 shows projections of solutions for $\phi_1 = 0.1$, $0.5, 1, 2, 4$, $\phi_1/\phi_2 = 0.2$, and $X = 2.9$. It is clear that the trajectories $u_1(x)$

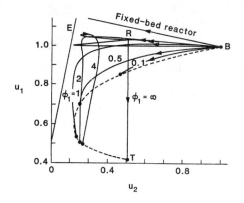

Figure 7.16

Chemical Reaction in a Countercurrent Reactor   Chap. 7

and $u_2(x)$, $0 \le x \le X$ (Fig. 7.17) are monotonic for small ϕ_1 and later develop shallow extrema before tucking themselves into the discontinuous solution for $\phi_1 \to \infty$. Sundaresan and others (1987) have recently shown that in the case of no reaction, the solution depends on the finite ratio ϕ_2/ϕ_1 as both tend to infinity. This is not the case for a linear isotherm and no reaction, and it has not yet been properly tested for the present situation, as the calculations are extremely difficult to push much beyond $\phi_1 = 4$.

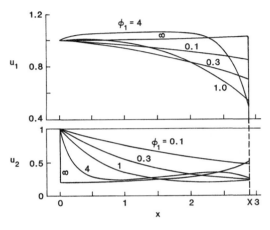

Figure 7.17

7.6 Further Developments

The models proposed here for countercurrent reactors are rich enough to be developed in several directions. Petroulas continued the general procedure of mapping out families of solutions on the lines laid out in this chapter. It would not be appropriate to reproduce this in detail here and the reader is referred to the references at the end of this chapter.

Another development is the use of a switched column instead of a moving bed. The essence of countercurrency is that the solid moves past the feed point in one direction and the carrier fluid in the other. This could be accomplished by keeping the solid fixed and moving the feed point in the same direction as the carrier but more slowly. However, it would be as difficult to do this in practice as to move the bed. On the other hand, it would be possible to have a number of fixed feed points and to switch the feed from one point to the next at prescribed intervals, thus achieving a discontinuous countercurrency. Then it might be well to build the sections between feed points as separate columns, which, with a flexible switching schedule, would provide versatile separation processes. Processes of this kind have been developed industrially under such names as Sorbex and Parex (for hydrocarbons) and Sarex (for glucose and fructose).

It is not easy to build nonlinear models of these processes for they are essentially made up from solutions of the transient fixed-bed model with careful bookkeeping to represent the switching. It has yet to be proved that the transients settle down in a cyclic pattern, although numerical experiments and results for the linear case suggest that this will be the case. Wankat has used the method of characteristics with linear isotherms to give a local equilibrium analysis of the simulated moving bed (see the references at the end of the chapter) and demonstrates clearly the advantages of the moving feed. He and his coworkers found improvements of 50% in reduced bandwidth and greater resolution. He accommodates dispersion by an error function approximation that is good for long columns. Ruthven and his coworkers have done extensive experimental work and supported it with calculations using two linear models. On the one hand, they have considered an equivalent continuous countercurrent column with longitudinal dispersion in the steady state, a model that is well treated by Ruthven in his monograph (1984). On the other, a cell model with transients is used to develop the cyclic solution. The concentration profiles of the latter solution at a time halfway between switchings are found to agree well with the steady-state profiles from the continuous models. This is interesting and appropriate modelling but rather outside the scope of first-order partial differential equations.

REFERENCES

7.1–7.5. The material in these sections is based on the work of Cho and Petroulas and is given in

B. K. CHO, R. ARIS, and R. W. CARR, "The mathematical theory of a counter-current catalytic reactor," *Proc. Roy. Soc. London* **A383,** 147–189 (1982).

T. PETROULAS, R. ARIS, and R. W. CARR, "Analysis of the counter-current moving bed chromatographic reactor," *Comput. Math. Appl.* **11,** 5–34 (1985).

T. PETROULAS, R. ARIS, and R. W. CARR, "Analysis and performance of a counter-current moving-bed chromatographic reactor," *Chem. Engrg. Sci.* **40,** 2233–2240 (1985).

Earlier work is presented in

R. C. MAKINO and E. ROGERS, "Sedimentation of chemically reacting nondiffusing macromolecules: I. Theory," *Arch. Biochem. Biophys.* **109,** 499–510 (1965).

R. C. MAKINO and E. ROGERS, "Sedimentation of chemically reacting nondiffusing macromolecules: II. Some special solutions" *Arch. Biochem. Biophys.* **109,** 560–570 (1965).

K. TAKEUCHI and Y. URAGUCHI, "Separation conditions of the reactant and the product with a chromatographic moving-bed reactor," *J. Chem. Engrg. Japan* **9,** 164–166 (1976).

K. TAKEUCHI and Y. URAGUCHI, "Basic design of chromatographic moving-bed reactors for product refining," *J. Chem. Engrg. Japan* **9,** 246–248 (1976).

K. TAKEUCHI and Y. URAGUCHI, "The effect of the exhausting section on the performance of a chromatographic moving-bed reactor," *J. Chem. Engrg. Japan* **10,** 72–74 (1977).

K. TAKEUCHI, T. MIYAUCHI, and Y. URAGUCHI, "Computational studies of a chromatographic moving bed reactor for consecutive and reversible reactions," *J. Chem. Engrg. Japan* **11,** 216–220 (1978).

K. TAKEUCHI and Y. URAGUCHI, "Experimental studies of a chromatographic moving-bed reactor," *J. Chem. Engrg. Japan* **10,** 455–460 (1977).

See also

S. SUNDARESAN, J. K. WONG, and R. JACKSON, "Limitations of the equilibrium theory of countercurrent devices," *AIChE J.* **33,** 1466–1472 (1987).

7.6. Wankat's analysis and experimental findings are given in

P. C. WANKAT, "Improved efficiency in preparative chromatographic columns using a moving feed," *Indust. Engrg. Chem. Fund.* **16,** 468–472 (1977).

P. C. WANKAT and P. ORTIZ, "Moving feed point gel permeation chromatography: an improved preparative technique," *Indust. Engrg. Chem. Process Des. and Dev.* **21,** 416–420 (1982).

R. S. MCGARY and P. C. WANKAT, "Improved preparative chromatography: moving port chromatography," *Indust. Engrg. Chem. Fund.* **23,** 256–260 (1984).

Ruthven's work is to be found in

C. B. CHING and D. M. RUTHVEN, "Experimental study of a simulated counter-current adsorption system: I," *Chem. Engrg. Sci.* **40,** 877–885 (1985).

C. B. CHING and D. M. RUTHVEN, "Experimental study of a simulated counter-current adsorption system: II," *Chem. Engrg. Sci.* **40,** 887–891 (1985).

C. B. CHING, D. M. RUTHVEN, and K. HIDAJAT, "Experimental study of a simulated counter-current adsorption system: III," *Chem. Engrg. Sci.* **40,** 1411–1417 (1985).

See also his paper

D. M. RUTHVEN, "The axial dispersed plug flow model for continuous counter-current adsorbers," *Canad. J. Chem. Engrg.* **61,** 881–883 (1983).

and book

D. M. RUTHVEN, *Principles of Adsorption and Adsorption Processes,* Wiley, New York (1984), Chap. 12.

Author Index

Subject Index

S

Sarex process, 537
Saturation, 114, 135–37, 283, 284, 290, 461
Semilinearity, 9, 233
Semishocks, 2, 53, 57, 108, 432, 492
Separation factor, 116, 253
Shock, 2, 47, 53, 89, 90, 125, 152, 268, 272, 283
  accelerated, 156
  decay of, 153, 304
  entropy change across, 274
  interactions of, with two simple waves, 164
  point of inception of, 163
Shock condition, 50, 55, 61, 90, 210, 268, 272, 374, 422, 429
Shock layer, 114, 207
Shock-like discontinuity, 431
  confluence of, 155, 156, 308
  definition of, 108
  formation of, 43, 114, 144
  transmission between two, 170, 313, 395
Shock path, 318
Shock wave, viscous, 446
Simple wave, 25, 27, 133, 152, 240, 243, 263, 283
  centered, 39, 163, 279
Simple wave and shock:
  cancellation of, 157
  transmission between, 172, 315, 396
Simple wave region, 25, 40
Simple waves:
  cancellation of, 155
  transmission between two, 177, 321, 399
Singularities, development of, 84
Solute:
  adiabatic adsorption of a single, 456
  adiabatic adsorption of two, 471
Solutions:
  asymptotic behavior of, 439
  decay of, 112
  discontinuities in, 48
  existence and uniqueness, 96
  progressive wave, 102
  standing wave, 102
  weak, 84, 88, 93, 429, 504
Sorbex process, 537
Stanton number, 208
State, region of constant, 25, 27, 34, 134, 243, 280

Stationary shock, 524
Steady state, 518
  attainable in finite time, 406
  method of mapping, 533
Straight-line characteristics (*See Characteristics*)
Strip conditions, 15
Surface coverage, 128
  total, 472
Systems, hyperbolic (*See Hyperbolic systems*)

T

Terms, dissipative and dispersive, 102
Thermal wave, pure, 459
Transformation, hodograph, 20, 24, 516
Transmission, among and between waves and shocks, 170, 172, 306, 313, 315, 395, 396

V

Viscosity, artificial, 92

W

Watershed point, 124
Wave:
  compression, 39, 43, 44, 145, 165
  expansion, 39, 44, 145
  rarefaction, 39
  sawtooth, 166
  shallow-water, 79, 81
  simple (*See Simple wave*) (*See also Approximation, long-wave*)
Waves:
  combined, interactions between, 151, 156, 301, 390, 432
  of different families, interactions between, 170, 313, 394
  pairs of interacting, 152
  of the same family, interactions between, 308, 391
  simple, 133, 152, 240, 243, 263, 283, 333
  centered, 279
  transmission between, 155, 305
Weak solutions (*See Solutions, weak*)

A CATALOG OF SELECTED
DOVER BOOKS
IN SCIENCE AND MATHEMATICS

A CATALOG OF SELECTED

DOVER BOOKS

IN SCIENCE AND MATHEMATICS

Astronomy

BURNHAM'S CELESTIAL HANDBOOK, Robert Burnham, Jr. Thorough guide to the stars beyond our solar system. Exhaustive treatment. Alphabetical by constellation: Andromeda to Cetus in Vol. 1; Chamaeleon to Orion in Vol. 2; and Pavo to Vulpecula in Vol. 3. Hundreds of illustrations. Index in Vol. 3. 2,000pp. 6⅛ x 9¼.
23567-X, 23568-8, 23673-0 Pa., Three-vol. set $46.85

THE EXTRATERRESTRIAL LIFE DEBATE, 1750–1900, Michael J. Crowe. First detailed, scholarly study in English of the many ideas that developed between 1750 and 1900 regarding the existence of intelligent extraterrestrial life. Examines ideas of Kant, Herschel, Voltaire, Percival Lowell, many other scientists and thinkers. 16 illustrations. 704pp. 5⅜ x 8½. 40675-X Pa. $19.95

A HISTORY OF ASTRONOMY, A. Pannekoek. Well-balanced, carefully reasoned study covers such topics as Ptolemaic theory, work of Copernicus, Kepler, Newton, Eddington's work on stars, much more. Illustrated. References. 521pp. 5⅜ x 8½.
65994-1 Pa. $15.95

AMATEUR ASTRONOMER'S HANDBOOK, J. B. Sidgwick. Timeless, comprehensive coverage of telescopes, mirrors, lenses, mountings, telescope drives, micrometers, spectroscopes, more. 189 illustrations. 576pp. 5⅝ x 8¼. (Available in U.S. only.)
24034-7 Pa. $13.95

STARS AND RELATIVITY, Ya. B. Zel'dovich and I. D. Novikov. Vol. 1 of *Relativistic Astrophysics* by famed Russian scientists. General relativity, properties of matter under astrophysical conditions, stars, and stellar systems. Deep physical insights, clear presentation. 1971 edition. References. 544pp. 5⅝ x 8¼.
69424-0 Pa. $14.95

Chemistry

CHEMICAL MAGIC, Leonard A. Ford. Second Edition, Revised by E. Winston Grundmeier. Over 100 unusual stunts demonstrating cold fire, dust explosions, much more. Text explains scientific principles and stresses safety precautions. 128pp. 5⅜ x 8½. 67628-5 Pa. $5.95

THE DEVELOPMENT OF MODERN CHEMISTRY, Aaron J. Ihde. Authoritative history of chemistry from ancient Greek theory to 20th-century innovation. Covers major chemists and their discoveries. 209 illustrations. 14 tables. Bibliographies. Indices. Appendices. 851pp. 5⅜ x 8½. 64235-6 Pa. $24.95

CATALYSIS IN CHEMISTRY AND ENZYMOLOGY, William P. Jencks. Exceptionally clear coverage of mechanisms for catalysis, forces in aqueous solution, carbonyl- and acyl-group reactions, practical kinetics, more. 864pp. 5⅜ x 8½.
65460-5 Pa. $19.95

THE HISTORICAL BACKGROUND OF CHEMISTRY, Henry M. Leicester. Evolution of ideas, not individual biography. Concentrates on formulation of a coherent set of chemical laws. 260pp. 5⅜ x 8½. 61053-5 Pa. $8.95

A SHORT HISTORY OF CHEMISTRY, J. R. Partington. Classic exposition explores origins of chemistry, alchemy, early medical chemistry, nature of atmosphere, theory of valency, laws and structure of atomic theory, much more. 428pp. 5⅜ x 8½. (Available in U.S. only.) 65977-1 Pa. $12.95

GENERAL CHEMISTRY, Linus Pauling. Revised 3rd edition of classic first-year text by Nobel laureate. Atomic and molecular structure, quantum mechanics, statistical mechanics, thermodynamics correlated with descriptive chemistry. Problems. 992pp. 5⅜ x 8½. 65622-5 Pa. $19.95

Engineering

DE RE METALLICA, Georgius Agricola. The famous Hoover translation of greatest treatise on technological chemistry, engineering, geology, mining of early modern times (1556). All 289 original woodcuts. 638pp. 6¾ x 11. 60006-8 Pa. $21.95

FUNDAMENTALS OF ASTRODYNAMICS, Roger Bate et al. Modern approach developed by U.S. Air Force Academy. Designed as a first course. Problems, exercises. Numerous illustrations. 455pp. 5⅜ x 8½. 60061-0 Pa. $12.95

DYNAMICS OF FLUIDS IN POROUS MEDIA, Jacob Bear. For advanced students of ground water hydrology, soil mechanics and physics, drainage and irrigation engineering and more. 335 illustrations. Exercises, with answers. 784pp. 6⅛ x 9¼. 65675-6 Pa. $24.95

ANALYTICAL MECHANICS OF GEARS, Earle Buckingham. Indispensable reference for modern gear manufacture covers conjugate gear-tooth action, gear-tooth profiles of various gears, many other topics. 263 figures. 102 tables. 546pp. 5⅜ x 8½. 65712-4 Pa. $16.95

MECHANICS, J. P. Den Hartog. A classic introductory text or refresher. Hundreds of applications and design problems illuminate fundamentals of trusses, loaded beams and cables, etc. 334 answered problems. 462pp. 5⅜ x 8½. 60754-2 Pa. $13.95

MECHANICAL VIBRATIONS, J. P. Den Hartog. Classic textbook offers lucid explanations and illustrative models, applying theories of vibrations to a variety of practical industrial engineering problems. Numerous figures. 233 problems, solutions. Appendix. Index. Preface. 436pp. 5⅜ x 8½. 64785-4 Pa. $15.95

STRENGTH OF MATERIALS, J. P. Den Hartog. Full, clear treatment of basic material (tension, torsion, bending, etc.) plus advanced material on engineering methods, applications. 350 answered problems. 323pp. 5⅜ x 8½. 60755-0 Pa. $11.95

ANALYTICAL FRACTURE MECHANICS, David J. Unger. Self-contained text supplements standard fracture mechanics texts by focusing on analytical methods for determining crack-tip stress and strain fields. 336pp. 6⅛ x 9¼. 41737-9 Pa. $19.95

A HISTORY OF MECHANICS, René Dugas. Monumental study of mechanical principles from antiquity to quantum mechanics. Contributions of ancient Greeks, Galileo, Leonardo, Kepler, Lagrange, many others. 671pp. 5⅜ x 8½.
65632-2 Pa. $18.95

STATISTICAL MECHANICS: Principles and Applications, Terrell L. Hill. Standard text covers fundamentals of statistical mechanics, applications to fluctuation theory, imperfect gases, distribution functions, more. 448pp. 5⅜ x 8½.
65390-0 Pa. $14.95

THE VARIATIONAL PRINCIPLES OF MECHANICS, Cornelius Lanczos. Graduate level coverage of calculus of variations, equations of motion, relativistic mechanics, more. First inexpensive paperbound edition of classic treatise. Index. Bibliography. 418pp. 5⅜ x 8½. 65067-7 Pa. $14.95

THE VARIOUS AND INGENIOUS MACHINES OF AGOSTINO RAMELLI: A Classic Sixteenth-Century Illustrated Treatise on Technology, Agostino Ramelli. One of the most widely known and copied works on machinery in the 16th century. 194 detailed plates of water pumps, grain mills, cranes, more. 608pp. 9 x 12.
28180-9 Pa. $24.95

ORDINARY DIFFERENTIAL EQUATIONS AND STABILITY THEORY: An Introduction, David A. Sánchez. Brief, modern treatment. Linear equation, stability theory for autonomous and nonautonomous systems, etc. 164pp. 5⅜ x 8¼.
63828-6 Pa. $6.95

ROTARY WING AERODYNAMICS, W. Z. Stepniewski. Clear, concise text covers aerodynamic phenomena of the rotor and offers guidelines for helicopter performance evaluation. Orignially prepared for NASA. 537 figures. 640pp. 6⅛ x 9¼.
64647-5 Pa. $16.95

INTRODUCTION TO SPACE DYNAMICS, William Tyrrell Thomson. Comprehensive, classic introduction to space-flight engineering for advanced undergraduate and graduate students. Includes vector algebra, kinematics, transformation of coordinates. Bibliography. Index. 352pp. 5⅜ x 8½. 65113-4 Pa. $10.95

HISTORY OF STRENGTH OF MATERIALS, Stephen P. Timoshenko. Excellent historical survey of the strength of materials with many references to the theories of elasticity and structure. 245 figures. 452pp. 5⅜ x 8½. 61187-6 Pa. $14.95

CONSTRUCTIONS AND COMBINATORIAL PROBLEMS IN DESIGN OF EXPERIMENTS, Damaraju Raghavarao. In-depth reference work examines orthogonal Latin squares, incomplete block designs, tactical configuration, partial geometry, much more. Abundant explanations, examples. 416pp. 5⅜ x 8¼.
65685-3 Pa. $10.95

Mathematics

HANDBOOK OF MATHEMATICAL FUNCTIONS WITH FORMULAS, GRAPHS, AND MATHEMATICAL TABLES, edited by Milton Abramowitz and Irene A. Stegun. Vast compendium: 29 sets of tables, some to as high as 20 places. 1,046pp. 8 x 10½. 61272-4 Pa. $32.95

FUNCTIONAL ANALYSIS (Second Corrected Edition), George Bachman and Lawrence Narici. Excellent treatment of subject geared toward students with background in linear algebra, advanced calculus, physics and engineering. Text covers introduction to inner-product spaces, normed, metric spaces, and topological spaces; complete orthonormal sets, the Hahn-Banach Theorem and its consequences, and many other related subjects. 1966 ed. 544pp. 6⅛ x 9¼. 40251-7 Pa. $18.95

ASYMPTOTIC EXPANSIONS OF INTEGRALS, Norman Bleistein & Richard A. Handelsman. Best introduction to important field with applications in a variety of scientific disciplines. New preface. Problems. Diagrams. Tables. Bibliography. Index. 448pp. 5⅜ x 8½. 65082-0 Pa. $13.95

FAMOUS PROBLEMS OF GEOMETRY AND HOW TO SOLVE THEM, Benjamin Bold. Squaring the circle, trisecting the angle, duplicating the cube: learn their history, why they are impossible to solve, then solve them yourself. 128pp. 5⅜ x 8½. 24297-8 Pa. $6.95

VECTOR AND TENSOR ANALYSIS WITH APPLICATIONS, A. I. Borisenko and I. E. Tarapov. Concise introduction. Worked-out problems, solutions, exercises. 257pp. 5⅜ x 8¼. 63833-2 Pa. $10.95

THE ABSOLUTE DIFFERENTIAL CALCULUS (CALCULUS OF TENSORS), Tullio Levi-Civita. Great 20th-century mathematician's classic work on material necessary for mathematical grasp of theory of relativity. 452pp. 5⅜ x 8¼. 63401-9 Pa. $14.95

AN INTRODUCTION TO ORDINARY DIFFERENTIAL EQUATIONS, Earl A. Coddington. A thorough and systematic first course in elementary differential equations for undergraduates in mathematics and science, with many exercises and problems (with answers). Index. 304pp. 5⅜ x 8½. 65942-9 Pa. $9.95

FOURIER SERIES AND ORTHOGONAL FUNCTIONS, Harry F. Davis. An incisive text combining theory and practical example to introduce Fourier series, orthogonal functions and applications of the Fourier method to boundary-value problems. 570 exercises. Answers and notes. 416pp. 5⅜ x 8½. 65973-9 Pa. $13.95

COMPUTABILITY AND UNSOLVABILITY, Martin Davis. Classic graduate-level introduction to theory of computability, usually referred to as theory of recurrent functions. New preface and appendix. 288pp. 5⅜ x 8½. 61471-9 Pa. $12.95

ASYMPTOTIC METHODS IN ANALYSIS, N. G. de Bruijn. An inexpensive, comprehensive guide to asymptotic methods–the pioneering work that teaches by explaining worked examples in detail. Index. 224pp. 5⅜ x 8½ 64221-6 Pa. $9.95

ESSAYS ON THE THEORY OF NUMBERS, Richard Dedekind. Two classic essays by great German mathematician: on the theory of irrational numbers; and on transfinite numbers and properties of natural numbers. 115pp. 5⅜ x 8½.
21010-3 Pa. $7.95

APPLIED COMPLEX VARIABLES, John W. Dettman. Step-by-step coverage of fundamentals of analytic function theory–plus lucid exposition of five important applications: Potential Theory; Ordinary Differential Equations; Fourier Transforms; Laplace Transforms; Asymptotic Expansions. 66 figures. Exercises at chapter ends. 512pp. 5⅜ x 8½.
64670-X Pa. $14.95

INTRODUCTION TO LINEAR ALGEBRA AND DIFFERENTIAL EQUATIONS, John W. Dettman. Excellent text covers complex numbers, determinants, orthonormal bases, Laplace transforms, much more. Exercises with solutions. Undergraduate level. 416pp. 5⅜ x 8½.
65191-6 Pa. $12.95

MATHEMATICAL METHODS IN PHYSICS AND ENGINEERING, John W. Dettman. Algebraically based approach to vectors, mapping, diffraction, other topics in applied math. Also generalized functions, analytic function theory, more. Exercises. 448pp. 5⅜ x 8¼.
65649-7 Pa. $12.95

CALCULUS OF VARIATIONS WITH APPLICATIONS, George M. Ewing. Applications-oriented introduction to variational theory develops insight and promotes understanding of specialized books, research papers. Suitable for advanced undergraduate/graduate students as primary, supplementary text. 352pp. 5⅜ x 8½.
64856-7 Pa. $9.95

COMPLEX VARIABLES, Francis J. Flanigan. Unusual approach, delaying complex algebra till harmonic functions have been analyzed from real variable viewpoint. Includes problems with answers. 364pp. 5⅜ x 8½.   61388-7 Pa. $10.95

AN INTRODUCTION TO THE CALCULUS OF VARIATIONS, Charles Fox. Graduate-level text covers variations of an integral, isoperimetrical problems, least action, special relativity, approximations, more. References. 279pp. 5⅜ x 8½.
65499-0 Pa. $10.95

CATASTROPHE THEORY FOR SCIENTISTS AND ENGINEERS, Robert Gilmore. Advanced-level treatment describes mathematics of theory grounded in the work of Poincaré, R. Thom, other mathematicians. Also important applications to problems in mathematics, physics, chemistry and engineering. 1981 edition. References. 28 tables. 397 black-and-white illustrations. xvii + 666pp. 6⅛ x 9¼.
67539-4 Pa. $17.95

INTRODUCTION TO DIFFERENCE EQUATIONS, Samuel Goldberg. Exceptionally clear exposition of important discipline with applications to sociology, psychology, economics. Many illustrative examples; over 250 problems. 260pp. 5⅜ x 8½.
65084-7 Pa. $10.95

NUMERICAL METHODS FOR SCIENTISTS AND ENGINEERS, Richard Hamming. Classic text stresses frequency approach in coverage of algorithms, polynomial approximation, Fourier approximation, exponential approximation, other topics. Revised and enlarged 2nd edition. 721pp. 5⅜ x 8½.   65241-6 Pa. $17.95

INTRODUCTION TO NUMERICAL ANALYSIS (2nd Edition), F. B. Hildebrand. Classic, fundamental treatment covers computation, approximation, interpolation, numerical differentiation and integration, other topics. 150 new problems. 669pp. 5⅜ x 8½. 65363-3 Pa. $16.95

THE FUNCTIONS OF MATHEMATICAL PHYSICS, Harry Hochstadt. Comprehensive treatment of orthogonal polynomials, hypergeometric functions, Hill's equation, much more. Bibliography. Index. 322pp. 5⅜ x 8½.
65214-9 Pa. $12.95

THREE PEARLS OF NUMBER THEORY, A. Y. Khinchin. Three compelling puzzles require proof of a basic law governing the world of numbers. Challenges concern van der Waerden's theorem, the Landau-Schnirelmann hypothesis and Mann's theorem, and a solution to Waring's problem. Solutions included. 64pp. 5⅜ x 8½.
40026-3 Pa. $6.95

CALCULUS REFRESHER FOR TECHNICAL PEOPLE, A. Albert Klaf. Covers important aspects of integral and differential calculus via 756 questions. 566 problems, most answered. 431pp. 5⅜ x 8½. 20370-0 Pa. $10.95

THE PHILOSOPHY OF MATHEMATICS: An Introductory Essay, Stephan Körner. Surveys the views of Plato, Aristotle, Leibniz & Kant concerning propositions and theories of applied and pure mathematics. Introduction. Two appendices. Index. 198pp. 5⅜ x 8½. 25048-2 Pa. $8.95

INTRODUCTORY REAL ANALYSIS, A.N. Kolmogorov, S. V. Fomin. Translated by Richard A. Silverman. Self-contained, evenly paced introduction to real and functional analysis. Some 350 problems. 403pp. 5⅜ x 8½. 61226-0 Pa. $14.95

APPLIED ANALYSIS, Cornelius Lanczos. Classic work on analysis and design of finite processes for approximating solution of analytical problems. Algebraic equations, matrices, harmonic analysis, quadrature methods, much more. 559pp. 5⅜ x 8½.
65656-X Pa. $16.95

AN INTRODUCTION TO ALGEBRAIC STRUCTURES, Joseph Landin. Superb self-contained text covers "abstract algebra": sets and numbers, theory of groups, theory of rings, much more. Numerous well-chosen examples, exercises. 247pp. 5⅜ x 8½.
65940-2 Pa. $10.95

SPECIAL FUNCTIONS, N. N. Lebedev. Translated by Richard Silverman. Famous Russian work treating more important special functions, with applications to specific problems of physics and engineering. 38 figures. 308pp. 5⅜ x 8½.
60624-4 Pa. $12.95

QUALITATIVE THEORY OF DIFFERENTIAL EQUATIONS, V. V. Nemytskii and V.V. Stepanov. Classic graduate-level text by two prominent Soviet mathematicians covers classical differential equations as well as topological dynamics and ergodic theory. Bibliographies. 523pp. 5⅜ x 8½. 65954-2 Pa. $14.95

NUMBER THEORY AND ITS HISTORY, Oystein Ore. Unusually clear, accessible introduction covers counting, properties of numbers, prime numbers, much more. Bibliography. 380pp. 5⅜ x 8½. 65620-9 Pa. $12.95

THEORY OF MATRICES, Sam Perlis. Outstanding text covering rank, nonsingularity and inverses in connection with the development of canonical matrices under the relation of equivalence, and without the intervention of determinants. Includes exercises. 237pp. 5⅜ x 8½. 66810-X Pa. $8.95

INTRODUCTION TO ANALYSIS, Maxwell Rosenlicht. Unusually clear, accessible coverage of set theory, real number system, metric spaces, continuous functions, Riemann integration, multiple integrals, more. Wide range of problems. Undergraduate level. Bibliography. 254pp. 5⅜ x 8½. 65038-3 Pa. $11.95

MODERN NONLINEAR EQUATIONS, Thomas L. Saaty. Emphasizes practical solution of problems; covers seven types of equations. ". . . a welcome contribution to the existing literature...."–*Math Reviews.* 490pp. 5⅜ x 8½. 64232-1 Pa. $13.95

MATRICES AND LINEAR ALGEBRA, Hans Schneider and George Phillip Barker. Basic textbook covers theory of matrices and its applications to systems of linear equations and related topics such as determinants, eigenvalues and differential equations. Numerous exercises. 432pp. 5⅜ x 8½. 66014-1 Pa. $12.95

MATHEMATICS APPLIED TO CONTINUUM MECHANICS, Lee A. Segel. Analyzes models of fluid flow and solid deformation. For upper-level math, science and engineering students. 608pp. 5⅜ x 8½. 65369-2 Pa. $18.95

ELEMENTS OF REAL ANALYSIS, David A. Sprecher. Classic text covers fundamental concepts, real number system, point sets, functions of a real variable, Fourier series, much more. Over 500 exercises. 352pp. 5⅜ x 8½. 65385-4 Pa. $11.95

AN INTRODUCTION TO MATRICES, SETS AND GROUPS FOR SCIENCE STUDENTS, G. Stephenson. Concise, readable text introduces sets, groups, and most importantly, matrices to undergraduate students of physics, chemistry, and engineering. Problems. 164pp. 5⅜ x 8½. 65077-4 Pa. $7.95

SET THEORY AND LOGIC, Robert R. Stoll. Lucid introduction to unified theory of mathematical concepts. Set theory and logic seen as tools for conceptual understanding of real number system. 496pp. 5⅜ x 8¼. 63829-4 Pa. $14.95

TENSOR CALCULUS, J.L. Synge and A. Schild. Widely used introductory text covers spaces and tensors, basic operations in Riemannian space, non-Riemannian spaces, etc. 324pp. 5⅜ x 8¼. 63612-7 Pa. $13.95

ORDINARY DIFFERENTIAL EQUATIONS, Morris Tenenbaum and Harry Pollard. Exhaustive survey of ordinary differential equations for undergraduates in mathematics, engineering, science. Thorough analysis of theorems. Diagrams. Bibliography. Index. 818pp. 5⅜ x 8½. 64940-7 Pa. $19.95

INTEGRAL EQUATIONS, F. G. Tricomi. Authoritative, well-written treatment of extremely useful mathematical tool with wide applications. Volterra Equations, Fredholm Equations, much more. Advanced undergraduate to graduate level. Exercises. Bibliography. 238pp. 5⅜ x 8½. 64828-1 Pa. $8.95

CATALOG OF DOVER BOOKS

FOURIER SERIES, Georgi P. Tolstov. Translated by Richard A. Silverman. A valuable addition to the literature on the subject, moving clearly from subject to subject and theorem to theorem. 107 problems, answers. 336pp. 5⅜ x 8½. 63317-9 Pa. $11.95

POPULAR LECTURES ON MATHEMATICAL LOGIC, Hao Wang. Noted logician's lucid treatment of historical developments, set theory, model theory, recursion theory and constructivism, proof theory, more. 3 appendixes. Bibliography. 1981 edition. ix + 283pp. 5⅜ x 8½. 67632-3 Pa. $10.95

CALCULUS OF VARIATIONS, Robert Weinstock. Basic introduction covering isoperimetric problems, theory of elasticity, quantum mechanics, electrostatics, etc. Exercises throughout. 326pp. 5⅜ x 8½. 63069-2 Pa. $12.95

THE CONTINUUM: A Critical Examination of the Foundation of Analysis, Hermann Weyl. Classic of 20th-century foundational research deals with the conceptual problem posed by the continuum. 156pp. 5⅜ x 8½. 67982-9 Pa. $8.95

CHALLENGING MATHEMATICAL PROBLEMS WITH ELEMENTARY SOLUTIONS, A. M. Yaglom and I. M. Yaglom. Over 170 challenging problems on probability theory, combinatorial analysis, points and lines, topology, convex polygons, many other topics. Solutions. Total of 445pp. 5⅜ x 8½. Two-vol. set.
Vol. I: 65536-9 Pa. $9.95
Vol. II: 65537-7 Pa. $8.95

A SURVEY OF NUMERICAL MATHEMATICS, David M. Young and Robert Todd Gregory. Broad self-contained coverage of computer-oriented numerical algorithms for solving various types of mathematical problems in linear algebra, ordinary and partial, differential equations, much more. Exercises. Total of 1,248pp. 5⅜ x 8½. Two volumes.
Vol. I: 65691-8 Pa. $16.95
Vol. II: 65692-6 Pa. $16.95

INTRODUCTION TO PARTIAL DIFFERENTIAL EQUATIONS WITH APPLICATIONS, E. C. Zachmanoglou and Dale W. Thoe. Essentials of partial differential equations applied to common problems in engineering and the physical sciences. Problems and answers. 416pp. 5⅜ x 8½. 65251-3 Pa. $13.95

THE THEORY OF GROUPS, Hans J. Zassenhaus. Well-written graduate-level text acquaints reader with group-theoretic methods and demonstrates their usefulness in mathematics. Axioms, the calculus of complexes, homomorphic mapping, *p*-group theory, more. Many proofs shorter and more transparent than older ones. 276pp. 5⅜ x 8½. 40922-8 Pa. $12.95

DISTRIBUTION THEORY AND TRANSFORM ANALYSIS: An Introduction to Generalized Functions, with Applications, A. H. Zemanian. Provides basics of distribution theory, describes generalized Fourier and Laplace transformations. Numerous problems. 384pp. 5⅜ x 8½. 65479-6 Pa. $13.95

Math–Decision Theory, Statistics, Probability

ELEMENTARY DECISION THEORY, Herman Chernoff and Lincoln E. Moses. Clear introduction to statistics and statistical theory covers data processing, probability and random variables, testing hypotheses, much more. Exercises. 364pp. 5⅜ x 8½. 65218-1 Pa. $12.95

STATISTICS MANUAL, Edwin L. Crow et al. Comprehensive, practical collection of classical and modern methods prepared by U.S. Naval Ordnance Test Station. Stress on use. Basics of statistics assumed. 288pp. 5⅜ x 8½. 60599-X Pa. $8.95

SOME THEORY OF SAMPLING, William Edwards Deming. Analysis of the problems, theory and design of sampling techniques for social scientists, industrial managers and others who find statistics important at work. 61 tables. 90 figures. xvii +602pp. 5⅜ x 8½. 64684-X Pa. $16.95

STATISTICAL ADJUSTMENT OF DATA, W. Edwards Deming. Introduction to basic concepts of statistics, curve fitting, least squares solution, conditions without parameter, conditions containing parameters. 26 exercises worked out. 271pp. 5⅜ x 8½. 64685-8 Pa. $9.95

LINEAR PROGRAMMING AND ECONOMIC ANALYSIS, Robert Dorfman, Paul A. Samuelson and Robert M. Solow. First comprehensive treatment of linear programming in standard economic analysis. Game theory, modern welfare economics, Leontief input-output, more. 525pp. 5⅜ x 8½. 65491-5 Pa. $17.95

DICTIONARY/OUTLINE OF BASIC STATISTICS, John E. Freund and Frank J. Williams. A clear concise dictionary of over 1,000 statistical terms and an outline of statistical formulas covering probability, nonparametric tests, much more. 208pp. 5⅜ x 8½. 66796-0 Pa. $8.95

PROBABILITY: An Introduction, Samuel Goldberg. Excellent basic text covers set theory, probability theory for finite sample spaces, binomial theorem, much more. 360 problems. Bibliographies. 322pp. 5⅜ x 8½. 65252-1 Pa. $11.95

GAMES AND DECISIONS: Introduction and Critical Survey, R. Duncan Luce and Howard Raiffa. Superb nontechnical introduction to game theory, primarily applied to social sciences. Utility theory, zero-sum games, n-person games, decision-making, much more. Bibliography. 509pp. 5⅜ x 8½. 65943-7 Pa. $14.95

FIFTY CHALLENGING PROBLEMS IN PROBABILITY WITH SOLUTIONS, Frederick Mosteller. Remarkable puzzlers, graded in difficulty, illustrate elementary and advanced aspects of probability. Detailed solutions. 88pp. 5⅜ x 8½. 65355-2 Pa. $5.95

PROBABILITY THEORY: A Concise Course, Y. A. Rozanov. Highly readable, self-contained introduction covers combination of events, dependent events, Bernoulli trials, etc. 148pp. 5⅜ x 8¼. 63544-9 Pa. $8.95

STATISTICAL METHOD FROM THE VIEWPOINT OF QUALITY CONTROL, Walter A. Shewhart. Important text explains regulation of variables, uses of statistical control to achieve quality control in industry, agriculture, other areas. 192pp. 5⅜ x 8½. 65232-7 Pa. $8.95

THE COMPLEAT STRATEGYST: Being a Primer on the Theory of Games of Strategy, J. D. Williams. Highly entertaining classic describes, with many illustrated examples, how to select best strategies in conflict situations. Prefaces. Appendices. 268pp. 5⅜ x 8½. 25101-2 Pa. $9.95

Math–Geometry and Topology

ELEMENTARY CONCEPTS OF TOPOLOGY, Paul Alexandroff. Elegant, intuitive approach to topology from set-theoretic topology to Betti groups; how concepts of topology are useful in math and physics. 25 figures. 57pp. 5⅜ x 8½. 60747-X Pa. $4.95

COMBINATORIAL TOPOLOGY, P. S. Alexandrov. Clearly written, well-organized, three-part text begins by dealing with certain classic problems without using the formal techniques of homology theory and advances to the central concept, the Betti groups. Numerous detailed examples. 654pp. 5⅜ x 8½. 40179-0 Pa. $18.95

EXPERIMENTS IN TOPOLOGY, Stephen Barr. Classic, lively explanation of one of the byways of mathematics. Klein bottles, Moebius strips, projective planes, map coloring, problem of the Koenigsberg bridges, much more, described with clarity and wit. 43 figures. 210pp. 5⅜ x 8½. 25933-1 Pa. $8.95

CONFORMAL MAPPING ON RIEMANN SURFACES, Harvey Cohn. Lucid, insightful book presents ideal coverage of subject. 334 exercises make book perfect for self-study. 55 figures. 352pp. 5⅝ x 8¼. 64025-6 Pa. $11.95

THE GEOMETRY OF RENÉ DESCARTES, René Descartes. The great work founded analytical geometry. Original French text, Descartes's own diagrams, together with definitive Smith-Latham translation. 244pp. 5⅜ x 8½. 60068-8 Pa. $9.95

THE THIRTEEN BOOKS OF EUCLID'S ELEMENTS, translated with introduction and commentary by Sir Thomas L. Heath. Definitive edition. Textual and linguistic notes, mathematical analysis. 2,500 years of critical commentary. Unabridged. 1,414pp. 5⅜ x 8½.. Three-vol. set. Vol. I: 60088-2 Pa. $11.95
Vol. II: 60089-0 Pa. $11.95
Vol. III: 60090-4 Pa. $12.95

GEOMETRY OF COMPLEX NUMBERS, Hans Schwerdtfeger. Illuminating, widely praised book on analytic geometry of circles, the Moebius transformation, and two-dimensional non-Euclidean geometries. 200pp. 5⅝ x 8¼. 63830-8 Pa. $8.95

DIFFERENTIAL GEOMETRY, Heinrich W. Guggenheimer. Local differential geometry as an application of advanced calculus and linear algebra. Curvature, transformation groups, surfaces, more. Exercises. 62 figures. 378pp. 5⅜ x 8½. 63433-7 Pa. $11.95

CURVATURE AND HOMOLOGY: Enlarged Edition, Samuel I. Goldberg. Revised edition examines topology of differentiable manifolds; curvature, homology of Riemannian manifolds; compact Lie groups; complex manifolds; curvature, homology of Kaehler manifolds. New Preface. Four new appendixes. 416pp. 5⅜ x 8½. 40207-X Pa. $14.95

TOPOLOGY, John G. Hocking and Gail S. Young. Superb one-year course in classical topology. Topological spaces and functions, point-set topology, much more. Examples and problems. Bibliography. Index. 384pp. 5⅜ x 8¼. 65676-4 Pa. $13.95

LECTURES ON CLASSICAL DIFFERENTIAL GEOMETRY, Second Edition, Dirk J. Struik. Excellent brief introduction covers curves, theory of surfaces, fundamental equations, geometry on a surface, conformal mapping, other topics. Problems. 240pp. 5⅜ x 8½. 65609-8 Pa. $9.95

Math–History of

A SHORT ACCOUNT OF THE HISTORY OF MATHEMATICS, W. W. Rouse Ball. One of clearest, most authoritative surveys from the Egyptians and Phoenicians through 19th-century figures such as Grassman, Galois, Riemann. Fourth edition. 522pp. 5⅜ x 8½. 20630-0 Pa. $13.95

THE HISTORICAL ROOTS OF ELEMENTARY MATHEMATICS, Lucas N. H. Bunt, Phillip S. Jones, and Jack D. Bedient. Fundamental underpinnings of modern arithmetic, algebra, geometry and number systems derived from ancient civilizations. 320pp. 5⅜ x 8½. 25563-8 Pa. $9.95

GAMES, GODS & GAMBLING: A History of Probability and Statistical Ideas, F. N. David. Episodes from the lives of Galileo, Fermat, Pascal, and others illustrate this fascinating account of the roots of mathematics. Features thought-provoking references to classics, archaeology, biography, poetry. 1962 edition. 304pp. 5⅜ x 8½. (Available in U.S. only.) 40023-9 Pa. $9.95

HISTORY OF MATHEMATICS, David E. Smith. Nontechnical survey from ancient Greece and Orient to late 19th century; evolution of arithmetic, geometry, trigonometry, calculating devices, algebra, the calculus. 362 illustrations. 1,355pp. 5⅜ x 8½. Two-vol. set.
Vol. I: 20429-4 Pa. $13.95
Vol. II: 20430-8 Pa. $14.95

A CONCISE HISTORY OF MATHEMATICS, Dirk J. Struik. The best brief history of mathematics. Stresses origins and covers every major figure from ancient Near East to 19th century. 41 illustrations. 195pp. 5⅜ x 8½. 60255-9 Pa. $8.95

THE HISTORY OF THE CALCULUS AND ITS CONCEPTUAL DEVELOPMENT, Carl B. Boyer. Origins in antiquity, medieval contributions, work of Newton, Leibniz, rigorous formulation. Treatment is verbal. 346pp. 5⅜ x 8½. 60509-4 Pa. $9.95

Physics

OPTICAL RESONANCE AND TWO-LEVEL ATOMS, L. Allen and J. H. Eberly. Clear, comprehensive introduction to basic principles behind all quantum optical resonance phenomena. 53 illustrations. Preface. Index. 256pp. 5⅜ x 8½.
65533-4 Pa. $10.95

ULTRASONIC ABSORPTION: An Introduction to the Theory of Sound Absorption and Dispersion in Gases, Liquids and Solids, A. B. Bhatia. Standard reference in the field provides a clear, systematically organized introductory review of fundamental concepts for advanced graduate students, research workers. Numerous diagrams. Bibliography. 440pp. 5⅜ x 8½.
64917-2 Pa. $11.95

QUANTUM THEORY, David Bohm. This advanced undergraduate-level text presents the quantum theory in terms of qualitative and imaginative concepts, followed by specific applications worked out in mathematical detail. Preface. Index. 655pp. 5⅜ x 8½.
65969-0 Pa. $16.95

ATOMIC PHYSICS (8th edition), Max Born. Nobel laureate's lucid treatment of kinetic theory of gases, elementary particles, nuclear atom, wave-corpuscles, atomic structure and spectral lines, much more. Over 40 appendices, bibliography. 495pp. 5⅜ x 8½.
65984-4 Pa. $14.95

AN INTRODUCTION TO HAMILTONIAN OPTICS, H. A. Buchdahl. Detailed account of the Hamiltonian treatment of aberration theory in geometrical optics. Many classes of optical systems defined in terms of the symmetries they possess. Problems with detailed solutions. 1970 edition. xv + 360pp. 5⅜ x 8½.
67597-1 Pa. $10.95

THIRTY YEARS THAT SHOOK PHYSICS: The Story of Quantum Theory, George Gamow. Lucid, accessible introduction to influential theory of energy and matter. Careful explanations of Dirac's anti-particles, Bohr's model of the atom, much more. 12 plates. Numerous drawings. 240pp. 5⅜ x 8½. 24895-X Pa. $8.95

ELECTRONIC STRUCTURE AND THE PROPERTIES OF SOLIDS: The Physics of the Chemical Bond, Walter A. Harrison. Innovative text offers basic understanding of the electronic structure of covalent and ionic solids, simple metals, transition metals and their compounds. Problems. 1980 edition. 582pp. 6⅛ x 9¼.
66021-4 Pa. $19.95

HYDRODYNAMIC AND HYDROMAGNETIC STABILITY, S. Chandrasekhar. Lucid examination of the Rayleigh-Benard problem; clear coverage of the theory of instabilities causing convection. 704pp. 5⅜ x 8¼. 64071-X Pa. $17.95

INVESTIGATIONS ON THE THEORY OF THE BROWNIAN MOVEMENT, Albert Einstein. Five papers (1905–8) investigating dynamics of Brownian motion and evolving elementary theory. Notes by R. Fürth. 122pp. 5⅜ x 8½.
60304-0 Pa. $7.95

THE PHYSICS OF WAVES, William C. Elmore and Mark A. Heald. Unique overview of classical wave theory. Acoustics, optics, electromagnetic radiation, more. Ideal as classroom text or for self-study. Problems. 477pp. 5⅜ x 8½.
64926-1 Pa. $14.95

PHYSICAL PRINCIPLES OF THE QUANTUM THEORY, Werner Heisenberg. Nobel Laureate discusses quantum theory, uncertainty, wave mechanics, work of Dirac, Schroedinger, Compton, Wilson, Einstein, etc. 184pp. 5⅜ x 8½.
60113-7 Pa. $8.95

ATOMIC SPECTRA AND ATOMIC STRUCTURE, Gerhard Herzberg. One of best introductions; especially for specialist in other fields. Treatment is physical rather than mathematical. 80 illustrations. 257pp. 5⅜ x 8½.   60115-3 Pa. $11.95

AN INTRODUCTION TO STATISTICAL THERMODYNAMICS, Terrell L. Hill. Excellent basic text offers wide-ranging coverage of quantum statistical mechanics, systems of interacting molecules, quantum statistics, more. 523pp. 5⅜ x 8½.
65242-4 Pa. $14.95

THEORETICAL PHYSICS, Georg Joos, with Ira M. Freeman. Classic overview covers essential math, mechanics, electromagnetic theory, thermodynamics, quantum mechanics, nuclear physics, other topics. First paperback edition. xxiii + 885pp. 5⅜ x 8½.   65227-0 Pa. $24.95

PROBLEMS AND SOLUTIONS IN QUANTUM CHEMISTRY AND PHYSICS, Charles S. Johnson, Jr. and Lee G. Pedersen. Unusually varied problems, detailed solutions in coverage of quantum mechanics, wave mechanics, angular momentum, molecular spectroscopy, more. 280 problems plus 139 supplementary exercises. 430pp. 6½ x 9¼.   65236-X Pa. $14.95

THEORETICAL SOLID STATE PHYSICS, Vol. 1: Perfect Lattices in Equilibrium; Vol. II: Non-Equilibrium and Disorder, William Jones and Norman H. March. Monumental reference work covers fundamental theory of equilibrium properties of perfect crystalline solids, non-equilibrium properties, defects and disordered systems. Appendices. Problems. Preface. Diagrams. Index. Bibliography. Total of 1,301pp. 5⅜ x 8½. Two volumes.   Vol. I: 65015-4 Pa. $16.95
Vol. II: 65016-2 Pa. $16.95

A TREATISE ON ELECTRICITY AND MAGNETISM, James Clerk Maxwell. Important foundation work of modern physics. Brings to final form Maxwell's theory of electromagnetism and rigorously derives his general equations of field theory. 1,084pp. 5⅜ x 8½. Two-vol. set.   Vol. I: 60636-8 Pa. $14.95
Vol. II: 60637-6 Pa. $14.95

OPTICKS, Sir Isaac Newton. Newton's own experiments with spectroscopy, colors, lenses, reflection, refraction, etc., in language the layman can follow. Foreword by Albert Einstein. 532pp. 5⅜ x 8½.   60205-2 Pa. $13.95

THEORY OF ELECTROMAGNETIC WAVE PROPAGATION, Charles Herach Papas. Graduate-level study discusses the Maxwell field equations, radiation from wire antennas, the Doppler effect and more. xiii + 244pp. 5⅜ x 8½.
65678-0 Pa. $9.95

INTRODUCTION TO QUANTUM MECHANICS With Applications to Chemistry, Linus Pauling & E. Bright Wilson, Jr. Classic undergraduate text by Nobel Prize winner applies quantum mechanics to chemical and physical problems. Numerous tables and figures enhance the text. Chapter bibliographies. Appendices. Index. 468pp. 5⅜ x 8½. 64871-0 Pa. $13.95

METHODS OF THERMODYNAMICS, Howard Reiss. Outstanding text focuses on physical technique of thermodynamics, typical problem areas of understanding, and significance and use of thermodynamic potential. 1965 edition. 238pp. 5⅜ x 8½. 69445-3 Pa. $8.95

TENSOR ANALYSIS FOR PHYSICISTS, J. A. Schouten. Concise exposition of the mathematical basis of tensor analysis, integrated with well-chosen physical examples of the theory. Exercises. Index. Bibliography. 289pp. 5⅜ x 8½. 65582-2 Pa. $13.95

RELATIVITY IN ILLUSTRATIONS, Jacob T. Schwartz. Clear nontechnical treatment makes relativity more accessible than ever before. Over 60 drawings illustrate concepts more clearly than text alone. Only high school geometry needed. Bibliography. 128pp. 6⅛ x 9¼. 25965-X Pa. $7.95

THE ELECTROMAGNETIC FIELD, Albert Shadowitz. Comprehensive undergraduate text covers basics of electric and magnetic fields, builds up to electromagnetic theory. Also related topics, including relativity. Over 900 problems. 768pp. 5⅝ x 8¼. 65660-8 Pa. $19.95

GREAT EXPERIMENTS IN PHYSICS: Firsthand Accounts from Galileo to Einstein, edited by Morris H. Shamos. 25 crucial discoveries: Newton's laws of motion, Chadwick's study of the neutron, Hertz on electromagnetic waves, more. Original accounts clearly annotated. 370pp. 5⅜ x 8½. 25346-5 Pa. $12.95

RELATIVITY, THERMODYNAMICS AND COSMOLOGY, Richard C. Tolman. Landmark study extends thermodynamics to special, general relativity; also applications of relativistic mechanics, thermodynamics to cosmological models. 501pp. 5⅜ x 8½. 65383-8 Pa. $15.95

LIGHT SCATTERING BY SMALL PARTICLES, H. C. van de Hulst. Comprehensive treatment including full range of useful approximation methods for researchers in chemistry, meteorology and astronomy. 44 illustrations. 470pp. 5⅜ x 8½. 64228-3 Pa. $14.95

STATISTICAL PHYSICS, Gregory H. Wannier. Classic text combines thermodynamics, statistical mechanics and kinetic theory in one unified presentation of thermal physics. Problems with solutions. Bibliography. 532pp. 5⅜ x 8½. 65401-X Pa. $14.95

Prices subject to change without notice.

Available at your book dealer or online at **www.doverpublications.com**. Write for free Dover Mathematics and Science Catalog (59065-8) to Dept. GI, Dover Publications, Inc., 31 East 2nd St., Mineola, NY 11501. Dover publishes more than 400 books each year on science, elementary and advanced mathematics, biology, music, art, literature, history, social sciences, and other subjects.